Trace Minerals in Foods

FOOD SCIENCE AND TECHNOLOGY

A Series of Monographs, Textbooks, and Reference Books

Other Volumes in Preparation

Trace Minerals in Foods

edited by

KENNETH T. SMITH
Miami Valley Laboratories
The Procter & Gamble Company
Cincinnati, Ohio

CRC Press
Taylor & Francis Group
Boca Raton London New York

CRC Press is an imprint of the
Taylor & Francis Group, an **informa** business

CRC Press
Taylor & Francis Group
6000 Broken Sound Parkway NW, Suite 300
Boca Raton, FL 33487-2742

First issued in paperback 2019

ISBN-13: 978-0-8247-7835-4 (hbk)
ISBN-13: 978-0-367-40336-2 (pbk)

Library of Congress Cataloging-in-Publication Data

Trace minerals in foods / edited by Kenneth T. Smith.--1st ed.
        p. cm. -- (Food science and technology ; 28)
    Includes index.
    ISBN 0-8247-7835-9
    1. Trace elements. 2. Food--Analysis. I. Smith, Kenneth Thomas.
    II. Series: Food science and technology (Marcel Dekker, Inc.) ; 28.
    TX553.T7T73 1988
    613.2'8--dc19   88-20377
                                                                        CIP

Visit the Taylor & Francis Web site at
http://www.taylorandfrancis.com

and the CRC Press Web site at
http://www.crcpress.com

Preface

Our current understanding of trace elements in general and trace element nutrition in particular has witnessed tremendous growth over the past several decades. The advent of new and improved methodology for the detection and quantitation of trace elements has provided much of the impetus to this growth. Of course, growth in the area of detection in foods or foodstuffs is just the first step toward eventual applications to human health. Knowledge in the intermediate steps of distribution, absorption, metabolism, and excretion must grow at similar rates. The identification of elements in terms of their essentiality and the definition of their requirements for growth, maintenance, pregnancy, and lactation must logically follow their appearance or detection in plants, animals, and man. Thus, trace element nutrition, as does nutrition in and of itself, touches upon a variety of disciplines. Advances in this science depend on the ability to integrate and coordinate the information in biochemical, physiological, pharmacological, epidemiological, and clinical observations alike. One can readily envision the diversity and complexity of problems that evolve in such a scenario. Despite such diversity, there have been significant advances in this rapidly growing field and these advances constitute the subject of this handbook.

One of the aims of this handbook is to begin to integrate the thoughts and hypotheses of a number of disciplines. As mentioned above, the field of nutrition touches upon such disciplines as medicine, biochemistry, physiology, epidemiology, pharmacology, chemistry, microbiology, and toxicology along with the subspecialties of these various groups. Our understanding of the effects of minerals in biology comes from a blend of information from these disciplines.

However, we must be careful to keep foremost in our minds that *Food* is the single most unifying component in the field of trace element nutrition. Our quest for in-depth knowledge about minerals must always return to our ability to translate this information into its applicability to food and its various relationships to health. Thus, lines of disciplinary endeavor must often be crossed to integrate the information into a cohesive framework that food scientists, nutritionists, biochemists, dieticians, and physicians can all use.

As you examine the contents of this book, I encourage you to keep such thoughts of scientific integration in mind. My personal experiences have taught that we can learn a great deal about trace elements by examining a variety of disciplinary lines of evidence. The food scientist concerned with the organoleptic properties of certain products needs also to consider such things as the chemistry of trace minerals and their potential bioavailability to the consumer. Similarly, the dietician making recommendations on food choices must also consider such items as palatability, mineral interactions, bioavailability, and the like. The clinical investigator developing protocols for experimental programs needs to carefully consider the chemistry, biology, and perhaps even pharmacokinetics of particular trace elements in order to make maximum use of the data obtained. Finally, epidemiologists are faced with a myriad of interrelationships among foods and food customs that affect their ability to interpret data on various disease relationships. These examples, as well as many others, stress the need for integration.

This handbook attempts to pull together information from a number of individuals with diverse backgrounds. Some of the topics covered include a definition of the problems in trace element nutrition from a global perspective, the chemistry and chemical fate of certain trace elements in food systems, the use of tracers in methodological aspects of trace element investigations, and the interaction among trace elements. Some chapters are devoted to specific trace elements. The authors are well known in their particular areas of expertise and bring their own unique perspectives.

In all, the message I wish to convey here is one of variety and cross-fertilization. Rather than travel down the path that progressively narrows to the point of singularity, take time to step back, examine the allied fields that affect your particular specialty, and see whether some new information can be incorporated into your hypothesis, theory, or data base. A better understanding of the processes that influence trace element nutrition, from soil through foods to human health, will surely serve to foster a more complete understanding of the relationships of trace elements to overall health.

Kenneth T. Smith

Contents

Contributors

Richard A. Anderson Vitamin and Mineral Nutrition Laboratory, United States Department of Agriculture, Beltsville Human Nutrition Research Center, Beltsville, Maryland

Fergus M. Clydesdale Department of Food Science and Nutrition, University of Massachusetts, Amherst, Massachusetts

Janet L. Greger Department of Nutritional Sciences, University of Wisconsin, Madison, Wisconsin

Lois Kramer Metabolic Section, Veterans Administration Hospital, Hines, Illinois

April C. Mason Department of Foods and Nutrition, Purdue University, West Lafayette, Indiana

Forrest H. Nielsen United States Department of Agriculture, Agriculture Research Service, Grand Forks Human Nutrition Research Center, Grand Forks, North Dakota

Jean A. T. Pennington Center for Food Safety and Applied Nutrition, Food and Drug Administration, Washington, D.C.

John G. Reinhold[*] Research Center for Study of Nutrition, University of Guadalajara, Guadalajara, Mexico

Ruth Schwartz College of Human Ecology, Cornell University, Ithaca, New York

---

[*]Present affiliation: University of Pennsylvania, Philadelphia, Pennsylvania

Kenneth T. Smith   Miami Valley Laboratories, The Procter & Gamble Company, Cincinnati, Ohio

Herta Spencer   Metabolic Section, Veterans Administration Hospital, Hines, Illinois

Donald B. Thompson   Department of Food Science, The Pennsylvania State University, University Park, Pennsylvania

Connie Weaver   Department of Foods and Nutrition, Purdue University, West Lafayette, Indiana

Trace Minerals in Foods

1
Problems in Mineral Nutrition: A Global Perspective

JOHN G. REINHOLD / University of Guadalajara, Guadalajara, Mexico

INTRODUCTION

Defects in mineral nutrition are capable of producing severe impairment of health. Three elements are mainly involved: calcium, iron, and zinc. Others known to be indispensable may be added to the list, but insufficient information is available to permit an evaluation on a worldwide basis. Calcium malnutrition in the form of rickets affects infants, children, and adolescents. A similar disturbance, osteomalacia, is a potent source of pain and disability of women in many developing countries, although both rickets and osteomalacia intrude into developed countries in the form of unexpected and frequently unrecognized epidemics. Rickets is an ancient illness, having been depicted in tomb paintings in ancient Egypt dating to 4000 B.C.(1). Anemia due to iron deficiency was recognized in ancient Greece. However, zinc deficiency has been identified as an important cause of faulty nutrition in animals and humans only within the past half century. Calcium, iron, and zinc, together with others of the trace metal group, share physical and chemical properties, including solubilities and the ability to react with complexants, that affect their availability as nutrients. For this reason deficiencies may involve more than one element. Low concentrations of the nutritionally important metals may exist in soils (2) and produce regional deficiencies. Yet, calcium, iron, and zinc are among the most abundant and widely distributed elements. It is surprising, therefore, that low concentrations of calcium in drinking water may unfavorably affect cardiovascular status (3). Plants have an extraordinary capability to extract metals from soils. As a result the content of minerals in a plant growing in deficient soil may not differ much from that of a plant growing in soils well supplied with nutrients (2). However, a

*Present affiliation: University of Pennsylvania, Philadelphia, Pennsylvania

1

peasant engaged in subsistence farming on a small tract is more vulnerable to soil deficiencies than is an urban dweller who derives food from many sources.

This chapter will be devoted to consideration of calcium, iron, and zinc deficiencies. Together, they currently afflict large numbers of children and adults. Rickets and osteomalacia can produce disabling deformities, and rickets may bring about death from pneumonia. The consequences of zinc deficiency are less obvious, yet recent work suggests that zinc deficiency may in some circumstances contribute to the development of the rickets syndrome. Indeed, one of the earliest effects of zinc deficiency recognized was a failure of long bones to develop in fowl. Iron deficiency may impair physical work productivity. Both zinc and iron are involved in immunocompetence.

RICKETS AND OSTEOMALACIA

Rickets and osteomalacia are among the more common syndromes resulting from malnutrition. Twenty-five years ago, rickets was said to be a disease of the past, conquered by supplementation of the milk supplies with vitamin D. The claim proved to be premature, because the claimants were not informed of the prevalence of rickets in many developing countries where supplementation was not feasible.

Historically, rickets grew in importance as the industrial revolution brought about migration of laborers from countryside to city. Urbanization was especially extensive in Britain. Smoke from tens of thousands of coal fires combined with cloud and fog and the dimmed light of northern latitudes deprived infants and children of the sunlight necessary for vitamin D synthesis. Epidemics of rickets resulted. Some relief was provided by medication with cod liver oil, its efficacy having been established in the seventeenth century. Sunlight was also known to produce remission, although it was not until 1890 that Palm (4) provided convincing evidence that lack of sunlight could cause rickets and that exposure to sunlight was curative. The discovery of vitamin D, and especially the successful prophylaxis of rickets by fortification of milk with vitamin D in the United States and elsewhere, led to the premature announcements of its disappearance. However, rickets and osteomalacia remain prevalent in the subtropics, including the Middle East, northern India, and the northern tier of Africa. Immigration from regions of abundant sunshine to Britain, especially, grew in volume and brought new tenants to the darkness of the slums. The diets and housebound ways of life of the mother countries were retained by many. Rickets again became endemic.

Definitions

Rickets: An impairment of calcification in which epiphyseal cartilage is thickened and excessive amounts of cartilage are formed within the epiphysis and areas of bone growth (5). It occurs most frequently in infancy. Late rickets develops during the rapid growth period of adolescence.

Osteomalacia: Rickets occurring in the adult female as a result of failure of newly formed osteoid to calcify and replace bone lost by resorption.

Osteoporosis: A condition in which bone contains a decreased amount of calcified tissue. Although it is mainly a problem in the elderly, it may occur at other ages.

There are said to be more than 30 causes of rickets and osteomalacia (6). This chapter is concerned with rickets and osteomalacia produced by nutritional deficiencies of calcium and vitamin D and by the action of antimetabolites, such as dietary phytate and fiber.

Dietary Requirements For Calcification

The Recommended Dietary Allowance (RDA) of calcium in the United States is 450 mg/day for infants between six months and one year of age and 800 mg/day after one year. FAO/WHO recommendations are the same for infants, but considerably lower for children, 400 to 500 mg/day and for adolescents, 600 to 700 mg/day. The FAO/WHO RDA figures are more attainable in the developing countries than are the U.S. RDAs. Gopalan and Narasinga Rao (7) obtained positive calcium balances for Indian women at an intake of 388 mg/d, but one third of the men examined remained in negative balance with an intake of 500 mg/d. Begum and Pereira (8) found that preschool children established a positive minimal balance of calcium at 200 mg/day, but that they retained considerably more calcium when intake increased to 280 mg/day. Moreover, it appears that calcium requirements established at minimal levels may not yield sufficient calcium to support more than minimal growth and calcification or to forestall eventual development of osteoporosis.

Ca/P ratios in diets are believed to be optimal between 1:1 and 2:1. Calcium intakes should be sufficient to maintain serum calcium concentrations at or above 9.0 mg/dl at all ages, and serum phosphorus concentrations above 4.0 mg/dl in children and 2.5 mg/dl in adults. Lowered concentration of phosphorus in serum is the most consistent change in blood associated with rickets and osteomalacia (9), and the downward movement is a consequence of increased parathyroid activity. Calcium concentrations in serum are decreased less frequently. Activity of alkaline phosphatase in serum is generally elevated but may remain within normal limits (10), particularly in the presence of protein-energy malnutrition (PEM). The product of the concentrations of calcium and phosphorus in serum is a useful, although empirical, expression of the capability for calcification (11, 12).

Causes of Calcium Depletion

These include (a) low calcium intake, (b) poor bioavailability of calcium, (c) lack of vitamin D, and (d) malfunction of the gastrointestinal tract. These factors may operate separately, but the likelihood of overt clinical expression is increased when more than one is operative.

Low Calcium Intake

Rickets resulting solely from low intake of calcium is uncommon. It has been described by Maltz et al. (13), Kooh et al. (14), and Pettifor et al. (15-20). The reports of Pettifor have special interest because the rickets they describe occurred in a region of South Africa where calcium intakes of children averaged 125 mg/day. Intakes of phosphorus were normal. Bone deformities closely resembled those of clinical rickets due to other causes. Concentrations of 1, 25-dihydroxycholecalciferol (1,25-dihydroxy D_3) and of parathyroid hormone were elevated. Healing followed an increase in calcium intake to 1200 mg/day (20). Findings in these calcium-deficient children suggested that vitamin D becomes less effective when dietary intakes of calcium are exceptionally low.

This view is supported by a study done by Pettifor and colleagues (18), in which they used baboons to compare the effects of a diet (a) having a low calcium content but sufficient vitamin D, with a diet (b) containing cornmeal with 100 mg of phytate and 1.6 gm of crude fiber per 100 gm but deficient in vitamin D. Both diets produced rickets or osteomalacia with lowered serum calcium concentrations, which the vitamin D in diet (a) failed to prevent. However, superior growth occurred in the animals fed this diet.

Low levels of calcium in diets occur in India, Thailand, Taiwan, and Morocco (See tables 1 to 4). More information concerning other countries and localities should be sought. Ackroyd and Kirshnan (22) showed that addition of calcium to the diets of nursery school children in India produced a striking increase in growth. Spencer and Kramer (23) found no adaptation by human adults to a diet providing 200 mg/day of calcium.

Vitamin D Deficiency

The primary function of vitamin D is to promote the absorption of calcium. The most important source of the vitamin is biosynthesis in the skin; very little is supplied by unfortified diets in developing countries. Biosynthesis is dependent upon the extent of exposure to sunlight. When this is diminished by shortened days at higher latitudes or other factors that decrease the intensity of the ultraviolet component of the solar spectrum, deficiency results. Smog has become an important source of interference in tropical and subtropical countries as industrial development proceeds. For example, rickets has appeared in Mexico City (123–125), where smog control is ineffective, in part because of the reluctance of the government to impose extra burdens upon industry.

Religious and folk traditions continue to be important determinants of life patterns in developing countries. This is true especially of the Moslem group. Strict seclusion of women (purdah) persists. Loomis (27) reported that the incidence of rickets and osteomalacia in India among women and children from prosperous Moslem families was seven times the rate observed in Hindu families of similar economic status. Deprivation of sunlight among the Moslems explains the difference. The Moslem women and children not only were secluded indoors, but ventured out of doors only when enshrouded in opaque garments. As a result the highest incidence of rickets and osteomalcia prevails in countries that are predominantly Moslem. Among the poor, life is lived in semidarkness in sunless slums. Even in villages sunlight is shunned. The aversion is found not only among Moslems, for Jensen (28) discovered children in non-Moslem Ethiopia to be deprived of sunshine to avoid the sun's evil eye. Nomads, living in tents, are more likely to be exposed to sunlight, but Guggenheim and Kaufman (29) encountered Bedouin women living in tents who received insufficient solar radiation to prevent osteomalacia. In many countries in which rickets is prevalent, infants are wrapped in swaddling garments that permit exposure of only a bit of face. Carpet weavers in the Middle East, many of them adolescents, work in darkened rooms or black tents and are particularly vulnerable to rickets. The advent of television in the Mexican village in which the author was living attracted children of all ages into darkened rooms from sunny streets and must be considered yet another cause of deprivation of sunlight.

Wilson and Widdowson (30) concluded from studies in India that vitamin D deficiency alone was unlikely to produce rickets or osteomalacia. It was only when it was superimposed upon calcium deficiency that the rickets and osteomalacia syndromes became fully developed. Such combinations of causes commonly exist. Studies by Lund et al. (30a) supported a direct involvement of 1,25-dihydroxy D_3 in the healing of bone lesions in rickets, while 25-hydroxy D_3 levels were too low to be measured. 24,25-Dihydroxy D_3 appeared to make no contribution. Magnesium depletion impaired biosynthesis of 1,25-dihydroxy D_3 (30b).

Complexants of Calcium

Phytate. Mellanby (31) in 1926 described interference by cereals with calcification. Bruce and Callow (32) presented evidence that the substance responsible was inositol hexaphosphate (phytate), and this was confirmed by Harrison and Mellanby (33). Concern about the nutritive value of wartime brown bread (extraction rate, 92%) (with added calcium carbonate and calcium monohydrogen phosphate) led

McCance and Widdowson (34) to perform balance studies on human subjects who ate the bread over periods up to nine months. Bioavailabilities of calcium and magnesium were decreased. A further finding was that phytate depressed the absorption of calcium and magnesium. Vitamin D failed to counteract the action of phytate. These observations were confirmed by Krebs and Mellanby (35) and McCance and Walsham (36). Mellanby (37), reviewing studies that extended over many years in which puppies served as experimental animals, concluded that phytate was the component of cereals most responsible for decreased deposition of calcium in bones. The action of phytate was not reversed by vitamin D. Phytate added to the diets of infants and children in amounts calculated to combine with 50 to 100% of the dietary calcium caused a considerable decrease in calcium absorption (38, 39).

Walker et al. (40) challenged the evidence supporting the rachitogenic action of phytate, claiming the occurrence of adaptation, but the conditions used were inapplicable to those prevailing in developing countries (25). Rajalakshmi (41) observed that calcium supplements to diets of Indian schoolboys promoted bone development more effectively than did vitamin D, indicating that the vitamin failed to overcome the retardant effect of phytate. The role of intestinal phytase is uncertain. Wise et al. (176) found no phytate hydrolysis of the upper gastrointestinal tract in rats.

Bread as a Source of Rachitogens

The phytate of bread is an important source of rachitogenic activity in developing countries where lightly sifted flours and omission of leaven or fermentation result in high intakes of phytate. The predominance of bread in the diets of many developing countries introduces high intakes of this potent complexant of metals. In Iran, where much of the rural population subsists on unleavened bread, this has important consequences for nutrition. A brief description of the several types of bread consumed in Iran will be helpful. *Bazari* is a leavened flatbread made by a sourdough process from flours of 80 to 85% extraction in urban bakeries. It is baked on the walls of an oven. Its phytate concentration was (mean and SD) 333 ± 104 mg/100 gm (41a,156). *Sangak*, another bread baked in urban bakeries, is also made with leaven from flours of somewhat higher extraction rate than *bazari*. Its phytate concentration was 388 ± 85 mg/100 gm. *Tanok* is a paper-thin rural bread made without leaven or fermentation from wholemeals of 90 to 100% extraction. It is baked briefly on sheet metal over a fire of twigs or straw. *Tanok* provides from 50 to 90% of the energy requirement of the village population in the Shiraz region. It contains 746 ± 131 mg/100 gm of phytate.

Kouhestani et al. (158) describe the preparation of various Iranian breads but omit mention of *tanok*. They describe another bread, *lavosh,* identical in appearance to *tanok* but made with leaven from flours of lower extraction. Phytate may remain high despite use of leaven (157). The *chapati,* also know as *rotti,* is the main staple of the diet of central and northwestern India. It is made without leaven from high-extraction wheaten flours and is eaten by both urban and rural populations. The *Chapati* is equal to *tanok* in phytate concentrations and resemble a pancake in appearance (287).

Dietary Fiber

A study designed to compare the action of purified phytate added to *bazari* with that of an equivalent amount of phytate in the form of *tanok* showed the latter to be far more potent in its effects on calcium and zinc balances (25) than phytate plus *bazari.* The discrepancy suggested that some other component of *tanok* was producing effects similar to those of phytate. Dietary fiber, present in large amounts in *tanok,* seemed to be a likely prospect. Accordingly, the ability of fiber to combine with calcium, zinc, and iron, and so to diminish their bioavailability, was examined. Studies done in vitro (44,44a) showed that dephytinized *tanok* bound as much or more calcium and zinc as did untreated *tanok.* An examination of cellulose, several hemicelluloses, and lignin showed that all were capable of binding calcium, iron, and zinc at pH 6.5 and of releasing bound metal at pH less than 6 (44a).

Experiments in which cellulose was added to human diets containing either *bazari* or *tanok* decreased balances of calcium, magnesium, and zinc (45). Several other reports showed that less calcium was retained when diets with high fiber content were eaten than from diets with low or moderate fiber content. Cummings (46) evaluated bran; Kelsay et al. (47), fruit and vegetables; Cummings et al. (48), wheat fiber; Slavin and Marlett (49), cellulose; Pathak et al., (50), Punjabi diet plus cellulose; Reinhold et al. (51), *tanok;* Rhada and Gurvani (52), Bengal gram; and Mahalka et al. (53), five fiber sources. Other studies have failed to demonstrate an effect: Southgate et al. (54), bran; Cummings et al. (55), pectin; Kelsay et al. (56), fruits and vegetables; King et al. (57) cellulose, xylan, pectin, corn bran; Ellis and Morris (58), dephytinized bran; van Dokkum et al. (59), fiber-rich bread, neutral detergent fiber; and Andersson et al. (60), reconstituted dephytinized bran bread. The lack of agreement regarding the effect of fiber on metal retention has no simple explanation. Fiber contents of the experimental diets may have been insufficient. Synergistic effects of phytate=, peptide=, or phosphorus-containing components may be involved. Rendleman (61) found that cellulose, starch, hemicellulose, and pectin components of bran made little contribution to binding of calcium. Water-soluble components of bran, mainly phytate, were responsible for 50% of its binding ability.

Examination of the mineral content of bran that had been passed through the human gastroinestinal tract disclosed that undigested residues recovered had affinities for copper, iron, and zinc that increased contents of these metals in the residues two to four times (63). However, calcium contents increased at least tenfold. Dietary fiber bound calcium in proportion to its uronic acid content (64).

Dietary fiber may decrease the effectiveness of vitamin D. Batchelor and Compston (65) reported that the time required for disappearance of ^{3}H-labeled 25-hydroxyvitamin D_3 was significantly shortened in human subjects by the addition of 20 g/day of fine bran to the diet. Vitamin D and its metabolites undergo enterohepatic circulation (66,67). Thus, fiber present in the tract may sequester portions of the vitamin and its hormonal products.

Mineral Depletion in Iranian Villagers

If the diets of iranian villagers supplied inadequate amounts of bioavailable calcium and other divalent metals, bodily reserves should become depleted. This hypothesis was examined by measuring the retention of calcium, phosphorus, zinc, and nitrogen by villagers when a nutritious hospital diet was eaten. Thirteen villagers—five females and eight males aged 17 to 47—were admitted to a metabolism ward. After three to five days, during which they ate the equivalent of their village diets, dietary balances of the four elements were measured under close supervision for six or seven days while they ate the nutritious diet. Strongly positive balances of calcium, zinc, and phosphorus were found as anticipated. Balances of nitrogen were the exception. Being close to zero, they indicated that protein nutrition was adequate. Females had more positive calcium and phosphorus retentions than males (69).

Rickets in Wartime Dublin

The opinion that rickets can be equated to vitamin D deficiency persists, despite evidence that dietary rachitogens are also important. The epidemic of rickets among infants and children in Dublin in 1942 and 1943 is instructive. Eire, while not a belligerent, faced a shortage of wheat, and a decision was made to extend the supply by increasing the rate of extraction from 70% in 1940 to 100% in 1942. The incidence of rickets rose rapidly after the change became effective (70). Subsequently, the extraction rate was lowered and the incidence returned to its 1940 level. Robertson et al. (71), after a review of the evidence, concluded that the epidemic could not be explained by changes in the proportion of children protected by vitamin D administration or by exposure to sunlight. Instead, an increase in rachitogen intake must have been responsible.

Bowel Malfunction

Diarrhea produced by invasion of the upper bowel by pathogens ranks high among the major disturbances of nutrition in developing countries in infancy and childhood. Poor hygiene is mainly responsible. Wastage of nutrients arrests growth. In these circumstances, signs of rickets may not be apparent despite depletion of bone materials. Sandstead et al. (10) detected evidence of rickets in half of the Egyptian children with protein-energy malnutrition (PEM). Three-fourths of the children with PEM in India examined by Faridi et al. (72) displayed decreased bone density. Other examples of rickets associated with PEM are cited in Table 1.

Association of Zinc Malnutrition with Rickets

Early studies of zinc deficiency showed that among its manifestations was faulty development of the long bones that led to crippling lesions of legs and feet of domestic fowl. Luecke (74) explains that substitution of soybean meal for animal meals in the rations was responsible. This practice introduced phytate into the ration in an amount sufficient to lower the bioavailability of zinc. The lesions were corrected by increasing the zinc intake.

Rhesus monkeys marginally deprived of zinc during the first year of life showed delayed skeletal maturation and defective mineralization resembling that in the human rachitic syndrome (75). In this connection, the unexplained elevation of alkaline phosphatase activities of serum exhibited by most of the hypogonadal zinc-deficient dwarfs originally studied by Prasad et al. (76) in Shiraz may have been manifestations of a delayed or faulty bone development. This speculation presumes that their zinc deficiency was not sufficiently severe to depress alkaline phosphatase activity.

Strause et al. (77) observed that moderate deficiency of manganese copper and iron decreased concentrations of calcium and phosphorus in bones of rats.

Rickets and Osteomalacia in Developing Countries

Introduction

Rickets and ostemalacia have a nearly worldwide distribution, and a survey of their geographical occurrence, the role of the principal etiologic factors, and incidence among various populations and ethnic groups is informative. (A compilation of such data is presented in Tables 1 to 4).

In the tables, staple cereals listed are those that predominate in the diets of the economically deprived social strata. All cereals are important sources of phytate, dietary fiber, and possibly other

Table 1. Rickets and Osteomalacia in India

Location	Year	Staple cereals	Ca	P	Sunlight	Rickets; Osteomalacia	Age group	Rate %	Reference
Amritsar Central	1931	Wheat Wm, Un	–	–	Deprived	Osteomalacia	Women	54	(79)
Outskirts			–	–	Partly deprived	Osteomalacia	Women	18	
Punjab Rural	1939	Wheat Wm, Un	Low	–	Exposed	Rickets	Children	25	(80,81)
		Wm, Un	Low	–	Exposed	Osteomalacia	Women	50	
Punjab Rural	1961	Wheat Wm, Un	Low	–	Exposed	Rickets	Children	5.3	(84)
Kangra Valley		Wheat Wm, Un	Low	–	–	Rickets	Children	56	
Orissa	1961	Rice	Low	–	Partly deprived	Rickets	Children	1.6	(84)
Hyderabad	1961	Rice	Low	–	Partly deprived	Rickets	Children	1.2	(84)
New Delhi	1962	Wheat Wm, Un	Low	–	Deprived	Rickets	Children	35	(82)
Maharashtra Village	1971	Wheat Wm, Un	Low	–	Exposed	Rickets	Preschool	3.5	(86)

Location	Year	Diet	Serum Ca[a]	Serum P[a]	Exposure	Condition	Group		Ref.
New Delhi	1971	Wheat Wm, Un	Low	—	—	Rickets, PEM	Children	12	(73)
Amritsar	1972	Wheat Rice	Low	—	Deprived	Rickets	Children	1	(87)
Lhudiana	1973	Wheat Wm, Un	Low	—	Exposed	Rickets Osteomalacia	Children Women	1	(88)
New Delhi	1973	Wheat mainly	<9 mg/dl 73%	<3 mg/dl 86%	Deprived	Osteomalacia	Women	—	(89)
Poona Rural	1974	Wheat Wm, Un	Low	—	Partly deprived	Rickets	Preschool	6.5	
Calcutta	1975	Rice	248[c]	—	Deprived	Rickets	Preschool	1.8	(90)
Hyderabad	1984	Rice	248[c]	—	Partly deprived	Rickets	Children	—	
Haryana	1980	Wheat Wm, Un	—	—	Oral vit. D	Osteomalacia	Women	1.0	
Jaipur	1985	—	—	—	—	Atrophic rickets	Infants	—	(91)

[a] Calcium and phosphorus concentrations in serum.
[b] % abnormal.
[c] mg/day
Wm = wholemeal.   Un = unleavened.

11

Table 2. Rickets and Osteomalacia in the Middle and Near East

Location	Year	Staple cereals	Ca	P	Sunlight	Rickets; Osteomalacia	Age group	Rate (%)	Reference
Pakistan	1961	Wheat Barley	Low	—	Deprived	Rickets	Early; late	High	(93)
Pakistan	1976	Wm, Un	Low	—	Deprived	Osteomalacia	—	12% non pregnant 33% pregnant	(94)
Iran	1961	Wheat Rice	349–642[c]	—	Deprived	Rickets	Infants	High	( )
Iran	1971	Wheat Wm, Un	Low	—	Deprived	Osteomalacia	Carpet Weavers	—	(95)
Village	1972	Wheat Wm, Un	Low	—	Deprived	Rickets	Infants	2.6	(96)
Mainly village	1972	Wheat	—	—	Exposed	Rickets	School	10	(97)
Village	1973	Wheat	9 mg/dl[a] 22/25	—	Deprived	Rickets	Children	—	(98)
Tehran	1975	Wheat	—	—	Deprived	Rickets	3 mos.– 6 yrs.	16	(99)
Village	1974	Wheat Wm, Un	—	—	Deprived	Rickets	6 mos.– 2 yrs	Overt 5 Subclin 36	(100)
Afghanistan	1961	Wheat barley	—	—	Deprived	Rickets Osteomalacia	Children Young Women	75	(93)
Iraq	1961	Wheat	327–722[c]	—	Deprived	Rickets	Children	—	(93)
Mosul	1972	Wheat	9 mg/dl[b] 43%	3mg/dl 60%	Deprived	Rickets + PEM	Infants	—	(101)

Country	Year	Diet		Exposure	Condition	Population	Percentage[b]	Ref.
Saudi Arabia	1961	(Dates) Wheat	—	Deprived	Rickets	Children	"Prevalent"	(93)
Saudi Arabia	1984	—	—	Deprived	Rickets / Osteomalacia	Children Newborn Pregnant	— / —	(102)
Turkey	1961	Wheat barley	—	Deprived	Rickets	Children	"Prevalent"	(93)
Syria Urban	1961	Wheat	—	Deprived	Rickets	Children	14	(93)
Israel (Haifa)	1953	Wheat	—	—	Rickets	Infants	16/98 necropsies	(103)
Isreal	1961	Wheat Wm, Un	—	Deprived	Rickets	Children Beduin	9.1	(93)
Israel	1962	Wheat	—	Exposed	Rickets	Children Israeli	4.6	(104)
Israel	1976	Wheat Wm, Un	—	Deprived	Ostecmalacia	Women Nomadic	—	(29)
Israel	1976	Wheat Wm, Un	Lowered 25(OH)D	Deprived	No signs	Women Beduin	—	(106)
Israel	1979	Wheat Wm, Un	Lowered 25(OH)D	Deprived	No signs	Women in labor Newborn	—	(107)
Israel	1973	Wheat Wm, Un	500[a]	Deprived	Osteomalacia	Women	—	(108)

[a] Mg/day.
[b] Calcium and phosphorus concentrations in serum with percentage abnormal.
[c] Deprived during preschool period.
Wm = Wholemeal. Un = unleavened.

Table 3. Rickets and Osteomalacia in the Far East

Location	Year	Staple cereals	Ca mg/day	P mg/day	Sunlight	Rickets; Osteomalacia	Age group	Rate (%)	Reference
China	1961	Wheat(in north) Rice	—	—	Exposed	Osteomalacia	Women	5	(93)
Taiwan	1961	Rice	219	987	Exposed	nil	Children	—	(93)
Manila	1958	Rice	—	—	Deprived	Rickets	Infants	—	(109)
Vietnam	1961	Rice	Low	—	Exposed	Rickets Osteomalacia	"Common"	—	(93)
Thailand	1961	Rice	132	—	Exposed	Not stated	—	—	(93)
Sri Lanka	1964	Rice	—	—	Exposed	Rickets	Children	—	(110)

complexants of nutritionally important metals. Rate of extraction of flours is the main determinant of the amounts of complexants consumed. Fermentation improves bioavailability of metals, and the omission of leaven and fermentations in breadmaking is noted. Cereals are also eaten as porridges and in soups or stews. Rice may be eaten in the phytate- and fiber-rich brown form or in the white (hulled) form, which contains diminished but appreciable amounts of complexants. The two are seldom distinguished when diets are described. Tea has not received sufficient consideration as a source of antimetabolites. Great quantities are consumed in Iran and countries of central and east Asia.

Exposure to sunlight is very important as an etiological factor. Deprivation is prevalent in Moslem countries. Evaluation is based on the author's observations and information in the literature. Although women provide agricultural labor in Turkey and are exposed to sunlight, this appears to be exceptional for a Moslem population.

Early rickets is that occurring before age ten; late rickets is that affecting adolescents.

India

Rickets and osteomalacia, long prevalent in India, have been the subjects of many studies (Table 1). Wilson in 1931 (79) collected 1000 cases in a survey of prepartition Punjab and Kashmir. All sects, Hindu, Moslem, Sikh, and Christian, were susceptible. One noteworthy finding was that Hindu female agricultural workers regularly exposed to sunshine were prone to develop osteomalacia. Hence, diet alone could produce the syndrome. However, Wilson observed a direct relationship between urban location and incidence of late rickets. More than half of the female adolescents in crowded central Amritsar developed rickets. In well-planned suburbs only one-sixth were affected. In 1940, Taylor and Day (80) and Taylor (81) confirmed the existence of osteomalacia in the female field workers exposed to sunlight. Both authors were interested primarily in the relationship of rickets and dental caries, and they were suprised to find perfect teeth in both rickets and osteomalacia.

In New Delhi in 1962, 35% of a sample of 28,400 outpatients were found to have rickets or osteomalacia (82). Nine years later, several surveys yielded lower incidences, 12% for New Delhi (73) and 6.5% of preschool children in Poona. However, reports describing rickets continue to appear, including one describing atrophic rickets, an unusually severe form associated with severe malnutrition and atypical radiological findings (92). Most of the surveys in Table 1 were made in chapati-consuming regions, some with deprivation of sunlight, some without. Thus, both etiologic factors were operational. The absence of reports from south India are in agreement with the conclusion of Wilson (79) that rickets seldom occurred south of Bombay.

Middle and Near East

Pakistan and Afghanistan rank high among the countries surveyed in incidence or rickets and osteomalacia (Table 2). Wheat (as *chapatis*) and barley predominate in the diets with intakes of phytate and fiber that are elevated. Infants, girls and women are deprived of sunlight.

In Iran (Table 2), 10% of children attending schools in Isfahan showed signs of active rickets. Another 10% presented deformities attributed to previous attacks of rickets (97). Many of these children were from villages in the vicinity where *tanok* was an important component of their diets. Mostly boys, they had been exposed to sunlight during preschool years, so that diet must have been predominant in etiology. Zorel et al. (97) attributed the rickets and deformities to high intakes of phytate.

Cannell et al. (100) examined 78 of 94 infants aged between 6 and 18 months in a moderately prosperous village located near Persepolis. Four had overt rickets. Twenty-eight showed at least two signs suggesting presence of rickets. The presence of active or subclinical rickets was confirmed in nearly all of these by radiological and chemical studies. In a study of fully developed rickets in village children, Amirhakimi (98) found low concentrations of serum calcium in 90% of a selected non-random hospitalized sample. Salimpour (99) found 200 cases of rickets in Tehran, a rate of 16%. These included infants and young children, many with PEM. He noted that radiological and chemical changes characteristic of rickets did not appear until the PEM was successfully treated. Severe anemia also was prevalent.

An incidence of 14% was reported for rickets in children in Damascus (93).

Reports of rickets from Israel are mainly from the south, where among the Bedouin (Table 2), the incidence was 9% in children (93), double that for Jewish children. Lowered concentrations of 25-hydroxyvitamin D were found in plasma of Bedouin women, who showed no physical signs of osteomalacia (106,107). These concentrations were approximately one-half to one-third those of Jewish women of Sephardic origin. Comparison of 25-hydroxyvitamin D in cord blood of newborn Bedouins yielded values that were about one-third those of Jewish newborn. Bedouin newborn also had a higher incidence of neonatal rickets and hypocalcemia.

East Asia

Osteomalacia was prevalent in the four northern provinces of China, Shansi, Shensi, Kansu, and Manchuria, at the time of May's survey in 1961 (93). May estimated that 5% of postpubertal females were affected (Table 3). No mention was made by May of rickets or osteomalacia in the rice-eating regions of China or Taiwan. Wheat was the staple grain in the north where rickets was reported. A recent study of breastfed infants in Beijing showed 11% to have radiological evidence of rickets (296).

May (93) stated that rickets was common in Vietnam. De Silva (110) observed rickets among the children admitted to hospitals in Colombo, Sri Lanka.

North Africa

The prevalence of rickets in North Africa is among the highest anywhere (Table 4). The main cause is insufficient exposure to sunlight among the predominantly Moslem populations. Poor and crowded housing, together with observance of purdah, are responsible. Wheat is the predominant staple. Calcium intakes are well below RDA. May (111) pointed out that rickets does not occur in the southern regions of the north African countries. These areas are populated by nomads who do not deprive themselves of sunlight.

In Cairo and northern Egypt, the incidence of rickets was 13% in a survey made by Abdou et al. (113) in 1965 (Table 4). A national survey in 1978 yielded a lower figure: 5%. Frequencies were high in Middle and Upper Egypt according to May (111): 45% of infants under two years and 80% of those under ten showed unmistakable signs. Maize, together with grain sorghum, were used for preparation of unleavened bread in rural areas where the incidence of rickets was 53%. In Upper Egypt, millet and some wheat were used as well. Avoidance of sunlight by women and its deprivation in infancy undoubtedly made important contributions to development of rickets in Egypt as a whole.

East Africa

Rickets is common in Ethiopia. May (111) estimated that one-third of all infants and children were affected (Table 4). Hojer et al. (116) developed criteria for recognition of rickets in children afflicted with PEM. A combination of the two occurred frequently. Detection of rickets in this form was important, because the rickets responded rapidly to treatment by injection of large dosages of vitamin D. Pulmonary disease often developed as a complication of the rickets-PEM syndrome (115). *Teff*, a staple cereal in Ethiopia is converted into bread by a sour-dough process that involves lengthy fermentation. However, the fermented bread retains much of its phytate (J. G. Reinhold, and O. J. Waslien, unpublished, 1975). Yet, the principal cause of rickets in Ethiopia is deprivation of sunlight due to mothers' fears of evil effects of exposure upon their infants. Exposure to sunlight for a total of 12 hours spread over two or three weeks sufficed to cure rickets (116a).

Uganda and Kenya produced no reports of rickets according to May (111). Tanzania did, although no particulars are cited. No rickets was reported from the countries of west or central Africa (111), probably because it had not been sought. When looked for, rickets proved to be common (295).

Table 4. Rickets and Osteomalacia in Africa

Location	Year	Staple cereals	Ca (mg/day)	P (mg/day)	Sunlight	Rickets; Osteomalacia	Age Group	Rate (%)	Reference
Algeria	1967	Wheat	—	—	Deprived	Rickets Osteomalacia	Infants Women	Prevalent	(111)
Nomads	1967	Wheat	—	—	Exposed	—	—	Absent	(111)
Morocco	1967	Wheat	263–475	1500	Deprived	Rickets	Children	Prevalent	(111)
Tunisia	1967	Wheat	—	—	Deprived	Rickets Osteomalacia	Children Women	50	(111)
North Africa	1967	Wheat	Low	—	Deprived	Children	0 to 5 yr	45–60	(112)
Egypt (cairo)	1965	Maize Sorghum Millet Wheat	450	—	Deprived	Rickets	Children	13	(113)
(Cairo)	1963	Same	—	—	Deprived	Rickets	4 mos– 4 yr	26.6 Inpatients 14.6 Clinic Pts.	(114)
(Middle)	1964	Same	—	—	Deprived	Rickets	<2 yr	45	(111)
(Upper)	1964	Same	—	—	Deprived	Rickets	<10 yr	80	(111)
All-Egypt	1978	Same	—	—	Deprived	Rickets	Preschool	5	(111)
Guinea	1962	(Manioc)	—	—	Exposed	—	Children	nil	(111)

Country	Year	Staple foods			Exposure	Condition	Age group	Number	Ref
Gambia	1962	Rice Millet Sorghum	—	—	Exposed	—	Children	nil	(111)
Liberia	1962	Rice	250	—	Exposed	—	Children	nil	(111)
Zambia	1962	Maize	—	—	Exposed	—	Children	Regional	(111)
Zimbabwe	1962	Maize Millet Sorghum	—	—	Exposed	Rickets	Children	Regional	(111)
Ethiopia	1962	Teff	—	—	Deprived	Rickets	Infants	13	(111)
Ethiopia	1973	Same	—	—	Deprived	Rickets	2 yr	High Mortality	(111)
Ethiopia	1977	Same	—	—	Deprived	Rickets + PEM	18 mo	65	(116)
Sudan (North)	1962	Sorghum (fermented) Millet	—	—	Deprived (mainly)	—	Children	nil	(111)
Tanzania	1962	Maize Millet Sorghum	—	—	Exposed (mainly)	Rickets (severe)	—	—	(111)
Kenya	1962	Maize Millet Sorghum	398	—	Exposed	—	Children	nil	(111)
Uganda	1962	Maize Millet Sorghum	—	—	Exposed	—	Children	nil	(111)

Table 4. (Continued)

Location	Year	Staple cereals	Ca	P	Sunlight	Rickets; Osteomalacia	Age Group	Rate (%)	Reference
			mg/day						
South Africa	1960	Maize Sorghum	—	—	Deprived	Rickets	<12 mos.	25	(118)
Urban	1961 -	Same	<300	—	Deprived	Rickets	<24 mos.	7	(119)
Urban	1966	Same	250—400	—	Deprived	Rickets Osteomalacia	Children Women	"Common" None	(120)
Rural	1966	Same	—	—	Exposed	Rickets	Children	"Rare"	
Urban	1969	Same	—	—	Deprived	Rickets	<6 mos.	14	(121)
Rural	1979	Same	125	—	Exposed	Rickets	4 to 13 yrs.	—	(122)

South Africa

Doncaster and Jackson (118) found that of 100 infants aged three months to two years with rickets, 30% had been exposed to sunlight less than ten minutes a day. Calcium intakes ranged from 250 to 500 mg/day and did not differ from those of control infants. As previously mentioned, Pettifor et al. (16) discovered rickets in village boys whose calcium intakes approximated 125 mg/day. In Cape Town, one infant in seven developed rickets before the age of six months (121).

Western Hemisphere

Rickets and osteomalacia are rarely reported from the developing countries of Latin America or the Caribbean. Recent exceptions, however, are descrived in three papers detailing the occurrence of rickets in Mexico City mainly in infants under 12 months (123—125). Heavy smog is believed to contribute. May (126) mentions reports of rickets in Panama affecting 4% of infants in their first year and 3% in their second. He also mentions that rickets occurs among Amerindian children in Guiana (127) at a rate of 8.5% for children under 12 years. Many children in Latin America receive supplemental calcium in the form of residual calcium hydroxide in maize tortillas. Maize is softened by heating in a suspension of calcium hydroxide, appreciable amounts of which remain in the dough (nixtamal) from which tortillas are prepared. Dwyer et al. (122) call attention to the risk of rickets among children consuming vegetarian diets.

Morbidity From Rickets in Terms of Population

Conversion of incidence expressed as percentages into numbers of children affected is instructive. Using Iran as an example, with a population of 30 million, conservatively estimated, 17% would be children under four years (128), a cohort of 5,100,000. With an incidence of rickets of 5%, the number of infants and children afflicted nationwide totals 255,000. If Salimpour's estimate (99) of an incidence of 16% for rickets in Tehran is used, the number of children, 16% of 2,500,000 (estimated population), is 625,000, and the number with rickets would be 100,000.

Rickets in Asian Immigrants to Britain

Around 1960, increased numbers of immigrants from Pakistan living in Glasgow were diagnosed as having rickets or osteomalacia (129). They included four groups: (a) congenital rickets, (b) prematurely newborn, (c) children who had been inadequately treated with vitamin D, and (d) adolescents (247). The reappearance of rickets in Glasgow, where it had been a grave problem for centuries, followed the termination of a program for supplementation of milk with vitamin D. This was done because of the sporadic appearance of hypercalcemia in children consuming the milk. The rickets was associated in most patients with lowered concentrations of calcium and phosphorus in serum and elevated alkaline phosphatase activities (130). Administration of vitamin D produced remission (131,132). Concentrations of the vitamin D metabolite 25-hydroxyvitamin D in plasma were lower in Asians than in white Glaswegians, lower in Asians with symptoms of rickets than in healthy Asians, and undetectable in those with overt rickets or osteomalacia (131,132). Stamp (133) reported that 25-hydroxyvitamin D concentrations in plasma of asymptomatic Asians were about half those of white controls, while Mawer and Holmes (134) showed that the ability of Asians to convert vitamin D into its hormonal metabolites was unimpaired. A high incidence of rickets was reported from many other centers where Asians had settled. Thus, in Manchester 30% of 168 Asians examined presented overt signs of rickets or osteomalacia (130). Vitamin D therapy corrected the physical, chemical, and radiographic lesions.

The Asian immigrants brought with them and retained the mode of life they had known in their native countries. This included diets in which *chapatis* were the principal source of energy. Together with legumes, high intakes of phytate and fiber were associated with decreased retention of calcium, magnesium, and zinc as a result. Wills et al. (134a) treated rickets successfully in a 15-year-old boy by lowering his consumption of *chapatis*. Ford et al. (131) replaced *chapatis* with a leavened bread made from flour of low rate of extraction and in this way restored serum calcium and phosphorus concentrations to normal and produced remission of rickets and osteomalacia. *Chapatis* consumption and phytate intake correlated significantly with serum alkaline phosphatase activity. Robertson et al. (136), using the data of Hunt et al. (137), calculated multiple regressions which showed that 25-hydroxyvitamin D levels, outdoor exposure, and vitamin D intake accounted for only 15.5% of a relationship to rickets. Inclusion of phytate raised the correlation to 34.7%, a significant gain. Singleton and Tucker (138) found an association between consumption of Asian diets and symptoms of rickets. According to Wills and Farney (139), lowered serum calcium concentrations were a response to increased phytate intakes. The decrease was corrected only partially by increased intake of vitamin D.

Some dissent was expressed from the view that high phytate intakes shared responsibility for the development of rickets. Dent et al. (140) removed *chapatis* from the diet of a 14-year-old boy with rickets without promoting remission. Resumption of the *chapatis*-containing diet combined with exposure to ultraviolent irradiation led to rapid healing of the rickets. O'Hara-May and Widdowson (141) compared *chapati* intakes of Asian boys and found no difference in the quantity consumed between those with and those without signs of rickets. The flour used was supplemented with calcium. Gupta et al. (142) described spontaneous correction of subnormal concentrations of calcium in serum during the summer months with increased exposure to sunlight. Single large doses of ergocalciferol provided protection for a year (143).

The episode of rickets in Asian immigrants in Britain produced an abundance of excellent investigations. It underscored the existence of two etiologic agents, phytate and other complexants of calcium that interfered with calcium absorption by decreasing bioavailability, and deficiency of vitamin D resulting from insufficient exposure to sunlight.

Rickets Elsewhere

Osteomalacia in guest workers from Turkey was reported from West Germany (144) and in an Asiatic in Switzerland by way of Uganda (145). In Australia, rickets occurs among immigrants from the Mediterranean region and also among aboriginals (147). A high incidence of rickets has been reported from Greece (146,146a). In the Soviet Union rickets is sufficiently widespread to justify publication of a book (148).

ZINC DEFICIENCY

Human Zinc Deficiency in Iran

Only two decades have passed since Prasad, Halsted, and Nadimi (149) associated the interrupted growth, hypogonadal status, and delayed adolescence of a group of dwarfs in the vicinity of Shiraz with a deficiency of zinc. Subsequently, a similar syndrome was found to exist in Egypt (150,152).

In Iran, the syndrome appeared only to occur among the rural population settled in villages. Migratory tribespeople indigenous to the area were unaffected. The evaluation of the zinc status of villagers was at first limited to examination of zinc concentrations in hair. These were significantly lower in village women and children than in their Shiraz conterparts (153). Hair analysis of zinc concentrations in men did not differ. Subsequently, Eminians et al. (153a) reported that preschool children from villages whose heights and weights were low for their age also had lower plasma zinc concentrations than did urban pairmates. Further support for the existence of a factor in the nutritional environment in villages detrimental to growth was obtained by Ronaghy (154), who surveyed the cadre of conscripts rejected for military service because of small stature. This group included 3% of the 19-year-olds. All lived in villages.

Definitive evidence that zinc deficiency was responsible for the attenuated growth and sexual development of village youths in Iran was supplied by Halsted et al. (155). Fifteen stunted males aged 19 years who had been rejected for military service because of subminimal stature, together with two stunted women of similar ages, were randomly assigned to three groups, one received a balanced nutritious diet plus a daily supplement of 27 mg/day of zinc as sulfate. Members of a second group received the diet plus a placebo. Those in the third group, received the diet and placebo for six months, after which the placebo was replaced by 27 mg of zinc/day. Growth resumed in all of the subjects. However, after six months the mean of the increment of growth of those receiving the zinc supplement from the beginning was 10.6 cm. This exceeded by a substantial margin the growth registered by the group receiving the placebo, 3.8 cm. The third group also showed a spurt of growth after the placebo was replaced by zinc. The difference between the gains of those receiving zinc supplementation and those receiving placebo was highly significant statistically ($P < 0.001$). Sexual maturation also occurred more rapidly in the zinc-supplemented groups, with a mean of 10.3 weeks before first seminal emissions occurred, compared with 29 weeks for those receiving diet plus placebo.

Tanok and Zinc Deficiency—Phytate and Fiber

An evaluation of the village diet made to identify the factor(s) responsible for the zinc deficiency syndrome established that its zinc content exceeded the RDA for zinc. Yet, the association of the syndrome with the rural environment and the preponderance of *tanok* with its high concentrations of phytate in the village diet strongly implicated *tanok* as the source of impaired bioavailability of zinc. The experiment previously described (25) designed to compare the effects of phytate and of *tanok* upon mineral metabolism proved that both phytate and *tanok* had detrimental effects upon zinc economy. During the control period, during which *bazari* provided 60% of the energy intake, zinc balances were positive in each of three subjects. Purified phytate added to the diet decreased zinc balances significantly. Consumption of an equivalent amount of phytate in the form of *tanok* produced further deterioration in zinc balances. *Tanok* not only was clearly detrimental but more so than phytate plus *bazari*.

Since the amount of phytate consumed was equivalent to that of the preceding period, the additional losses of zinc suggested that some other agent was enhancing the action of phytate upon zinc economy. Fiber appeared to be a likely possibility. Studies in vitro showed that cellulose, hemicellulose, and other fiber components, as well as dephytinized *tanok*, bound zinc at pH 6.5 (44,44a). However, it has been difficult to distinguish the complexant effects of fiber from those of phytate, since concentrations of both are dependent upon extraction rate. Phytate and fiber also combine with intrinsic zinc entering the intestine in salivary, gastric, biliary, pancreatic, and intestinal secretions. A portion of the zinc so combined is diverted to the large intestine and is lost in the feces. An enteropancreatic circulation of zinc has been postulated by Matseshe et al.

(159). Neutral detergent fiber added to human diets decreased zinc turnover and reduced its bioavailability (294).

Phytate/Zinc Ratios

The ability of phytate to depress bioavailability of zinc has been related to the molar ratios of phytate to zinc. Oberleas (174) has reviewed the relationship. Ratios of less than 12 appear to be compatible with adequate availability. Oberleas calculated ratios for Iranian flatbreads and reported a ratio of 10.8 for *bazari*, 12.8 for *sangak*, and 22.6 for *tanok*. These values appear to be in agreement with the behavior of *bazari* and *tanok* in human balance studies. (*sangak* has not been investigated.) The studies of Morris and Ellis (175) supported the ratio of 12 as a limiting value in human subjects. However, phytate/zinc ratios do not account for the greater effect of intrinsic phytate in *tanok* when compared with phytate added to *bazari* to produce an equivalent total concentration (25) observed in human balance studies.

No "intrinsic" phytase activity was detected in the gastrointesttinal secretions of the rat (176). It was concluded that cecal flora are the agents for phytate destruction occurring in the tract. Wise et al. (176) also reported that phytate combined with calcium was not degraded by phytase. The extent of the inhibition of phytase activity depended upon the quantity of calcium in the diet, which in rats far exceeded that in human diets. Spencer et al. (178) confirmed previous observations that availability of zinc to humans is not altered by a wide range of dietary calcium and phosphorus concentrations. Dietary calcium intake had no effect on retention of nitrogen, magnesium, copper, or zinc by infants (179). Ballam et al. (179a) suggested that dietary fiber source may affect the hydrolysis of dietary phytate in rats. Wise and Gilburt (160a) observed that concentration of amino acids in digesta was an important determinant of the solubility and retention of zinc and copper in the presence of calcium phytate.

Dietary Protein and Zinc Bioavailability

Consumption of protein from animal sources increased the bioavailability of zinc (161—168). Meat intakes of most villagers in Iran are low and less than those in cities (68). Yogurt is usually included in the village diet, but quantities may be limited and availability is seasonal. A number of studies designed to evaluate the extent of interference by dietary fiber with zinc absorption in Western-type diets containing generous allowances of meat and other proteins from animal sources have yielded negative findings (170,175). Such studies probably have limited applicability to impoverished societies, such as that of rural Iran and other developing countries. High fiber intakes yielded by vegetarian diets more nearly reproduce conditions existing in these regions.

Using a diet rich in fiber from fruit and vegetables, Kelsay and Prather (171) found that balances of zinc, calcium, and magnesium were decreased by significant margins as a result of diminished bioavailability. Obizoba (172) measured the response of zinc and copper balances to diets containing cereal and legumes together with milk in human subjects. Zinc balances became negative when phytate (and fiber) were elevated, but balances of copper remained unchanged. Nitrogen intakes were less than half those provided by Western diets. Vegetarian diets may bring about adaptive changes (173). Adaptation, however, did not overcome wastage of zinc produced by Iranian diets rich in phytate and fiber except in two Americans (24).

A comparison of zinc intakes of schoolboys aged 13 and 14 living in villages with those of city boys in Maleki's study (68) showed that the intakes of the rural boys, 19.0 ± 6.2 (mean and SD) mg/day were somewhat lower than those of the urban boys, 22.3 ± 7.3 mg/day. Both exceeded the RDA for zinc. Bioavailability was undoubtedly higher than in urban diets because of the use of leaven in preparation of urban breads (160).

Avid retention of zinc by villagers eating a nutritious Western-type diet, as described in preceding section, testified to their zinc-depleted status (69).

Geographical Distribution of Zinc Deficiency

The fully developed syndrome of interrupted growth, hypogonadism, and delayed adolescence associated with impaired zinc nutrition described in Iran and northern Egypt occurs also in the Anatolian region of Turkey. As in Iran, it was characterized by geophagia and severe microcytic anemia (180). Treatment with zinc was effective (181). Hambidge et al. (182) found low concentrations of zinc (and six other heavy metals) in the hair of residents of Chandigar, India, and Bangkok, Thailand. Hambidge and Walravens (183) listed Cairo, Pretoria, Cape Town, and Hyderabad as locations where zinc deficiency (low plasma zinc concentrations) were associated with generalized malnutrition and response to zinc therapy. Subnormal zinc concentrations were found in 20% of more than 2,000 plasma samples examined in Tunisia by Jacobs et al. (184), the lowest being found in pregnant or lactating women. In the People's Republic of China, low concentrations of zinc in plasma and hair of children aged one to six were attributed to low dietary intakes (185). Treatment with a zinc supplement corrected pica, which was prevalent, and stimulated resumption of growth. Holt et al. (186) reported low concentrations of zinc in serum in 24% of aboriginals examined in western Australia, particularly in preadolescent and adolescent age groups. Serum iron concentrations were depressed in over 50%, but those of serum copper were often elevated. Zinc intakes were inadequate in Papua New

Guinea, where less than 3% of dietary zinc was derived from foods of animal origin (187). Lowered concentrations of zinc in serum of children of poor families were reported from Brazil (188). Solomons (189, 190) showed that the two main staples of the Guatemalan rural diet, maize and beans, decreased absorption of zinc in both inorganic and organic form. A survey by Iyengar (191) confirmed the occurrence of low serum zinc concentrations in blood in Bangladesh, India, and Turkey and in hair in Turkey, Kenya, Egypt, South Africa (Bantu), India, Bangladesh, and in parts of the United States. Kew et al. (192) had previously reported that zinc concentrations in the plasma of Bantu were the same as those of South African Caucasoids. Hambidge et al. (193) encountered subnormal concentrations of zinc in plasma and hair of children enrolled in the Denver Head Start Program. Pica in a two-year-old infant was corrected within three days by administration of zinc (194). Zinc concentrations in plasma of Thai infants tended to be lower than those established in Western countries (198). Mexican-American Children with retarded growth had low concentrations of zinc in serum and hair associated with low concentrations of vitamin A in serum (196).

Zinc and Protein-Energy Malnutrition

Zinc deficiency often complicates PEM as a result of lowered zinc intake and increased losses due to diarrhea. Zinc metabolism in PEM has been studied in Eqypt (10), South Africa (197,198), India (199, 200), Cuba (201), and Jamaica (202). The latter observed that low plasma zinc concentrations were strongly associated with nutritional edema. Absence of correlation between albumin and zinc in plasma indicated that the low concentrations of zinc observed were not secondary to lowered albumin concentrations.

Zinc and Pregnancy

Pregnancy is associated with declining concentrations of zinc in hair and plasma. The decrease in plasma zinc is only partly accounted for by expansion of plasma volume and lowered albumin concentrations. The fall in plasma zinc is physiological and is not necessarily a result of lowered zinc consumption (203,204). Maternal plasma zinc concentrations are inversely correlated with fetal weight. An association between fetomaternal complications of pregnancy was detected in mothers with the lowest concentrations of zinc and albumin in plasma (205−207). Mexican women living in Mexico have poorer zinc nutrition than those living in the United States (208). The incidence of pregnancy-related hypertension was decreased by dietary supplements of zinc (209). Infant birth weight was not correlated with zinc levels in plasma and hair in South Lebanon (210). Dietary intakes of zinc by Mexican-Americans were 67% of the RDA (204). Intakes of zinc and plasma zinc concentrations of women who ate vegetarian diets did not differ from those of nonvegetarians (211), although intakes of zinc by both groups were low. In a review of zinc and reproduction, Apgar (212) reports that malformations tended to be associated with low concentrations of plasma zinc. Prenatal iron supplementation may adversely affect maternal zinc status (203).

Metabolic Functions of Zinc

Certain functions of zinc may be deduced from the defects of metabolism associated with severe zinc deficiency. Retarded growth may result from a critical need for zinc to accomplish transcription and translation. Nearly 20 nucleotide polymerases are zinc-dependent metalloenzymes, as are other enzymes involved in production of mRNAs and their translation. Vallee and Falchuck (213) postulated a role of zinc in gene activation and/or repression. In the absence of zinc, chemically altered forms of some enzymes are produced. Lowered incorporation of thymidine into DNA is involved (214).

Disturbed lipid metabolism may result from impaired incorporation of mucosal lipids into chylomicrons due to zinc deficiency. Abnormal metabolism of fatty acids is associated with disturbed prostaglandin activity (215). Evans and Johnson (216) observed that aspirin, which inhibits prostaglandin synthesis, also inhibited zinc absorption.

Hypogonadism, a conspicuous response in human zinc deficiency, was manifested by decreased testosterone in serum of rats fed zinc-deficient diets. However, weight-restricted rats showed similar changes compared with pair-fed rats or rats fed ad libitum (217). Zinc-deficient rats did not respond to injection of human chorionic gonadotropin in contrast to weight-restricted rats, which exhibited a significant testosterone response (217). The difference was attributed to Leydig cell failure.

Immunochemical Functions of Zinc

Zinc-deficient animals and children develop thymic atrophy, which is reversed by zinc supplements of 2 mg/kg/day (218). Children with PEM also show thymic atrophy and increased susceptibility to infection. Such children, when skin tested with Candida antigen, showed a highly significant increase in response if the injection site was treated with zinc sulfate (219). Chandra (220) states that zinc uptake is an important factor modulating immune response. Duchateau et al. (221) fed normal subjects zinc supplements daily for one month. Tests of lymphocyte reaction to phytohemagglutinin and concanavalin A showed an enhanced response. This behavior, however, was not related to pretreatment concentration of zinc or to zinc deficiency. Haynes et al. (222) showed that zinc-deprived rhesus monkey infants exhibited numerous alterations in immunocompetence.

Other Functions of Zinc

The skin becomes vulnerable in zinc deficiency. Acrodermatitis enteropathica, a rare human zinc deficiency disease, is completely corrected by treatment with zinc (223,224). Lesions of the skin developing as a complication of PEM also respond to treatment with zinc (225).

Zinc is essential for vitamin A metabolism (226). Zinc deficiency leads to impaired synthesis of vitamin A-binding protein and decreased release into the blood stream of vitamin A stored in the liver. In addition, retinol dehydrogenase, like other dehydrogenases, is a zinc dependent metalloenzyme (227).

Prasad et al. (228) produced zinc deficiency experimentally in human volunteers and observed adverse effects on connective tissue including decreased synthesis of collagen, RNA and activity of deoxythymidine kinase. Activities of zinc-dependent enzymes in plasma were decreased. Ammonia concentrations in blood increased. Zinc deficiency induced in humans produced changes in alkaline phosphatase (AP), and ALA dehydratase activities in serum and AP in leucocytes (229). Impaired glucose tolerance and leucocyte hemotaxis also were observed. Prasad et al. (230) reported that treatment with zinc stimulated growth in growth-retarded sickle cell anemics.

Teratoxic effects associated with zinc deficiency have been reviewed by Hurley (231).

The importance of adequate zinc nutrition for bone development has been discussed in a preceding section.

IRON DEFICIENCY

Introduction

Iron deficiency is prevalent throughout the developing countries. The causes are multiple, but poor bioavailability of nonheme iron in food, particularly the iron of foods derived from cereal grains, is among the more important. Fleming (232) ranked the iron of rice as having the lowest bioavailability, about 1% of the total iron, followed by maize at 3% and wheat at 5%. A survey made by an International Nutritional Anemia Consultative Group (232a) ranked the iron of maize as having the lowest bioavailability, 3.7%, followed by rice at 6.3% and wheat, 31%. Narasinga Rao (232a) rated wheat lower than rice and also found the iron of grain sorghum and millet to be poorly available. Iron contents of whole meals prepared from these or other grains may be artefactually elevated by iron-rich soil (232b). Soil iron is poorly soluble in water or digestive secretions. Bioavailability of iron in the form of heme is superior to that of nonheme iron, but heme intakes of many populations are low to the point of being negligible.

Morris et al. (233) found that about 70% of the iron of wheat bran consisted of monoferric phytate, a water-soluble substance with

bioavailability equal to that of ferrous sulfate (234). Until this discovery, the poor availability of iron in wheat had been attributed to
the low solubility of iron phytate. Ellis et al. (235) assayed the
response of hemoglobin production by rats fed tricalcium iron phytate
and found it to be inferior to that produced by iron phytate. Such
a complex possibly could form in calcium enriched foods, e.g. maize
tortillas in Mexico. Little knowledge exists regarding the nature of
the intrinsic iron of cereals other than wheat.

Dietary fiber forms complexes with ferrous iron that are stable
at pH > 6 (44,44a). Iron so bound is released in increasing proportions as pH falls below 6 (236). Iron also is bound by dietary protein, e.g., zein, with similar pH dependence (237). Both ferrous and
ferric iron form poorly soluble hydroxypolymers at pH > 4 and pH >
3, respectively. pH values that would be attained only fleetingly in
the duodenal contents. Reversible binding of iron by fiber, protein,
and phytate may stabilize iron in the small intestine by protecting
against polymerization. Leigh and Miller (243) showed that fiber
stabilized iron solutions in the course of digestion in vitro.

The occurrence and behavior of iron bound to fiber has been
reviewed recently (239,240). Dietary fiber decreased uptake of iron
by intestinal segments of rat intestine in vitro (242), from rat intestine in vivo (238), and from dog intestine in vivo (241). Simpson et
al. (244) reported that wheat bran decreased iron absorption in human
subjects by 51 to 74%. A soluble phosphate-rich fraction separated
from dephytinized bran was more inhibitory than an insoluble high-
fiber fraction. Phosphorus in various forms was found by Peters et
al. (244a) to decrease iron absorption by human subjects. Consumption of 36 g/day of wheat bran by human subjects maintained positive
iron balances. However, the iron intakes were high, and the diets
had high contents of adjuvants of iron absorption, such as ascorbic
acid and proteins of animal origins (245). Iron retention by young
men in Iran decreased when fiber-rich *bazari* bread made up 60% of
the energy intake of their diet (246).

Populations at risk for rickets and osteomalacia also tend to be
at risk for anemia due to iron deficiency. Such an association has
been found for Asian immigrants in Britain (247–249), in India, (250),
in South Africa (251), and in Egypt (252). The frequency of the association suggests that it is more than coincidental. Although an action of
phytate appears to be unlikely because of the preponderance of ferric
monophytate, at least in the diets of wheat-consuming populations,
consumption of *chapatis* was associated with iron deficiency (253).
Complexes of iron, phosphate, protein, and fiber have been described
(255). The low uptake of iron from cow's milk or iron-fortified infant
formulas as compared with human milk should not be overlooked (254).
Vegetarians utilized the iron of a vegetarian diet better than did
omnivores (256).

The occurrence of anemia due to iron deficiency was disguised
in PEM until treatment brought about repletion of body fluids (256a).

There was an inverse relationship between protein depletion and iron deficiency (256b) when the two states coexisted.

Modification of Iron Bioavailability

Ascorbic acid (AsA) enhances the availability of iron of foods, an observation that appears to have been made originally by Heilmeyer and van Mutius (257) and often confirmed since. The effects of AsA can be spectacular. Thus, addition of 50 or 100 mg of AsA to maize porridge increased the bioavailability of its iron tenfold (258). The adjuvant action of AsA has been so often demonstrated that reports of its failure are noteworthy, yet Mathan et al. (259), in a study made in India under WHO sponsorship, found no beneficial effect when 500 mg of AsA daily was given to pregnant women. Instead, they found that casein, 15 g/day, produced a superior hematologic response. No explanation was offered for the action of casein, but enhanced secretion of gastric acid stimulated by the protein may explain the improvement (260). Clydesdale and Nadeau (263) showed that milk improved the solubility of iron of cereals and, as a result, bioavailability. Flesh foods rank with AsA in effectiveness as promoters of iron utilization (264). AsA was less effective in the presence of soybean products compared with meals based on egg white (261). AsA acted by releasing iron from dietary complexants but did not stimulate iron absorption (238).

Calcium chloride appeared to inhibit absorption of nonheme iron (262). Calcium hydroxide residues remaining in tortillas after steeping in heated lime water may affect iron solubility and so decrease iron availability (237). Geophagia decreased iron (and zinc) absorption by Turkish children (267).

Consumption of tea with a meal lowered availability of iron by about 50%, an action that was largely overcome by increasing the quantity of AsA given as an adjuvant (258). Tea consumption by infants lowered hemoglobin concentrations in blood (265). Tannin and associated polyphenols were held responsible, and their effect upon iron availability probably contributes substantially to the iron depletion prevalent in the tea-drinking populations that abound in Asia. The tannin content of foods and beverages of the Indian diet has been measured (266) and its effect on ionizable iron concentrations calculated. Phytoferritin in soy flour decreased iron absorption (292).

Iron Deficiency in Pregnancy

The relationship between anemia, mainly due to iron deficiency, and the outcome of pregnancy is a source of concern. Its effects have been evaluated by Garn et al. (268) and by Prema et al. (269). Lowered birth weights and other risk factors were associated with maternal hemoglobin concentrations below 110 gm/L (Garn) or 80 gm/L (Prema). Royston (270) surveyed the prevalence of lowered hemoglobin concentrations among the developing countries throughout the world. A condensed version of his findings is shown in Table 5. He confirms that the anemia of pregnancy is a nutritional syndrome that is the result of low nutrient intake, poor absorption of iron due to low bioavailability, increased nutrient losses and/or poor utilization of iron and perhaps of other mineral elements. Table 5 compares the mean hemoglobin concentrations in pregnancy found in the major developing regions of the world and also shows the percentage of women whose hemoglobin concentrations were below the lower limit for healthy, well-nourished pregnant women. The developing countries of Asia have the highest proportion of lowered hemoglobin concentrations, followed by those of Africa and the Americas. Table 5 also shows the large disparity between the values found in developing countries and those in the United States.

Iron Supplementation

Correction of iron deficiency by means of dietary iron supplementation is beneficial in most circumstances. However, it may produce adverse effects (271,272,273). Masawe et al. pointed out that infectious diseases occurred infrequently in the presence of chronic iron deficiency. They noted that attempts to correct the deficiency by administration of iron supplements to patients with quiescent malarial infections was followed by activation of the illness. Murray et al. (272) found that the incidence of infections among iron-deficient Somali nomads was increased sevenfold by supplementation of the diet with iron medication. This response included activation of quiescent brucellosis, malaria, or tuberculosis. Controls who received placebos were unaffected. The authors suggested that host defenses were more effective during iron deficiency than during iron repletion and that iron deficiency may be regarded as an ecological compromise between host and infectious agent. Another example of a detrimental effect of iron supplementation was provided by milk-drinking nomads with an unexpected freedom from infection with *Entamoeba histolytica* (274). Administration of iron to correct iron deficiency increased susceptibility to infection by amoeba. It appeared that the lactoferrin in milk competed successfully for iron with the pathogen and inhibited multiplication of the latter.

Gracey et al. (275,276) stressed the importance of the microbic ecology of the small intestine in malnutrition. Diarrhea resulting from contaminated food, water, and environment were among the major causes of debilitating malnutrition, particuarly PEM. Since iron is essential for the multiplication of organisms, including pathogens, in the small intestine, an abundance of iron may facilitate invasion of the upper intestinal tract with devastating effects on the host.

Table 5. Anemia in Pregnancy

Region	Hb < Normal Mean gm/L	%	Region	Hb < Normal Mean gm/L	%
Africa			America		
northern	110	47	middle	124	25
western	107	47	Caribbean	117	37
eastern	119	44	tropical, south	125	44
southern	118	48	temperate, south	118	47
Mean	110	47	Mean	121	38
Asia			U.S., all women	138	6
			U.S., poor women	132	12
southwestern	115	54			
middle	113	56			
southeastern	108	53			
eastern	107	61			
Mean	111	56			

Source: Adapted from Ref. 270.

Iron and Immunocompetence

Iron deficiency is associated with impaired immunocompetence that affects cell-mediated mechanisms particularly (277,278). It appears, however, that no major changes occur in plasma immunoglobulin concentrations or other humoral factors (279). Bhaskaram and Reddy (280) reported finding decreased percentages of T lymphocytes in anemic children (hemoglobin concentrations between 40 and 70 g/L) whose anemia responded to iron medication. Decreased ^{3}H-thymidine incorporation was also observed but was not corrected by the iron.

Treatment of iron deficiency, therefore, presents a dilemma: Unless it is treated with iron supplementation, impaired immunocompetence may fail to provide resistance to pathogens. However, administration of iron may encourage invasion by new pathogens and rejuvenate those that are quiescent. The state of hygiene in the environment may govern the decision whether or not to use iron medication. Anemic infants treated en masse with iron in developed countries did not exhibit similar side effects (293).

Iron Nutrition and Productivity

Productivity of adult workmen in Indonesia was related to their iron nutrition (281). Work performance was impaired when hemoglobin concentrations were less than 100 gm/L. As a result, earnings of rubber plantation workers were related to their hemoglobin concentrations.

DISCUSSION

Multiple Etiologies of Rickets and Osteomalacia

The syndromes of rickets and osteomalacia result from two primary etiologic factors: deprivation of sunlight and lowered bioavailability of calcium. The latter follows the excessive intake of complexants, such as dietary phytate and fiber. Alone, either factor may ordinarily produce mild symptoms that commonly escape detection. Acting in combination, they produce the fully developed clinical syndromes of rickets or osteomalacia. It is in the predominantly Moslem countries of the Middle East and north Africa, together with non-Moslem northern and central India that the fully developed syndrome currently appears most frequently. Treatment directed toward restoration of Vitamin D levels, either by exposure to sunlight or administration of the vitamin, possibly in large dosage, to produce a depot that will ensure a supply for a period of months is usually, but not always, effective (26).

Treatment directed at lowering of the consumption of metal complexants is more difficult to accomplish yet may be required because complexants alone may cause depletion (26). Incorporation of leaven, already in use for production of many breads, should be extended to those, like *tanok* in Iran, from which it is now omitted. Introduction of starter and fermentation for two or more hours would provide only a partial solution because of the resistance of wholemeals to action of leaven (160,282,288). However, as the pH of the sponge falls, phytase in wheat would become more active and aid in phytate degradation.

A better method for reducing phytate content, which would also lower that of fiber, would be the use of more effective sifting of grist, particularly, the avoidance of recycling offal into the flour (a practice which the author has witnessed). The ultimate solution for eliminating excessive intakes of mineral complexants is the introduction of modern milling equipment to lower rates of extraction. Lowered consumption of bread by the introduction of more varied diets must await an improvement in economic status of the populations at risk.

Failure to provide adequate incentives to production of food coupled with crop failures has led to dependence upon imported grains in many developing countries. Purchase of imported grains is costly and attempts are being made to extend grains by incorporation of bran into wheat or triticale flours. This practice has been a traditional response to famine as well. Halim and Lorenz (288) have found that bran in the amount of 10% added to flour and allowed to ferment for a day or more, as practiced in the Sudan, yielded an acceptable bread. The extent of phytate destruction under these conditions and the effect of such treatment upon mineral metabolism remains to be evaluated.

Multiple Mineral Deficiencies

Excessive intakes of phytate, fiber and perhaps other complexants of nutritionally indispensable mineral elements leads to the simultaneous depletion of calcium, magnesium, iron and zinc. Of particular interest in this connection is the likelihood that depletion of zinc may contribute to the disturbances of bone metabolism characteristic of rickets and osteomalacia. On the other hand, depletion of calcium undoubtedly shares with zinc a role in the stunting of growth that characterized the syndrome of zinc deficiency described by Prasad et al. in Iran (149) and Egypt (150-152).

The association of iron deficiency anemia with rickets and osteomalacia is well established. Despite the ready availability, of ferric monophytate demonstrated by Lipschitz et al. (224), Faraji et al. (246) found that consumption of the Iranian bread, *bazari*, tended to increase excretion of iron in feces, decrease iron balances and decrease serum iron concentrations. These tendencies were enhanced by consumption of cellulose in applesauce (45). Also to be watched is the phosphate-rich, soluble fraction of wheat bran described by Morris and Ellis (175) as having the ability to decrease iron bioavailability.

Absence of Adaptation to Dietary Phytate

A change from negative to positive calcium balances observed by Walker et al. (40) when national wartime bread was eaten for several weeks was attributed to adaptation to dietary phytate. A similar improvement in calcium availability was described by Bhaskaram and Reddy (42) during periods of *chapati* consumption by Indian children. However, numerous calcium balance studies conducted in Iran during which *bazari* or *tanok* served as the principal source of energy for periods of 16 to 96 days failed to demonstrate improvement in calcium availability (24-26,45,51,297). James (291) claims to have perceived changes in these studies that suggested adaptation. This is questionable. The absence of adaptation in the Iranian studies may be related to a substantial difference in content of phytate in the diets employed. These are estimated to be ten times greater in the Iranian diets than in those of Walker. Zinc balances of the two American volunteers did improve with time, a change that could be attributed to adaptation. None of the Iranian subjects showed similar changes. Addition of vitamin D produced no improvement in calcium retention in vitamin D-replete subjects (26). The mineral-depleted state of Iranian villagers supplied additional evidence of failure to adapt (69), as did the subnormal serum calcium concentrations of 21% of adults and 34% of children among randomly selected villagers (284).

Sunlight

The prevalance of rickets and osteomalacia in Moslem countries is largely a result of deprivation of sunlight. Women of childbearing age are far more susceptible than males because religious dicta require women to wear veils. The practice of seclusion of women probably has diminished in recent years, although it may again be rising in countries that have turned toward strict interpretation of Islamic law. Aside from religion, many Moslem women prefer to wear veils because of convenience and as protection from intense sunlight. Children may be deprived of sunlight to preserve light complexions. However, avoidance of sunlight is clearly being overdone.

The effects of total deprivation of sunlight upon young men have been investigated (285). Calcium absorption began to fail after only three to four weeks. Concentrations of 25-hydroxyvitamin D in serum declined by 63% between 14 and 63 days of deprivation. Increased skin pigmentation lengthens the exposure needed to produce a given amount of vitamin D (286,289).

COMMENTS

Rickets also occurs in developed countries but in mild form without lasting effects. However, rickets as it occurs in rural Iran may produce deformities that persist over the life span. According to Zarel et al. (97) one in five boys attending school in Isfahan suffered from rickets or showed deformities from previous attacks. Severely crippled young women were found in villages in the Shiraz region, some unable to walk without assistance (P. Abadi, 1974, unpublished). According to A. Lahimgarzadeh (1973, unpublished), the most frequent complaint of village women of all ages was bone pain associated with osteomalacia or osteoporosis. It is evident that bone disease of nutritional origin presents an important threat to public health in many developing countries. Besides assistance directed toward prevention or correction of calcium, iron, and zinc depletion, the need to avoid pregnancy by woman with active or incipient osteomalacia must be emphasized. Otherwise, the stresses imposed by pregnancy upon mineral metabolism added to those from faulty diet may lead to serious complications for both mother and infant.

ACKNOWLEDGMENTS

The author's work in Shiraz on mineral metabolism was made possible largely by the facilities of the Institute of Nuclear Medicine in the Nemazee Hospital. These included a metabolism ward. The Institute was created by Dr. Russell Barakat, at that time Associate Professor of Medicine, who found financial support, designed the laboratories, assembled equipment and recruited the technical staff.

Studies of villagers became possible as a result of a program of health care initiated in rural areas by the Department of Community Medicine of the University in Shiraz under the direction of Dr. Hosain Ronaghy. As a result, the attitude of indifference and suspicion first encountered was replaced by willing cooperation.

REFERENCES

1. V. N. Patwardhan and W. J. Darby, *The State of Nutrition of the Arab Middle East*, Vanderbilt University Press, Nashville, TN, 1972.

2. T. S. West, Soil as a source of trace elements, *Phil. Trans. R. Soc. London (B)*, *294*: 19–39 (1981).

3. M. Parrott-Garcia and D. A. McCarron, *Nutr. Rev.* *42*: 205–213 (1984).

4. T. A. Palm, Geographical distribution and etiology of rickets, *Practitioner*, *45*: 270 (1890).

5. R. J. C. Stewart, Bone pathology in experimental malnutrition, *World Rev. Nutr. and Diet.*, *21*: 1–74 (1975).

6. C. E. Dent and T. C. B. Stamp, Vitamin D, rickets and osteo-malacis, in *Metabolic Bone Disease* (L. V. Avioli and S. M. Krane, eds.) Academic Press, New York, 1977, pp. 237–320.

7. C. Gopalan and B. S. Narasingo Rao, Dietary allowances for Indians, in *Indian Council of Medical Research, Special Report Series 60*, Hyderabad-7 (1968).

8. A. Begum and S. M. Pereira, Calcium balance studies on children accustomed to low calcium intakes, *Br. J. Nutr.*, *23*: 905–916 (1969).

9. R. Smith, Rickets and osteomalacia, *Hum. Nutr. Clin. Nutr.*, *36*: 115–133, (1982).

10. H. H. Sandstead, A. S. Shukry, A. S. Prasad, et al., Kwash-iorkor in Egypt. *Am. J. Clin. Nutr.*, *17*: 15–26 (1965).

11. J. Howland and B. Kramer, Factors concerned in the calcification of bone, *Tr. Am. Pedr. Soc.*, *34*: 204–208 (1922).

12. F. C. McLean and M. R. Urist, *Bone*, 2nd ed., University of Chicago Press, Chicago, 1961.

13. H. E. Maltz, M. B. Fish, and M. A. Holliday, Calcium deficiency rickets and the renal response to calcium infusion, *Pediatrics*, *46*: 865–870 (1970).

14. S. W. Kooh, D. Fraser, B. J. Reilly, et al., Rickets due to cal-cium deficiency, *N. Eng. J. Med.*, *297*: 1264–1266 (1977).

15. J. M. Pettifor, F. P. Ross, J. Wong et al., Rickets in rural blacks in South Africa. Is dietary calcium a factor? *J. Pediatr.*, *92*: 320–324 (1978).

16. J. M. Pettifor, F. P. Ross, G. F. Moodley, and F. Shuenyane, Calcium deficiency in rural black children in South Africa, *Am. J. Clin. Nutr.*, *32*: 2477–2483 (1979).

17. J. M. Pettifor, F. P. Ross, G. Moodley, et al., Calcium deficien-cy rickets associated with elevated 1,25 dihydroxyvitamin D con-centrations in a rural black population. *Proc. Workshop Vitamin D, 4th*, 1125–1127 (1979); *Chem. Abst.*, *92*: 40205 (1980).

18. J. M. Pettifor, W. A. DeKlerk, M. R. Sly et al., Effects of varied calcium, phosphorus and vitamin D contents on mineral and bone metabolism in baboons, a preliminary report. *S. Afr. J. Sci.*, *77*: 136—139 (1981).

19. J. M. Pettifor and F. B. Ross, Low dietary calcium intakes and its role in pathogenesis of rickets, *S. Afr. Med. J.*, *63*: 179—184 (1983).

20. J. M. Pettifor, F. P. Ross, R. Travers, M. Glorieux, and H. F. DeLuca, Dietary calcium deficiency: a syndrome associated with bone deformities and elevated 1,25 dihydroxyvitamin D concentrations, *Metab. Bone Dis. Related Res.*, *2*: 301—305 (1981).

21. M. R. Sly, W. H. Van der Walt, H. Willem, et al., Exacerbation of rickets and osteomalacia by maize; a study of bone histomorphometry and composition in young baboons, *Calcified Tissue Int.*, *36*: 370—379 (1984).

22. W. R. Ackroyd and B. G. Krishnan, The effect of calcium lactate on children in nursery school, *Lancet*, *235*: 153—155 (1939).

23. H. Spencer and D. B. Kramer, Factors influencing calcium balances in man, in *Calcium in Biological Systems* (K. P. Rubin, G. B. Weiss, and J. W. Putney, Jr., eds.), Plenum, New York, 1985.

24. B. J. Campbell, J. G. Reinhold, J. J. Cannell, and I. Nourmand The effects of prolonged consumption of wholemeal bread upon metabolism of calcium, magnesium, zinc and phosphorus of two young American adults, *Pahlavi Med. J. (Shiraz)*, *7*:1—17 (1976).

25. J. G. Reinhold, K. Nasr, A. Lahimgarzadeh, and H. Hedayati, Effects of purified phytate and phytate-rich breads upon metabolism of zinc, calcium, phosphorus and nitrogen in man, *Lancet*, *1*: 283—288 (1973).

26. J. G. Reinhold, B. Faraji, P. Abadi, and F. Ismail-Beigi, An extended study of the effect of Iranian village and urban flatbreads on mineral balances of two men before and after supplementation with vitamin D., *Ecol. Food Nutr.*, *10*: 169—178 (1981).

27. W. F. Loomis, Rickets, *Sci. Amer.*, *223*: 77—91 (1970).

28. G. R. Jansen, The nutritional status of preschool children in Egypt, *World Rev. Nutr. Diet*, *45*: 42—67 (1985).

29. K. Guggenheim and N. A. Kaufman. Nutritional health in a changing society—studies from Istael, *World Rev. Nutr. Diet.*, *24*: 219—240 (1976).

30. D. C. Wilson and E. M. Widdowson, *Indian Res. Mem.*, *34* (1942).

30a. B. Lund, P. Charles, C. Egsmose, C. Lund, et al., Changes in vitamin D metabolites and bone histology in rats, *Calc. Tissue Int.*, *37*: 478—483 (1985).

30b. R. K. Rude, J. S. Adams, E. Ryzen, et al., Low serum concentrations of 1,25 dihydroxycholecalciferol in human Mg deficiency, *J. Clin. Endocrin. Metab.*, *61*: 933—940 (1985).

31. E. Mellanby, The presence in foodstuffs of substances having

specific harmful effects under certain conditions, *J. Physiol.*, *61*: xxiv–xxvi (1926).

32. H. M. Bruce and R. K. Callow, Cereals and rickets. The role of inositol hexaphosphoric acid, *Biochem. J.*, *28*: 527–537 (1934).

33. D. C. Harrison and E. Mellanby, Phytic acid and the rickets-producing action of cereals, *Biochem. J.*, *33*: 1660–1680 (1938).

34. R. A. McCance and E. M. Widdowson. Mineral metabolism of healthy adults on white and brown bread dietaries, *J. Physiol*, *101*: 44–85 (1942).

35. H. A. Krebs and K. Mellanby, The effect of National Wheatmeal on the absorption of calcium, *Biochem. J.*, *37*: 466–468 (1943).

36. R. A. McCance and C. M. Walsham, Digestibility and absorption of calories, protein, purines, fat and calcium in wholemeal wheaten bread, *Br. J. Nutr.*, *2*: 26–34 (1948).

37. E. Mellanby, The rickets-producing and anticalcifying action of phytate, *J. Physiol*, *109*: 488–533 (1949).

38. E. Hoff-Jorgensen, O. Andersen, H. Begtruo, and G. Nielsen, The effect of phytic acid on the absorption of calcium and phosphorus. 2. In Infants, *Biochem. J.*, *40*: 453–454 (1946).

39. E. Hoff-Jorgensen, O. Andersen, and G. Nielsen, The effect of phytic acid on the absorption of calcium and phosphorus, *Biochem J.*, *40*: 555 (1946).

40. A. R. P. Walker, F. W. Fox, and J. T. Irving, Studies in human calcium metabolism. The effect of bread rich in phytate phosphorus on the metabolism of certain mineral salts with special reference to calcium, *Biochem. J.*, *42*: 452 (1948).

41. R. Rajalakshmi, Preschool child malnutrition. Patterns, prevalence, prevention, *Baroda J. Nutr.*, *3*: 1–129 (1976).

41a. J. G. Reinhold, Phytate concentrations of leavened and unleavened Iranian breads, *Ecol. Food Nutr.*, *1*: 187–192 (1971).

42. C. Baskaram and V. Reddy, Role of dietary phytate in etiology of nutritional rickets, *Indian J. Med. Res.*, *69*: 265–270 (1979).

43. J. B. Eastwood, H. C. De Wardener, R. W. Gray, and J. L. Lemann, Jr., Normal plasma 1,25 dihydroxy-vitamin D concentrations in nutritional osteomalacia, *Lancet*, *1*: 1377–1378 (1979).

44. J. G. Reinhold, F. Ismail-Beigi, and B. Faraji, Fiber vs. phytate as determinant of the availability of calcium, zinc and iron of breadstuffs, *Nutr. Rept. Int.*, *12*: 75–85 (1975).

44a. F. Ismail-Beigi, B. Faraji, and J. G. Reinhold. Binding of zinc and iron to wheat bread, wheat bran, and their components, *Am. J. Clin. Nutr.*, *30*: 1721–1725 (1977).

45. F. Ismail-Beigi, J. G. Reinhold, B. Faraji, and P. Abadi, Effects of cellulose added to diets of low and high fiber content upon the metabolism of calcium, magnesium, zinc and phosphorus by man, *J. Nutr*, *107*: 510–518 (1977).

46. J. H. Cummings, Nutritional implications of dietary fiber, *Am. J. Clin. Nutr.*, *31*: s21–s29 (1978).

47. J. L. Kelsay, K. M. Behall, and E. S. Prather, Effect of fiber

from fruits and vegetables on metabolic responses of human subjects. II. Calcium, magnesium, iron and silicon balances, *Am. J. Clin. Nutr., 32*: 1876—1880 (1979).

48. J. H. Cummings, M. J. Hill, H. Houston, W. J. Branch, and D. J. A. Jenkins, Effect of meat protein and dietary fiber on colonic function and metabolism. Changes in bowel habit, bile acid excretion and calcium absorption, *Am. J. Clin. Nutr., 32*: 2086—2093 (1979).

49. J. L. Slavin and J. A. Marlett, Influence of refined cellulose upon human bowel function and calcium and magnesium balances, *Am. J. Clin. Nutr., 33*: 1932—1939 (1980).

50. V. Pathak, B. L. Kwatra, and S. Bajaj, Effect of dietary fiber on the absorption of calcium and zinc by human beings. *J. Research Punjab Agric. Univ.* (Lhudiana) *18*:216—220 (1982).

51. J. G. Reinhold, P. Abadi, B. Faraji, F. Ismail-Beigi, and R. M. Russell, Depletion of calcium, magnesium, zinc and phosphorus associated with consumption of wholemeal wheaten bread rich in fiber and phytate, *Baroda J. Nutr., 7*:55—62 (1981).

52. V. Rhada and P. Gurvani. Utilization of protein and calcium in adult women on cereal-legume diets containing various amounts of fiber, *Nutr. Rept. Int., 30*: 859—864 (1984).

53. J. R. Mahalko, F. R. Dintzis, L. K. Johnson et al., Effect of wheat bran, soy hulls, corn bran, apple or carrot on calcium, cooper, iron, phosphorus and zinc balances in adult men, *Fed. Proc., 44*: 1851 (1985).

54. D. A. T. Southgate, W. J. Branch, M. J. Hill, et al., Metabolic responses to dietary supplements of bran, *Metabolism, 25*: 1129—1135 (1975).

55. J. H. Cummings, D. A. T. Southgate, W. J. Branch, et al., The digestion of pectin in the human gut and its effect on calcium absorption and large bowel function, *Br. J. Nutr., 41*: 477—495 (1979).

56. J. L. Kelsay, W. M. Clark, B. J. Herbst and E. S. Prather, Nutrient utilization by human subjects consuming fruits and vegetables as sources of fiber, *J. Agric. Food Chem., 29*: 461—465 (1981).

57. J. C. King, F. M. Costa and N. F. Butts, Fecal mineral excretion of young men fed diets high in fiber components or phytate, *Fed. Proc. 41*: 712 (1982).

58. R. Ellis and E. R. Morris, Relation between phytic acid and trwce metals in wheat bran and soybean, *Cereal Chem., 58*: 367—370 (1981).

59. W. van Dokkum, A. Wesstra, and F. A. Schippers, Physiological effects of fiber-rich types of bread. 1. The effect of dietary fiber from bread on the mineral balances of young men, *Br. J. Nutr. 47*: 451—460 (1982).

60. H. Andersson, B. Navert, S. Bingham, et al., The effects of breads containing similar amounts of phytate but different amounts of wheat bran on calcium, zinc, and iron balances in man, *Br. J. Nutr.*, *50*: 503—516 (1983).

61. J. A. Rendleman, Cereal complexes: binding of calcium by bran and components of bran, *Cereal Chem.*, *59*: 303—309 (1982).

62. Withdrawn.

63. F. R. Dintzis, P. R. Watson, and H. H. Sandstead, Mineral content of brans passed through the human gastrointestinal tract, *Am. J. Clin. Nutr.*, *41*: 901—908 (1985).

64. W. P. James, W. J. Branch, and D. A. T. Southgate, Calcium binding by dietary fiber, *Lancet*, *1*: 638—639 (1978).

65. A. J. Batchelor and J. E. Compston, Reduced plasma half-life of radio-labeled 25-hydroxyvitamin D_3 in subjects receiving a high fibre diet, *Br. J. Nutr.*, *49*: 213—216 (1983).

66. S. B. Arnaud, R. S. Goldsmith, P. W. Lambert and V. L. W. Go, 25(OH)D_3, Evidence of an enterohepatic circulation in man, *Proc. Soc. Exp. Biol. Med.*, *149*: 570—572 (1975).

67. I. H. Rosenber, M. D. Sitrine and J. G. Bolt, in *Vitamin D. Basic Research and its Clinical Applications*, de Gruyter, Berlin (1979).

68. M. Maleki, Food consumption and nutritional status of 13-year old village and city schoolboys in Fars Province, Iran, *Ecol. Food Nutr.*, *2*: 39—43 (1973).

69. J. G. Reinhold, H. Hedayati, A. Lahimgarzaden, and K. Nasr, Zinc, calcium, phosphorus and nitrogen balances of Iranian villagers following change from phytate-rich to phytate-poor diets, *Ecol. Food Nutr.*, *2157*—2162 (1973).

70. J. E. Jessup, Results of rickets surveys in Dublin, *Br. J. Nutr.*, *4*: 289—293 (1950).

71. I. Robertson, J. A. Ford, W. B. McIntosh and M. G. Dunnigan, The role of cereals in the aetiology of nutritional rickets: the lesson of the Irish National Survey 1943—48, *Br. J. Nutr.*, *45*: 17—22 (1981).

72. M. M. Faridi, Z. Anseri and S. K. Bhavgari. Imprints of protein energy malnutrition on the skeleton of children. *J. Trop. Pediatr 30*: 150—153 (1984).

73. A. K. Pramanik, S. Gupta, and P. S. Agarwal, Rickets in protein calorie malnutrition, *Indian Pediatr.*, *8*: 195—199 (1971).

74. R. W. Luecke, Domestic animals in the elucidation of zinc's role in nutrition, *Fed. Proc.*, *43*: 2823—2828 (1984).

75. J. C. Leek, J. O. Vogler, M. E. Gershwin, et al., Studies of marginal zinc deprivation in rhesus monkeys. Fetal and skeletal effects, *Am. J. Clin. Nutr.*, *40*: 1203—1212 (1984).

76. A. S. Prasad, J. A. Halsted, and M. Nadimi, Syndrome of iron deficiency anemia, hepatosplenomegaly, hypogonadism, dwarfism and geophagia, *Am. J. Med.*, *31*: 532—546 (1961).

77. L. Strause, P. Saltman, and M. Miller, The role of trace elements in the etiology of osteoporosis: results with an animal model, in *Osteoporosis, Proc. Copenhagen Int. Symp.* (C. Christianson, ed.) 1984, *Chem. Abst., 103*: 159458.

78. J. G. Reinhold, Nutritional osteomalacia in immigrants in an urban community, *Lancet, 1*: 386 (1972).

79. D. C. Wilson, The incidence of osteomalacia and late rickets in Northern India, *Lancet, 2*: 10–12 (1931).

80. G. F. Taylor and C. D. M. Day, Osteomalacia and dental caries, *Br. Med. J., 2*: 221–222 (1940).

81. G. F. Taylor, Osteomalacia and calcium deficiency, *Br. Med. J., 1*: 960 (1976).

82. S. Ghosh, S. Sarin, and S. K. Shegal, Study of rickets, *J. Indian Ped. Soc., 1*: 253 (1962).

83. Withdrawn.

84. V. N. Patwardhan. Incidence of rickets in India, *Nutrition in India, 2nd ed.*, Bombay, 1961.

85. Anon., Studies on pre-school children. Report of the Working Party of the Indian Council of Med. Research, *ICMR Technical Report Series No. 26*, New Delhi, 1974.

86. M. V. Phadke, N. S. Deodhar, H. O. Kulkarni, The rural pre-school child, his growth and nutritional status, *Ind. J. Med. Res., 59*: 748–755 (1971).

87. S. S. Manchanda and Harbaus Lal., The challengy of rickets in the Punjab, *Indian J. Pediatr., 39*: 52–57 (1972).

88. D. Hodgkin, G. H. Kug, P. M. Hine, et al., Vitamin D deficiency in Asians at home and in Britain, *Lancet, 2*: 167 (1973).

89. H. P. Vaishnava and S. N. A. Rizvi, Rickets in India, *Br. Med. J., 1*:112 (1967) Letter. *Lancet, 2*: 1147 (1971) Letter. *Lancet, 2*: 621 (1973). Letter.

90. M. Chandhura, Nutritional profile of Calcutta pre-school children, *Indian J. Med. Res., 63*: 189–195 (1975).

91. A. K. Marye, A. S. Saini, S. Rathae, and S. R. Arora, Osteomalacia in a Hindu population of Haryana, *Indian J. Med. Res., 73*: 756–780 (1980).

92. L. Jain, S. K. Chatorvedi, S. Saxena, and M. Udawat, Atrophic rickets: A pattern to be reckoned with in tropical countries, *J. Trop. Pediatr., 31*: 167–169 (1985).

93. J. M. May and I. S. Jarcho, *The Ecology of Malnutrition in the Far and Near East (Food Resources, Habits and Deficiencies)*, Hafner Publishing Co., New York, 1961.

94. S. M. Rab and A. Baseer, Occult osteomalacia amongst healthy and pregnant women in Pakistan, *Lancet, 2*: 1211–1213 (1976).

95. K. Chapman, Osteomalacia in Iran, *J. Obst. Gynecol. Br. Commonw., 78*: 857–860 (1971).

96. J. Eminians, G. H. Amirhakimi, and M. Mahloudji, Health status in a small village community near Shiraz, Iran, *J. Trop. Pediatr., 18*: 11–21 (1972).

97. J. Zarel, H. Khayambashi, R. Emami, et al., High phytic acid content of wheat flour. Possible factor in prevalence of bone deformities in Isfahan school children, *Acta Biochim. Iran, 9*: 74–78 (1972).

98. G. H. Amirhakimi, Rickets in a developing country. Observations of general interest from Southern Iran, *Clin. Pediatr., 12*: 88–92 (1973).

99. R. Salimpour, Rickets in Tehran. Study of 200 cases, *Arch. Dis. Childh., 50*: 63–66 (1975).

100. J. J. Cannell, P. Abadi, Daniel Craig, B. J. Campbell, et al., Rickets: Clinical, subclinical and healing in an Iranian village, unpublished studies, 1975.

101. N. A. Nagi, Vitamin D deficiency rickets in malnourished children, *J. Trop. Med. Hyg., 75*: 251–254 (1972).

102. F. Serenius, A. H. T. Eldressy, and P. Dandona, Vitamin D nutrition in pregnant women at term and in newborn babies in Saudi Arabia, *J. Clin. Pathol., 37*: 444–447 (1984).

103. B. Griffith and S. T. Winter, The prevalence of rickets in subtropical Israel, *J. Trop. Pediatr., 4*: 13 18 (1958).

104. H. Costeff and Z. Breslaw, Rickets in Southern Israel. Some epidemiological observations, *J. Pediatr., 61*: 919–924 (1962).

105. Withdrawn.

106. S. Shany, J. Hersh, and G. M. Berlyne, 25-Hydroxycholecalciferol levels in Bedouins in the Negev, *Am. J. Clin. Nutr., 29*: 1104–1107 (1976).

107. Y. Biale, S. Shany, M. Levi, et al., 25-Hydroxycholecalciferol levels in Bedouin women in labor and in cord blood of their infants, *Am. J. Clin. Nutr., 31*: 2380–2382 (1979).

108. G. M. Berlyne, J. BenAri, E. Nord, et al., Bedouin osteomalacia due to calcium deprivation caused by high phytic acid content of unleavened bread, *Am. J. Clin. Nutr., 26*: 910–911 (1973).

109. E. Stransky and P. O. Dizon, Clinical rickets in the Phillipines, *J. Trop. Pediatr., 4*: 17–22 (1958).

110. C. C. deSilva. Common nutritional diseases of childhood in the tropics, *Adv. Pediatr., 13*: 253–264 (1964).

111. J. M. May and D. L. McLellan, *Ecology of Nutrition of Eastern Africa*, Hafner Publ. Co., New York, 1970.

112. Joint FAO/WHO Expert Committee on Nutrition, *WHO Techn. Report Series No. 377* (1967).

113. I. A. Abdou, N. P. Ali, and Lebshtein, Rickets in Cairo, Cited by G. R. Jansen, *Wld. Rev. Nutr. Diet., 45*: 42–67 (1985).

114. Y. W. Aboul-Dahab, Clinical studies of rickets in Cairo, *J. Egypt. Pub. Health Assoc., 38*: 203 (1963).

115. T. W. Mariam and G. Sterky, Severe rickets in infancy and childhood in Ethiopia, *J. Pediatr., 82*: 876–878 (1973).

116. B. Hojer, M. Gebre-Medhin, G. Sterky, et al., Combined vitamin D deficiency rickets and protein energy malmutrition in Ethiopian infants, *J. Trop. Pediatr., 23*: 73–79 (1977).

116a. B. Hojer and M. Gebre-Medhin, Rickets and exposure to sunshine, *J. Trop. Pediatr.*, *21*: 88–91 (1975).

117. K. E. Knudsen, M. Bach and L. Munch, Dietary fiber contents and compositions of sorghum and sorghum-based foods, *J. Cereal Sci.*, *3*: 153–164 (1985).

118. C. P. Doncaster and W. F. U. Jackson, South African studies in rickets in the Cape Peninsula, *S. Afr. Med. J.*, *34*: 776–780 (1960).

119. C. P. Doncaster and W. P. U. Jackson, Studies in rickets in the Cape Peninsula. II Aetiology, *S. Afr. Med. J.*, *35*: 890–894 (1961).

120. A. R. P. Walker, Nutritional, biochemical and other studies on South African populations, *S. Afr. Med. J.*, *40*: 814–852 (1966).

121. I. Robertson, A survey of clinical rickets in the infant population in Cape Town in 1967 to 1968, *S. Afr. Med. J.*, *43*: 1072 (1960).

122. J. T. Dwyer, W. H. Diets, Jr., G. Hass, et al., Risk of nutritional rickets among vegetarian children. *Am. J. Dis. Child.* *133*: 134–148 (1979).

123. J. Larracilla Alegre, L. Perez-Arteage, A. Juarez Fausto, et al., Raquitismo por carencia de vitamin D. Analisis de 70 casos. *Cuad. Nutr.*, *1*: 55–66 (1976).

124. A. Cuellar R., J. Luengas, and R. Ruiz Raquitismo carencial, *Cuad. Nutr.*, *2*: 115–128 (1977).

125. B. Luengas, S. Manzano, and M. M. Sanchez, Diagnostico de raquitismo en ninos con desnutricion, *Cuad. Nutr.*, *4*: 37–46 (1979).

126. J. M. May and D. L. McClellan, *Ecology of Nutrition. Mexico and Central America*, vol. 11, Hafnet Publ. Co., New York, 1972.

127. J. M. May, *Ecology of Nutrition. South America*, vol. 13, Hafner Publ. Co., New York, 1974.

128. N. Keyfitz and W. Flieger, *World Population. An Analysis of Vital Data*, University of Chicago Press, Chicago, 1968.

129. M. G. Dunnigan, J. J. Paton, S. J. Haase, et al., Late rickets and osteomalacia in the Pakistan Community in Glasgow, *Scottish Med. J.*, *7*: 159–167 (1961).

130. A. M. Holmes, B. A. Enoch, J. L. Taylor, and M. E. Jones, Adult rickets and osteomalacia among the Asian immigrant population, *Quart. J. Med.*, *42*: 125–149 (1973).

131. J. A. Ford, E. M. Colhoun, W. B. McIntosh, et al., Biochemical response of late rickets and osteomalacia to a chupatti-free diet, *Br. Med. J.*, *3*: 446 (1972).

132. M. A. Preece, J. A. Ford, W. B. McIntosh, et al., Vitamin-D deficiency among Asian immigrants to Britain, *Lancet*, *1*: 907–910 (1973).

133. T. C. B. Stamp, Factors in human vitamin D nutrition and in the production and cure of classical rickets, *Proc. Nutr. Soc.*, *34*: 119–130 (1975).

134. E. B. Mawer and A. M. Holmes, Rickets in Glasgow Pakistanis, *Br. Med. J.*, *3*: 177 (1972).

134a. M. R. Wills, J. B. Phillips, R. C. Day, et al., Phytic acid and nutritional rickets in immigrants, *Lancet*, *1*: 971 (1972).

135. M. G. Dunnigan, W. B. McIntosh, and J. A. Ford, Rickets in Asian immigrants, *Lancet*, *1*: 1346 (1976).

136. I. Robertson, A. Kelman, and M. G. Dunnigan, Chapatty intake, vitamin D status and Asian rickets, *Br. Med.*, *1*: 229 (1977).

137. S. D. Hunt, J. L. H. O'Riordan, J. Windo, and A. S. Truswell, Vitamin D status in different subgroups of British Asians, *Br. Med. J.*, *2*: 1351–1354 (1976).

138. N. Singleton and S. M. Tucker, Vitamin D status of Asian infants, *Br. Med. J.*, *1*: 607–610 (1978).

139. M. R. Wills and A. Farney, Effect of increased dietary phytate on cholecalciferol requirements in rats, *Lancet*, *2*: 406 (1972).

140. C. E. Dent, J. M. Round, D. G. F. Rowe, and T. C. D. Stamp, Effect of chapattis and ultraviolet rediation on nutritional rickets in Indian immigrants, *Lancet*, *1*: 1282 (1973).

141. J. O'Hara-May and E. M. Widdowson, Diets and living conditions of Asian boys in Coventry with and without signs of rickets, *Br. J. Nutr.*, *36*: 23–36 (1976).

142. M. M. Gupta, J. M. Round, and T. C. B Stamp, Spontaneous cure of vitamin D deficiency during summer in Britain, *Lancet*, *1*: 586–588 (1974).

143. W. P. Stephens, B. S. Klimiuk, J. L. Berry, and J. L. Mauser, Annual high dose vitamin D prophylaxis in Asian immigrants, *Lancet*, *2*: 1199 (1981).

144. G. Opperman, Ostomalacia of immigrants in Germany, *Dtsch. Med. Wochensch.*, *103*: 1387–1388 (1978).

145. A. Gallino and A. Radvila, Immigrant osteomalacia, *Schweiz Med. Wochenschr.*, *112*: 163–165 (1982).

146. P. Lapatsanis, V. Deleyanni, and S. Doxiades, Rickets in Greece, *J. Pediat.*, *73*: 195 (1968).

146a. P. Lapatsanis, G. Makaronis, C. Vreto, et al., Two types of nutritional rickets in infants, *Am. J. Clin. Nutr.*, *29*: 1222–1226 (1976).

147. V. Mayne and D. McCredie, Rickets in Melbourne, *Med. J. Austral.*, *2*: 873–875 (1972).

148. K. Svyatkina, A. Khvul, and M. Rassolova, *Rickets* (P. Ponomareva, ed.), Mir Publishers, Moscow, 1968.

149. A. S. Prasad, J. A. Halsted, and M. Nadimi, Syndrome of iron deficiency anemia, hepatosplenomegaly, dwarfism, hypogonadism and geophagia, *Am. J. Med.*, *31*: 532–546 (1961).

150. A. S. Prasad, A. Miale, Z. Farid, H. H. Sandstead, A. R. Schulert, and W. J. Darby, Biochemical studies on dwarfism, hypogonadism and anemia, *Arch. Int. Med.*, *111*: 407—428 (1963).

151. A. S. Prasad, A. Miale, Z. Farid, et al., Zinc metabolism in patients with the syndrome of iron deficiency anemia, hepatosplenomegaly, dwarfism and hypogonadism, *J. Lab. Clin. Med.*, *61*: 537 (1963).

152. A. S. Prasad, A. R. Schulert, A. Miale, et al., Zinc and iron deficiencies in male subjects with dwarfism and hypongonaism but without ancylosiomiasis, schystosomiasis or severe anemia, *Am. J. Clin. Nutr.*, *12*: 437—444 (1963).

153. J. G. Reinhold, G. A. Kfoury, M. A. Ghalambor, and J. C. Bennett, Zinc and copper concentrations in the hair of Iranian villagers, *Am. J. Clin. Nutr.*, *18*: 294 (1966).

153a. J. Eminians, J. G. Reinhold, G. A. Kfoury, et al., Zinc nutrition of children in Fars Province of Iran, *Am. J. Clin. Nutr.*, *20*: 734 (1967).

154. H. A. Ronaghy, Growth retardation as a factor in rejection from military service, *Pahlavi Med. J. (Shiraz)*, *1*: 29 (1970).

155. J. A. Halsted, H. A. Ronaghy, P. Abadi, et al., Zinc deficiency in man. The Shiraz experiment, *Am. J. Med.*, *53*: 277—284 (1972).

156. J. G. Reinhold, High phytate content of rural Iranian bread: a possible cause of human zinc deficiency, *Am. J. Clin. Nutr.*, *24*: 1204—1208 (1971).

157. N. Ter-Sarkissian, M. Azar, H. Ghavifekr, et al., High phytic acid in Iranian breads, *J. Am. Diet. Assoc.*, *65*: 651—653 (1974).

158. A. Kouhestani, H. Ghavifekr, M. Rahmanian, et al., Composition and preparation of Iranian Breads, *J. Am. Diet. Assoc.*, *55*: 262—266 (1969).

159. J. W. Matseshe, J. F. Phillips, J. R. Malegelade, et al., Recovery of dietary iron and zinc from proximal intestine of healthy man: studies of different meals and supplements, *Am. J. Clin. Nutr.*, *33*: 1946—1953 (1980).

160. J. G. Reinhold, A. Parsa, N. Karimian, et al., Availability of zinc in leavened and unleavened wholemeal wheaten breads as measured by solubility and uptake by rat intestine in vitro, *J. Nutr.*, *104*: 976—982 (1974).

160a. A. Wise and D. J. Gilburt, In vitro competition between calcium phytate and the soluble fraction of rat small intestinal contents, *Toxicol. Lett.*, *11*: 49—54 (1982).

161. B. Sandstrom, Food components of significance for the intestinal absorption of zinc, *Naeringsforskning*, *22*: 224—228 (1978).

162. B. Sandstrom, B. Arvidsson, U. Cederbled, et al., Zinc absorption from composite meals. The significance of wheat extraction

rate, zinc, calcium and protein content in meals based on bread, *Am. J. Clin. Nutr.*, *33*: 739–745 (1980).

163. J. R. Greger and S. M. Snedeker, Effect of dietary protein and phosphorus levels on the utilization of zinc, copper and maganese by adult males., *J. Nutr.*, *110*: 2243–2253 (1980).

164. S. M. Snedeker and J. L. Greger, Metabolism of zinc, copper and iron as affected by dietary protein, cystine and histidine, *J. Nutr.*, *113*: 644–652 (1983).

165. W. Frolich and B. Sandstrom, Zinc absorption from composite meals, in *Nutritional Bioavailability of Zinc* (G. E. Inglett, ed.), American Chemical Society, Washington, 1983, pp. 211–221.

166. S. J. Ritchey and L. J. Taper, Utilization of zinc by humans, in *Nutritional Bioavailability of Zinc* (G. E. Inglett, ed.), American Chemical Society, Washington, 1983, pp. 108–126.

167. N. T. Davies, Effects of dietary phytate on mineral availabilty, in *Dietary Fiber in Health and Disease* (G. F. Vahouny and D. Kritchevsky, eds.), Plenum, New York, 1982, pp. 108–116.

168. B. Pederson and B. O. Eggum, Interrelations between protein and zinc utilization in rats, *Nutr. Rept. Int.*, *27*: 441–453 (1983).

169. Withdrawn.

170. H. H. Sandstead, H. Klevay, J. Munoz, et al., Zinc and copper balances in humans fed fiber, *Am. J. Clin. Nutr.*, *31*: 180–184 (1978).

171. J. L. Kelsay and E. S. Prather, Mineral balances of human subjects consuming spinach in a low fiber diet and in a diet containing fruits and vegetables, *Am. J. Clin. Nutr.*, *38*: 12–19 (1983).

172. I. C. Obizoba, Zinc and copper metabolism of human adults fed combinations of corn, wheat, beans, rice and milk containing various levels of phytate, *Nutr. Rept. Int.*, *24*: 203–210 (1981).

173. J. H. Freeland-Graves, M. L. Ebangit and P. J. Hendrickson, Alteration in zinc absorption and salivary sediment zinc after lacto-ovo-vegetarian diet, *Am. J. Clin. Nutr.*, *33*: 157 (1980).

174. D. Oberleas, Role of phytate in zinc biovailability and homeostatis, in *Nutritional Bioavailability of Zinc* (G. F. Inglett, ed.), American Chemical Soceity, Washington, D.C., 1983, pp. 145–158.

175. E. F. Morris and R. Ellis, Dietary phytate/zinc molar ratio and zinc balance in humans, in *Nutritional Bioavailability of zinc* (G. F. Inglett, ed.), American Chemical Society, Washington D.C., 1983, pp. 159–183.

176. A. Wise, C. P. Richards, and M. A. Trimble, Phytate hydrolysis in the gastrointestinal tract of the rat followed by phosphorus-31 Fourier transform nuclear magnetic resonance spectroscopy, *Applied Environmental Microbiol.*, *45*: 313–314 (1983).

177. A. Wise, Dietary factors determining the biological activity of phytate, *Nutr. Abst. Rev. Rev. Clin. Nutr.*, *53*: 791–806 (1983).

178. H. Spencer, L. Kramer, C. Norris, and D. Osis, Effect of calcium and phosphorus on zinc metabolism in man, *Am. J. Clin. Nutr.*, *40*: 1213–1218 (1984).

179. B. De Vizia, S. J. Fomon, S. E. Nelson, et al., Effect of dietary calcium on metabolic balances of normal infants, *Pediatr. Res.*, *19*: 800–806 (1985).

179a. G. H. Ballem, T. S. Nelson, and L. K. Kirby, The effect of phytate and fiber source on phytate hydrolysis and mineral availability in rats, *Nutr. Rept. Int.*, *30*: 1089–1100 (1984).

180. A. O. Cavdar, A. Arcasy, and S. Cin, Zinc deficiency in Turkey, *Am. J. Clin. Nutr.*, *30*: 833–834 (1977).

181. A. O. Cavdar, Zinc deficiency and geophagia, *J. Pediatr.*, *100*: 1003 (1982).

182. K. M. Hambidge, B. A. Walravens, V. Kamen, et al., Chromium, zinc, manganese, copper, nickel, iron and cadmium concentrations in hair of residents of Chandigarh, India, and Bangkok, Thailand, in *Trace Substances in Environmental Health* (D. Hemphill, ed.), Columbia, MO, 1974, pp. 31–44.

183. K. H. Hambridge and B. A. Walravens, Zinc deficiency in infants and pre-adolescent children, in *Trace Elements in Human Health and Disease*, vol. 1 (A. S. Prasad and D. Oberleas, ed.) Academic Press, New York, 1976, pp. 23–32.

184. R. M. Jacobs, R. Riza, Z. Kallal, et al., Zinc nutrition of Tunisians, *Fed. Proc.*, *39*: 897 (1980).

185. C. Xue-Cun, Y. Tai-An, H. Jin-Shen, et al., Low levels of zinc in hair and blood, pica, anorexia and poor growth in Chinese preschool children, *Am. J. Clin. Nutr.*, *42*: 694–700 (1985).

186. A. B. Holt, R. M. Sparg, J. B. Ineson, et al., Serum and plasma zinc copper and iron concentration in aboriginal communities in Northwestern Australia, *Am. J. Clin. Nutr.*, *33*: 119–132.

187. R. S. Gibson, J. S. Rose, and J. H. Sabry, Trace element status of households from the Wosera, Papua, New Guinea, *Fed. Proc.*, *44*: (Abst.) 1507 (1985).

188. C. M. Donangelo, C. L. Azevido, Serum zinc levels in Brazilian children of low socioeconomic status, *Arch. Latinoam. Nutr.*, *39*: 290–297 (1984).

189. N. W. Solomons, Biological availability of zinc in humans, *Am. J. Clin. Nutr.*, *35*: 1048–1075 (1982).

190. N. W. Solomons, Interaction between zinc and dietary factors, *Arch. Latinoam. Nutr.* *32*: 26–31 (1982).

191. G. V. Iyengar, Reference values for elemental concentrations in some human samples of clinical interest. A preliminary evaluation., *Sci. Total Environ.*, *38*: 125–131 (1984).

192. H. C. Kew, R. C. Mallett, J. A. Dunn et al. Serum, whole blood and erythrocyte zinc levels in South African Bantu, *S. Afric. J. Med. Sci.*, *35*: 57—59 (1970)

193. K. M. Hambidge, D. A. Walravens, R. M. Brown, et al., Zinc nutrition of preschool children in the Denver Head Start Program, *Am. J. Clin. Nutr.*, *29*: 734—738 (1976).

194. K. M. Hambidge and A. Silverman, Pica with rapid improvement after dietary zinc supplementation, *Arch. Dis. Child.*, *48*: 567—568 (1973).

195. W. Varavithya, P. Porenanent, S. Srianojate, et al., Zinc status in normal Thai infants and children, *J. Trop. Med. Pub. Health*, *10*: 534 (1979).

196. H. P. Chase, H. M. Hambidge, S. E. Barnett, et al., Low vitamin A and zinc concentrations in Mexican-American migrant children, *Am. J. Clin. Nutr.*, *33*: 2346—2349 (1980).

197. L. M. Smit and P. J. Pretorius, Serum zinc levels and urinary zinc excretion in South African Bantu kwaashiorkor patients, *J. Trop. Pediatr.*, *9*: 105—122 (1964).

198. J. D. Hansen, B. H. Lehman, et al., Serum zinc and copper concentrations in children with protein calorie malnutrition. *S. Afr. Med. J.*, *43*: 1248—1251 (1969).

199. S. Kumar, K. S. Rao, et al., Plasma and erythrocyte zinc levels in protein energy malnutrition, *Nutr. Metab.*, *15*: 364—371 (1973).

200. B. Sharda and B. Bhandari, Serum zinc in protein calorie malnutrition., *Indian Pediatr.*, *14*: 195—196 (1977).

201. A. Amador, M. Hermelo, R. Fernandez-Regelado, et al., Concentration of zinc in hair of children with protein energy malnutrition, *Revist. Cuba Pediatr.*, *49*: 629—663 (1976); *Trop. Dis. Bull.*, *75*: 591 (1978).

202. B. E. Golden and M. H. N. Golden, Plasma zinc and clinical features of malnutrition, *Am. Clin. Nutr.*, *32*: 2490—2494 (1979).

203. K. M. Hambidge, N. F. Krebs, M. A. Jacobs, et al., Zinc nutritional status during pregnancy: a longitudinal study, *Am. J. Clin. Nutr.*, *37*: 429—442 (1983).

204. I. F. Hunt, N. J. Murphy, J. Gomez, and J. C. Smith, Dietary zinc intake of low-income pregnant women of Mexican descent, *Am. J. Clin. Nutr.*, *32*: 1511—1518 (1979).

205. M. D. Mukherjie, H. H. Sandstead, M. Rhanaparkhi, et al., Maternal zinc, iron, folic acid and protein nutriture and outcome of human pregnancy, *Am. J. Clin. Nutr.*, *40*: 496—507 (1984).

206. S. Jameson, Effects of zinc deficiency in human reproduction, *Acta. Med. Scand. Suppl.*, *593*: 1—99 (1976).

207. C. M. Donangelo, Zinc status of nursing mothers and weight adequacy of their newborn babies, *Nutr. Rept. Int.*, *30*: 1157—1163 (1984).

208. I. F. Hunt, A. Sanchez, D. Martner, et al., Zinc nutriture in pregnant women in Mexico and in California, *Fed. Proc.*, *44*: 1507 (1985).

209. I. F. Hunt, N. J. Murphy, A. E. Cleaver, et al., Zinc supplementation during pregnancy: effects on selected blood constituents and on progress and outcome of pregnancy in low income women of Mexican descent, *Am. J. Clin. Nutr.*, *40*: 508–521 (1984).

210. J. R. Turnlund, G. C. King, C. J. Wahben, et al., Zinc status and pregnancy outcome of pregnant Lebanese women, *Nutr. Res.*, *3*: 309 (1983).

211. J. C. King, T. Stein, and M. Doyle, Effect of vegetarianism on the zinc status of pregnant women, *Am. J. Clin. Nutr.*, *34*: 1049–1055 (1981).

212. J. Apgar, Zinc and reproduction, *Ann. Rev. Nutr.*, *5*: 43–68 (1985).

213. B. L. Vallee and K. H. Falchuk, Zinc and gene expression, *Trans. R. Soc. London. Ser B*, *294*: 185–197 (1981).

214. R. M. Forbes, Use of laboratory animals to define physiological functions and bioavailability of zinc, *Fed. Proc.*, *43*: 2835–2839 (1984).

215. M. K. Song and N. F. Adhoun, Evidence for an important role of prostaglandins E_2 and F_2 in the regulation of zinc transportation in the rat, *J. Nutr.*, *109*: 2152–2159 (1979).

216. G. W. Evans and P. E. Johnson, Defective prostaglandin synthesis in acrodermatitis enteropathica, *Lancet*, *1*: 52 (1977).

217. C. J. McClain, J. S. Garaler and D. H. Van Thiel, Hypogonadism in the zinc deficient rat: localization of the functional abnormalities, *J. Lab. Clin. Med.*, *104*: 1007–1015 (1984).

218. M. H. N. Golden, A. A. Jackson, and B. E. Golden, Effect of zinc on the thymus of recently malnourished children, *Lancet*, *2*: 1057–1059 (1977).

219. M. H. N. Golden, B. E. Golden, P. S. E. G. Harland, et al., Zinc and immunocompetence in protein energy malnutrition, *Lancet*, *1*: 1226– 1228 (1978).

220. R. K. Chandra, Single nutrient deficiency and cell-mediated immunoresponses, *Am. J. Clin. Nutr.*, *33*: 736–738 (1980).

221. J. Duchateau, G. Delespesse, and P. Verercke, Influence of oral zinc supplementation on the lymphocyte response to initogens of normal subjects, *Am. J. Clin. Nutr.*, *42*: 252–262 (1985).

222. D. L. Haynes, M. E. Gershwin, M. S. Golub, et al., Studies of marginal zinc deprivation in thesus monkeys. VI. Influence on the immunohematology of infants in the first year, *Am. J. Clin. Nutr.*, *42*: 252–262 (1985).

223. E. J. Moynahan, Acrodermatitis enteropathica: a lethal inherited human zinc disorder, *Lancet*, *2*: 399–400 (1974).

224. P. A. Walraven, M. K. Hambidge, K. H. Neldnet, et al., Zinc metabolism in acrodermatitis enteropathica, *J. Pediatr.*, *93*: 71–73 (1978).

225. M. H. N. Golden, B. E. Golden, and A. A. Jackson, Skin breakdown in kwashiorkor: response to zinc, *Lancet*, *1*: 1276 (1980).

226. J. C. Smith, Jr., E. G. McDaniel, E.G. Fan, et al., A trace element essential in vitamin A metabolism, *Science*, *131*: 954–956 (1973).

227. N. W. Solomons and R. M. Russell, The interaction of vitamin A and zinc. Implications for human nutrition, *Am. J. Clin. Nutr.*, *33*: 2031–2040 (1980).

228. A. S. Prasad, E. Rabbani, A. Abbasli, et al., Experimental zinc deficiency in humans, *Ann. Int. Med.*, *89*: 483–490 (1978).

229. M. T. Baer, J. C. King, T. Tamura, et al., Nitrogen utilization enzyme activity, glucose intolerance and leucocyte chemotaxis in human experimental zinc deficiency, *Am. J. Clin. Nutr.*, *41*: 1220 (1985).

230. A. S. Prasad, Zinc deficiency in sickle cell disease. *Prog. Clin. Biol. Res. 165*: 49–58 (1984).

231. L. S. Hurley, The role of trace elements in foetal and neonatal development, *Phil. Trans. Roy. Soc. London, Ser. B, 294*: 145–152 (1981).

232. A. F. Fleming, Iron deficiency in the tropics, *Clin. Hematol.*, *11*: 365–383, (1980).

232a. B. S. Narasinga Rao, Studies on iron deficiency anemia, *Indian. J. Med. Res.*, *68* (Suppl.): 58–69 (1978).

232b. Y. Hofwander, Hematologic investigations on Ethiopia with special reference to high iron intake, *Acta Med. Scand. Supp.*, *484*: 1–74 (1968).

233. E. R. Morris, F. E. Greene, and A. C. Marsh, On the chemistry and bioavailability to the rat of iron in wheat, in *Trace Element Metabolism in Animals*, 2nd ed. W. G. Hoekstra, J. W. Suttie, H. E. Ganther and W. Mertz, eds. Iniv. Park Press, Baltimore, 1974, pp. 506–508.

234. B. A. Lipschitz, K. M. Simpson, J. D. Cook, and E. Morris, Absorption of monoferric phytate by dogs, *J. Nutr.*, *104*: 1154–1160 (1979).

235. R. Ellis, E. R. Morris, and A. D. Hill, Bioavailability to rats of iron and zinc in calcium-iron phytate, *Nutr. Res. (New York)*, *2*: 319–322 (1983).

236. J. G. Reinhold, J. S. Garcia L., and P. Garzon, Binding of iron by fiber of wheat and maize, *Am. J. Clin. Nutr.*, *34*: 1384–1391 (1981).

237. J. G. Reinhold, P. Garcia L., and P. Garzon, Solubility of ferrous and ferric iron as affected by constituents of the maize tortilla, *Nutr. Rept. Int.*, *30*: 603–615 (1984).

238. J. G. Reinhold, J. Garcia-Estrada, P. Garcia L., et al., Retention of iron by rat intestine in vivo as affected by dietary fiber, ascorbate and citrate, *J. Nutr.*, *116*: 1007–1018 (1986).

239. R. Ali, H. Staub, G. Coccodrilli, Jr., et al., Nutritional significance of dietary fiber: effect on nutrient bioavailability and selected gastrointestinal functions, *J. Agric. Food Chem.*, *29*: 465–472 (1981).

240. J. G. Reinhold, Dietary fiber and the bioavailability of iron, in *Nutritional Bioavailability of iron* (C. Kies, ed.), American Chemical Society, Washington, D.C., 1982, pp. 143–161.

241. R. Fernandez and S. F. Phillips, Components of fiber impair iron absorption in the dog, *Am. J. Clin. Nutr.*, *35*: 107–112 (1982).

242. J. G. Reinhold, P. Garcia L., L. Arias-Amado, and P. Garzón, Dietary fiber-iron interactions. Fiber-modified uptakes of iron by segments of rat intestine, in *Dietary Fiber in Health and Disease*, (G. Vahouny and D. Kritchevsky, eds.), Plenum Publ. Co., New York, 1982, pp. 117–132.

243. M. J. Leigh and D. D. Miller, Effects of pH and chelating agents on iron binding by dietary fiber: implications for iron bioavailability, *Am. J. Clin. Nutr.*, *38*: 202–213 (1983).

244. K. M. Simpson, E. R. Morris, and J. D. Cook, The inhibitory effect of bran on iron absorption in man, *Am. J. Clin. Nutr.*, *34*: 1469–1478 (1981).

244a. T. Peters, Jr., L. Apt, and J. F. Ross, Effect of phosphates upon iron absorption studied in normal human subjects and in an experimental model using dialysis, *Gastroenterol*, *66*: 315–322 (1971).

245. R. R. Morris and R. Ellis, Phytate, wheat bran and bioavailability of dietary iron, in *Nutritional Bioavailability of Iron* (C. Kies, ed.), Amer. Chemical Society, Washington, D.C., 1982, pp. 121–141.

246. B. Faraji, J. G. Reinhold, and P. Abadi, Human studies of iron absorption from fiber-rich Iranian flatbreads, *Nutr. Rept. Int.*, *23*: 267–278 (1981).

247. G. C. Arneil, The return of infantile rickets to Britain, *Wld. Rev. Nutr. Diet.*, *10*: 239–261 (1969).

248. M. W. Moncrieg, L. T. H. Arthur, and H. R. W. Hunt, Nutritional rickets at puberty, *Arch Dis. Child.*, *48*: 221–224 (1973).

249. J. Black, Pediatrics among ethnic minorities. Asian families II. Conditions that may be found in the children, *Br. Med. J.*, *290*: 830–833 (1985).

250. S. Swarup-Mitra and A. K. Bhattacharya, Iron nutrition in rickets, *Trans. Roy. Soc. Trop. Med. Hyg.*, *72*: 444–445 (1978).

251. G. F. Kirsten, H. Oev Heese, S. DeVilliers, et al., Prevalence

of iron deficiency in apparently healthy Cape colored infants, *S. Afric. Med. J.*, *65*: 378–380 (1984).

252. M. Abdel-Pattah, S. Shalabyand, S. El Ashmavi, An epidemiological study of iron deficiency anemia in early infancy, *Gaz. Egypt. Paed. Ass.*, *22*: 143–147 (1974).

253. I. A. Abdou, H. E. Ali, and A. K. Lebshtein, Cited by G. R. Jansen, *World Rev. Nutr. Diet*, *45*: 42–67 (1985).

254. C. W. Weber, L. A. Vaughan, and W. A. Stini, Bioavailability of iron and other trace minerals from human milk, in *Nutritional Bioavailability of Iron* (C. Kies, ed., Amer. Chemical Soc., Washington, D.C., 1982, pp. 173–181.

255. M. I. Schnepf and L. D. Satterlea, Partial characterization of an iron soy protein complex, *Nutr. Rept. Int.*, *31*: 371–380 (1985).

256. C. Kies and L. McEndree, Vegetarianism and the bioavailability of iron, in *Nutritional Bioavailability of Iron* (C. Kies, ed.), Amer. Chemical Soc., Washington, D.C., 1982, pp. 183–198.

256a. G. MacDougall, G. Moodley, C. Eyberg, et al., Mechanisms of anemia in protein energy malnutrition in Johannesberg, *Am. J. Clin. Nutr.*, *35*: 229–235 (1982).

256b. J. L. Beard, H. A. Huebers, and C. A. Finch, Protein and iron deficiency in rats, *J. Nutr.*, *114*: 1396–1340 (1984).

257. L. Heilmeyer and I. van Mutius, Release of iron from foods by acids, *Z. ges. exp. Med.*, *112*: 192–207 (1943).

258. D. Derman, M. Sayers, S. R. Lynch, et al., Iron absorption from a cereal-based meal containing cane sugar fortified with ascorbic acid, *Br. J. Nutr.*, *38*: 261–269 (1977).

259. V. Mathan, S. J. Baker, S. K. Sood, et al., WHO sponsored collaborative studies on nutritional anemia in India, *Br. J. Nutr.*, *42*: 391–298 (1979).

260. H. W. Davenport, *A Digest of Digestion*, 2nd, Year Book Medical Publishers, Chicago, 1978, pp. 45–60.

261. B. R. Schricker, D. D. Miller and D. Van Campen, In virto estimation of iron availability in meals containing soy products, *J. Nutr.*, *112*: 1696–1705 (1982).

262. J. C. Barton, M. E. Conrad, and R. T. Parmley, Calcium inhibition of inorganic iron absorption in rats, *Gastroenterol*, *84*: 90–100 (1983).

263. F. M. Clydesdale and D. R. Nadeau, Solubilization of iron in cereals by milk fractions, *Cereal Chem.*, *61*: 350–355 (1984).

264. E. Bjorn-Rasmussen and L. Hallberg, Effect of animal proteins on absorption of food iron in man, *Nutr. Metab.*, *23*: 192–202 (1979).

265. H. Merhav, Y. Amitei, H. Palti, et al., Tea drinking and microcytic anemia in infants, *Am. J. Clin. Nutr.*, *41*: 1210–1213 (1985).

266. B. S. Narasinga Rao and T. Prahanathi, Tannin content of

foods commonly consumed in India and its influence on ionizable iron, *J. Sci. Food Agric.*, *33*: 89—96 (1982).

267. H. Arcasoy and A. O. Cavdar, Decreased iron and zinc absorption in Turkish children with iron deficiency and geophagia, *Acta Haematol.*, *60*: 76—84 (1978).

268. S. M. Garn, M. T. Keating and F. Falkner, Hematological status and pregnancy outcome, *Am. J. Clin. Nutr.*, *34*: 115—117 (1981).

269. K. Prema, B. Ramalakshmi, R. Mahdasapeda, et al., Changes in hemoglobin levels during different periods of gestation, *Nutr. Rept. Int.*, *23*: 629 (1981).

270. E. Royston, The prevalence of nutritional anemia in developing countries. A critical review of available information, *World Health Statistics Quart.*, *35*: 52—91 (1982).

271. A. E. G. Masawe, J. M. Muindi, G. B. I. Swai, Infection in iron deficiency and other types of anemia, *Lancet, 2*: 314—317 (1974).

272. M. J. Murray, A. B. Murray, M. B. Murray, and C. J. Murray, The adverse effect of iron repletion on the course of certain infections, *Br. Med. J.*, *2*: 1113—1115 (1978).

273. G. V. Mann, Food intake and resistance to disease, *Lancet, 1*: 1238—1239 (1980).

274. M. J. Murray, A. Murray, and C. J. Murray, The salutory effect of milk on amoebiasis and its reversal by iron, *Br. Med. J., 1*: 1351 (1980).

275. M. Gracey, J. B. Iveson, Sunoto, and Suharyono, Human *Salmonella* carriers in a tropical urban environment, *Trans, Roy. Soc. Trop. Med. Hyg.*, *74*: 479—482 (1981).

276. M. S. Gracey, Nutrition, bacteria and the gut, *Br. Med. Bull.*, *37*: 71—75 (1981).

277. R. K. Chandra, Immunocompetence as functional index of nutritional status, *Br. Med. Bull.*, *37*: 89—94 (1981).

278. L. G. MacDougall, R. Anderson, G. B. Nichols, J. Katz, Immune response in iron deficient children: impaired cellular defense mechanisms with altered humoral components, *J. Pediat.*, *86*: 833 (1975).

279. R. K. Chandra and A. K. Saraya, Impaired immunocompetence associated with iron deficiency, *J. Pediatr.*, *86*: 897—902 (1975).

280. C. Bhaskaram and V. Reddy, Cell-mediated immunity in iron and vitamin-deficient children, *Br. Med. J.*, *3*: 522 (1975).

281. S. S. Basta, M. S. Soaherna, D. Karyadi, et al., Iron deficiency and productivity of adult males in Indonesia, *Am. J. Clin. Nutr.*, *32*: 916—925 (1979).

282. J. G. Reinhold, Phytate destruction by yeast fermentation in whole wheat meals, *J. Am. Diet. Assoc.*, *66*: 38—41 (1975).

283. B. F. Harland and J. Harland, Fermentative destruction of

phytate in rye, white and whole wheat breads, *Cereal. Chem.*, *51*: 226–227 (1980).

284. J. G. Reinhold, Zinc and mineral deficiencies in man. The phytate hypotheses, in *Proc. 9th Int. Cong. Nutr.* vol. 1 (A. Chavez, H. Bourges, and S. Basta, eds.), S. Karger, Basel, 1975, pp. 115–122.

285. D. M. Davies, Calcium metabolism of healthy men deprived of sunlight, *Ann. N. Y. Acad. Sci.*, *453*: 21–27 (1985).

286. T. L. Clements, J. A. Adams, S. L. Henderson, et al., Increased skin pigment reduces the capacity of the skin to synthesize vitamin D, *Lancet*, *1*: 74–76 (1982).

287. S. K. Dutta, R. M. Russell and B. Chowdury, Folate content of North Indian breads, *Nutr. Rept. Int.*, *21*: 251–256 (1980).

288. M. Halim and K. Lorenz, High fiber Sudanese breads, *Nutr. Rept. Int.*, *31*: 91–100 (1985).

289. C. W. Low, P. W. Paris, and W. F. Holick, Indian and Pakistani immigrants have the same capacity as Caucasians to produce vitamin D in response to ultraviolet irradiation, *Am. J. Clin. Nutr.*, *44*: 683–685 (1986).

290. C. J. Lee, L. Y. Panemusjalore, and E. Wilson, Effect of dietary energy restriction on bone mineral content of mature rats, *Nutr. Res.*, *6*: 51–59 (1986).

291. W. P. T. James, Dietary fiber and mineral nutrition, in *Medical Aspects of Dietary Fiber* (G. A. Spiller and H. M. Kay, eds.), Plenum Books, New York, 1980, pp. 239–286.

292. S. R. Lynch and A. M. Cavell, Iron in soybean flour is bound to phytoferritin, *Am. J. Clin. Nutr.*, *45*: 865 (1987).

293. W. B. Andelman and S. R. Sered, Utilization of dietary iron by term infants, *Am. J. Dis. Child.*, *111*: 45–55 (1966).

294. M. R. Wastney, R. L. Aamodt, W. F. Rumble et al., Effect of fiber on zinc kinetics in humans, *Fed. Proc.*, *46*: 884 (1987).

295. F. Okonofra, S. Houlder, J. Bell, et al., Vitamin D Metabolism in pregnant Nigerian women and their newborn infants, *J. Clin. Path.*, *39*: 650–653 (1986).

296. M. L. Ho, H. C. Yen, R. C. Tsang, et al., Randomized study of sunshine exposure and serum 25(OH)D in breastfed infants in Beijing, China, *J. Pediatr.*, *107*: 128–131 (1985).

297. J. G. Reinhold, B. Faraji, P. Abàdi, et al., Decreased absorption of calcium, magnesium, zinc and phosphorus by humans due to increased fiber and phosphorus consumption as wheat bread, *J. Nutr.*, *106*: 493–503 (1976).

2
Minerals: Their Chemistry and Fate in Food

FERGUS M. CLYDESDALE / University of Massachusetts, Amherst, Massachusetts

INTRODUCTION

Concern about the interaction of minerals in food due to their chemical reactiviey is relatively recent. Indeed, studies in the areas of food science and nutrition have often treated the minerals as inert materials. Thus availability was thought to be affected only by processes such as canning or milling, where leaching or physical separation might occur. An excellent review of the influence of processing on vitamin-mineral content and biological availability in processed foods written in 1974 stated that "the losses of mineral salts in processing are solely due to leaching during processing steps" (1). Another stated that "the major inorganic elements in foods are the base-forming calcium, magnesium, sodium, and potassium; and the acid-forming phosphorus, chlorine, and sulfur. None of the minerals are altered significantly during the storage of foods canned in either metal or glass containers" (2). This is not to say that interactions were unknown, since animal nutritionists have used their knowledge of trace element reactions with dietary constituents to practical advantage for some time (3).

However, since the 1960s there has been an increasing amount of research in human mineral nutrition. In fact, Mertz (4) has stated that research in the last decade has changed our concepts of mineral nutrition. He attributed this to three major areas of research emphasized during this time:

1. The "new trace elements" as biologically active substances.
2. The investigation of mineral and trace element interactions and the recognition of their importance as determinants of metabolism and nutritional status.

57

3. Better definition of human requirements for minerals and trace elements.

Such work has been described in a number of publications and symposia, many of which are cited in several recent reviews (4–6). Most of these studies are concerned with interactions that take place throughout the small intestine or within the body but not with interactions that take place during processing and/or storage. However, the end result is often the same—either a positive or negative effect on bioavailability. For instance, Solomons (5) has suggested that when minerals are ingested simultaneously in a meal they may compete in the process of absorption within the gastrointestinal (GI) tract in any of the following ways:

1. Displacement of one mineral by another from the molecules necessary for their uptake from the lumen into the intestinal wall.
2. Competition for pathways through the intestinal wall or into the bloodstream.
3. Interaction between two minerals or a third substance to form an insoluble complex, thereby impairing the absorption of both minerals. However, even when minerals are not ingested simultaneously, high intakes of one mineral can condition the reception and handling of a second mineral not necessarily consumed at the same time.

Interestingly, from the point of view of this chapter, Solomons (5) also indicated that (a) the mechanism of biological mineral-mineral interactions can be related to the similar orbital configuration of electrons in the ions of several minerals of biological importance in man, and (b) it has been hypothesized that elements having similar physical and chemical properties will be biologically antagonistic toward each other. Thus it is apparent that the chemistry of the minerals and their subsequent interactions in the GI tract is extremely important. Moreover, the same chemical characteristics that influence interactions within the GI tract will also control interactions within a food during harvesting, preparation, processing, and storage and will ultimately effect bioavailability.

It is the aim of this chapter to attempt to describe the chemistry of minerals in such a way that reactions that might occur during processing and storage will be more apparent. Obviously each mineral cannot be treated in detail, as that would require a separate chapter for each. Therefore, their chemistry and most important potential reactions will be discussed in the hope of alerting the reader to the effects of postharvest-preconsumption treatments on mineral bioavailability.

CHEMISTRY

Definition and Classification

Definition

Taber's Cyclopedic Medical Dictionary (7) defines a mineral as "1. An inorganic element or compound occurring in nature, esp. one that is solid. 2. Inorganic; not of animal or plant origin. 3. Impregnated with minerals, as mineral water. 4. Pert. to minerals." This same reference refers to a mineral acid as "an acid containing no carbon atoms" and mineral compounds as "compounds of mineral elements, excepting carbon, constitute the mineral constituents of the body." However, the Handbook of Chemistry and Physics (8) states that "mineralogical names should be used only to designate actual minerals and not to define chemical composition; thus the name calcite refers to a particular mineral (contrasted with other minerals of similar composition) and is not a term for the chemical compound whose composition is properly expressed by the name calcium carbonate." The minerals are elements, and an element may be defined as a substance that cannot be decomposed by the ordinary types of chemical change or made by chemical union (8). (This excludes the compounds recently named electrides, which are made of alkaline metal elements and electrons. Under this definition electrons would be considered elements.) Therefore, minerals may be classified chemically just as any other element. In this classification an important observation is that the chemical properties of an atom are primarily associated with the electron configuration of its outermost shell. Thus the elements of Group IA in the Periodic Table contain a single outermost S electron. All of them are soft, with low melting points. They all react violently with water, evolving hydrogen gas and forming a basic solution containing a hydroxide of the formula MOH. Nearly all compounds formed by these metals are soluble in water, and the formulas are related. Because of their similarities, they are called alkali metals, of which Na and K are a part.

As we move from left to right in the Periodic Table, there is an increase in the number of outermost electrons along with a concurrent change in chemical properties and, therefore, reactivity. For instance, among the A group elements, the first three in the third period, Na(group IA), Mg(group IIA), and Al(group IIIA), have typical metallic properties. Of the three, Na gives up electrons most readily and therefore is the most reactive. After Al, the next four elements, Se(4A), P(5A), S(6A), and Cl(7A), increasingly show the characteristics of nonmetals (by definition metals tend to lose electrons and form positively charged cations, while nonmetals tend to gain electrons and form negatively charged anions). Of these four elements, chlorine has the greatest tendency to gain electrons and is the most reactive towards metals.

If, instead of proceeding across the Periodic Table, we move down through each individual group, we will find that the metallic character of A group elements increases. Therefore, when dealing with chemical reactivity, those A group elements to the far left and along the bottom of the Periodic Table are most likely to have metallic characteristics, while those fewer number in the top right will be nonmetals—a valuable consideration when evaluating mineral reactivity in foods. Thus, location in the Periodic Table is the first general classification step for predicting mineral reactivity. However, it should be kept in mind that the properties of any element will vary gradually and smoothly and will be an average of adjacent elements.

Classification

Certainly, the best way to classify the minerals chemically would be through their individual electron configurations as is done in the Periodic Table. However, it should be remembered that the Periodic Table was not developed through a knowledge of the theoretical electron configurations of the elements. It was developed on the basis of observations that there was a periodic recurrence of similar properties when the elements were organized as to atomic weight, resulting in families of elements that shared similar chemical properties. Therefore, it might be of value to classify the minerals of biological importance in terms of their families and list some of the general chemical characteristics of these families. Such characteristics will define their behavior in a food environment and thus their potential for absorption.

Alkali Metals: Sodium and potassium belong to the family of elements known as the alkali metals (Group IA). Each atom of this series contains one loosely bound electron and they are thus known as monovalent or univalent metals, generally reacting as a monovalent cation (Na^+ or K^+). These ions are extremely stable, and higher valencies are most unlikely, with the exception of compounds such as sodium peroxide, Na_2O_2, potassium superoxide, KO_2, and potassium triiodide, KI_3 (9).

The alkali metals are powerful reducing agents, which allows them to displace less electropositive metals from their salts. In general, this will occur with a more electropositive element handing over its electrons to a less electropositive metal. For instance, sodium will replace alumium from its salts, and aluminum will replace iron.

The electropositive nature of the alkali metals also make them ready to donate their spare electrons to nonmetals.

The alkali metals will form oxides and peroxides with oxygen as well as displacing hydrogen from water to form hydroxides. The degree of reactivity with water and the strength of the base formed increases with the atomic weight of the metal.

Alkaline Earths: The metals of the alkaline earths, being in Group 2A, follow those of the alkali metals in the classification of the elements. Magnesium and calcium are members of this group and, as such, have two free electrons and are therefore divalent. They form colorless salts much like the alkali metals, but generally these salts are less soluble. They are electropositive but less so than the alkali metals and therefore have similar properties in terms of reduction as well as reaction with nonmetals and oxygen. Therefore, they form salts with nonmetals and bases in water but less vigorously than the alkali metals.

Transition Metals: Vanadium, chromium, manganese, iron, cobalt, nickel, copper, and zinc are members of the group of ten elements in the fourth period of the Periodic Table, often referred to as the transition metals. They start with scandium (atomic number 21) and end with zinc (atomic number 30). In these elements the 3d sublevel is being filled and, since there are ten 3d electrons, ten electrons are needed to fill it. Since they differ in structure only in an inner sublevel, they resemble one another in some properties. Due to their electronic structure they show less sharply defined electrochemical activity than the alkali metals or alkaline earths. They are, therefore, relatively inert, but in the presence of electronegative elements such as chlorine they lose electrons and form positively charged cations. However, they do not react with oxygen with the same alacrity as the previous groups.

The transition elements will also form covalent bonds to produce metallic carbonyls as well as oxy-acids. Perhaps, the most important characteristic of the transition metals is their ability to form stable complexes. The relative abilities of different ligands to coordinate with metal ions depend upon a great many factors. One of these is the basicity of the ligand. Molecules which have a strong attraction for protons are among the better coordinating agents. The species NH_3, OH^-, and CN^-, which are strong bases in the Brönsted-Lowry or Lewis sense, form stable complexes with a wide variety of transition metal ions.

Halogens: Chlorine, fluorine, and iodine are members of the electronegative group of elements known as the halogens, which occur in group VIIA. The name halogen, or "salt producer," was introduced by Berzelius in 1825 to describe those nonmetals which form salts by direct combination with metals. The degree to which they do this is based on their electronegative character, which decreases from F to Cl to Br to I. Therefore, F will displace Cl from its salts, Cl will displace Br and Br will displace I. Iodine, although the least electronegative, will still accept electrons from nearly all the metals to form a salt.

All four halogens combine directly with hydrogen, with their respective activity declining from F to I, as would be expected from

their position in the Periodic Table. The four hydrides thus formed
are known as hydrofluoric, hydrochloric, hydrobromic, and hydroiodic
acids, since they dissolve in water and produce strongly acid solutions.

The affinity of chlorine and iodine for oxygen is greater than
that of fluorine and bromine.

A few comments about fluorine might be in order since it has
some properties that are not typical of the other halogens (9), and
which might be of significance in food. Calcium fluoride is insoluble
in water, while the chloride, bromide, and iodide are so soluble that
they form deliquescent salts. Hydrogen fluoride forms a much weaker
acid than the other halogens, and fluorine does not react with alkalis
to form oxy-salts but instead a fluoride and difluorine monoxide (OF_2)
or oxygen are produced.

Groups IVA, VA, VIA, and Posttransition Minerals: The remaining
essential minerals are members of groups IVA, VA, and VIA of the
periodic classification with the exception of molybdenum, which can be
classified as a posttransition element. Silicon and tin are IVA elements, phosphorus a VA element, and selenium and sulfur VIA elements. Their chemical characteristics are similar to other elements
surrounding them in the Periodic Table.

Having discussed a loose classification scheme for the essential
minerals, it would be of interest to now discuss selected chemical
characteristics which might be of primary importance to the reactions these minerals undergo alone, or in the presence of one another,
in food. Of necessity, only a few chemical properties will be discussed, and these will represent a rather arbitrary selection of those
considered important from a personal interpretation of both the literature and studies done in this laboratory. Apologies are given in advance to those who feel that important areas have been missed, but
be assured, brevity and not disinterest is the culprit.

Electrochemistry: Oxidation and Reduction

Many of the reactions discussed thus far involve the transfer of electrons. In such reactions, if electrons are lost from an atom producing a positive ion and electrons, the process is known as an oxidation
half-reaction. If, on the other hand, an atom gains electrons to form
an ion, the process is known as a reduction half-reaction and the atom
is said to be reduced. When the oxidation and reduction occur simultaneously in the same reaction, it is called an oxidation-reduction reaction or, more simply, a redox reaction.

It is also possible to broaden the meaning of these terms to cases
where no ionic species exist. That is, electrons are not lost or gained
but shared as in a covalent bond. In this latter case the term oxidation number is used to denote the charge that an atom would have if

the shared bonding electrons were assigned to the more electronegative element. In the hydrogen fluoride (HF) molecule, the hydrogen is said to have an oxidation number of +1 and the fluorine -1. In water the bonding electrons are assigned to the more electronegative oxygen atom, giving it an oxidation number of -2 and hydrogen an oxidation number of +1.

This leads to a working definition of the terms oxidation and reduction. Oxidation is defined as an increase in oxidation number and reduction as a decrease. Reactions in which one element increases in oxidation number at the expense of another are referred to as oxidation-reduction reactions. Further, the phrases oxidizing agent and reducing agent are frequently used to designate the species responsible for oxidation and reduction (10).

Since the ease of oxidation or reduction is an indication of the ease of removal or addition of electrons, a measure of this characteristic provides a valuable tool in assessing not only reactivity but preferential reactivity in the presence of other compounds. Such a measure may be obtained by measuring the standard voltage of a cell, which is the algebraic sum of the standard oxidation potential of the species being oxidized in the cell reaction and the standard reduction potential of the species being reduced. At a given temperature (25°C) this depends upon the nature of the cell reaction and the concentrations of the ions or molecules taking part in the reaction. The standard voltage ($E°$) corresponds to a cell reaction when all ions or molecules are at a concentration of 1 M and all gases are at a partial pressure of 1 atmosphere.

Just as an oxidation-reduction reaction may be considered in two parts, the standard cell voltage may also be considered in two parts. The potential representing the oxidation half-reaction is known as the standard oxidation potential and that of the reduction half-reaction is the standard reduction potential. Table 1 lists some standard reduction potentials, SRP (volts), for the essential elements and water. Not every reaction of each essential element is shown, but those of some interest were arbitrarily chosen.

The SRP for these elements allows one to predict their relative ease of oxidation and reduction and thus the tendency for the reactions shown in Table 1 to occur. Under a given set of conditions, the more positive the potential, the more spontaneous is the corresponding half-reaction shown in Table 1. Conversely, the more negative the potential, the less spontaneous is the reaction shown but the more spontaneous is the reverse reaction. Later we shall consider how this affects such factors as solubility and binding potential, which will obviously affect the bioavailability of these minerals in the GI tract.

The reason for the sign of $E°$ effecting the spontaneity of a reaction is easily understood from the fact that a spontaneous reaction

Table 1. Standard Reduction Potentials (SRP) for the Essential Elements and Water

Element	Reduction half-reaction	SRP (E, volts)
K	$K^+ + e^- = K$	-2.92
Ca	$Ca^{+2} + 2e^- = Ca$	-2.76
Na	$Na^+ + e^- = Na$	-2.71
Mg	$Mg^{+2} + 2e^- = Mg$	-2.38
Mn	$Mn^{+2} + 2e^- = Mn$	-1.03
P	$P + 3H_2O + 3e^- = PH_3(g)$	-0.87
H_2O	$2H_2O + 2e^- = H_2 + 2OH$	-0.83
Se	$Se + 2e^- = Se^{-2}$	-0.78
Zn	$Zn^{+2} + 2e^- = Zn$	-0.76
Cr	$Cr^{+3} + 3e^- = Cr$	-0.74
Cr	$Cr^{+2} + 2e^- = Cr$	-0.56
S	$S + 2e^- = S^{-2}$	-0.51
S	$S + H_2O + 2e^- = HS^- + OH^-$	-0.48
Fe	$Fe^{+2} + 2e^- = Fe$	-0.44
Cr	$Cr^{+3} + e^- = Cr^{+2}$	-0.41
Co	$Co^{+2} + 2e^- = Co$	-0.28
P	$P + 3H^+ + 3e^- = PH_3(g)$	-0.04
Fe	$Fe^{+3} + 3e^- = Fe$	-0.04
H	$2H^+ + 2e^- = H_2$	0.00
Mo	$H_2MoO_4 + 6H^+ + 6e^- = Mo + 4H_2O$	0.0
S	$S^2 + 2H^+ + 2e^- = H_2S$	0.14

Table 1. Continued.

Element	Reduction half-reaction	SRP (E, volts)
Cu	$Cu^{+2} + e^- = Cu^+$	0.16
Cu	$Cu^{+2} + 2e^- = Cu$	0.34
Cu	$Cu^+ + e^- = Cu$	0.52
I	$I_2 + 2e^- = 2I^-$	0.54
Fe	$Fe^{+3} + e^- = Fe^{+2}$	0.77
Cl	$Cl_2 + 2e^- = 2Cl^-$	1.36
F	$F_2 + 2e^- = 2F^-$	3.03

Source: Ref. 8.

must have a negative free energy change ($\Delta G < 0$), and the relation-ship between G and E may be written as:

$$\Delta G = -nFE$$

where ΔG is the free energy change in calories, n is the number of moles transferred, F is the faraday (23,060 cal/volt), and E repre-sents the voltage or the SRP. Thus, if E is positive, ΔG must be negative, and the reaction will occur spontaneously.

Hard and Soft Metals

Both metals and ligands may be classified as hard or soft (11). Hard metals will most likely react to form complexes with hard ligands, whereas soft metals will react with soft ligands. Anglici (11) has characterized the metals in the following manner: hard metals have small radii, low polarization, high positive charges, no unshared electrons in the valence shell. Hard metals complex most strongly with very electronegative atoms, such as O, N, and F, which are also difficult to oxidize. Bonding is largely ionic between hard metals and hard ligands. Soft metals form stable complexes with atoms, such as P, S, and I, which have high polarizabilities, low electronegativ-ities, and are readily oxidized.

Of the characteristics used to differentiate between hard and soft metals, the least self-evident is the atomic radius. If the atom

is viewed as a series of concentric circles around a nucleus, then
the force between this set of spheres and the nucleus is proportional
to the algebraic sum of all the charges included in the set of spheres.
Therefore, the nuclear charge, +Ze, is shielded from an outermost
electron by the charges -qx, -qy, etc. of the inner electron shells.
The force of attraction, then, between the outermost electron and the
nucleus is:

$$F = \frac{e(Ze - qx - qy)}{\phi^2}$$

where e is the magnitude of the electronic charge and ϕ is the inter-
particle distance.

Applying this to the elements it can be seen that those in Group
1A all have a single S electron outside a closed shell or a closed p
subshell. The full subshells are closer to the nucleus than the outermost electron and act to decrease the net attractive force on the latter.
Therefore, since the average distance of the electron from the hydrogen nucleus increases rapidly with the principal quantum number, n,
the size of the atoms should also increase within this, and any other,
group (10). For predictive purposes it is worth noting that within a
period, the number of electrons and configuration in the inner shells
is the same. Therefore, the attraction for the outer electron will
vary with Z, the atomic number, being equal to Z -e, where e represents the total change on the inner electron shells. Since Z increases
from left to right across any period, then the attraction (Z -e) on the
outermost electron will increase and atomic size will decrease.

Bonding

The chemical properties of molecules are governed in large part by
the strength of the bonds which hold their atoms together, while the
physical properties, such as melting point and solubility, are governed
by intermolecular bonding. Therefore, it is important to briefly review the general types of atomic and molecular binding.

There are three different types of chemical bonds which account
for most interatomic forces. These are (1) ionic bonds, (2) covalent
bonds, and (3) metallic bonds.

Ionic compounds, which are joined by ionic bonds, consist of
positively and negatively charged ions held together by strong electrostatic forces. Such compounds are formed between metals and
nonmetals.

Covalent bonds are formed by the sharing of an electron pair
rather than by the attraction of opposite charges. In this case
stability is conferred to the molecule formed because the increased

volume available to an electron when it is associated with two nuclei, rather than one, lowers its energy.

The metallic bond explains properties such as electrical and thermal conductivity, luster, ductility, and emission of electrons, which is peculiar to metals. This bond is often explained on the basis of the "electron-sea model" where the metallic lattice is visualized as an array of positive ions anchored in a sea of electrons which can wander through the lattice.

In the case of intermolecular forces there are also three types of bonds: (1) dipole forces, (2) hydrogen bonds, and (3) dispersion (London or van der Waals) forces.

Dipole forces exist between polar molecules because such molecules tend to line up in an electric field. The forces involved are similar in kind to those in ionic bonds but are an order of magnitude weaker. For instance, dipolar forces will exist in solid iodine chloride, ICl, where the ICl molecules are arranged so that the positive end of one ICl molecule (iodine atom) is adjacent to the negative end (chlorine atom) of another ICl molecule. The polarity is due to differences in electronegativities between two electronegative atoms in the same molecule producing partial positive and negative charges at each end of the molecule. However, in ionic bonds, a complete transfer of electrons is involved between atoms so that ions with full positive and negative charges are formed, as is the case with the mineral salts.

Hydrogen bonds are stronger than dipole forces and exist mainly with hydrogen fluoride, water, and ammonia. The reason for the strength of this bond may be found by examining the electronegative forces and relative size of the atoms involved. The difference in electronegativity between hydrogen (2.1) and fluorine (4.0), oxygen (3.5), or nitrogen (3.0) is great enough to cause the bonding electrons in HF, H_2O, and NH_3 to be pulled away from the hydrogen atom to such an extent that it acts almost like a positively charged bare proton in its interaction with any adjacent molecule. Since the strength of the hydrogen bond is directly related to the difference in electronegativity within the molecule, it follows that HF would have the strongest hydrogen bond (greatest difference in electronegativity) and NH_3 the weakest since the difference is relatively small. As well as the establishment of a charge in these molecules, the small size of the hydrogen atom allows the fluorine, oxygen, or nitrogen atoms, which carry an apparent negative charge, of another molecule to approach it very closely. It is interesting to note for predictive purposes that hydrogen bonding is mostly limited to compounds containing these three elements, all of which have comparatively small atomic radii. Larger atoms, such as chlorine and sulfur, with electronegativity (3.0, 2.8) similar to that of nitrogen, show little tendency to form hydrogen bonds in such compounds as HCl and H_2S (10).

Dispersion or London forces are somewhat different in origin from those discussed thus far which exist between polar molecules. While hydrogen bonds and dipole forces arise from an attraction between permanent dipoles, dispersion forces are due to what might be called temporary or instantaneous charge separations. These come about due to the fact that while, on the average, the two bonding electrons in a molecule such as hydrogen are as close to one nucleus as another, there still may be a concentration of the electron cloud in some part of the molecule at any given instant. This will set up a temporary or instantaneous dipole, which produces a small attractive force known as the dispersion force. The strength of this instantaneously produced force depends upon how easily the electron cloud in a molecule can be distorted or polarized by another temporary dipole in an adjacent molecule. This in turn will depend upon the distance of the electrons from the nucleus such as occurs in large molecules, as discussed previously. Since molecular size and molecular weight parallel each other, it may be concluded that dispersion forces often increase in magnitude with molecular weight.

It should be noted that dispersion forces are often called van der Waals forces. However, in a strict sense van der Waals forces include all the intermolecular forces discussed as well as those from an induced dipole. It is the sum of these forces that determines the magnitude of the deviation of real gases from ideal behavior as described in the van der Waals equation(10).

Chelation

It could be argued that, of all the interactions that biologically important minerals undergo, chelation is the most important. The following quotation (11) illustrates this fact. "It can be no great exaggeration to describe chelation as a seminal concept in modern biochemical theory and practice. . . . By now, several thousand papers dealing with chelation phenomena in living systems have appeared in the literature, and no modern textbook of biochemistry or pharmacology is without numerous examples of chelation reactions between metals such as calcium, magnesium, iron, zinc, copper, manganese, molybdenum, cobalt, or chromium, and organic metabolites ranging in complexity from glycine to coenzyme-mediated enzymes." I would like to add that chelation is no less important when considering mineral reactions in food that occur postharvest and preconsumption than it is within the living organism.

The metals most often involved in chelation reactions are the alkaline earths, transition metals, and molybdenum, a posttransition metal whose other chemical reactions have been discussed previously.

The metal atom (cation) in a complex ion or chelation reaction is referred to as the central atom. The molecules (anions) that have an

unshared electron pair(s), which can be donated to the central atom
to form a coordinate covalent bond, are known as coordinating groups
or ligands.

The number of bonds formed by the central atom is called its
coordination number, and this along with the geometry of its complex
are characteristic for the various minerals. However, it should be
noted that some of the minerals have only one coordination number
and form complex ions with only one type of geometry, whereas others
may have more than one coordination number and/or have different
geometries.

A chelating agent is a molecule that is capable of seizing and
holding a metal ion in a clawlike grip (*chele* is the Greek word for
claw). Like a claw, the clutching structure forms a ring in which the
ion is held by a pair of pincers. The pincers consist of "ligand"
atoms (usually nitrogen, oxygen, or sulfur), which donate two electrons to form the coordinate covalent bond mentioned previously. In
most cases the central atom can be grasped by more than one molecule,
so that the ion is held in a set of rings. Each ring is usually com-
posed of five or six members consisting of single atoms or groups of
atoms (12). Lindenbaum (13) warns of the common erroneous practice
of referring to most metal-organic complexes as chelates. The terms
chelate and chelation should be applied only to a special form of complex formed between a metal ion and a ligand.

In working with chelation reactions the following five factors
should be considered (13):

1. *Ring Formation*: What distinguishes chelation as a special
   form of complexation is that the ligand molecule contains at
   least two electron-donating atoms; these atoms, usually oxygen, sulfur, or nitrogen, are spaced along the ligand molecule in such a way that binding to a metal results in the
   formation of a heterocyclic ring containing at least one covalent bond. Covalency generally limits the metal member of
   the ring to members of the alkaline earth, transition, and
   rare earth series. Due to the limitations imposed by bond
   angles, the most stable chelate rings contain five to seven
   atoms, although some stable four-membered rings are known,
   usually containing sulfur.
2. *Dentation*: The term bidentate refers to a ligand with two
   donor atoms. Other ligands may be multidentate (e.g.,
   ethylene-diaminetetreacetic acid, EDTA), although all donor
   atoms need not be involved. As we shall note, multivalency
   of the metal ions allows the formation of more than one chelate ring. Frequently the entire organic molecule is referred
   to as the ligand. Preferably, the term ligand should apply
   only to the portion of the molecule involved in chelation.

3. *Resonance*: Unsaturated chelate rings are stabilized by reson ance; these rings are planar, as compared to saturated rings which are puckered.

4. *pH*: In most organic ligands some of the oxygen, nitrogen, or sulfur atoms are already sharing electrons with hydrogen atoms. Thus, for a metal chelate to be formed, the pH of the medium must be sufficiently high to ensure that mass action will allow the metal to compete with protons bound to these ligand atoms. But it is important to note that in some cases a pH high enough to release protons from the ligand may result in the formation of insoluble metal hydroxides, thus lowering the concentration of metal ions available for chelation.

5. *Specificity*: From a manipulative point of view, it often would be desirable if a specific ligand were to form a chelate only with a specific cation. In practice, only relative specificity can be expected. Thus, it can be arranged, taking the above factors into account, that of two metal cations in solution with a ligand, a greater fraction of the cation in lower concentration is bound to the ligand, despite competition from the more abundant cation. It might be pointed out also that organic ligands containing sulfur tend preferentially to bind Cu^+ as well as Hg^{+2}, As^{+3}, and Sb^{+3}.

Chelating agents may be classified as to their degree of dentation i.e., the number of donor atoms they provide. They may be mono-dentate, bidentate, tridentate . . . or multidentate. For instance, a monodentate chelate provides one coordinate covalent bond to the cen-trol atom. Since iron has a coordination number of six, it would re-quire six of these monodentate ligands per iron molecule (14). Bi-dentate chelates provide two bonds and iron would require three of these, whereas it would require only two tridentate chelators to satis-fy its coordination number of six. It is interesting to note that since the favored geometry of ferric iron chelates is an octahedron, triden-tate compounds are the most effective in binding it (15).

As noted previously, ring size or the number of atoms in the chelate ring is also important to formation and stability. Van Campen (16) has also noted that tridentate chelators bind iron most strongly and that cysteine and lysine are both tridentate chelators. However, Sigel (17) has pointed out that the sulfhydryl group of cysteine, being a five-membered ring, has a greater tendency to chelate a metal ion than does the nitrogen group of lysine, which forms an eight-membered ring. This is an important consideration when predicting mineral reactions in food. Another important feature of chelation to be kept in mind when dealing with nutrient minerals in food is that the oxidation-reduction potential of a metal is always altered by chelation (12). A valence change resulting from chelation could alter the subsequent affinity of the metal for the ligand or could shift the location of the chelate as a result of charge reduction, increased liposolubility, and consequently increased membrane permeability. On the other hand, if chelation leads to polymerization (such as in iron hydroxide formation), the solubility or mobility of a metabolite could be restricted, or its biological function severely altered.

For more information on the chemistry of the metal chelate compounds, the reader is referred to the classic text of that name by Martell and Calvin (18) to begin their study.

Solubility

Many of the chemical properties discussed in the preceding sections affect solubility either directly or indirectly. Also, many of the factors which will be discussed in the subsequent section on food processing will owe their effects, in one way or another, to solubility. Nevertheless, there are some general comments that should be made about mineral solubility which do not easily fit into any of these sections.

The final determinant of absorption of a mineral from the intestine is in large part its solubility. The importance of this parameter in mineral availability cannot be overemphasized. For instance, because of the widespread agreement on the importance of solubility, most of the in vitro techniques for predicting the availability of dietary iron have focused on solubility in one form or another (19).

Many of the reactions which the minerals undergo involve the formation of ionic compounds which do not fall into neat categories of soluble and insoluble with a sharp dividing line. As a result, it is difficult to predict the solubility of compounds which might form from other compounds of known solubility. Therefore, it has been suggested (10) that the following facts would be useful for predicting the results of precipitation reactions:

All nitrates are soluble.
All chlorides are soluble except $AgCl$, Hg_2Cl_2, and $Pb\ Cl_2$.
All sulfates are soluble except $CaSO_4$, $SrSO_4$, $BaSO_4$, Hg_2SO_4, $HgSO_4$, $PbSO_4$, and Ag_2SO_4.
All carbonates are insoluble except those of the 1A elements (alkali metals) and $NH_4^{\pm}$.
All hydroxides are insoluble except those of the 1A elements (alkali metals), $Sr(OH)_2$, and $Ba(OH)_2$. $Ca(OH)_2$ is slightly soluble.
All sulfides except those of the 1A (alkali metals) and 2A (alkaline earths) elements and NH_4^+ are insoluble.

Unfortunately, the above observable facts are not readily explained by any simple theory. The difficulties involved in predicting the solubilities of ionic compounds results from the solution process in which water molecules are involved. Normally, when a solid dissolves in a liquid both the enthalpy change, ΔH, and entropy change, ΔS, are positive. This may not occur when an ionic solute and water are involved, since the formation of hydrated ions causes heat to be produced, thus making ΔH negative. Also water molecules are oriented around ions, which may lead to a decrease in entropy. Thus it is difficult to predict. In most cases it is found that the solubilities of ionic compounds are inversely related to the charge density of their ions. These containing ions of low charge density tend to be more soluble than those with high charge density. Thus, compounds containing K^+(radius = 1.33 Å) are often soluble in water, while many Ca^{2+}(radius = 0.99 Å) salts are not. As noted previously, salts containing NO_3^-, a large ion of low charge, are soluble, while those containing anions with a higher charge, such as CO_2^{2-}, or smaller size, such as OH^-, are insoluble in water.

To complicate the matter futher we must also deal with the common ion effect and solubility product. This simply means that the presence or addition of an ion, which is in the ionic compound of interest, may effect the solubility and thus the relative concentration of all ions in the compound of interest.

THE FATE OF MINERALS IN FOOD

The physicochemical determinants of mineral absorption may be defined as those reactions that take place prior to mucosal absorption and transfer. This includes the effect of the food itself and any process it might undergo, as well as the role of the changing environment of the stomach and the duodenum on the chemical and physical state of the minerals.

In a discussion of iron (19), which can be generalized to other minerals, it has been suggested that when the inorganic minerals are ingested alone, their absorption is a function of solubility and charge density, which are in turn affected by conditions in the GI tract. However, when the same sources are fed with food, their absorption is a function of not only their own inherent solubility and charge density, but also depends on their reactivity within the food environment as well as the presence of enhancers and inhibitors. These factors may be affected by conditions in the GI tract as well as by any preparation or processing the food undergoes. When food is cooked, processed, or stored, minerals may be lost, new compounds may be formed, the food environment may be changed, and new external factors may be introduced. These changes may decrease the concentration of minerals in the food as well as having dramatic effects on the physicochemical state of both endogenous and exogenous minerals.

Mineral Losses During Processing

Prior to discussing the processes that cause the major loss of minerals, a note of caution should be expressed concerning the variability of mineral levels among samples of the same type of food which may limit the validity of such data. In plants and vegetables this variability is thought to be due to such factors as genetics, agricultural practices, variations in the soil content of various elements, soil fertility and pH, and environmental factors and plant maturity (3,20). After reviewing the literature on variability, Rotruck (20) concluded that intercomparisons of analysis for different samples cannot be made validly for most elements and that studies on the effect of processing on trace element levels in food should be conducted on samples that have first been shown to be uniform. It should be added that this problem can be overcome to some extent by always measuring the endogenous mineral content. However, if the reaction causing the change is dependent upon mineral concentration, even this will not solve the problem. Other steps which can be taken to minimize the effects of variability are to verify the analytical technique used, employ sampling techniques that ensure uniformity, report the precise identification of the sample including species, variety, and origin of the food, and include a precise description of any processing or preparation effects studied (20).

With the hope that the reader will keep these limitations in mind, the following sections will review some of the studies on mineral losses.

Leaching

Canning, boiling, steaming, blanching and cooking are the processes most likely to cause leaching of minerals. However, the extent to which minerals are lost during these processes varies with the food product, type of processing, and properties of the minerals involved.

Schroeder (21) has reported that canning spinach caused losses of 81.7% of the manganese, 70.6% of the cobalt, and 40.1% of the zinc. In beans and tomatoes 60% and 83.8% of the zinc was lost, respectively. Canning carrots, beets, and green beans caused a loss in cobalt of 70%, 66.7%, and 88.9%, respectively. Interestingly, they reported an increase of 226.8% for manganese and 60% for zinc in canned beets, which could have been due to contamination and/or variability. Heintze et al. (22) observed that the loss was greatest for sodium in heat-treated spinach, followed by potassium, magnesium, and calcium. Bielig et al. (23) noted a similar pattern with carrots, but in beans the magnesium loss was less than that of potassium.

The mineral content of the water used for processing or cooking can also affect the mineral content of the processed food. Heintze et al. (22) reported an increase in the calcium content of cooked spinach due to the hardness of the water, while Marston et al. (24)

found that the content of both calcium and magnesium in the water affected the degree of change of these minerals during processing.

The amount of water used in the process has also been found to affect mineral losses. Some processes use inherently less water (steaming) than others (boiling), while at times the amount of water used in a given process may vary. Krehl and Winters (25) found that when assorted vegetables were processed with their wash water only, greater than 90% of the minerals were retained. Pressure cooking or the use of only ½ cup of water caused an 85% increase in retention, while immersion cooking caused a decrease of 75%. Similar results have been reported in other investigations (26,27).

Higgs et al. (28) reported that in the major sources of selenium (meats, seafoods, eggs, cereal products) there was not a decrease when cooked in the normal manner. However, some vegetables, such as asparagus and mushrooms, which are relatively high in selenium, lose significant amounts when cooked. This effect may be due to the presence of volatile selenium compounds.

Losses of calcium, manganese, and iron in cooked spaghetti were found to occur during the first ten minutes of cooking, while magnesium continued to decrease steadily and potassium continued to be lost but at a slower rate (29). Ranhotra et al. (30) reported changes in manganese, iron, copper, phosphorus, zinc, calcium, and magnesium ranging from 86.5 to 119.1% during the cooking of spaghetti. Values greater than 100% were attributed to losses in solids during cooking.

Klein and Mondy (31) and Mondy and Ponnampolam (32) reported a decrease in the cortical content of potassium and iron and an increase in the pith content during baking of potatoes. This suggested that the minerals migrated inward. Frying (32) reduced the content of these minerals throughout the tuber, while microwave processing resulted in only minimal losses (31).

Freezing, as would be expected, results in greater retention of minerals than canning (33).

Separation

The loss of minerals due to separation often occurs in such processes as milling, sugar refining, and processing of legumes or seeds into oils and/or proteins.

The loss of minerals due to the milling of cereals, in particular wheat, has been extensively studied (20). The iron content of cereals is greatly reduced by removing the germ and bran, since this is where iron is concentrated. However, a large part of this iron is restored through supplementation of flour. Refining of wheat to flour also results in a significant loss of copper, manganese, zinc, and cobalt as follows: 75.6% loss for iron, 67.9% for copper, 88.5% for manganese, 77.7% for zinc, and 67.9% for cobalt (20).

Although molybdenum is concentrated to some degree in the bran and germ of wheat, the extent of concentration is not as great as for the elements mentioned previously. The concentration of chromium in the bran and germ closely resembles molybdenum, with losses being 48.0% for molybdenum and 40.0% for chromium.

Milling results in a relatively small decrease in cadmium, indicating that it is distributed more evenly throughout the seed. Selenium shows a loss of only 15.9%, and data are not available on other minerals (20).

The processing of rice, corn, sugar, and soybeans also involves some loss of minerals due to separation. Polishing rice causes a 75% loss in chromium and zinc, while losses of manganese, cobalt, and copper are much less, ranging from 26 to 45%.

The processing of corn appears to result in little loss of trace minerals but in a substantial redistribution. Corn meal contains about one-half the chromium and zinc that dry corn contains, but it has higher levels of cobalt and copper. Corn starch contains less chromium, manganese, and zinc but slightly more cobalt than dry corn. Corn oil contains much of the original copper and chromium from dry corn, but manganese, cobalt, and zinc levels were 21%, 42%, and 9% of these in dry corn, respectively (20).

The data on sugar processing show an increase in the concentration of trace elements in molasses as compared to sugar cane. White sugar has less chromium, manganese, copper, zinc, and molybdenum than molasses, sugar cane, or raw sugar and less cobalt than molasses. However, in terms of human consumption these data probably do not have much significance.

Soybean processing does not appear to cause large losses of trace elements, with the exception of silicon. In fact, since the soybean process is one of concentration with respect to protein, it has been found that there is an increase of iron, zinc, aluminum, strontium, and selenium. In order to explain this increase, one would have to postulate that these minerals are either bound or intimately associated with the protein, thus following it through processing and concentration. Manganese, boron, copper, molybdenum, iodine, and barium showed variable changes without any trends. The large loss of silicon probably reflects a loss in soil on the soybean through washing and processing (20).

Effects of Environmental Changes

The fact that a mineral, or some percentage of it, is not lost during processing or storage does not mean that the mineral is in the same form that it was in the raw product. Indeed, chances are that it is not. It is interesting that a great deal of the literature refers almost exclusively to the presence or absence of enhancers or inhibitors of mineral absorption and largely ignores other effects of the food environment. It would seem more apt to refer to the total enhancement or inhibition of mineral bioavailability by food and/or processing.

This would include the presence of enhancers and inhibitors, as well as any other alterations in the state of the mineral due to environmental factors. Many minerals are extremely reactive, particularly in the presence of moisture and/or water activity. Such reactions might occur with free radicals, other reactive groups, oxygen, and compounds entering the food via diffusion from packing materials, to name a few. In turn, solubility, valence, charge density, ionization, dehydration, and rehydration might be affected.

Redox Potential and pH

It is obvious from Table 1, where the standard reduction potentials of the reduction half-reactions of the elements are shown, that the reduction potential of the food and/or components of the food in which the element resides will affect its valence and in so doing will affect its solubility. The importance of solubility in absorption has already been emphasized, but it should also be noted that the greater the charge on a molecule, the more difficult it is for that molecule to traverse cellular membranes (34).

In order to illustrate such effects it might be instructive to use an example. From Table 1 it can be seen that the standard reduction potential (SRP) for the conversion of ferric ion to ferrous ion ($Fe^{+3} + e^- \rightarrow Fe^{+2}$) is +0.77 V. This means that whenever ferric iron is present in a system with a redox potential of less than +0.77 V, reduction should be spontaneous as explained previously. Now, if we examine a number of common food items (Table 2) we find that most foods have potentials within ± .10 V of +0.40 V. This does not seem extraordinary when one considers that each of the foods contain at least a trace of dissolved oxygen (Table 2) and the SRP for the oxygen half-cell reaction ($O_2 + 2H_2O + 4e^- \rightarrow 4OH^-$) is +0.401 V.

Nojeim et al. (35) utilized an electrolytic cell model system at pH 4.2 and three foods of different redox potentials to evaluate this effect with four iron fortificants. In the model system the lower potentials favored the formation of Fe^{+2} in the case of elemental iron and ferrous sulfate but not in the case of ferric orthophosphate and sodium ferric EDTA. In the food systems Fe^{+2} was favored in tomato juice (0.24 V) and cranberry juice (0.40 V), but all the iron was found to be insoluble in the biscuit dough (0.34 V).

At first glance these results appear to contradict the theoretical changes which would be expected. This is because the effect of pH was not considered. The SRP of $Fe^{+3} \rightarrow Fe^{+2}$ is 0.77 V in an aqueous system. However, if we examine this reaction under basic conditions ($Fe(OH)_3(s) + e^- \rightarrow Fe(OH)_2(s) + OH^-$) it is found that the SRP is -0.56 V, indicating nonspontaneity and, in fact, precipitation, since, as was pointed out in a previous section, all hydroxides are insoluble with the exception of the alkali metals. Hydrolysis of the positive essential metal ions takes place readily under neutral or even slightly

Table 2. pH, Dissolved Oxygen and Oxidation Values of Some
Common Food Items

Item	pH	Dissolved oxygen (ppm)	Redox potential (mvSHE)
Apple cider vinegar	2.8	5.0	425
Chemically leavened biscuit dough	6.5	3.4	341
Cherry-flavored instant beverage	4.5	5.5	370
Commercially canned green beans	5.7	0.8	285
Commercially canned sauerkraut	4.5	0.4	235
Commercially canned spinach	5.5	0.6	341
Cranberry juice	2.7	2.0	401
Distilled water	6.3	4.0	400
Ginger ale	5.5	2.0	360
Skim milk	6.8	0.8	321
Tomato juice	4.3	0.3	241
Vitamin C-fortified orange drink	2.8	3.5	390

Source: Ref. 35.

acid conditions (36). Thus, Fe^{+2} and Fe^{+3} ions are converted to their
insoluble hydroxides in foods at or near neutrality, and the redox
potential at this pH favors the formation of the very insoluble
$Fe(OH)_3$. Hydrolysis is not limited to iron since Cu^{+2} ions are readily
hydrolyzed to insoluble basic salts in weakly acidic conditions, and
Mn^{+2}, Co^{+2}, and Zn^{+2} are converted to their insoluble hydroxides at
or near neutrality.

It should be noted that the insoluble hydroxides at any given
pH are in equilibrium with very small concentrations of the metal
ion according to the solubility product (Ksp) principle:

$$M^{+2} + 2 \ OH^- \rightleftarrows M(OH)_2$$

$$Ksp = [M^{+2}] \ [OH^-]^2$$

Therefore, since Ksp is constant, an increase in OH (i.e., increase in pH) results in a decrease in the concentration of the free
metal ion (M^{+2}). This decrease in free metal ion will follow the first,
second, and third powers of hydroxyl ion concentration with the
mono-, di-, and trivalent metal hydroxides, respectively (36). Therefore, small increases in pH cause a large decrease in solubility. This
is why such care must be taken in clinical studies to control and
specify the pH at which a mineral is fed.

The requirements of some essential elements, such as vanadium and molybdenum, are so small that the solubility for almost any compound is such that enough is provided. Therefore, solubility is not a problem.

The pH will also affect the solubility of salts formed from metal ions and anions commonly present in food, such as phosphates, carbonates, and oxalates. At high pH these salts have minimum solubility, maximum stability, and minimum bioavailability. A decrease in pH increases metal ion solubility by decreasing the concentration of the anion and shifting the equilibrium away from salt formation (36). However, at the pH of food some of these salts may be soluble, thus increasing the potential bioavailability of the metal involved. For instance, oxalic acid forms both soluble and insoluble salts with metal ions, which is of particular nutritional importance. It forms soluble salts with alkali metal ions (Li,Na,K) and with iron. All other oxalates are sparingly soluble in water, which would dramatically effect the bioavailability of the metal ions involved. Oxalate forms a practically insoluble salt at neutral or alkaline pH with calcium, being soluble to the extent of 0.67 mg per 100 ml of water at pH 7.0 and 13°C. Zinc also has limited solubility (0.79 mg/100 ml, 18°), while Fe^{+2} and Fe^{+3} show solubilities of 22.0 mg/100 ml and "very soluble," respectively (37). These chemical facts and their effect on iron absorption have recently been substantiated in a biological sense by Van Campen and Welch (38), who investigated the availability to rats of iron from two varieties of spinach. They also compared the absorption of iron between $FeCl_3$ and Fe-oxalate, as well as the effects of adding 0.75% oxalate to the diet. They found that absorption of iron from both varieties of spinach was comparable to that from $FeCl_3$ and that the iron was equally available from Fe-oxalate and $FeCl_3$. The addition of 0.75% oxalic acid to the diet did not depress iron absorption and, if anything, appeared to enhance iron utilization by rats.

Aging

Aging or length of storage is another factor that must be considered. Metal salts that are normally soluble under a given set of conditions, once precipitated and allowed to assume a stable structure over time, either can become insoluble or dissolve so slowly that they may be considered insoluble. This aging can proceed via hydrolysis from free ions in solution to various oxides.

In the case of iron, alkalinization, oxidation, and heating, which might occur in food processing and storage, tend to drive the reactants toward their final form of stable dehydrated polymers with increasingly ordered structures (39). It should be emphasized that there are many different stages, with varying degrees of solubility, which the iron undergoes on aging. Spiro and Saltman (40) have reviewed in detail the properties of colloidal ferric hydroxide polymers and their polynuclear iron chelates. Interestingly, these various stages which the iron undergoes may be correlated to bioavailability in several different studies. Derman et al. (41) have shown that aqueous suspensions of ferric hydroxide are poorly available to humans, but Berner et al. (42) found that a soluble species of ferric hydroxide polymers prepared by hydrolyzing a ferric nitrate solution with $KHCO_3$ was available to rats. Derman et al. (43) found that ferric hydroxide was not interchangeable in the nonheme iron pool of maize porridge, since it was absorbed only half as well as the intrinsic iron present. More fully oxidized and aged species, such as ferric orthophosphate and ferric pyrophosphate, are insoluble and have been found to be unavailable in many studies (44).

A technique utilizing visible/infrared spectrophotometry has been suggested as a possible technique to follow the aging process of iron in food, on storage, in order to predict potential bioavailability (39).

Enhancers and Inhibitors

The interaction of dietary components with minerals through some form of bonding or by chelation is the basis for enhancement or inhibition of minerals in the diet. If the food component forms a compound with the mineral which is soluble and can be absorbed through the mucosa and/or it has a kinetic stability constant (45) which allows the mineral to be transferred to a mucosal or serosal acceptor, it is known as an enhancer. If, on the other hand, the compound so formed is insoluble and/or it has a dissociation constant which does not allow transfer to a mucosal or serosal acceptor, it is known as an inhibitor. This means that an inhibitor may form a soluble mineral complex but not allow the mineral to transfer to biological acceptors within a physiologically reasonable time. This might explain, in part, why some in vitro studies based primarily on solubility might disagree with clinical findings.

It is very difficult to state with certainty whether compounds formed with minerals are chelates or are held in a complex by one of the bonds described in an earlier section. Often investigators do not chemically define the type of complex formed, nor is it necessary. Therefore, in the present discussions we may use the terms complex and chelate interchangeably with the knowledge that other types of bonding may be responsible for formation.

There are many excellent reviews available on the effect of mineral interactions on bioavailability (4—6, 46—51). In fact, many of these effects are discussed in other chapters of this volume. Therefore, I am going to attempt to limit this discussion to some of the reactions which may be responsible for these interactions in food, postharvest. This may include the effects of processing and/or storage.

The number of interactions possible are incredibly large. All the naturally occurring compounds, as well as compounds from processing-induced reactions, such as the Maillard reaction, could interact with minerals. Therefore, we have arbitrarily chosen to mention proteins, dietary fibers, carbohydrates, lignin, phytic acid, and organic acids.

Proteins and Amino Acids

A recent review has summarized the literature on the effects of dietary proteins on iron bioavailability (52). Many of the reactions discussed in this paper are also applicable to elements that have an electronic configuration similar to iron.

It has been pointed out that at least half of the twenty most commonly occurring amino acids in proteins contain side chain donor atoms that are capable of forming chelates with metal ions (53). As discussed previously chelates form from electron-donating atoms such as oxygen, sulfur, or nitrogen and contain five to seven atoms in a ring in their most stable form. Serine and threonine (oxygen), 2,3-diaminopropanoate (nitrogen), and cysteine, cystine, and penicillamine (sulfur) will form stable five-membered rings. Aspartate and asparagine (oxygen), 2,4 diaminobutanoate (nitrogen), and homocysteine (sulfur) form six-membered rings. Glutamate (oxygen) and ornithine (nitrogen) form seven-membered rings and lysine (nitrogen) a ring with eight atoms.

It has been noted that L-cysteine and D-penicillamine will form chelates with the transition metals (53). Flynn et al. (54) determined the effective stability constants for both cysteine and lysine with five different iron sources and ferric cysteine was the only one to show potential for the improvement of iron absorption in food. It would be unseemly to end this section without at least mentioning the enhancing effect of meat on the bioavailability of both heme and nonheme iron. This effect is extremely well documented and will be discussed elsewhere in this book. However, it is interesting to note that a recent in vitro study (55) has found that when the ability to solubilize iron was plotted versus time, during a pepsin pancreatin digestion of chicken muscle, a maximum was reached after approximately two hours. This indicates that the compounds responsible for the iron absorption-enhancing effect of meat (meat factor) may be specific peptides formed furing digestion.

Dietary Fiber

The effects of dietary fiber on the inhibition of mineral absorption are well documented (19,50,51)

Pectin. Pectin has long been known to form complexes with divalent metals (56), and Schweiger (57) has shown that both pectate and alginate form cross-linking metal chelates in which pairs of adjacent hydroxyl groups are required.

Significant binding of iron, calcium, magnesium, and zinc by both pectin and guar gum has been reported (58). Platt and Clydesdale (59) have shown that these fibers exhibit both specific and non-specific binding sites with Fe^{+2}. Interestingly guar gum showed the highest avidity for Fe^{+2} of several fibers evaluated (59).

Cellulose. Cellulose is a linear polymer of anydro-D-glucopyranose units joined in 1-4 glucosidic bonds, and is the basic structural material of plant cell walls. The effect of cellulose on the absorption of minerals is somewhat unclear due to some apparent contradictions in clinical studies as outlined by James (51). In vitro studies of binding would seem to indicate that it is unlikely that cellulose would cause problems in mineral absorption in a food material. James (51) cites unpublished data, which indicate that calcium binding to cellulose is negligible, and in another study (58) cellulose was found to have a low avidity for iron and showed nonspecific binding. Camire and Clydesdale (58) found that cellulose bound less than 6% of Fe, Ca, Mg, or Zn at pH 5.0 and a 24-hr incubation at 30°C. In this study the effects of two processing treatments, toasting and boiling, were also evaluated. There was no significant effect of the toasting treatment on the binding of Fe, Ca, Mg, and Zn to cellulose, and boiling caused a significant increase in the amount of iron bound at pH 6.0 and a significant increase in the amount of zinc bound at pH 6.0 and 7.0. Clydesdale (19) reviewed the available data on the binding capacity of cellulose to iron, which showed that it had negligible binding at low pH but greater binding at higher pH values. Several studies showed that compounds such as phytate, citrate, EDTA, and lignin were able to remove the iron from cellulose, indicating that its binding constant with iron was weak so that it would not be a significant factor in iron absorption in food systems (19).

Hemicellulose. The hemicelluloses are heterogeneous groups of poly-saccharides which may contain numerous kinds of hexose and pentose monosacchharide units and, in some instances, residues of glucoronic aicd. They are often referred to as xylans, arabinogalactans, and glucomannins, and often some of the polysaccharide gums are included in this group.

The hemicelluloses have been implicated in mineral bioavailability. This may be due to the fact that the uronic acid components can offer free binding sites for cations.

Mod et al. (60) reported that the sequence of binding to alkali soluble hemicellulose from rice was Cu > Zn > Fe, and for water-soluble hemicellulose Cu > Fe > Zn. Fernandez and Phillips (61) showed that the hemicellulose, psyllium mucilage, bound significant amounts of iron, and Platt and Clydesdale (58) found that guar gum had a very high avidity for Fe^{+2}.

Carbohydrates

Discussions of metal complexation often exclude the role of carbohydrates. However, there is the potential for complex formation between sugars and metal ions. The extent of this occurring in a food matrix has not been fully elucidated. It appears that the axial-equitorial-axial sequence of three oxygen atoms on a six-membered ring forms a good site for complex formation with cations and that a *cis-cis* sequence of three hydroxyl groups on a five-membered ring is also suitable for formation of complexes with metal ions (62). Complexation is also possible with polysaccharides through coordination of the metal ion with more than three oxygen atoms. Alginic acid, for example, is a polysaccharide of industrial importance because its calcium salt forms gels useful to the food industry. Rendleman (62) has discussed the chemistry of these complexes in detail.

Charley and Saltman (63) reported that reducing sugars and polyhydroxy compounds form stable complexes of low molecular weight with metal ions and that these stimulate uptake from the diet. In fact, it has been suggested (64) that ferric fructose be used as an iron fortificant due to its stability and apparent bioavailability.

Calcium binding to lactose is well documented (63,65) and is thought to be responsible for the reported enhanced absorption of calcium from milk (48). Also, glucose polymers are reported to have increased calcium absorption from a liquid formula diet or an aqueous solution (66), but this enhancement has not yet been shown in a mixed meal.

Lactose has been found to combine with Ca, Ba, Sr, Mg, Mn, Zn, Na, and Li in a ratio of one to one over the pH range 2.0−6.5 (65). This means that such reactions are likely in a wide variety of foods, both processed and fresh. Various polysaccharide gums, such as carageenan, agar, alginates, guar gum, locust bean gum, and carob bean gum may react with minerals through some of the mechanisms discussed with variable results on bioavailability. Since these gums may be used in foods as thickeners, it would be wise to keep these potential effects in mind.

Lignin

Lignin is a high molecular weight polymer containing aromatic phenolic residues and is usually found in the food matrix associated with other fiber constituents. The chemical nature of lignin is responsible for its ion exchange properties, and lignin can act as an absorbant for many organic compounds. Of all the fiber components lignin has most consistently been shown to bind minerals with high affinity (58,59, 61,68,69) as well as decreasing absorption of iron in clinical studies.

Platt and Clydesdale (69) postulated that with iron the following order of binding strength exists: lignin (type A group), phytate, lignin (type B), beta-glucan, and cellulose. In a later study (59) lignin was found to offer two binding sites for Cu and Zn as well as Fe. The high affinity binding sites of lignin were shown to bind Fe approximately twice as strongly as copper, and 7.5 times more strongly than zinc. However, the high affinity groups bound more Cu than either Fe or Zn. The low affinity groups of lignin were shown to bind Cu approximately twice as strongly as Fe and 6 times more strongly than Zn, however, all complexes were weak at this binding site.

It should be noted that in most of these studies purified forms of lignin were used. Therefore, it is questionable if these results can be translated directly to food. Nonetheless, indications are that lignin complexes once formed in a food will remain and may decrease availability.

Phytic Acid

Although both the in vitro and in vivo evidence for the deleterious effects of lignin on mineral absorption is very sound, phytic acid still seems to generate more adverse publicity. This should not suggest that phytic acid is blameless, but simply that the evidence for its effects is much more controversial than that with lignin.

Chemically phytic acid is myo-inositol 1,2,3,4,5,6-hexakis (dihydrogen phosphate). Obviously with this structure it will easily form complexes with other food components, including the minerals. In fact this is probably the reason for the controversial results that have been obtained on its effect on the bioavailability of minerals. Phytic acid will form complexes with minerals that include a range of molar ratios from 1:1 to 6:1. Also, it can complex with more than one mineral simultaneously and in turn bind to protein. These various possibilities, all affected by pH, produce an almost limitless array of combinations and permutations, each of which might form a complex of vastly different solubility and chemical reactivity, thus influencing bioavailability.

For instance, Vohra et al. (70), using a titrimetric method, found that the capacity of sodium phytate to form a complex with a metal is pH dependent and that at pH 7.4 it bound metal in the following decreasing order: Cu, Zn, Ni, Co, Mn, Fe, Ca. In other studies it was found that water washing did not remove iron from wheat bran (71) nor from soy isolates at pH 4.3 (72), although Ca, Mg and Zn were removed. Since this doesn't agree with the order found for individual minerals alone (70), it would appear that phytic acid is bound in a protein-cationic-phytic acid complex in these foods. Another study (73) has shown that the association between phytic acid and protein occurs in the pH 1—2 range in rice bran, indicating the effect of substrate, pH, and minerals present. When both Zn and Fe were added to a wheat bran fraction, a significant decrease was seen in phytic acid, phosphorus, protein, endogenous Ca, and soluble Fe as compared to the addition of Fe alone. However, when Fe and

Zn were added to Na phytate alone, they all remained soluble, indicating the need for another component, such as protein or endogenous Ca present in the wheat bran, to cause precipitation (74).

Other workers (75,76) have shown the effect of pH on the solubility of metal phytic acid complexes. Jackman and Black (76) found the Mg and Ca salts to be soluble at low and insoluble at high pH. Al salts were soluble at extremes of pH with a minimum solubility at pH 3.5, and ferric salts were insoluble at pH 1.0—3.5, but became progressively more soluble as the pH was raised. It should be remembered that the molar ratio of iron to phytate in the complex is an extremely important determinant of solubility. Morris and Ellis (77) isolated soluble monoferric phytate from wheat bran and found it to be available to rats.

Champagne et al. (73) has shown that the solubility of Ca in rice bran decreases above pH 5, which would be in agreement with the other studies if this was due to a Ca-phytate complex. Consideration of pH is therefore vital before drawing any conclusions from either in vivo or in vitro studies. For this reason Platt and Clydesdale (74) incorporated a sequential pH treatment simulating gastrointestinal changes in their study of mineral interactions on the solubility of phytic acid complexes. Using this technique with phytate-rich wheat bran containing added Fe, it was found that both Ca and Zn caused a decrease in the solubility of phytate and iron, with Zn also causing a decrease in protein. This indicated that Ca does not require protein in its phytate-metal complex to cause precipitation. Cu had no effect on the solubility of Fe, phytate, or protein, indicating that a soluble complex is formed. This is in accord with previous work (70) on single metal-phytate complexes and could explain why phytate inhibits zinc but not copper absorption (78).

Calcium and magnesium also react with phytic acid to produce a mixed salt (79), and other trace metals interact together with phytate both in vitro and in vivo (80).

Mills (50) has suggested that the convincing argument made for the inhibitary effect of phytate on zinc utilization does not consider the restricted physiological relevance of the simplified aqueous systems within which such evidence was first obtained. The major variables interfering with extrapolations are the stability of phytic acid to hydrolysis both in the food and in vivo, the presence of other intraluminal ligands which bind Zn and prevent it from reacting with phytic acid (we found this phenomenon to occur in food systems with iron (69)), and the high concentration of Ca in rat bioassay diets as compared to human diets. In fact, it is suggested (50) that the rate of weight gain of rats (Δ gm/day) can be related to the (molar) concentration of dietary Ca, Zn, and phytate by the expression:

$$\Delta W = 6.213 - 0.331 \ X$$

where X = [Ca][phytate]/[Zn] moles/kg diet; and the critical threshold ratio of [Ca][phytate]/[Zn] above which available zinc supply was inadequate to maintain growth was approximately 3.5. Mills (50) also makes a telling statement concerning research on minerals which is worth repeating: "This relationship is clearly of greater value for predicting zinc availability from the diets of rodents; its predictive value for human diets cannot yet be assessed because of the startling infrequency with which concentrations of all relevant variables are reported."

Although the predictive value of the [Ca][phytate]/[Zn] ratio may prove to be superior to the more commonly used phytate/zinc ratio, it must be realized that either one will have limitations in diets which differ in respect to amino acids, proteins, organic acids, pH, and redox potential as discussed elsewhere. All of these factors will effect the formation and stability of Ca-phytate-Zn complexes, since some compounds will bind zinc prior to its uptake by Ca phytate, while others may remove it from the complex. For instance there is evidence that histidine, cysteine, and methionine readily desorb zinc from the particulate, insoluble Zn-Cu-phytate complex (81). It was further shown that the yield of Cu or Zn from phytic acid complexes was proportional to the amino-N content of the intestinal soluble phase, and the effectiveness of release of a range of metals from their complexes with phytate was typically Cu > Cd > Mn > Zn > P. This fact, along with the differences in solubility of Zn- versus Cu-phytate complexes mentioned previously, might be used to explain the bioavailability of Cu, as compared to the decreased bioavailability of Zn, in the presence of phytate. Mg deficiency can also be induced by a high fiber diet, and the Mg phytate complex formed may be as effective in inhibiting Zn absorption as Ca phytate (50).

With this degree of complexity the various arguments about the inhibitory effects of bran being due to fiber or phytate (82) may or may not be relevant in a mixed diet.

Although the inhibitory effect of phytate on Ca absorption is well documented in the literature, the numerous chemical reactions involved with its interaction with phytate (80) must be considered. Graf and Eaton (83) have shown that phytate does not impair the absorption of Ca when fed to mice by gavage, indicating that the reactions which occur both in vitro and in vivo should be considered in this phenomenon.

There are a great many more studies on phytate which could be mentioned. However, a discussion of phytate mineral interactions would not be complete without acknowledging the studies of McCance and Widdowson, which set the foundation for later investigations.

Organic Acids

Organic acids occur naturally in foods as well as being important
food additives. In this latter role they are employed in several dozen
functions, some of which are: enhancing and modifying flavor, aiding
in the preservation of food by lowering pH, simplifying processing
operations, acting as gelling agents for pectin, serving as a source of
acidity in leavening, catalyzing the inversion of sucrose, stabilization,
buffers, and numerous other roles.

However, if the chemical structures of these acids are examined,
it becomes apparent that many of them are excellent ligands for chelation reactions as described in an earlier section. As such, these
organic acids are able to complex free metal ions or desorb them from
other complexes, such as those with phytic acid. Therefore, these
can act to either enhance metal absorption or inhibit it depending
upon the solubility of the complex they form and its dissociation
constant.

The best known enhancer of this type is ascorbic acid, which
has been shown to increase nonheme iron absorption in countless
studies.

The ability of ascorbic acid to increase iron absorption is related
both to its role as a reducing agent and its ability to form soluble
iron-ascorbate complexes. However, complexation of iron with ascorbate occurs much faster than reduction (45). Further, the thermodynamic stability constant of the iron ascorbate complex reported by
Gorman and Clydesdale (84) and its kinetic stability constant (45)
are important considerations in its mode of action. When ascorbic
acid solubilizes iron from another complex it appears that the mechanism involves a reductive release (85,86). This is chemically sound
since Fe^{+3} binds more tightly than Fe^{+2} to many acceptor molecules
(45). By reducing Fe^{+3} to Fe^{+2} the bond strength of a complex may
be lessened, shifting the equilibrium to the more soluble Fe^{+2}.

Unfortunately, the enhancing effect of ascorbic acid is destroyed
by cooking and baking due to the oxidative destruction of ascorbic
acid by high temperatures (87,88). Lynch and Cook (89) found that
boiling oxidized only 30% of the ascorbic acid present in a meal, while
baking resulted in very little of the ascorbic acid being in the reduced
state.

However, the instability of ascorbic acid is well known to the
food technologist and problems will occur whenever a food is exposed
to oxygen, high relative humidity, low acidity (the pH of most foods),
heat, other trace metals, and, to a lesser extent, other factors (19).
Therefore, ascorbic acid is unable to function effectively over time
unless it is in a room-temerature acid environment or in a dry food
packaged to prevent the introduction of moisture and oxygen. This
rules out its use in many applications. As a result, studies were

initiated to evaluate ascorbate, citrate, malate, and lactate as possible enhancers of iron in processed soy systems (90–92). In all of these studies it was evident that the ability of each ligand to enhance iron solubility was dependent upon pH, type of iron source, process, and ligand. In the fried soy meat system (91), combinations of hydrogen-reduced iron (HRI) with either ascorbic, lactic, citric, or malic acid produced significant increases in soluble iron. Supplementation of a baked corn-soy-milk (CSM) food blend with HRI and either ascorbate or citrate was best while ferric orthophosphate and cysteine, ascorbate, citrate, or malate was superior in the boiled product (92).

There have been several studies indicating that both malate and citrate increase the bioavailability of iron, but there has not been as much work on the other minerals. Chemically, there would seem to be the potential for chelation with all the minerals as described by Martell and Calvin (18). Warner and Weber (93) have investigated the cupric and ferric citrate complexes as a function of pH and in each case found a chelate with a citrate: metal ratio of 1:1 in which four protons have been displaced by the citric acid. However, this was achieved at a pH of 3 in the case of the iron and at 6 to 7 in the case of the copper. They also found that at pH values below these, other 1:1 complexes were present but fewer than four protons were displaced.

Mineral-Mineral Interactions in Food

As noted in the beginning of this chapter, a great deal of research has been carried out to investigate the competition between minerals for physiological absorption sites. However, little has been done to describe quantitatively the competition between minerals for the binding sites of food components. Such competition could have an enhancing or inhibiting effect on absorption of a given mineral, depending upon whether it competed favorably for a site or a ligand which formed soluble (bioavailable) complexes or insoluble (unavailable) complexes. This could be a particularly important consideration in determining the amounts, ratios, and order of addition of minerals in fortified foods.

Many formulated foods contain alginates, and the fact that Ca has been found to compete successfully with Fe for binding sites (93) may affect the relative availability of these minerals.

Garcia-Lopez and Lee (94) reported that Cu and Zn increased the amount of soluble iron in the presence of neutral detergent fiber (NDF) from pinto beans, but Cu decreased it in acid detergent fiber (ADF). Mg and Ca did not alter the solubility of Fe with either fraction.

Platt and Clydesdale (95) found an effect on competitive mineral binding to lignin due to the order of addition of the minerals involved.

When Zn, Mg, or Ca was added at a concentration equivalent to the RDA to a lignin sample containing 9 mg iron, there was an increase in percent soluble iron. However, this increase was significantly greater if the iron was added after the minerals. In a further study (96) with wheat brans it was shown that the addition of Fe and Zn caused an increase in soluble Fe but a decrease in Mg compared to Fe alone. Fe and Cu in the same system did not affect Fe solubility but increased that of Mg. Fe and Ca caused a decrease in Fe and an increase in Mg, suggesting that there are mineral specific binding sites in bran.

From these studies it is apparent that the competitive displacement of one mineral by other minerals at the various cation exchange sites of food will affect mineral solubility. In turn the solubility may affect bioavailability and is therefore an important consideration in formulating foods.

CONCLUSIONS

It is a somewhat frustrating endeavour to try to cover such a vast topic as this in a meaningful way. Throughout this effort I have been haunted by the number of topics and references which had to be omitted due to limited space. However, it is hoped that, if nothing else, this chapter will bring to the reader an appreciation of the importance of mineral chemistry and its role in the ultimate bioavailability of the mineral in foods.

It is obvious that the clinician will have to keep these interactions in mind when designing a technique or study to measure bioavailability. In order to fully understand the results it is necessary to control and record such simple parameters as pH, cooking time and temperature, purity of ingredients used, utensils employed, ratios of minerals used, and their order of addition. If this is done, we will be able to analyze the chemical reasons for observed bioavailabilities and hopefully begin to understand the nature of the process.

In this way, we can contribute to the design and production of better food vehicles to carry those most important nutrients, the minerals.

ACKNOWLEDGMENT

Paper No. 2762, submitted by the Massachusetts Agricultural Experiment Station, University of Massachusetts at Amherst. This work was supported in part by Experiment Station Project No. NE116.

REFERENCES

1. R. E. Hein and I. J. Hutchings, in *Nutrients in Processed Foods: Vitamins-Minerals*, American Medical Association, Publ. Sci. Group Inc., Acton, MA., 1974.

2. I. I. Somers, R. P. Farrow, and J. M. Reed, in *Nutrients in Processed Foods: Vitamins-Minerals*, American Medical Association, Publ. Sci. Group Inc., Acton, MA., 1974.

3. E. J. Underwood, *Trace Elements in Human and Animal Nutrition*, 4th ed., Academic Press, New York, 1977.

4. W. Mertz, Mineral elements: New perspectives, *J.A.D.A.*, 77: 258 (1980).

5. N. W. Solomons, Mineral interactions in the diet, *Contemporary Nutrition*, 7 (7): 1 (1982).

6. H. H. Sandstead, Trace element interactions. *J. Lab. Clin. Med.*, 98: 457 (1981).

7. *Taber's Cyclopedic Medical Dictionary*, 7th printing, F. A. Davis Co. Philadelphia, 1984.

8. *Handbook of Chemistry and Physics*, 63rd ed., CRC Press Inc., Boca Raton, FL, 1982–83.

9. T. M. Lowry and A. C. Cavell, *Intermediate Chemistry*, 6th ed., MacMillan and Co., New York, 1954.

10. W. L. Masterton and E. J. Slowinski, *Chemical Principles*, 5th ed, W. B. Saunders Co., Philadelphia, 1981.

11. R. J. Anglici, Stability of coordination compounds, in *Inorganic Chemistry* vol. I (G. L. Eichorn, ed.) Elsevier Publ. Co., New York, 1973.

12. A. Lindenbaum, A survey of naturally occurring chelating ligands. Metal ions in biological systems: studies of some biochemical and environmental problems, *Adv. Exp. Med. Biol.*, 40: 67 (1973).

13. J. Schubert, Chelation in medicine, *Scientific American, 214*: 40 (1966).

14. W. Forth and W. Rummel, Iron absorption, *Phys. Rev.*, 53: 724 (1973).

15. F. A. Cotton and G. Wilkinson, *Advanced Inorganic Chemistry: A Comprehensive Text*, John Wiley and Sons, New York, 1962.

16. D. Van Campen, Enhancement of iron absorption from ligated segments of rat intestine by histidine, cysteine, and lysine. Effect of removing ionizing groups and stereoisomerism, *J. Nutr., 103*: 139 (1973).

17. H. Sigel, Amino acids and derivatives as ambivalent ligands in metal ions, in *Biological System Series* vol. 9, Marcel Dekker, Inc., New York, 1979.

18. A. E. Martell and M. Calvin, *Chemistry of the Metal Chelate Compounds*, Prentice-Hall, Inc., New York, 1952.

19. F. M. Clydesdale, Physicochemical determinants of iron bioavailability, *Food Technol.*, 37 (10): 133 (1983).

20. J. T. Rotruck, Effect of processing on nutritive value of food: trace elements, *Handbook of Nutritive Value of Processed Food Vol. I. Food for Human Use*, (Miloslav Recheigled) CRC Press, Boca Raton, FL, 1982.

21. H. A. Schroeder, Losses of vitamins and trace minerals resulting from preservation and processing of foods, *Am. J. Clin. Nutr.*, *24*: 562 (1971).

22. H. Heintze, H. Zohm, A. Fricker, and K. Paulus, Influence of heat treatment of spinach at temperatures from 100 to 130°C on major constituents. III Changes in content of minerals, protein, nitrate, and oxalic acid, *Lebens. Wiss. und Tech.*, *11*: 301 (1978).

23. H. J. Bielig, H. J. Hofsommer, and E. Valldorf, Effects of processing technology on mineral contents of canned vegetables, *Industrielle Obst und Gemseverwertung*, *66*: 611 (1981).

24. E. V. Marston, E. A. Davies, and J. Gordon, Mineral retention in vegetables as affected by phosphates in cooking water, *Home Ec. Res. J.*, *2*: 147 (1974).

25. W. A. Krehl and R. W. Winters, Effect of cooking method on retention of vitamins and minerals in vegetables, *J. Amer. Diet Assoc.*, *26*: 966, (1950).

26. A. J. Leinert, D. P. Becker, J. C. Somogyi, and D. Hutzel, Effect of cooking method on mineral losses, *Ernahrungs Umschau.*, *28*(1):12 (1981).

27. R. S. Harris and E. Karmas, in *Nutritional Evaluation of Food Processing*, AVI Publ. Co., Westport, CT, 1965.

28. D. J. Higgs, V. C. Morris, and O. A. Levander, Effect of cooking on selenium content of foods, *J. Agric. Food Chem.*, *20*: 678 (1972).

29. A. H. Y. Abdel-Rahman, Effect of cooking time on the quality, minerals, and vitamins of spaghetti produced from two Italian duram wheat varieties, *J. Food Sci.*, *17*: 349 (1982).

30. G. S. Ranhotra, J. A. Gelroth, F. A. Novak, and M. A. Bock, Retention of selected minerals in cooked pasta products, *Nutr. Rep. Int.*, *26*: 821 (1982).

31. L. B. Klein and N. I. Mondy, Comparison of microwave and conventional baking of potatoes in relation to nitrogenous constituents and mineral composition, *J. Food Sci.*, *46*: 1874 (1981).

32. N. I. Mondy and R. Ponnampolam, Effect of baking and frying on nutritive value of potatoes: minerals, *J. Food Sci.*, *48*: 1475 (1983).

33. H. A. Schnitt and C. M. Weaver, Effects of laboratory scale processing on chromium and zinc in vegetables, *J. Food Sci.*, *47*: 1693 (1982).

34. T. Emery, Iron metabolism in humans and plants, *Am. Scientist.*, *70*: 626 (1982).

35. S. J. Nojeim, F. M. Clydesdale, and O. T. Zajicek, Effect of redox potential on iron valence in model systems and foods, *J. Food Sci.*, *46*: 1265 (1981).
36. S. Chaberek and A. E. Martell, *Organic Sequestering Agents*, John Wiley and Sons, Inc., New York, 1959.
37. F. M. Clydesdale, The effects of physicochemical properties of food on the chemical status of iron, in *Nutritional Bioavailability of Iron* (C. Kies, ed.), Sym. Series No. 203, American Chemical Society, Washington, D.C., 1982.
38. D. Van Campen and R. M. Welch, Availability to rats of iron from spinach: effects of oxalic acid, *J. Nutr.*, *110*: 1618 (1980).
39. L. S. Eyerman, F. M. Clydesdale, R. Huguenin, and O. T. Zajicek, Characterization of solution properties of four iron sources in model systems, *J. Food Sci.*, *52*: 197 (1987).
40. G. Spiro and P. Saltman, Polynuclear complexes of iron and their biological implications, in *Structure and Bonding*, vol. 6 (P. Hemmerich, C. K. Jorgensen, J. B. Neilands, Sir R. S. Nyholm, D. Reinen, and R. J. P. Williams, ed.), Springer-Verlag, Berlin, 1969.
41. D. P. Derman, T. H. Bothwell, J. D. Torrance, A. P. Macphail, W. R. Bezwoda, R. W. Charlton, and F. G. H. Mayet, Iron absorption from ferritin and ferric hydroxide, *Scand. J. Haematol*, *29*: 18 (1982).
42. L. A. Berner, D. D. Miller, and D. Van Campen, Availability to rats of iron on ferric hydroxide polymers, *J. Nutr*, *115*: 1042 (1985).
43. D. P. Derman, M. Sayers, S. R. Lynch, R. W. Charlton, T. H. Bothwell, and F. Mayet, Iron absorption from a cereal diet containing cane sugar fortified with ascorbic acid, *Brit. J. Nutr.*, *38*: 261 (1977).
44. *International Nutritional Anemia Consultative Group*, The effects of cereals and legumes on iron availability, The Nutrition Foundation, Washington, D.C., 1982.
45. J. E. Gorman and F. M. Clydesdale, Thermodynamic and kinetic stability constants of selected carboxylic acids and iron, *J. Food Sic.*, *49*: 500 (1984).
46. K. Lee, Iron chemistry and bioavailability in food processing, in *Nutritional Bioavailability of Iron*, (C. Kies, ed.), Sym. Series No. 203, American Chemical Society, Washington, D.C., 1982.
47. H. Spencer and L. Knomer, Effect of certain minerals on the bioavailability of calcium in adult males, in *Nutritional Bioavailability of Calcium* (C. Kies ed.), Sym. Series No. 275, American Chemical Society, Washington, D.C. 1985.
48. I. H. Rosenberg and N. W. Solomons, Physiological and pathophysiological mechanisms in mineral absorption, in *Absorption and Malabsorption of Mineral Nutrients* (N. W. Solomons and I. H. Rosenberg, eds.), Alan R. Liss, Inc., New York, 1984.

49. O. A. Levander and L. Cheng, eds., *Micronutrient interactions: Vitamins Minerals and Hazardous Elements*, Annals of the New York Academy of Science, New York, 1980.

50. C. F. Mills, Dietary interactions involving the trace elements, in *Annual Review of Nutrition* (R. E. Olson, ed.), Annual Rev. Inc., Palo Alto, CA, 1985.

51. W. P. T. James, Dietary fiber and mineral absorption, in *Medical Aspects of Dietary Fiber* (G. A. Spiller and R. McPherson Kay, eds.), Plenum, New York, 1980.

52. L. A. Berner and D. D. Miller, Effects of dietary proteins on iron bioavailability—a review, *Food Chem*, *18*: 47 (1985).

53. H. Sigel, Amino acids and derivatives as ambivalent ligands in metal ions, in *Biological System Series*, vol 9, Marcel Dekker, Inc., New York, 1979.

54. S. M. Flynn, F. M. Clydesdale, and O. T. Zajicek, Complexation, stability and behaviour of L-cysteine and L-lysine with different iron sources, *J. Food Prot.*, *47*: 36 (1984).

55. C. A. Bonner and F. M. Clydesdale, The effects of proteolytic digestion products of chicken breast muscle on iron solubility, *Am. J. Clin. Nutr.*, in press (1987).

56. Z. I. Kertesz, *The Pectic Substances*, Interscience Publ., Inc., New York, 1951.

57. R. G. Schweiger, Metal chelates of pectate and comparison with alginate, *Kolloid-Z.*, *Z. X. Polymere.*, *208*: 28 (1966).

58. A. L. Camire and F. M. Clydesdale, The effect of pH on the binding of Ca, Mg, Zn, and Fe to wheat bran and fractions of dietary fiber, *J. Food Sci.*, *46*: 548 (1981).

59. S. R. Platt and F. M. Clydesdale, Mineral binding characteristics of lignin, guar qum, cellulose, pectin and neutral detergent fiber under simulated duodenal conditions, *J. Food Sci.*, *52*: 1414 (1987).

60. R. R. Mod, R. L. Ory, N. M. Morris and F. L. Normand, Chemical properties and interactions of rice hemicellulose with trace minerals in vitro, *J. Agric. Food Chem.*, *29*: 449 (1981).

61. R. Fernandez and S. F. Phillips, Components of fiber bind iron in vitro, *Am. J. Clin. Nutr.*, *35*: 100 (1982).

62. J. A. Rendleman, Interaction of alkali and alkali earth metals with carbohydrates, *Advances in Carbohydrate Chemistry*, *21*: 209 (1966).

63. P. Charley and P. Saltman, Chelation of calcium by lactose: its role in transport mechanisms, *Science*, *139*: 1205 (1963).

64. G. W. Bates, J. Boyer, B. A. Hegenauer, and P. Saltman, Facilitation of iron absorption by ferric fructose, *Am. J. Clin. Nutr.*, *25*: 983 (1972).

65. C. E. Bugg and W. J. Cook, Calcium ion binding to uncharged sugars: crystal structures of calcium bromide complexes of lactose, galactose and inositol, *A. C. S. Chem. Comm.*, *561*: 727 (1972).

66. Anon., Glucose polymers enhance the intestinal absorption of calcium, *Nutr. Rev.*, *43*: 200 (1985).
67. M. L. Swartz, R. A. Bernhard, and T. A. Nickerson, Interaction of metal ions with lactose, *J. Food Sci.*, *43*: 93 (1978).
68. P. J. Van Soest, Some physical characteristics of dietary fibers and their influence on the microbial ecology of the human colon, *Proc. Nutr. Soc.*, *43*: 25 (1984).
69. S. R. Platt and F. M. Clydesdale, Binding of iron by cellulose, lignin, sodium phytate and beta glucan, alone and in combination, under simulated gastrointestinal pH conditions, *J. Food Sci.*, *49*: 531 (1984).
70. P. Vohra, G. A. Gray, and F. H. Kratzer, Phytic acid-metal complexes, *Proc. Soc. Exp. Med.*, *120*: 447 (1965).
71. R. Ellis and E. R. Morris, Effect of sodium phytate on the stability of monoferric phytate complex and the bioavailability of the iron to rats, *Nutr. Rept. Inter.*, *20*: 739 (1979).
72. P. N. Davies, L. C. Norris, and F. H. Kratzer, Interference of soybean proteins with the utilization of trace minerals, *J. Nutr.*, *77*: 217 (1962).
73. E. T. Champagne, R. M. Rao, J. A. Liuzzo, J. W. Robinson, R. J. Gale, and F. Miller, Solubility behaviour of the minerals, proteins and phytic acid in rice bran with time, temperature, and pH, *Cereal Chem.*, *62*: 218 (1985).
74. S. R. Platt and F. M. Clydesdale, Interactions of iron alone and in combination with Ca, Zan, and Cu with a phytate-rich fiber-rich fraction of wheat bran under gastrointestinal pH conditions, *Cereal Chem.*, *64* (21): 102 (1987).
75. R. Hill and C. Tyler, The effect of decreasing acidity on the solubility of calcium, magnesium, and phosphorous in bran and certain pure salts, *J. Agric. Sci.*, *44*: 311 (1954).
76. R. H. Jackman and C. A. Black, Solubility of iron, aluminum, calcium, and magnesium inositol phosphates at different pH values, *Soil Sci.*, *72*: 179 (1951).
77. E. R. Morris and R. Ellis, Isolation of monoferric phytate from wheat bran and its biological value as an iron source to the rat, *J. Nutr.*, *106*: 53 (1976).
78. J. R. Turnlund, J. C. King, B. Gong, W. R. Keyes, and M. C. Mitchell, A stable isotope study of copper absorption in young men: effect of phytate and alpha-cellulose, *Am. J. Clin. Nutr.*, *42*: 18 (1985).
79. M. Cheryan, Phytic acid interactions in food systems, *CRC Crit. Rev. Food Sci. Nutr.*, *13*: 297 (1980).
80. A. Wise, Dietary factors determining the biological activities of phytate, *Nut. Abs. Rev: Rev. Clin. Nutr.*, *53*: 791 (1983).
81. A. Wise and D. Gilburt, In vitro competition between calcium phytate and the soluble fraction of rat small intestine contents for cadmium, copper and zinc, *Toxicol. Lett.*, *11*: 49 (1982).

82. J. G. Reinhold, F. Ismail-Beigi, and B. Faradji, Fiber vs. phytate as determinant of the availability of calcium, zinc, and iron of bread stuffs, *Nutr. Rept. Inter.*, *12*: 75 (1975).

83. E. Graf and J. W. Eaton, Dietary phytate and calcium bioavailability, in *Nutritional Bioavailability of Calcium* (C. Kies ed.), Sym. Series No. 275, American Chemical Society, Washington, D.C., 1985.

84. J. E. Gorman and F. M. Clydesdale, The behaviour and stability of iron ascorbate complexes in solution, *J. Food Sci.*, *48*: 1217 (1983).

85. N. Kojima, D. Wallace, and G. W. Bates, The effect of chemical agents, beverages, and spinach on the in vitro solubilization of iron from cooked pinto beans, *Am. J. Clin. Nutr.*, *34*: 1392 (1981).

86. F. M. Clydesdale and D. B. Nadeau, Effect of acid pretreatment on the stability of ascorbic acid iron complexes with various iron sources in a wheat flake cereal, *J. Food Sci.*, *50*: 1342 (1985).

87. L. Hallberg, Bioavailability of dietary iron in man, *Ann. Rev. Nutr.*, *1*: 123 (1981).

88. S. Rathee and K. Pradham, Effect of ascorbic acid on availability of iron from an egg based whole day diet of college girls, *Ind. J. Nutr. Diet.*, *17*: 90 (1980).

89. S. R. Lynch and J. D. Cook, Interaction of vitamin C and iron, *Annals N. Y. Acad. Sci.*, *355*: 32 (1980).

90. S. W. Rizk and F. M. Clydesdale, Effectiveness of organic acid to solubilize iron from a wheat soy drink, *J. Food Prot.*, *48*: 648 (1985).

91. S. W. Rizk and F. M. Clydesdale, Effect of organic acids in the in vitro solubilization of iron from a soy-extended meat patty, *J. Food Sci.*, *80*: 577 (1985).

92. S. W. Rizk and F. M. Clydesdale, Effects of baking and boiling on the ability of selected organic acids to solubilize iron from a corn-soy-milk food blend fortified with exogenous iron sources, *J. Food Sci.*, *50*: 1088 (1985).

93. R. C. Warner and I. Weber, The cupric and ferric citrate complexes, *J. Am. Chem. Soc.*, *75*: 5086 (1953).

94. J. S. Garcia-Lopez and K. Lee, Iron binding by fiber is influenced by competing minerals, *J. Food Sci.*, *50*: 424 (1985).

95. S. R. Platt and F. M. Clydesdale, Binding of iron by lignin in the presence of various concentrations of calcium, magnesium and zinc, *J. Food Sci.*, *50*: 1322 (1985).

96. S. R. Platt and F. M. Clydesdale, Effects of iron alone and in combination with Ca, Zn, and Cu on the mineral binding capacity of wheat bran, *J. Food Prot.*, *49*: 37 (1986).

3
Calcium, Phosphorus, and Fluoride

HERTA SPENCER and LOIS KRAMER / Veterans Administration
Hospital, Hines, Illinois

INTRODUCTION

This presentation will deal with the metabolism of calcium, fluoride,
and phosphorus, as well as with the effect of other agents, dietary
as well as medications, on the metabolism of these three elements.

The major organ of deposition of calcium is bone. Another important mineral associated with calcium in bone is phosphorus, the
calcium/phosphorus ratio in bone being 2:1. Bone is also the major
organ of deposition of fluoride. These three elements, in addition
to magnesium, appear to play an important role in maintaining the
normal structure of the skeleton. The continued maintenance of the
integrity of the skeletal structure throughout life is of importance,
as there is bone loss with aging, particularly in females. The bone
disorder osteoporosis is a major public health problem, as it may
cause serious complications leading to incapacitation and requiring
costly medical care. The ratio of the incidence of osteoporosis in
females to males is about 2:1, and the onset of osteoporosis is highest in women after the menopause. It has been estimated that there
are 15 to 20 million women in the United States who have osteoporosis
(1). In addition to hormonal deficiency and aging (2), certain factors
can play an important role in intensifying the loss of bone that occurs
with advancing age. Although calcium deficiency has not been proven
to be the etiologic factor in osteoporosis, calcium intake over the
years may play a major role in maintaining the normal skeletal structure (3), and the fact remains that a considerable loss of calcium has
occurred once osteoporosis is present.

Although little is known about the relationship of dietary phosphorus intake in relation to the phosphorus content of bone, one may

safely assume that both calcium and phosphorus are needed to protect the normal bone structure. Certain food items, such as dairy products, contain both minerals, while other foods, such as red meat, have a high phosphorus but a low calcium content.

Fluoride is a bone seeker and has been shown to improve the crystal structure of bone (4). A survey in the United States showed that the incidence of osteoporosis was lower in areas naturally high in fluoride than in localities where the fluoride content of the soil and of water is low (5). This observation indicates that a certain concentration of fluoride in bone over the years may be of importance for the maintenance of the normal skeletal structure.

CALCIUM

Calcium Requirement

The recommended dietary intake of calcium in the United States is 800 mg/day (6). However, one must consider the wide variability of intestinal absorption of calcium of different individuals. In addition, there may be certain factors that may interfere with the utilization of an adequate intake of calcium, such as fiber, phytates, and oxalates. Also, certain dietary factors and various drugs may affect the intestinal absorption and retention of calcium. The lifelong intake of an adequate calcium supply throughout adult life appears to play an important role in maintaining the normal skeletal structure and may thus minimize the extent of bone loss with aging, especially in females. It has been shown that the peak bone mass is achieved relatively early in life, at age 30—35 years, and declines thereafter. The decrease in bone mass is then accelerated in females after the menopause (7). Therefore, the question arises whether the inevitable bone loss due to increased bone resorption in later life can be lessened by increasing calcium intake in early life and by building a strong skeleton during younger years. In agreement with this assumption are statements in the literature that individuals who have been diagnosed with osteoporosis have allegedly had a low dietary calcium intake for many years (3). There is a general belief that there is an adaptation mechanism to long-term low calcium intake (8), however, there are differences of opinion on this subject. Lack of adaptation to a low calcium intake has been reported (7,9). Our own extensive experience in long-term calcium balance studies has shown that there is no adaptation to a low calcium intake with time, irrespective of the duration of an inadequate and very low dietary calcium intake (10). This was shown by the persistent negativity of the calcium balance after several months of a low calcium intake. We have also observed that the intestinal absorption of calcium did not increase with time during a prolonged low calcium intake. Therefore, it appears that there is no compensation for the long-term insufficient intake of calcium and for its effect on bone. This situation prevails despite the fact that our observations have shown that elderly persons with or without osteoporosis have the ability to decrease the urinary calcium excretion when the calcium intake is changed from a high to a low intake level.

With regard to the recommended dietary calcium intake of 800 mg/day (6), extensive studies carried out in this research unit have shown that a large percentage of the calcium balances of adults were negative during this calcium intake, so that variable amounts of calcium continue to be lost daily at this calcium intake (11). However, when the calcium intake was increased to 1200 mg/day the calcium balance improved. Furthermore, when the calcium intake was increased beyond this intake level, up to 2200 mg/day, the calcium balance did not become more positive than during a calcium intake of 1200 mg/day, indicating a plateau at the 1200-mg level of calcium. These data indicate that a calcium intake greater than 800 mg/day would be desirable. This aspect is empha- because all the negative calcium balances at the 800-mg calcium intake level were those of individuals who had either asymptomatic or symptomatic osteoporosis. Bone loss with aging or osteoporosis is a very gradual process extending over many years before this condition can be clinically diagnosed, and one cannot predict who will develop osteoporosis or is in the process of developing this bone condition.

Table 1 shows data of comparative calcium balances determined in 76 studies during different calcium intakes in control subjects and in patients with osteoporosis. During a low calcium intake of approximately 200 mg/day, the calcium balance was more negative in patients with osteoporosis than in the control subjects. Increasing the calcium intake four-fold to about 800 mg/day resulted in a slightly positive calcium balance in control subjects, while the calcium balance of the patients with osteoporosis was just in equilibrium, -3 mg/day when this calcium intake was given as whole milk, and -12 mg/day when the source of calcium was calcium lactate. These values have to be compared with the positive calcium balances of the control subjects, which were +99 and +82 mg/day during these two calcium intakes, respectively. Increasing the calcium intake to levels higher than 800 mg/day and adding another 400 mg calcium per day, so that the calcium intake was now about 1200 mg/day, resulted in a more positive calcium balance in both groups of patients whether the source of calcium was milk or calcium lactate. However, the calcium balances of the patients with osteoporosis were less positive than those of the control subjects. These balances ranged from +74 to +120 mg/day for the patients with osteoporosis, while the calcium balances of the control subjects ranged from +177 to +188 mg/day. When the calcium intake was increased further from 1200 to 1400 mg and to 2200 mg/day, the calcium balance did not become more positive than during the 1200-mg calcium intake. At these calcium intakes the calcium balance of patients with osteoporosis was also less positive than in control subjects. It should be emphasized that the calcium balances presented here are maximal

Table 1. Calcium Balances in Normals and in Patients with Osteoporosis

Type of patient	Number of studies	Calcium Intake (mg/day)	Calcium supplement	Calcium (mg/day)[a]			
				Intake	Urine	Stool	Balance
Normal	12	230	None	233	87	228	- 82
Osteoporosis	9	230	None	234	90	265	-121
Normal	3	800	Milk	812	84	629	+ 99
Osteoporosis	3	800	Milk	816	43	776	- 3
Normal	7	800	Calcium lactate	834	124	628	+ 82
Osteoporosis	5	800	Calcium lactate	830	104	738	- 12
Normal	5	1200	Milk	1239	131	931	+177
Osteoporosis	4	1200	Milk	1237	157	960	+120
Normal	2	1200	Calcium lactate	1210	92	930	+188
Osteoporosis	4	1200	Calcium lactate	1237	224	939	+ 74
Normal	4	1400	Calcium lactate	1386	137	1066	+183
Osteoporosis	3	1400	Calcium lactate	1466	212	1251	+ 3
Normal	5	2200	Calcium lactate	2182	264	1715	+203
Osteoporosis	10	2200	Calcium lactate	2251	183	1996	+ 72

[a]Values are averages for the entire study period for each group of patients.

balances, as dermal losses of calcium were not considered, and therefore the actual calcium balance may either be more negative or less positive than that determined from the urinary and fecal calcium excretions.

Dairy products are very good sources of calcium. Should these not be tolerated because of lactase deficiency or for other medical reasons, a great variety of calcium supplements are available over the counter. Commonly available calcium supplements are calcium gluconate, calcium lactate, and calcium carbonate. The first two compounds have a considerably lower calcium content, approximately 10%, than calcium carbonate, which contains 40% calcium. Therefore, the use of calcium carbonate requires fewer calcium tablets to achieve a specific calcium intake than the other two calcium supplements. All three types of calcium products are well tolerated and are not toxic if given in reasonable dosage and do not cause adverse reactions. However, these would have to be used with great caution for patients who either have or had kidney stones. Among the calcium carbonate-containing compounds are calcium-containing antacids, such as Tums and Titralac.

It is well known that vitamin D is essential for the intestinal absorption of calcium, and the recommended intake is 400 I.U. per day. This amount of vitamin D is apparently sufficient to facilitate the absorption of calcium. Larger amounts are not more effective in increasing the absorption of calcium and can lead to adverse effects. High doses of vitamin D can cause excessive bone breakdown as well as soft tissue calcifications in the presence or absence of hypercalcemia. One of the adverse effects of large doses of vitamin D is a very marked increase in urinary calcium, most probably due to a direct effect of vitamin D on bone causing increased bone resorption. In fact, studies carried out in this unit have shown that large doses of vitamin D, i.e., 25,000 and 50,000 I.U. per day, increased the uninary calcium excessively (12). In view of the fact that the intestinal absorption of calcium did not increase during the intake of these doses of vitamin D, the excess calciuria is most likely due to increased bone resorption and/or a renal effect of vitamin D inducing calcium loss.

Effect of Other Nutrients on Calcium Metabolism

Phosphorus

A great deal of emphasis is being placed on the relationship between the intake of calcium and phosphorus, particularly with regard to the effect of the dietary Ca/P ratio because of the alleged undesirable effects of dietary phosphorus on calcium metabolism and on bone (13, 14). Large amounts of phosphorus used in animal experiments have been reported to cause bone loss and osteoporosis (15,16). On the other hand, in humans phosphate has been shown to decrease the urinary calcium excretion and this is a desirable effect. The mechanism of the decrease during the high phosphate intake is not clear. Some investigators have ascribed the decrease in urinary calcium to stimulation of parathyroid hormone secretion and to subsequent increased tubular reabsorption of calcium. Others suggested that the decrease in urinary calcium during a high phosphorus intake is due to deposition of calcium in soft and hard tissues, to a renal effect, or to decreased intestinal absorption of calcium. Phosphate (14) and phosphate additives (17) have been reported to decrease the serum calcium level, and the decreased serum calcium level can lead to stimulation of parathyroid activity. Phosphate has been used therapeutically to decrease elevated serum calcium levels in patients with neoplasia (18). The mechanism of this decrease in hypercalcemia due to neoplasia is not clear.

Studies carried out in this Unit on calcium-phosphorus interrelations have shown that the intestinal absorption of calcium did not change during the addition of phosphorus to the diet, and therefore the decrease in urinary calcium is not due to decreased intestinal absorption of calcium. The studies have also shown that the main effect of phosphorus is a decrease in urinary calcium (19,20). In view of the fact that the intestinal absorption is not altered during the high phosphorus intake, the dietary calcium/phosphorus ratio does not appear to play an important role in the utilization of calcium (20).

Table 2 shows data on the effect of phosphorus on the calcium balance. Increasing the phosphorus intake from 800 to 2000 mg/day during calcium intakes of 800 and 2000 mg/day did not result in a change of the calcium balance or of the intestinal absorption of calcium, determined with Ca^{47}. The high phosphorus intake led to a decrease in urinary calcium. The latter effect has also been reported by others (18).

Table 2 also shows data of the effect of added phosphorus on the phosphorus balance. Urinary and fecal phosphorus increased when phosphate supplements, given as glycerophosphate, were added to the diet. The increase in urinary phosphorus was observed when phosphorus was added to all calcium intakes, the increase in stool phosphorus was less than the increase in urinary phosphorus, and the phosphorus balance became more positive during the addition of phosphate.

Protein Intake

The usual dietary intake of protein is approximately 1 gm/kg body weight per day. A high protein intake has been thought to have deleterious effects on the calcium status and therefore on bone by

Table 2. Effect of Phosphorus on the Calcium Balance During Different Calcium Intakes*

Number of patients Studied	Study	Days	Calcium (mg/day)				Phosphorus (mg/day)				Ca^{47} Absorption (% dose)
			Intake	Urine	Stool	Balance	Intake	Urine	Stool	Balance	
			Normal calcium intake								
8	Control	40	828 ±12†	211 ±33	621 ±40	- 4 ±34	845 ±25	532 ±20	293 ±29	+ 20 ±11	41 ±5
	Phosphate	46	823 ±10	143[d] ±24	656 ±39	+ 24 ±31	1977 ±13	1349[c] ±69	535[c] ±40	+ 93 ±37	38 ±5
			High calcium intake								
6	Control	33	2018 ±42.6	190 ±51.3	1662 ±104.5	+166 ±33.7	805 ±53.3	338 ±59.0	382 ±30.3	+ 85 ±48.5	30 ±3
	Phosphate	31	2019 ±43.2	106[b] ±30.8	1743 ±69.8	+170 ±29.8	2003 ±9.5	1028[c] ±86.6	805[d] ±73.5	+170[a] ±42.8	23 ±2

*Values are averages ± Sem for the calcium and phosphorus balances for the number of patients studied.

†Significance among means is indicated as: [a]$P < 0.05$, [b]$P < 0.01$, [c]$P < 0.001$, [d]$P < 0.005$.

inducing an increase in urinary calcium. However, only specific proteins cause this increase, for instance, purified proteins such as casein, lactalbumin, gelatin, egg white, and a variety of amino acids (21,22). These purified protein fractions are usually not consumed in isolated form in the human diet but are in combination with other nutrients, which apparently counteract the effect of the isolated proteins of increasing the excretion of calcium. The main sources of dietary protein are meat and dairy products. In studies carried out in this research unit it has been conclusively shown that large amounts of protein given as red meat did not have any adverse effect on the calcium status. Specifically, the urinary excretion of calcium did not increase during this high protein intake (23,24). Table 3 shows examples of the urinary calcium excretion determined in strictly controlled studies carried out during a normal and high protein intake given as meat. When the meat intake was more than doubled for time periods ranging from 36 to 84 days, there was no increase in urinary calcium, and in fact the urinary calcium even tended to be lower in the high protein studies than in the control studies. It should be mentioned that there was also no significant change in fecal calcium, and the calcium balances and the intestinal absorption of calcium remained unchanged during the high protein intake (23,24).

Effects of Other Minerals

Magnesium and calcium may compete for common absorption sites and the intestinal absorption of calcium may thereby be reduced. Experimental studies in animals have shown that magnesium has an adverse effect on calcium metabolism. Another mineral that may interfere with the absorption of calcium is zinc. Very little is known about the effect of zinc on calcium metabolism in humans, and a few animal and in-vitro studies indicate competition of zinc and calcium. Fluoride has been reported to affect the metabolism of calcium, namely, to improve the calcium balance and to increase the intestinal absorption of calcium in patients with osteoporosis (25).

This section will briefly describe the effects of magnesium, zinc, and fluoride on calcium metabolism.

The effect of added magnesium on the calcium balance was investigated in adults in this research unit. The dietary magnesium intake was approximately 250 mg/day in the control study, and the high magnesium intake was due to the addition of magnesium oxide to the diet, providing an additional 600 mg elemental magnesium.

Table 4 shows data of the effect of added magnesium on the calcium balance. When magnesium was added to the diet during a low calcium intake of approximately 300 mg/day, there was no change of the urinary or fecal calcium excretion and the calcium balance remained unchanged (26). Similarly, there were no changes in the

Table 3. Effect of Meat Protein on Urinary Calcium

Patient	Study days		Protein intake (gm/day)		Meat intake (gm/day)		Urinary calcium (gm/day)	
	Control	High protein	Control	High protein	Control	High protein	Control	High protein
1	48	42	116	166	255	660	170	149
2	60	36	70	136	200	510	173	157
3	24	84	75	131	200	475	119	105

Table 4. Effect of Other Minerals and Trace Elements on Calcium
Metabolism

	Calcium (mg/day)			
Study	Intake	Urine	Stool	Balance
Control	306	185	133	- 12
Magnesium[a]	298	183	135	- 20
Control	866	134	685	+ 47
Magnesium[a]	868	129	705	+ 34
Control	1410	359	1017	+ 34
Magnesium[a]	1403	368	985	+ 50
Control	230	93	299	-162
Zinc[b]	233	45	350	-163
Control	1972	114	1665	+193
Fluoride[c]	1949	66	1688	+195

[a] An average of 550 mg magnesium was given per day as magnesium
oxide. Magnesium intake in control study = 250 mg/day.

[b] 150 mg zinc was given per day as $ZnSO_4$. Zinc intake in control
study = 15 mg/day.

[c] 45 mg fluoride was given per day as NaF. Fluoride intake in control
study = 4.2 mg/day.

calcium excretions or in the calcium balance when magnesium was added
to a normal calcium intake of 800 mg/day or 1400 mg/day.

Zinc supplements are freely available over the counter and are
widely used by the public. Studies carried out in this research unit
have shown that a daily dose of 140 mg of zinc as the sulfate or as
the gluconate significantly decreases the intestinal absorption of calcium when these zinc supplements are taken during a low calcium intake. However, this reduction in calcium absorption does not occur
when these zinc supplements are taken during a normal calcium intake
of 800 mg/day. These observations have particular relevance for the
elderly who may be consuming an inadequate calcium intake as the use
of this dose of zinc may further intensify the bone loss that occurs
with aging. Table 4 shows that added zinc does have a modest effect
on the metabolism of calcium. When the zinc intake was increased
approximately ten-fold from 15 mg to 150 mg per day, there was a

modest increase in stool calcium and the urinary calcium decreased while the calcium balance did not change. However, calcium absorption studies have shown that the intestinal absorption of calcium, determined with Ca^{47}, decreased significantly during the high zinc intake (27). This was the case when zinc was given during a low calcium intake of approximately 200 mg/day but not during a normal calcium intake of 800 mg/day.

When fluoride was added to the constant diet as sodium fluoride and the fluoride intake was increased approximately ten-fold, the primary effect of fluoride on calcium metabolism was a decrease in urinary calcium while there was no change of the fecal calcium. The calcium balance did not change as there was a slight increase in fecal calcium (Table 4). The change of the calcium balance during the intake of fluoride depends primarily on the decrease in urinary calcium (28). It should also be mentioned that the intestinal absorption of calcium, which was determined with oral doses of Ca^{47}, did not change during the high fluoride intake.

Effects of Certain Drugs on Calcium Metabolism

Several commonly used medications can induce an increase in urinary calcium. Should this drug-induced calcium loss persist for prolonged periods of time, it can lead to demineralization of the skeleton or contribute to or intensify an already existing low calcium status, for instance in elderly females who tend to develop or already have osteoporosis. Among the drugs known to cause calcium loss are corticosteriods and thyroid medications. However, several other medications can also lead to an increase in urinary calcium and thereby induce calcium loss (29).

Antacids are commonly used for various gastrointestinal disorders. These drugs contain aluminum, which can cause significant calcium loss primarily due to an increase in urinary calcium. If these drugs are taken for prolonged periods of time, they can lead to bone loss and accelerate and/or intensify already existing demineralization of the skeleton (30,31). This is of particular relevance for persons who may already be in a low calcium status, for instance, due to hormonal deficiency or aging. These antacids do not inhibit the intestinal absorption of calcium, but they cause profound phosphorus depletion, and the loss of calcium is a secondary effect.

True phosphorus depletion occurs when the intestinal phosphorus content, which is due to ingestion of phosphorus with the diet, is complexed. The complexation of phosphate by aluminum practically abolishes the intestinal absorption of phosphate (30,31). It is usually assumed that phosphorus depletion is reflected by low serum levels of phosphorus. However, this is not always the case and depends on the time relation between the intake of the last dose of the phosphorus-depleting agent and procurement of the blood sample for phosphorus analysis. Very high fecal phosphorus excretion and low urinary phosphorus are more reliable indicators of phosphorus depletion than the plasma level of phosphorus. Most commonly phosphorus depletion is induced by aluminum-containing antacids, and even relatively small doses have this effect (31).

Table 5 shows examples of the effect of a small and a large dose of aluminum-containing antacids on the calcium balance. In both cases there was a distinct increase in urinary calcium, particularly during the intake of the larger dose. The fecal calcium increased somewhat and the calcium balance became more negative. This balance was particularly negative, -502 mg/day, during the commonly used therapeutic dose of antacids. These results have been previously published by this group and also by others (30,31).

PHOSPHORUS

It is generally alleged that the American diet has a very high phosphorus content and that the high phosphorus intake has adverse effects on calcium metabolism and on bone. Some of these effects have been described in the subsection Effect of Other Nutrients on Calcium Metabolism.

Although the recommended dietary intake (RDA) of phosphorus is 800 mg/day, it is safe to assume that the usual dietary phosphorus intake is approximately 1200 mg/day. Strictly controlled studies carried out in adults in this Unit have shown that increasing the phosphorus intake from 800 mg/day to various levels up to 2000 mg/

Table 5. Effect of Small and Large Doses of Antacids on Calcium Metabolism

Study	Study Days	Calcium (mg/day)[a]			
		Intake	Urine	Stool	Balance
Control	10	220	184	152	-116
Maalox (90 ml/day)	12	253	247	213	-197
Control	18	253	82	348	-177
Maalox (450 ml/day)	12	279	421	360	-502

[a]Values are averages for the number of study days.

Table 6. Effect of a High Phosphorus Intake on Urinary and Stool
Phosphorus

Phosphorus intake (mg/day)	Calcium intake (mg/day)			
	200	800	1400	2000
Urinary Phosphorus Excretion (mg/day)				
800	534	533	466	389
2000	1494	1347	1107	1137
Increase	960	814	641	748
Stool Phosphorus Excretion (mg/day)				
800	228	296	324	346
2000	422	547	758	687
Increase	194	251	434	341
Total Increase[a]	1154	1065	1075	1089

[a]Total increase = sum total of the increase in urinary and stool
phosphorus during the different calcium intakes.

day had no adverse effect on calcium metabolism even when a high
phosphorus intake of 2000 mg/day was given together with a high
calcium intake of 2000 mg/day. Data on the effect of phosphate on
calcium metabolism have already been shown in Table 2. The primary
effect of the high phosphorus intake (2000 mg/day) is a decrease in
urinary calcium, while there is no significant change of other parameters of calcium metabolism, such as the intestinal absorption of
calcium.

Table 6 shows data on the disposition of the ingested phosphorus
when the phosphorus intake is increased. A comparison of the average
increase in urinary and in fecal phosphorus during the addition of
phosphate is shown. On increasing the phosphorus intake by a factor
2.5 during different calcium intakes, the increase in urinary phosphorus was greatest during the low calcium intake, less during the
800-mg calcium intake, and remained in about the same range when
phosphorus was added to all higher calcium intakes. The fecal phosphorus increased progressively when phosphate was added to increasing calcium intakes. This increased excretion ranged from 194
mg/day during the low calcium intake to 434 mg/day during the 1400-
mg calcium intake and remained in the 400-mg range during the other
calcium intakes up to a calcium intake of 2000 mg/day. The increase

in urinary phosphorus together with the increase in fecal phosphorus accounted for approximately the entire amount of phosphorus which was added to the different calcium intakes. The total excess phosphorus excretion was approximately 1100 mg/day compared to an average increase in phosphorus intake of 1156 mg/day.

FLUORIDE

As fluoride enters the human food chain the metabolism of fluoride has become of important. The dietary fluoride content and the extent and the pathways of fluoride excretions were determined in strictly controlled studies in this research unit. These studies were carried out in fully ambulatory adult males. The dietary intake of fluoride and the excretions of fluoride in urine and stool were analyzed. As various minerals may affect the metabolism of fluoride, the interaction of fluoride with several inorganic elements, such as calcium, phosphorus, magnesium, and aluminum, was investigated.

Dietary Fluoride Intake of Adults

Prior to water fluoridation, dietary fluoride intake was low and was estimated to range from 0.5 to 1.5 mg total fluoride per day (32,33). In a survey of the fluoride contents of shopping baskets of four young men, the dietary fluoride intake ranged from 2.1 to 2.4 mg/day, including dietary beverages and fluoridated drinking water (34). Fluoride analyses of numerous diets used for metabolic studies at this hospital in the Chicago area averaged approximately 1.2 mg total fluoride (35), exclusive of the fluoride content of drinking water. In another study the fluoride content of diets from several areas in the United States was found to vary widely (36). The variability in dietary fluoride in both fluoridated and non-fluoridated areas appears to be due to the type of water used in the preparation of processed foods, namely to the varying fluoride content of water (37), so that the dietary fluoride content may even be relatively high in non-fluoridated areas. Other reports indicate a daily intake of fluoride by adults ranging from 0.85 to 1.44 mg/day, exclusive of the fluoride content of beverages, and from 1.46 to 2.57 mg/day including beverages (38). Certain dietary items, for instance, tea (32) and seafood (39), have a high fluoride content.

Excretion of Fluoride

The major pathway of fluoride excretion is via the kidney, and urinary fluoride excretion corresponds to 50—70% of the fluoride intake (40,41). In contrast, fecal fluoride excretion is very low, corresponding to about 5% of the fluoride intake. In view of the low

Table 7. Fluoride Excretions and Retention During Different Fluoride Intakes

No. of studies	Study days	Intake	Urine	Stool	Balance	Net Absorption of fluoride (% Intake)	Fluoride retention (%)
10	28	4.0	2.5	0.3	+1.2	85	30
10	32	14.0	7.5	0.8	+5.7	94	41
20	90	23.8	14.0	0.5	+9.3	98	39
4	87	46.0	26.8	4.9	+14.3	89	31

fecal fluoride excretion, the retention of fluoride in the body depends mainly on the magnitude of the urinary fluoride excretion. When supplemental fluoride is given, the main pathway of fluoride excretion is also via the kidney, while the fecal fluoride remains relatively low in relation to the fluoride intake.

An example of fluoride balances determined during different fluoride intakes is shown in Table 7. The urinary fluoride increased with increasing fluoride intake and accounted for approximately 50—60% of the intake. The fecal fluoride was very low during fluoride intakes up to 23.8 mg/day. Even during the highest fluoride intake of 46 mg/day, the fecal fluoride corresponded to approximately 10% of the fluoride intake. The fluoride balances were positive and increased with increasing intakes. These balances are maximal balances as dermal losses of fluoride were not determined. The intestinal absorption of fluoride was very high during all fluoride intakes and ranged from 85% during the lowest fluoride intake to 98% during a six-fold increase in fluoride intake. The fluoride retention calculated from fluoride balance data ranged from 30 to 41%.

Retention of Fluoride

The major portion of retained fluoride is deposited in bone. Fluoride is incorporated in the hydroxyapatite crystal of bone (4) and has been reported both to decrease bone resorption and to increase bone formation. Because fluoride is used for the treatment of osteoporosis, it is important to know whether fluoride which has been retained during fluoride therapy continues to be retained after its discontinuation or is released.

In a study performed in this research unit it was found that only a very small fraction of the retained fluoride is excreted after discontinuation of the high fluoride intake (42). This is in contrast to reports of other investigators, who stated that large amounts of fluoride are excreted for prolonged periods of time by persons who previously lived in high fluoride areas for many years and who subsequently moved to a low fluoride area (43). This would indicate that a large proportion of the previously retained fluoride is released from bone. The difference between the latter study and our study may be due to the fact that persons who resided in high fluoride areas may have ingested fluoride for many years, while in our studies the high fluoride intake was given for time periods of three months.

Interaction of Fluoride with Other Minerals

Several animal experiments have indicated that minerals such as calcium, phosphorus, magnesium, and aluminum interfere with the intestinal absorption and utilization of fluoride. Studies have been carried out in this Unit to examine the effect of these minerals on fluoride metabolism (44—47).

Table 8. Effect of Various Minerals on Fluoride Metabolism

Study	Fluoride (mg/day)			
	Intake	Urine	Stool	Balance
Low fluoride intake				
Control	4.0	2.9	0.13	+0.9
High Calcium[a]	3.9	2.9	0.20	+0.7
Control	4.0	2.9	0.13	+0.9
High phosphorus[b]	4.0	2.6	0.17	+1.2
Control	5.2	3.5	0.11	+1.6
High magnesium[c]	5.4	3.7	0.21	+1.5
Control	4.0	2.6	0.30	+1.1
Aluminum[d]	4.8	1.9	2.70	+0.2
High fluoride intake				
Control	13.1	7.8	0.31	+5.0
High calcium[a]	13.8	7.4	0.56	+5.8
Control	13.1	7.8	0.31	+5.0
High phosphorus[b]	13.2	7.8	0.31	+5.1
Control	24.9	14.0	0.31	+10.6
High magnesium[c]	25.1	14.2	0.93	+10.0
Control	49.7	34.1	2.20	+13.4
Aluminum[d]	50.6	17.8	20.30	+12.8

[a]Calcium intakes = 200 mg/day and 2300 mg/day in control and high calcium studies, respectively.

[b]Phosphorus intakes = 800 mg/day and 1400 mg/day in control and high phosphorus studies, respectively.

[c]Magnesium intakes = 300 mg/day and 800 mg/day in control and high magnesium studies, respectively.

[d]30 ml of aluminum hydroxide was given three times a day, containing 2 gm aluminum.

Table 8 shows data on the effect of added calcium, phosphorus, magnesium, and aluminum on fluoride metabolism during a low fluoride intake of approximately 4.0 mg/day and during various higher fluoride intakes. When the dietary calcium intake was increased from a low level of 200 mg/day to a high calcium intake of 2300 mg/day, there was no change in the urinary or fecal fluoride excretions or in the fluoride balance, no matter what the fluoride intake. Similar results were observed when the phosphorus intake was increased from 800 mg/day to 1400 mg/day. Increasing the magnesium intake from approximately 300 mg/day to 800 mg/day by the use of magnesium oxide, the urinary fluoride remained unchanged. The fecal fluoride barely increased during the low fluoride intake, while it increased from 0.31 to 0.93 mg/day during the higher fluoride intake of 25 mg/day. Despite this increase, the total fecal fluoride excretion was still very low and therefore the fluoride balance remained unchanged when magnesium was added to the low or high fluoride intake. When a relatively small amount of aluminum hydroxide, 90 ml/day, was added to the diet having a low fluoride content, the major change was an approximately nine-fold increase in fecal fluoride, the urinary fluoride decreased and the fluoride balance became less positive. Similar results were obtained when the same dose of aluminum hydroxide was added to the higher fluoride intake of 50 mg/day. These studies have shown that aluminum, mostly used in aluminum-containing antacids, has a marked inhibitory effect on the intestinal absorption of fluoride. This negative effect, which may contribute to a lower fluoride concentration in bone, in conjunction with the calcium-losing property of alumnium secondary to phosphorus depletion, may contribute to further demineralization of the skeleton in persons who are in a low calcium status.

REFERENCES

1. Office of Medical Applications of Research, National Institutes of Health: Consensus Conference—Osteoporosis, *JAMA, 252:* 799 (1984).
2. F. Albright and E. C. Reifenstein, Jr., *The Parathyroid Glands and Metabolic Bone Disease,* Williams & Wilkins, Baltimore, 1948.
3. B. E. C. Nordin, Osteoporosis and calcium deficiency, in *Bone as a Tissue* (K. Rodahl, J. T. Nicholson, and E. M. Brown, Jr., eds.), McGraw-Hill Book Co., Inc., New York, 1960, p. 46.
4. I. Zipkin, A. S. Posner, and E. D. Eanes, The effect of fluoride on the x-ray diffraction pattern of the apatite of human bone, *Biochim. Biophys., 59:*255 (1962).
5. D. Bernstein, H. Sadowsky, D. M. Hegsted, C. Guri, and F. Stare, Prevalence of osteoporosis in high- and low-fluoride areas in North Dakota, *JAMA, 193:*499 (1966).

6. Food and Nutrition Board, *Recommended Dietary Allowances,* ninth edition, National Academy of Sciences, National Research Council, Washington, D.C., 1980.

7. R. P. Heaney, J. C. Gallagher, C. C. Johnston, R. Neer, A. M. Parfitt, and G. D. Whedon, Calcium nutrition and bone health in the elderly, *Am. J. Clin. Nutr.,* 36:986 (1982).

8. H. H. Draper and C. A. Scythes, Calcium, phosphorus, and osteoporosis, *Fed. Proc.,* 40:2434 (1981).

9. L. H. Allen, Calcium bioavailability and absorption: A review, *Am. J. Clin. Nutr.,* 35:783 (1982).

10. H. Spencer and L. B. Kramer, Factors influencing calcium balance in man, in *Calcium in Biological Systems* (R. P. Rubin, G. B. Weiss, and J. W. Putney, Jr., eds.), Plenum Publishing Corporation, New York, 1985, p. 583.

11. H. Spencer, L. Kramer, M. Lesniak, M. DeBartolo, C. Norris, and D. Osis, Calcium requirements in humans: Report of original data and a review, *Clin. Orthop.,* 184:270 (1984).

12. H. Spencer, L. Kramer, C. Gatza, and D. Osis, Calcium absorption and calcium balances in man during vitamin D intake, in *Vitamin D: Biochemical, Chemical and Clinical Aspects Related to Calcium Metabolism* (A. W. Norman, K. Schaefer, J. W. Coburn, H. F. DeLuca, D. Fraser, H. G. Grigoleit, and D. V. Herrath, eds.), Walter de Gruyter, Berlin, 1977, p. 611.

13. G. H. Anderson and H. H. Draper, Effect of dietary phosphorus on calcium metabolism in intact and parathyroidectomized adult rats, *J. Nutr.,* 102:1123 (1972).

14. T. L. Sie, H. H. Draper, and R. R. Bell, Hypocalcemia, hyperparathyroidism and bone resorption in rats induced by dietary phosphate, *J. Nutr.,* 104:1195 (1974).

15. G. H. LaFlamme and J. Jowsey, Bone and soft tissue changes with oral phosphate supplements, *J. Clin. Invest.,* 51:2834 (1972).

16. L. Krook, Dietary calcium-phosphorus and lameness in the horse, *Cornell Vet. Suppl.,* 58:59 (1968).

17. R. R. Bell, H. H. Draper, D. Y. M. Tseng, H. K. Shin and G. R. Schmidt, Physiologic responses of human adults to foods containing phosphate additives, *J. Nutr.,* 107:42 (1977).

18. R. S. Goldsmith and S. H. Ingbar, Inorganic phosphate treatment of hypercalcemia of diverse etiologies, *N. Engl. J. Med.,* 274:1 (1966).

19. H. Spencer, J. Menczel, I. Lewin, and J. Samachson, Effect of high phosphorus intake on calcium and phosphorus metabolism in man, *J. Nutr.,* 86:125 (1965).

20. H. Spencer, L. Kramer, D. Osis, and C. Norris, Effect of phosphorus on the absorption of calcium and on the calcium balance in man, *J. Nutr.,* 108:447 (1978).

21. H. M. Linkswiler, C. L. Joyce, and C. R. Anand, Calcium retention of young adult males as affected by level of protein and of calcium intake, *Trans. NY Acad. Sci.*, *36*:333 (1974).

22. S. Margen, J. Y. Chu, N. A. Kaufmann, and D. H. Calloway, Studies in calcium metabolism. 1. The calciuretic effect of dietary protein, *Am. J. Clin. Nutr.*, *27*:584 (1974).

23. H. Spencer, L. Kramer, D. Osis, and C. Norris, Effect of a high protein (meat) intake on calcium metabolism in man, *Am. J. Clin. Nutr.*, *31*:2167 (1978).

24. H. Spencer, L. Kramer, M. DeBartolo, C. Norris, and D. Osis, Further studies of the effect of a high protein (meat) intake on calcium metabolism, *Am. J. Clin. Nutr.*, *37*:924 (1983).

25. C. Rich, J. Ensinck, and P. Ivanovich, The effects of sodium fluoride on calcium metabolism of subjects with metabolic bone diseases, *J. Clin. Invest.*, *43*:545 (1964).

26. H. Spencer, M. Lesniak, C. A. Gatza, L. Kramer, C. Norris, and J. Coffey, Magnesium-calcium interrelationships in man, in *Trace Substances in Environmental Health-XII* (D. D. Hemphill, ed.), University of Missouri, Columbia, 1978, p. 241.

27. H. Spencer, N. Rubio, L. Kramer, C. Norris, and D. Osis, Effect of zinc supplements on the intestinal absorption of calcium, *J. Am. Coll. Nutr. 6*: 47 (1987).

28. H. Spencer, I. Lewin, D. Osis, and J. Samachson, Studies of fluoride and calcium metabolism in patients with osteoporosis, *Am. J. Med.*, *49*:814 (1970).

29. H. Spencer, L. Kramer, and D. Osis, Factors contributing to calcium loss in aging, *Am. J. Clin. Nutr.*, *36*:776 (1982).

30. M. Lotz, E. Zisman, and F. C. Bartter, Evidence for a phosphorus-depletion syndrome in man, *N. Engl. J. Med.*, *278*:409 (1968).

31. H. Spencer, L. Kramer, C. Norris, and D. Osis, Effect of small doses of aluminum-containing antacids on calcium and phosphorus metabolism, *Am. J. Clin. Nutr.*, *36*:32 (1982).

32. F. J. McClure, Fluoride in foods-survey of recent data, *Pub. Health Rep.*, *64*:1061 (1949).

33. H. C. Hodge and F. A. Smith, in *Fluorine Chemistry*, vol. 4 (J. H. Simons, ed.), Academic Press, New York, 1965, p. 155 and p. 171.

34. F. A. San Fillippo and G. C. Battistone, The fluoride content of a representative diet of the young adult male, *Clin. Chim. Acta.*, *31*:453 (1971).

35. D. Osis, L. Kramer, E. Wiatrowski, and H. Spencer, Dietary fluoride intake in man, *J. Nutr.*, *104*:1313 (1974).

36. L. Kramer, D. Osis, E. Wiatrowski, and H. Spencer, Dietary fluoride in different areas in the United States, *Am. J. Clin. Nutr.*, *27*:590 (1974).

37. I. L. Shannon and W. B. Wescott, Fluoride levels in drinking water, *N. Carolina Dent. J.*, *58*:15 (1975).
38. L. Singer, R. H. Ophaug, B. F. Harland, and R. Marts, Fluoride content of adult foods, *J. Dent. Res.*, *56*:335 (1978).
39. F. J. McClure, H. H. Mitchell, T. S. Hamilton, and C. A. Kinser, Balances of fluorine ingested from various sources in food and water by five young men, *J. Indust. Hyg. Toxicol.*, *27*:159 (1945).
40. W. Machle, E. W. Scott, and E. J. Largent, The absorption and excretion of fluorides. 1. The normal fluoride balance, *J. Indust. Hyg. Toxicol.*, *24*:199 (1942).
41. H. Spencer, I. Lewin, E. Wiatrowski, and J. Samachson, Fluoride metabolism in man, *Am. J. Med.*, *49*:807 (1970).
42. H. Spencer, L. Kramer, D. Osis, and E. Wiatrowski, Excretion of retained fluoride in man, *J. Appld. Physiol.*, *38*:282 (1975).
43. R. C. Likins, F. J. McClure, and A. C. Steere, Urinary excretion of fluoride following defluoridation of a water supply, in *Fluoride Drinking Water* (F. J. McClure, ed.), U. S. Government Printing Office, Washington, D. C. Public Health Service Publication No. 825, 1962, p. 287.
44. H. Spencer, D. Osis, L. Kramer, E. Wiatrowski, and C. Norris, Effect of calcium and phosphorus on fluoride metabolism in man, *J. Nutr.*, *105*:733 (1975).
45. H. Spencer, L. Kramer, E. Wiatrowski, and D. Osis, Magnesium-fluoride interrelationships in man. II. Effect of magnesium on fluoride metabolism, *Am. J. Physiol.*, *234*:E343 (1978).
46. H. Spencer, L. Kramer, C. Norris, and E. Wiatrowski, Effect of aluminum hydroxide on fluoride metabolism, *Clin. Pharm.*, *28*:529 (1980).
47. H. Spencer, L. Kramer, C. Norris, and E. Wiatrowski, Effect of aluminum hydroxide on plasma fluoride and fluoride excretion during a high fluoride intake in man, *Toxicology and Applied Pharmocology*, *58*:140 (1981).

4
Magnesium

RUTH SCHWARTZ / Cornell University, Ithaca, New York

INTRODUCTION

Magnesium is the fourth most abundant cation in the body and, after potassium, the most abundant in the cell. As the essential metal in chlorophyll, magnesium has a central role in life processes. Its many additional roles in plant and animal organisms include all reactions that depend on synthesis or hydrolysis of ATP, stabilization of macromolecules such as DNA and ribosomes, and activation or regulation of a host of enzymes, primarily those involved in phosphate transfer (1–3). In combination with calcium, magnesium regulates the permeability and excitability of membranes. Its concentration in the extracellular fluids is critical for the integrity and function of the nervous system, the conductance of neural stimuli, and their transmission across myoneural junctions (2,4). That magnesium is an essential nutrient was unequivocally established over half a century ago by inducing a well defined and reproducible magnesium deficiency syndrome in rats (5,6). Yet, despite the recognized importance and ubiquity of magnesium in biological processes, its metabolism and functions are poorly understood.

From the first, research on the biological roles of magnesium has been difficult. For almost three decades after the demonstration of magnesium deficiency, investigations were held back by lack of adequately sensitive and accurate methods of analysis (7,8). Moreover, the only available radiotracer for magnesium is short-lived (9,10) and its stable minor isotopes are, at best, marginally useful as tracers in intact animals or man (11,12). Even the unquestioned importance and ubiquity of magnesium in biological processes have tended to deter research. It has been simpler for the biochemist to include magnesium routinely in the many in vitro systems and reaction mixtures that require its presence than to determine why it had to be included. Purely dietary magnesium deficiency is rare in human beings since magnesium is present in most foods and body magnesium is conserved by highly effective renal mechanisms. Thus incentives for nutritional and clinical research on magnesium have also been weak. Much is still to be learned about the metabolism and biological roles of magnesium.

MAJOR LANDMARKS IN THE ADVANCEMENT OF KNOWLEDGE OF MAGNESIUM

The history of the medicinal uses of magnesium salts and the production and identification of the metal have been reviewed in detail in two monographs (7,13). They will be covered only briefly here, with special emphasis on observations and events that revealed previously unrecognized facets of magnesium metabolism or led to methodological advances which opened up productive lines of research. Magnesium sulfate was recognized as the cathartic agent in Epsom salts by Dr. Nehemia Grey in 1695, 78 years after the discovery of the bitter-tasting water from the now famous Epsom Spring (14). In 1886 Colin correctly ascribed the cathartic effect of $MgSO_4$ to osmotic action in the lower gastrointestinal tract, an effect common to several salts, including sodium sulfate. Three years later, Jolyet and Cahours (15) reported that injected magnesium sulfate depressed central nervous system activity and respiration in dogs and appeared to mimic the effect of curare by blocking neuromuscular responses. Meltzer and Auer attempted to utilize the depressive action of magnesium sulfate on the central nervous system in a search for surgical anesthetics. They were unable to adjust the effective dosage of magnesium sulfate, however, and abandoned its use for anesthesia to avoid fatalities (16−18). Smaller doses of magnesium sulfate were shown later to ameliorate convulsions in patients with tetany (19) and in toxemia of pregnancy (20). Magnesium sulfate injections in moderate doses are still commonly used to treat pre-eclamptic convulsions (21).

In 1932, Kruse et al. described the now classical signs of magnesium deficiency, conclusively establishing magnesium as an essential nutrient in rats and dogs (6,22). Two earlier attempts at inducing magnesium deficiency in rats (23) and mice (24) had failed, presumably because the deficient diets contained too much magnesium. The report by Kruse et al. (22) was followed by a series of investigations on experimental magnesium depletion that confirmed the reproducibility of magnesium deficiency signs, with some variation among species. They showed, in addition, that the rate of onset and severity depended on the stage of growth as well as the calcium content of the diet (25,26). Overall, the effects of magnesium depletion were severe and far reaching including, in addition to disturbances in central nervous and neuromuscular function, derangements in calcium metabolism (27—30), bone development and structure, and calcification of renal and other soft tissues (30—32).

Attempts to identify magnesium deficiency in human beings were unsuccessful until the development of atomic absorption methods in the late 1950s made possible the accurate diagnosis of hypomagnesemia (8,33). Atomic absorption is still the method of choice for rapid, accurate, and sensitive magnesium analysis. Its use has led to the diagnosis of hypomagnesemia in association with a variety of pathological conditions (1,7).

One of the stumbling blocks in research on magnesium metabolism in man has been the lack of an adequate chemical indicator for magnesium status, since plasma magnesium levels do not appear to reflect consistently body or tissue magnesium stores (21). In 1953 Sheline and Johnson (10) reported the production and properties of ^{28}Mg($T1/_2$ = 21h), both a β and γ emitter (9). Hailed with enthusiasm as a new tool for research on the fate of magnesium in man and experimental animals (13,34—43), ^{28}Mg has proved somewhat disappointing. Sporadic availability, high cost, and the short half-life severely limits it uses, particularly in human beings in whom 80% or more of body magnesium is in compartments with slow turnover rates (35—37,44).

Recent efforts to substitute stable magnesium isotopes for ^{28}Mg have been only partly successful. The minor magnesium isotopes ^{25}Mg and ^{26}Mg have relatively high natural abundances. This factor, combined with the rapid turnover of extracellular magnesium and the relatively poor precision of currently available methods for stable magnesium isotope detection, limits their usefulness as tracers in man and intact animals (11,12,45). The current pace in the development of analytical mass spectrometry, however, has already broadened the scope for convenient and more sensitive detection of stable isotopes. Whether further technological developments can provide means for adequate utilization of stable magnesium tracers in man is not certain. They do, however, offer possibilities for isotopic studies in cell and tissue cultures to complement the more limited data obtainable in intact animals or human beings (46,47).

DISTRIBUTION OF MAGNESIUM IN THE BODY

The concentrations and distribution of magnesium in body tissues are summarized in Table 1. About 60% of the approximatley 28—40 gm of magnesium contained in an adult human body (51) is in the skeleton which also shows the highest tissue magnesium concentration (48). The remaining 40% is distributed in the muscles and soft tissues. One percent or less is in the plasma and extracellular fluids. With some variation between tissues, total tissue magnesium concentrations exceed those in extracellular fluids by a factor of about 5 to 10 (Table 1).

Table 1. Magnesium Concentration of Selected Tissues

Tissue	Magnesium, mmoles/Kg wet wt.[1] Man	Rat
Plasma	0.6-1.1	0.8-1.2[2]
Erythrocytes	2.1-3.1	—
Skeletal Muscle	6.9-8.9	10.6-12.0
Heart	6.6-8.2	10.0-11.0
Aorta	4.5	—
Kidney	4.0-7.0	7.0
Brain	4.0-7.0	6.0- 7.0
Liver	7.3-8.2	9.0-10.0[2]
Pancreas	12.0	10.0-14.0[3]
Bone	42.0	68.0[4]

[1]Walser (197), [2]Schwartz et. al. (165), [3]Gregor & Schwartz (68)
[4]Chou et. al. (36)

Table 2. Subcellular Distribution of Magnesium in Rat Liver and Pancreas

Cell Fraction	% of total Mg in Fresh Tissue Liver		Pancreas	
Supernatant	12.8[1]	19.2[2]	34.8[3]	34.0[4]
Microsomes	21.8	17.4	3.4	4.4
Mitochondria	48.0	13.7	13.2	32.7
Nuclei & Debris	13.0	47.8	42.5	25.4
Connective Tissue	—	2.0	—	—

[1]From Griswold & Pace (70), [2]Thiers & Vallee (186),
[3]Schwartz et. al. (165), [4]Greger and Schwartz (69).

The low magnesium concentrations of erythrocytes compared to other tissues (Table 1) reflect cellular differences in structure and function. Total cell magnesium content is related to the presence or absence of a nucleus and cellular content of ribosomes and other organelles. Measurements made on magnesium content of subcellular fractions show severalfold differences in magnesium concentration of the major cell compartments (52–54) (Table 2). The large error of such determinations and the pitfalls of cellular separation techniques for the measurement of cation distribution are well recognized. Nonetheless, the wide differences between the intracellular concentrations of calcium and magnesium (56), more recently confirmed by cell imaging techniques (46), suggest that the distribution of magnesium within cells is relatively uniform. Little is known at present about the forms or degree of binding of magnesium in subcellular compartments.

Plasma magnesium is normally 0.85–1.0 mM in man. Over 50% has been determined to be present as free Mg^{2+}. The remainder, about 35%, is nonspecifically bound to plasma albumin. Ten to fifteen percent is complexed to phosphate and small molecular weight organic ligands (48). Plasma magnesium levels remain remarkably constant over a wide range of magnesium intakes (48,57). Under conditions of severe magnesium depletion, plasma magnesium levels decrease before significant changes are seen in the red cells or soft tissues (58,59). Severe magnesium depletion has relatively little effect on soft tissue magnesium concentrations. The liver, skeletal, and heart muscle showing the greatest magnesium deficits. Bone magnesium content can be markedly reduced in growing animals. Smaller magnesium deficits are seen in the absence of growth (1,48,60,61).

MAGNESIUM HOMEOSTASIS

Most Western diets provide about 8–20 mmoles magnesium per day. These amounts appear to be adequate to maintain normal plasma magnesium levels in healthy individuals with normal kidney function. Moreover, since magnesium deficiency has not been reported in people in the absence of contributing diseases even with diets very low in magnesium (1,7), the body must have potent mechanisms for magnesium conservation. The nature of these mechanisms is incompletely understood.

Magnesium Absorption

About 40–70% of ingested magnesium is normally excreted in the feces (Fig. 1), only 1–2 mmoles per day being derived from endogenous sources (35–37). Aikawa (1) suggested that magnesium secreted into the gastrointestinal tract in digestive secretions is more completely absorbed than ingested magnesium. This would be possible only if

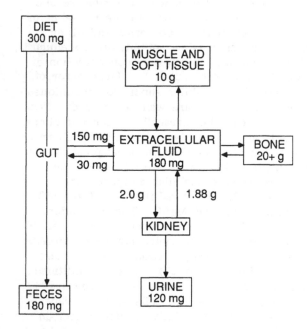

Fig. 1. Distribution of magnesium in major body compartments and magnesium homeostasis.

endogenous and exogenous magnesium did not become completely mis-cible in the gastrointestinal tract. The high solubility of dietary magnesium (62) and magnesium salts in general (63) does not support such an assumption. The few available data on food magnesium exchangeability in vivo have shown it to be close to 100% (64,65).

No single dietary factor has been conclusively shown to influence availability of dietary magnesium. Of those that have been impli-cated—proteins or amino acids, which may enhance magnesium absorp-tion (22,49,66), and phosphate, phytate, and fiber (65,67—70), which are thought to reduce it—only fiber appears to have been consistently found to reduce magnesium absorption (67,70). However, most of the data on factors affecting magnesium absorption were generated in human metabolic studies in which the inclusion of fiber and/or phytate involved addition or substitution of foods which must have introduced changes in dietary factors in addition to those of primary interest.

A major factor that affects magnesium absorption is the level of dietary magnesium (41,71). The repeatedly observed inverse relation-ship of fractional magnesium absorption to intake or luminal load in man (71,72) has led to the concept that gastrointestinal magnesium transport is saturable (73,74). The mechanisms of magnesium absorption have not been clearly defined, however. There is little doubt that magnesium in solution (presumably as free Mg^{2+}) can be transported by all segments of the intestine, including the colon (75). On the other hand, when magnesium is consumed orally, little if any absorption occurs below the small intestine in intact rats (70) or man (41,71,76). Apparently, magnesium interacts with gastro-intestinal contents in a way that progressively reduces its availability in passage through the GI tract (70). One of the possibilities suggested by following the fate of an isotopically labelled test meal along the gastro-intestinal tract in rats was uptake of soluble magnesium by microorganism in the cecum or large intestine (70).

Current understanding of epithelial magnesium transport is based on a small number of studies carried out with in vitro gut preparations (77–80) or with in situ ligated loops in animals (38,55,75,81–83). Data conflict in most major aspects relating to mechanisms of magnesium absorption. Areas of uncertainity include the primary intestinal sites of magnesium absorption, transport saturability, dependence on metabolic energy (77,79,81), interaction with calcium (78,82), and the influence of vitamin D (69,84,85). Based on a relatively recent in vivo study using ligated loops of rat ileum and colon, transport of magnesium appears to be saturable and to depend on sodium flux or gradient but not on metabolic energy (81,82). Magnesium absorption was found to resemble calcium absorption in the ileum, showing characteristics of facilitated diffusion, augmented by bulk flow. No specific carrier for magnesium transport has been identified, although the rare, but well-documented occurrence of a congenital defect in magnesium absorption in young infants suggests such a mechanism in man, at least during early development (74,86,87).

Overall, magnesium absorption is one of the least understood aspects of magnesium metabolism. Most of the scant data on magnesium absorption at the tissue or cellular level were obtained before the mid-1970s (38,75,77–83,109). Reexamination of magnesium transport in gastrointestinal mucosal cells and across isolated membranes would not only add to understanding of magnesium absorption mechanisms, it could also provide insight into mechanism of membrane transport of magnesium and the processes by which cells maintain magnesium homeostasis.

Renal Excretion

The kidney is well established as a primary site of magnesium homeostasis. In people with normal kidney function, the quantity of magnesium excreted in the urine closely reflects the amount absorbed over a wide range of magnesium intakes (29,59). Injected magnesium is rapidly cleared by the kidney, and renal excretion decreases to very low levels in response to a diet low or lacking in magnesium (29,88).

Tubular magnesium transport has been studied fairly extensively by micropuncture techniques (89—97). A portion of filtered magnesium, 20—25%, is reabsorbed in the proximal convoluted tubule (90,94,95). None appears to be transported by the descending limb of the loop of Henle where much of sodium and calcium is reabsorbed. Between 55—>60% of filtered magnesium is reabsorbed from the thick ascending limb of the loop of Henle (90,94). Only a negligible quantity (1—3%) can be reabsorbed in the distal tubule. There appears to be no tubular Tmax for renal magnesium transport (92,93). However, a feedback mechanism must exist at the contraluminal membrane of the thick ascending limb of the loop of Henle, since the fractional magnesium reabsorption rate decreases with increased Mg^{2+} in the extracellular fluids (93). It has been suggested, but not yet substantiated, that magnesium may be secreted into the renal tubule when plasma magnesium is elevated (29,94).

The Role of the Skeleton in Magnesium Homeostasis

Bone magnesium concentrations can be reduced by as much as 70—80% in growing magnesium depleted animals (9,48,51,60,148). However, the proportion of skeletal magnesium available for release to the extracellular fluids decreases with age (60,61,99) and can be very low in adult animals (61). Walser (48) questioned the lability of skeletal magnesium at any age, suggesting that the low concentrations seen in the bones of growing magnesium-deficient animals merely reflected low magnesium levels in extracellular and bone fluids during bone formation. The close relationship between plasma and bone magnesium in rapidly growing rats (9) and chicks (197) during magnesium depletion lends support to Walser's theory, but does not preclude the possibility that a fraction of skeletal magnesium is available for exchange with that in the surrounding fluids (60,101,102).

Incorporation of magnesium into hydroxyapatite has been studied under controlled conditions in vitro. Unlike calcium, magnesium does not appear to become part of the apatite crystal lattice but remains surface bound, either physically held on crystal surfaces or in adherent water of hydration (100). As synthetic apatite crystals age, they extrude magnesium toward the surfaces and hydration shells from which they can be rapidly eluted. More than 90% of the magnesium contained in aged synthetic apatite becomes exchangeable with that in the medium (100). Not all of the magnesium in bone mineral is exchangeable, and little is known of the nature of the nonexchangeable bone magnesium pool. The exchangeable portion, like that in synthetic apatite, is freely elutable or exchangeable with extraosseous, isotopically tagged magnesium (60,101). Both in vivo and in vitro, exchangeable bone magnesium is linearly related to magnesium concentration in the plasma or medium. Although the exchangeable magnesium pool decreases with age, and as a result of magnesium depletion (60), the linear relationship between the exchangeable magnesium pool and plasma or medium magnesium concentration has been demonstrated under various conditions, including magnesium deficiency and excess (101,102). Thus, magnesium is passively released or taken up by bone to augment or reduce plasma magnesium levels. It provides a buffer against fluctuations in extracellular magnesium concentrations, but is not a true store (102).

BIOLOGICAL AND PHYSIOLOGICAL ROLES

Before describing the functions of magnesium, it is useful to consider briefly some of its basic properties. Magnesium is the second element in group IIA of the Periodic Table. Like all elements in this group, magnesium has two outer shell electrons which are given up to form Mg^{2+}. The ionic radius of magnesium is 0.78Å (that of calcium is 1.06Å). Thus the major difference between the two ions in solution is their size and charge density. This accounts for the greater polarizing capability of magnesium compared to calcium and for its greater tendency to form hydrated and soluble salts with organic ligands. Magnesium forms predominantly electrostatic associations with hard anions. In biological systems it is complexed to strong electron donors such as the oxygen of carboxyacids, phosphates and the lone pair of electrons of the nitrogen atom (1,63).

Photosynthesis

The presence of magnesium in the chlorophyll molecule was first described by Willstatter and Stoll (1) following years of research on its characteristics and structure. Chlorophyll is the only known biologically functional magnesium chelate that is stable at physiological pH. It is a modified porphyrin structure with a central magnesium atom linked to four pyrrole groups. The stability of the magnesium complex has been demonstrated by its nonexchangeability with ^{28}Mg in aqueous organic solution. Harsher treatments, for instance, dilute acid, will remove magnesium from chlorophyll and cause loss of its characteristic green color and photochemical activity (1). Other metals can substitute for magnesium in chlorophyll but, with the exception of zinc, cannot render it photochemically active (103). Zinc, however, binds to chlorophyll more tightly than magnesium. It is unlikely to have been as plentifully available for the evolution of photosynthetic processes. Moreover, the zinc-chlorophyll complex has a weaker capacity to aggregate than magnesium chlorophyll, evidently a necessary attribute for efficient capture and utilization of light energy.

The structure of chlorophyll differs from that of heme in that it contains a fifth 5-carbon ring and a phytol side chain (104). The

hydrophobic character of the latter is important for the orderly stacking of chlorophyll units in the chloroplast grana, the units in which
energy capture takes place. The stacking of chlorophyll molecules
within layers of membranes is thought to optimize their function as
acceptors and transmitters of energy quanta which activate photosynthetic reactions (1,104).

In addition to its role as part of the chlorophyll molecule, magnesium participates in various stages of photosynthesis, including
electron transport, phosphate transfer, and transmembrane transport
in or out of chloroplasts. For instance, illumination has been shown
to increase chloroplast magnesium concentration by an energy-dependent process. Darkness reverses the magnesium flux (87,105—107).
For a more comprehensive review of the roles of magnesium in photosynthesis, the reader is referred to Aikawa (1).

Cellular Roles of Magnesium

The biochemical roles of magnesium are many and far reaching. Only
a cursory treatment of its biochemical functions can be included here.
More detailed reviews have been published by Wacker and his associates (2,3,4,108) and a thought-provoking section on the intracellular
role of magnesium is included in an article by Rasmussen et al. (56)
on ionic regulation of metabolism. A recent compilation of articles on
the role of metal ions in genetic information transfer edited by Eichhorn and Marzilli (72,109—113) contains up-to-date information on
involvements of magnesium in nucleic acid and protein synthesis.

Magnesium is a cofactor or activator for more than 300 enzyme
reactions. It is, in addition, an essential factor in the following
major processes: (1) stabilization of the plasma membrane and other
intracellular membranes, (2) stabilization and conformation of macromolecules, in particular nucleic acid and nucleoproteins, (3) generation
and hydrolysis of ATP, (4) nucleic acid and protein synthesis. Each
of these processes underlies numerous fundamental reactions in cellular
growth and metabolism. The stabilization of intracellular membranes
is essential for cellular compartmentalization and homeostasis. The
proper conformation of macromolecules such as the DNA double helix
(109,114), the ribosomes and ribosomal subunits (97,110) and the
binding of t-RNAs to ribosomes (97,115,116) are critical factors in
the transmission of the genetic code, cellular growth and multiplication.
The generation and utilization of ATP underlies all biosynthetic processes, glycolysis, and energy dependent transport across cell membranes and within the cell (47,112,117,118).

Enzymes that depend on magnesium as a cofactor or activator
include DNA and RNA polymerases (111,113,119), phosphatases,
kinases and phosphotransferases (2,7,115,120—123). Although manganese and sometimes other divalent cations have been shown to substitute for magnesium in vitro, either structurally in metal-nucleotide-

enzyme constellations (109,114,120), or in catalyzing specific reactions
such as nucleotide polymerization (109,119), magnesium appears to be
the metal utilized in the cell. Magnanese is not present in cellular
compartments in adequate concentrations. Moreover, if manganese is
substituted for magnesium during DNA or RNA polymerization in vitro,
the rate of error incorporation is significantly increased (97).

 Cytoplasmic magnesium concentration approximately equals that
of ATP. How much magnesium is complexed or exists as free Mg^{2+}
depends on intracellular pH and energy charge, since the binding
constants for Mg complexes of ADP and AMP are weaker than that for
Mg-ATP (104). Free cytoplasmic Mg^{2+} has been estimated to vary
between less than 0.1 to 1 mM (52−54,124−126), although 1 mM is
unlikely under normal circumstances. It has been suggested that
Mg^{2+} may be a regulator of intracellular metabolism (7,104,127).
Wacker has illustrated this possibility with reference to brain hexokinase, a key enzyme in brain glycolysis (7), which utilizes Mg-ATP
as a substrate and Mg^{2+} for allosteric activation. Consequently,
reaction rates are high when the ATP level is relatively low and Mg^{2+}
is high. The resultant formation of glucose-6-P stimulates glycolysis
and the formation of ATP. As the latter complexes to Mg^{2+}, thereby
reducing the concentration of free magnesium ions, the enzyme returns
to a less active conformation.

 Rubin and associates (127,128) posed the hypothesis that magnesium is central to coordinate control of cellular metabolism. Their
studies, carried out with cultured fibroblasts, showed the concentration of Mg^{2+} in the medium to affect simultaneously a number of fundamental cellular processes, including DNA and protein synthesis,
glycolysis, and uptake of ions. Recently, Walker and Duffus (129,
130) have proposed a role for magnesium in the regulation of the cell
cycle based on observations of cyclic reductions in cellular magnesium
content of the fission yeast Schizosaccharomyces Pombe as the organism
approaches cell division.

Magnesium and Neural Function

The disturbances in central nervous system and neuromuscular function that have been consistently shown to accompany both elevations
and reductions in plasma magnesium (5,6,22,48,58,59,131−133) leave
little doubt that magnesium is of vital importance to the functional
integrity of the nervous system. The exact roles of magnesium in
neural functions are still unclear, partly because they are intimately
linked to those of calcium, sodium, and potassium and difficult to
distinguish from them. What is currently known about the functions
of magnesium in the nervous system stems primarily from three lines
of investigation: studies on magnesium depletion in vivo, observations
of the pharmacological effects of injected magnesium salts, and experiments with in vitro nerve or nerve-muscle preparations.

Neural Effects of Magnesium Depletion

Neuromuscular hyperexcitability or weakness have been observed in all species with experimental or spontaneous magnesium deficiency (22,48,131,133—136). Other manifestations of alterations in central nervous or neuromuscular function are less consistent and may vary between species. For instance, the tonic-clonic convulsions first described by Kruse et al. (5,22) appear to be characteristic of the rat. Symptoms described in people with hypomagnesemia cover a range of neurologic and mental disturbances, including positive Chvostek and Trousseau signs, athetotic movements, gross muscular weakness and tremor, ataxia and tetany, as well as depression, confusion, and hallucinations (7,58,59,131,134,137,138). However, none or only a few of these signs may be present in an individual with hypomagnesemia. Among those that appear to be due to magnesium depletion rather than primary or secondary complications are positive Trousseau signs and myographic changes (88,121).

The free magnesium concentration of the spinal fluid is higher than that of the plasma (1,139—142), suggesting a specific mechanism for uphill transport of magnesium across the blood brain barrier. That such a mechanism must exist is strengthened by observations which indicate that the spinal fluid magnesium concentration appears to be buffered against both downward and upward fluctuations in plasma magnesium (1,143). The choroid plexus is thought to secrete magnesium into the spinal fluid at rates that vary in response to plasma levels (141). When the spinal fluid magnesium content was acutely reduced in young rats, lowered magnesium concentrations were also seen in the brain (139). These deficits were correlated with hypersensitivity to noise. Ninety-eight percent of brain magnesium was thought to be intracellular with 29—33 μg/gm dry weight in the grey matter and 17—22 μg/gm in white matter (133). Normal brain magnesium levels were restored within 2—6 hours after parenteral magnesium injections, which transitorily raised plasma magnesium levels 2—3-fold above normal. Reduction in sensitivity to tone stimuli appeared to reflect brain magnesium levels (133).

Pharmacological Effects of Magnesium

The observations of Jolyet and Cahours (15) and later those of Meltzer and Auer (17,18) seemed to suggest that parenteral administration of magnesium sulfate had two effects: depression of the central nervous system (CNS) and neuromuscular paralysis similar to that produced by curare. It now appears likely that the apparently direct effect on the CNS is the consequence of the action of excess mangesium on synaptic and neuromuscular transmission. As summarized in the preceding section, it is difficult to alter spinal fluid magnesium content significantly by raising plasma magnesium even up to 6—8-fold its normal levels (143). Moreover, analgesia or anesthesia are not predictable consequences of elevated plasma magnesium concentrations, and the difference between a level that induces anesthesia and one that leads to death by respiratory failure is small and difficult to define (16,18). It is possible, therefore, that what has been interpreted as a direct narcotic effect on the CNS was, in fact, secondary to depressions in neural and myoneural transmission resulting in cardiac and cardiovascular changes as well as in respiratory muscular paralysis (1).

Effects on Neural Function in Vitro

Numerous in vitro studies have shown that magnesium, like calcium, can stabilize axonal membrane potentials, but does so less effectively (2). The widely different concentrations of calcium and magnesium ions in the squid giant axon (2,144) (Ca: 0.1 μM, Mg: 2—3 mM) suggest different, if parallel, modes of action for the two cations in the maintenance of axonal membrane potentials. Details of the specific role of magnesium in this process are not known.

It is well established, however, that magnesium antagonizes calcium in the transmission of synaptic and myoneural impulses and that it does so by inhibiting the release of acetylcholine (4,140,145). Calcium promotes acetylcholine release from neurons by a process similar to others in which calcium mediates stimulus secretion coupling (127). The roles of magnesium in this process are still mostly conjecture, but are likely to involve a number of cellular processes (56). When these are more clearly defined it may become possible to reconcile the presently contradictory observations that high concentrations of magnesium promote acetylcholine synthesis in the brain but inhibit its release, while low concentrations promote acetylcholine release but do not appear to influence its synthesis (2).

MAGNESIUM DEPLETION

In Experimental Animals

Magnesium deficiency has been experimentally induced in many species including rats (5,22,25,26,28,40,133,139,146—148), mice (146), guinea pigs (149), rabbits (13,150), chicks (51,151), dogs (5,152), pigs (160), quail (153), ruminants (135,152), and monkeys (59,154). In all species hypomagnesemia is found associated with signs of nervous and neuromuscular disturbances (59), although the manifestations of the latter vary somewhat among species. The effects of magnesium depletion have been studied most intensively in rats, however, and much of current understanding of the roles of magnesium in animals rests on observations made in rats. Unfortunately, some responses of the rat to magnesium depletion differ from those shown in other species. The relevance of observations made in rats to other species may therefore not be valid. Nonetheless, the major effects of magnesium deficiency on organ function and pathology are similar across species, and these permit some generalizations regarding possible functions of magnesium. Hypomagnesemia and the attendant signs of magnesium deficiency develop most rapidly and acutely in young growing animals. As might be expected, the severity and rate of onset is directly related to the degree of dietary magnesium deprivation. In addition, high calcium (26), phosphorus, and protein (18, 50,52,194) have been shown to enhance susceptibility to magnesium depletion. A diet high in saturated fats and cholesterol has also been demonstrated to precipitate magnesium deficiency (156).

As diets used in different laboratories may differ in many constituents, it is not surprising that there is considerable variation in the symptoms reported. For instance, decreased growth has been accepted as a consequence of magnesium depletion since publication of Kruse, Orent, and McCollum's classic paper (22). Yet the relationship of magnesium depletion to growth is not clear cut. Severe magnesium restriction severely depresses feed consumption and growth (5,22,26). Less severe restriction which, nonetheless, results in rapid and severe hypomagnesemia may have only minor effects on feed comsumption. Under such conditions, initial depressions in weight gain appear solely to reflect feed intake (12,98). Moreover, stimulation of feed consumption by raising dietary protein will increase weight gain for a period while enhancing the rate of onset and severity of hypomagnesemia (12). Severe or prolonged magnesium depletion reduces feed consumption and feed efficiency (12,50,52,98).

Not all tissues reflect magnesium depletion by a decrease in magnesium concentrations (21,48). The largest deficits in magnesium concentration are seen in erythrocytes and bones of growing animals (48,51,60). Skeletal muscle may show a 10–15% decrease in magnesium concentration, liver and heart muscle from 5–8% (61). Other tissues including kidney and pancreas do not lose magnesium to a measurable degree.

Many investigators have confirmed Kruse et al. (22) in the observation of two distinct stages of magnesium deficiency: acute and chronic. The acute stage is particularly well defined in the rat, which shows, during early magnesium depletion, increasing hypomagnesemia, nervous irritability, hyperemia, peripheral vasodilation, and edema (48). This stage, which may last for up to three weeks, may or may not be accompanied by severe depression and weight gain. Hypomagnesemia and irritability, as well as neuromuscular disturbances, persist into the chronic stage. However, the hyperemia and peripheral vasodilation abate to be replaced by alopecia and trophic skin changes. If severely deprived of magnesium, the animals may die suddenly following convulsive seizure or coma (21,133). Less severely depleted animals can survive for several months (30,50,52).

Pathological Changes

One of the most consistent consequences of magnesium depletion is alteration in calcium utilization. In the rat, increased calcium retention and reduced urinary excretion may be accompanied by increased renal calcium concentrations, renal tubular casts and cellular calcium deposits (1,22,30). Other tissues that have been shown to develop calcified lesions are cardiac and skeletal muscle (21,31). While such lesions have been described primarily in rats, renal necrosis has been reported in ruminants also, in some cases with calcification (21). Apparently tissue calcification is not necessarily linked to the hypercalciuria that occurs only in rats (59,88).

The lesion most characteristic of magnesium deficiency in rats is degranulation of dermal and abdominal mast cells (50,134,157). It may occur in dogs (5) and possibly in mild forms in other species, but this has not been substantiated by reported observations. In rats, mast cell degranulation coincides with a severalfold increase in plasma histamine concentrations, which peak and later subside as the mast cells become exhausted (50,157). At certain sites such as the submucosa of the gastrointestinal and urinary tracts, immature mast cells proliferate, apparently deficient in the capacity to store and release histamine (134,157). Rats exhibiting this syndrome also show leucocytosis and eosinophilia as well as thymic enlargements and invasion by immature lymphocytes (1). The latter phenomenon, coupled with cell-mediated immune incompetence has been suggested to lead to a fatal type of lymphoma-leukemia seen in about 20−25% of magnesium deficient rats, but only in animals of the Sprague-Dawley strain (158). More general is a decrease in plasma immunoglobulins, which has been described in mice (146) as well as in rats. Whether such changes are specific to magnesium depletion or secondary to general malnutrition and depressed protein synthesis requires further investigation.

Among the early descriptions of tissue lesions are vascular lipid deposits that were thought to resemble those seen in atherosclerosis (22). Later Vitale et al. (190) observed that a diet high in atherogenic factors (20% saturated fats, 1% cholesterol and 0.3% cholic acid) produced signs of magnesium deficiency in rats fed 240 ppm magnesium, a level then considered adequate for this species. Moreover, very high dietary magnesium levels (1920 ppm) decreased the development of lipid deposits in the aorta and myocardium but did not lower the high plasma cholesterol that typically develops in rats fed this diet. Numerous studies have been carried out in rats, rabbits and cebus monkeys to clarify the complex relationship between dietary magnesium lipids and atherosclerosis as reviewed in detail by Seelig (21). The outcome is equivocal. One of the mechanisms by which magnesium supplements may ameliorate the effects of atherogenic diets may be by formation of insoluble and unabsorbable complexes with dietary lipids and fatty acids.

Effect of Magnesium Deficiency on the Skeleton

Watchorn and McCance (30) first described skeletal and dental changes during long term chronic, subacute magnesium deficiency in rats. The bones were brittle and weak, the teeth discolored, "chalky," and brittle. More severe magnesium depletion was later shown by Duckworth et al. (28) to impair bone formation and mineralization. Since then a series of reports have confirmed that magnesium depletion impairs bone formation in growing rats (98,159,160) and chicks (51,151) and may be associated with changes in bone density and mineralization (151,160). Inconsistent findings can probably be ascribed to species differences, the stage of growth of experimental animals, and the length and severity of magnesium depletion.

The decrease in bone magnesium content is most marked in rapidly growing bone. A concomitant increase in bone calcium content or denisty (51) has led to the suggestion that calcium may exchange for magnesium in apatite crystals. Bone growth per se may or may not be reduced by magnesium depletion (98,160). When it is, the width of the epiphyseal growth plate is markedly reduced as are the histologically and chemically identifiable manifestations of endochondral bone formation (51,159). Even when there is relatively little reduction in growth at the epiphyseal plate, cortical bone shows various abnormalities, including incompletely mineralized osteoid, cementing lines and small elongated osteocytes with apparently low osteolytic capacity (95,151). In the rat, but not in other species, subperiosteal exostoses, thought to consist of abnormally differentiated immature bone cells and tissue, occur in the shafts of long bones (98,159). These and other bone abnormalities are rapidly reversed by repletion with magnesium.

The mechanism by which magnesium depletion causes bone abnormalities are still conjectural. Presumably, the decrease in plasma magnesium is reflected in the bone fluid, although bone fluid composition has not been analyzed in magnesium-deficient bone. In vitro studies on apatite formation show that magnesium profoundly affects the equilibrium between amorphous calcium phosphate and apatite (161–163). It may also alter apatite crystal structure. In addition, both the development and function of bone cells and their precursors are likely to be influenced by the magnesium concentration of the bone fluid. A finding that is now accepted as a feature of the magnesium deficiency syndrome is end organ resistance to parathyroid hormone (PTH) stimulation (51,59,148,151,164–166).

The hypercalcemia which frequently accompanies magnesium deficiency in rats (29,98) has been interpreted as indicating excess secretion of PTH in response to decreased plasma magnesium concentrations (13,27). This view found support in the prevention of hypercalcemia in parathyrodectomized rats (29) and by observations of suppression of PTH secretion by increased plasma or medium magnesium in PTG tissue in vivo and in vitro (29,167). However, the potency of magnesium ions to influence PTG secretion is 2—3-fold less than that of calcium. Moreover, histological observations suggested decreased rather than increased activity of the PTG (98,160). Since the advent of reliable measurements, plasma PTH levels made in hypomagnesemic hypocalcemic human beings and dogs have been found apparently normal, elevated, or decreased but usually too low for the degree of observed hypocalcemia (59,88,168). According to Shils the sequence of events is probably (1) reduced calcium release from bone due to blunted endorgan sensitivity, (2) an initial PTG response to produce temporary hypersecretion of PTH and, (3) failure of the PTG to respond adequately to the degree of hypocalcemia until the hypomagnesemia is corrected by magnesium repletion. This sequence of events negates earlier hypotheses that the PTG regulates magnesium homeostasis (59).

In Ruminants (Grass Tetany)

Grass tetany was first shown to be associated with hypomagnesemia by Sjollema and Seekles, who noted similarities in the symptoms exhibited by dairy cows recently transferred from winter feed to new pasture to those described in magnesium depleted rats (136). Among the signs most prevalent in animals with grass tetany are hyperexcitability, muscle weakness, muscle twitchings and convulsions which may result in death. About 30% of cases with clinical signs have been estimated to have fatal outcomes (1) unless promptly treated with therapeutic doses of magnesium given orally or by injection and followed up with dietary magnesium supplements.

The etiology of grass tetany is complex. The disease is most prevalent in dairy cattle and in temperate regions. It has been of major economic importance in Australia, New Zealand, Holland and Britain (1,136) but also occurs in the United States with sufficient frequency to cause concern (169). In all locations it appears to be associated with spring pastures and cool weather. Although early grasses may be somewhat lower in magnesium than later crops, they cannot be considered magnesium deficient. Thus research on possible causes of grass tetany has been focused on factors likely to influence magnesium availability and absorption or to directly depress plasma magnesium.

Grass crops can vary in nitrogen content and in sodium, potassium, calcium, phosphate, and phytate, all of which have been shown to depress magnesium absorption (1,170). These factors vary in different grass varieties, and can be altered by fertilization and climatic conditions (169,171). One of the characteristics of spring grasses grown during cool cloudy weather is a high concentration of transaconitate (171). Transaconitate has been shown to disappear rapidly from the rumen and can, if maintained in the pasture at high concentrations, reduce plasma magnesium (169). Recently, transaconitate has been shown to be converted to tricarbalyllic acid (TCBA) by rumen bacteria (172). It can be absorbed but, unlike transaconitate or citrate, TCBA does not appear to be metabolized in mammalian tissues and may in fact compete with citrate or transaconitate for enzyme sites in Krebs cycle oxidation. Thus TCBA may contribute importantly to the sudden onset of grass tetany. This has not yet been demonstrated.

In Human Beings

Clinical magnesium deficiency not confounded by hypocalcemia was first described by Vallee et al. (138), who identified hypomagnesemia in patients with tetany following abdominal irradiation. In most instances, however, symptomatic hypomagnesemia is accompanied by hypocalcemia and/or other disturbances in electrolyte metabolism (58, 59).

Spontaneously occurring magnesium deficiency has not been described in the absence of other diseases or conditioning factors (48). The settings in which symptomatic hypomagnesemia has been detected fall mainly into two categories: conditions associated with excessive losses of body tissue, fluids, or electrolytes; or reduced gastrointestinal absorption or renal conservation. The latter include severe or prolonged diarrhea, nasogastric suctioning, protein-calorie malnutrition, sprue, celiac disease, or other malabsorption syndromes, gastroileal and bowel resections, alcoholism, renal dysfunction, and drug therapies, in particular thiazides and mercurial diuretics (1,3,7, 21,59,131,173−176). Magnesium depletion has also been diagnosed in association with a number of endocrine abnormalities, including hyperaldosteronism (48) and hyperparathyroidism, frequently following parathyroidectomy (177). The symptoms that have been reported to accompany hypomagnesemia vary considerably and cover a range of mental, neurological and neuromuscular disorders (59).

Several attempts have been made to induce symptomatic magnesium deficiency by dietary restriction in human volunteers (58,132,174, 178). Only one of these resulted in hypomagnesemia with symptoms (58). In this investigation six males and one female subject were depleted by consuming a diet which contained less than 0.8 Meq Mg per day. Although urinary and fecal magnesium fell rapidly to very low levels, plasma magnesium concentrations decreased continuously to levels comparable to those seen in experimentally magnesium depleted animals. Hypocalcemia and hypokalemia developed in six patients, all of whom showed symptoms that included anorexia, nausea, apathy, and a variety of neurological distrubances. Shels (58,59) concluded that magnesium depletion caused alterations in potassium and calcium homeostasis in human beings and that most of the symptoms appeared to be related to these changes in electrolyte metabolism. Some symptoms that were not seen in the experimentally depleted patients but

have been reported in clinically diagnosed magnesium deficient people
may have been the result of conditions or diseases that precipitated
the hypomagnesemia rather than magnesium deficiency per se.

ASSESSMENT OF MAGNESIUM STATUS

Numerous reports describing the progressive development of hypomagnesemia during experimental magnesium depletion suggest measurement of plasma magnesium as a simple and sensitive criterion of magnesium status (58,59). According to such data, a plasma magnesium
concentration significantly below the normal range would indicate
early magnesium depletion if erythrocyte magnesium is normal. Significant depression in erythrocyte magnesium would suggest magnesium
depletion of longer duration. However, erythrocyte magnesium levels
have not been measured with sufficient frequency to establish their
validity as criteria of magnesium status.

Plasma magnesium measurements have been useful in diagnosing
clinically unrecognized magnesium depletion, which responded positively to magnesium therapy. Unfortunately plasma magnesium levels
do not consistently appear to reflect whole body or tissue magnesium
status. Marked hypomagnesemia can occur in the presence of normal
magnesium concentration in muscle biopsy samples (21,61). Conversely,
plasma magnesium concentrations have been found to be normal or
only minimally depressed in patients with low muscle magnesium content
(61) or greater than normal retention of injected or orally administered
magnesium (176). Such discrepancies have prompted the development
of two alternative methods for measuring magnesium status: urinary
excretion of a parenteral magnesium load (179–183), and measurement
of leukocyte magnesium concentration (21). Both methods are more
time consuming than plasma and/or erythrocyte magnesium measurements. Their superiority as criteria of magnesium status is yet to
be established.

The magnesium load test has been tested in children with protein-
calorie malnutrition (180), infants (179,183) and postpartum women in
Thailand (181) and the United States (182). Although lower-than-
normal plasma magnesium levels were generally associated with greater
than normal retentions of intramuscular magnesium sulfate, the load
test appeared to show magnesium depletion more often than was
indicated by plasma magnesium levels. Moreover, plasma magnesium
did not predictably match magnesium retention in individuals (182).
Regrettably, these detailed and careful studies did not include an
objective criterion of body magnesium status against which the usefulness of the magnesium load test could be independently assessed.
Thus the conclusion that the magnesium load test is a more sensitive
detector of magnesium depletion than plasma magnesium concentration
has not yet been substantiated.

Some of the difficulties in establishing a diagnosis of magnesium depletion lie in the complex and still poorly understood interactions of magnesium with other electrolytes. This is illustrated by the results of a study carried out by Alfrey et al. (102) to probe the reasons for the frequently observed discrepancy between muscle, bone, and serum magnesium in patients with hypo- or hypermagnesemia. Correlations between serum and red cell magnesium (r = 0.69), erythrocyte and muscle magnesium (r = 0.62) were significant but less so than the correlation between serum and bone magnesium (r = 0.96). Muscle and bone magnesium did not correlate significantly (r = 0.12). On the other hand, muscle magnesium was highly correlated with muscle potassium (r = 0.82), indicating that potassium depletion can confound the diagnosis of magnesium deficiency. It is possible that factors such as undetected changes in potasssium or other electrolyte status occurring during the 48 or 56 hours needed for the parenteral magnesium load test can confound its outcome. In light of the information available to date, plasma or serum magnesium concentrations, if possible, supplemented by erythrocyte magnesium content are still the most convenient and reasonably reliable means of assessing magnesium status. A very low 24 hour urinary magnesium output determined before dietary changes or therapies are initiated can confirm a diagnosis of magnesium depletion (59).

HUMAN MAGNESIUM REQUIREMENTS

The current U.S. recommended dietary allowances (RDA) for magnesium (Table 3) are a compromise between recommended intakes for healthy adults as low as 200 mg (184) and as high as 600–700 mg per day (185). All estimations of magnesium requirements are based on data obtained in metabolic balance studies, mostly with subjects consuming constant controlled diets (184–188). A few measurements have been made in people eating self selected foods (67, 185). Since several food components are known or thought to influence magnesium absorption and utilization (49,66,68,70,184), some differences in findings from various laboratories are to be expected. However, the wide range in dietary magnesium intakes reported to have resulted in zero or positive balance (185) is unparalleled in similar studies with other nutrients and throws doubt on the validity of balance data for the determination of magnesium requirements.

The mechanisms for renal magnesium conservation are so effective that it is almost impossible to devise a combination of foods that does not provide enough magnesium to offset the very small urinary and endogenous fecal losses that have been reported in individuals eating diets designed to produce magnesium depletion (48,57,132,174,178).

Table 3. Recommended Dietary Allowance for Magnesium[1]

Category	Age (years)	Magnesium (mg)
Infants	0.0- 0.5	50
	0.5- 1.0	70
Children	1-3	150
	4-6	200
	7-10	250
Males	11-14	350
	15-18	400
	19-22	350
	23+	350
Females	11-14	300
	15-18	300
	19-22	300
	23+	300
Pregnant	—	450
Lactating	—	450

[1]From: Recommended Dietary Allowances, 9th Edition, 1980. National Research Council. National Academy of Sciences, Washington, D.C.

Under such conditions urinary losses were as low as 0.2-0.3 mmoles per day and daily fecal endogenous excretion 0.5 mmoles or less (58, 132). Yet negative magnesium balances of 3-4 mmoles have been reported in individuals whose diets contained 400-500 mg magnesium (185).

In 1964 Seelig (185) published a review of published data from magnesium balance studies. Daily magnesium intakes varied from <5 to >10 mg/Kg. Apparent magnesium retentions ranged widely between >200 mg negative or as much positive. Although Seelig was careful to exclude from her review studies of vary short duration, there was inevitably a wide spread in the lengths of balance periods and the numbers of periods that contributed to the data from different laboratories. Nonetheless, some of the generalizations that emerged from this review appeared to have stood the test of time. Men showed a greater tendency towards negative magnesium balances than women, particularly on low magnesium intakes. The highest proportion of negative balances occurred in people consuming less than 5 mg magnesium/Kg/day. Fewer negative balances were seen with diets supplying between 5−10 mg/Kg/day and none apparently at intakes exceeding 10 mg/Kg/day. Seelig concluded that dietary magnesium should be at least 6 and preferably 7−10 mg/Kg/day to ensure positive balance sufficient to replace average daily sweat and cutaneous losses which she estimated at about 15 mg (185,189). Her review and recommendations were influential in raising the previous RDA for magnesium from 200−250 mg in adults to the current level of 300−450 mg (Table 3).

The relatively few balance data published since 1964 tend to confirm Seelig's conclusion that dietary magnesium levels below 200−250 mg/day may be inadequate to maintain balance in adult males (12,65,137). It should be pointed out, however, that many of the data included in the review which prompted Seelig's conclusions were obtained with analytical techniques less specific or precise than atomic absorption spectrophotometry (33). The large apparent daily losses or gains of magnesium reported (185), amounting in some instances to cumulative balances of 3−4 g within a few weeks, suggest systematic error. In addition to possible analytical error, sources of contamination, for instance unaccounted contributions from tap water used for drinking (188), could result in erroneously negative balances. Whether a more current review of magnesium balance data would lead to quantitatively different estimates of magnesium requirements is uncertain. No good alternative for the metabolic balance is available for the estimation of magnesium requirements. A great need exists for the inclusion of blood magnesium measurements in clinical practice (190) and in the population at large, especially in conjunction with reliable dietary intake data.

Magnesium Intakes—Health Implications

Table 4 provides a list of food categories grouped in order of their relative magnesium contents and contributions to the diet. The most concentrated sources of magnesium are nuts, dried legumes, whole grains, and certain shellfish, notably winkles and conch. Foods somewhat less concentrated but still high in magnesium are clams, muscles, crabs, and dark green leafy vegetables. Seelig (185) justified a recommendation for high magnesium intakes by pointing out that the diets of preindustrialized man provided substantially more magnesium than the diets typical of industrialized societies. As Table 4 shows, the bulk of foods that make up the average U.S. diet contain

Table 4. Range of Magnesium Content in Major Food Categories[1]

Category	Selected Foods	Mg (mg/100g)
Dairy	—	4-30
	Whole milk	13
	Cottage cheese	4
	Cheddar cheese	28
Fish and	—	20-40
Seafood	Haddock	24
	Salmon	30
	Oysters	26
	Crab	40
Meat and	—	12-21
Poultry	Beef and lamb	13
	Pork	21
	Chicken, turkey	18-20
Vegetables:	—	15-90
Leafy	Cabbage, lettuce	15-25
	Collards, broccoli	
	Kale, spinach	88
Vegetables:	—	15-30
Root	Carrots	15
	Potatoes	21
	Winter squash	32
Vegetables:	—	11-95
Other	Eggplant	11
	Tomatoes	14
	Snapbeans	35
	Broad beans	95
Cereal Products:	—	90->500
Unrefined	Flour	109
	Bread	92
	Rolled Oats	148
	Wheat bran	490
Refined	—	
	Flour	24
	Bread	25
	Rice	28

Table 4. (continued)

Category	Selected Foods	Mg (mg/100g)
Dried legumes, nuts and seeds	—	100-300
	Navy beans	163
	Peanuts	180
	Pecans	120
	Almonds	296
	Sesame seeds	351

[1]Adapted from: Composition of Foods, Agriculture Handbook No. 8, ARS, USDA, 1963, and revised 1976.

moderate to low concentrations of magnesium ($<20-<50$ mg/100 gm). What, if any, are the consequences of such low magnesium intakes? The answers to this question can be only speculative. Little is known of the long-term effects of adaptation to low or high magnesium intakes. Nor is it possible to demonstrate with confidence by available methods whether an individual is marginally depleted of magnesium.

It has been suggested that chronic, subclinical magnesium deficiency may contribute to a number of degenerative diseases, including cardiovascular disease, renal calculi and osteoporosis (21,48,185). The basis for such suggestions is derived from (1) well-documented observations of tissue or organ damage in animals experimentally depleted of magnesium (6,22,30,31,156), (2) the high incidence of chronic diseases in societies whose diets tend to be low in magnesium (191), and (3) epidemiological studies showing lower myocardial magnesium concentrations at autopsy in populations with relatively high incidence of cardiovascular disease and sudden death (150,191,192). There is, however, little direct evidence that the tissue lesions seen in severely magnesium deficient animals are related to chronic marginal, clinically silent, magnesium deficiency in man. All that can be concluded from such observations is that some tissues and systems may be particularly susceptible to damage when magnesium is severely limited.

The epidemiological data are still too tenuous to allow conclusions regarding the possible role of magnesium status in the etiologies of diseases which are known to be complex and multidimensional. A major problem is the inadequacy of indicators of magnesium status, as has been discussed earlier. Although there are reports of low tissue magnesium levels in individuals with normal plasma magnesium (21), they are relatively few. Neither analyses of biopsied nor autopsied tissue samples are likely to produce reliable data unless carried out on large populations and under well controlled conditions. Even with adequate and accurate tissue data, the correlation with dietary magnesium is uncertain. Since magnesium levels have not been established for many commonly eaten foods, intakes which are based on dietary recalls or records, and calculated with available data tend to be low. Neglect of the contribution of drinking water (12,191) and possible nondietary sources of magnesium may further increase the difference between estimated and actual magnesium intakes.

Several features of contemporary diets and lifestyles have been shown experimentally to lower magnesium status or utilization. They include high levels of proteins (49,155,193,194), lipids (156) and calcium (26), ingestion of alcohol (1,7,21,61), and chronic use of drugs, in particular, diuretics (92,173,195). The magnesium content of diets typical of industrialized populations (185) is relatively low, in part, because they include substantial amounts of refined foods. Simple changes, such as reductions in sugar and fat, or substitution of whole grain for refined cereal products, can significantly increase dietary magnesium levels (Table 4) without resort to supplements. Until more is known definitely about human magnesium requirements and relationships of magnesium status to disease, it seems prudent to counsel dietary patterns that maximize magnesium intakes.

REFERENCES

1. J. K. Aikawa, *Magnesium: Its Biologic Significance*, CRC Press, Inc., Boca Raton, Fl., 1982.
2. D. M. Livingston, W. E. C. Wacker, Magnesium metabolism, in *Handbook of Physiology*, G. D. Aurbach, (Ed.), Vol. 7, American Physiological Society, Washington, DC, 1976, pp. 215—223.
3. W. B. Vernon, W. E. C. Wacker, Magnesium metabolism, in *Rec. Adv. Clin. Chem.*, K. G. H. H. Alberti (Ed.), Churchill Livington, New York, 1978.
4. O. F. Hutter, K. Kostial, Effect of magnesium and calcium ions on the release of acetyl choline, *J. Physiol.*, *124*: 234—241 (1954).
5. H. D. Kruse, E. R. Orent, E. V. McCollum, Studies on magnesium deficiency in animals. III. Chemical changes in the blood following magnesium deprivation, *J. Biol. Chem.*, *100*: 603—643 (1933).
6. E. R. Orent, H. D. Kruse, E. V. McCollum, Studies in magnesium deficiency in animals. II. Species variation in symptomatology of Mg deprivation, *Am. J. Physiol.*, *101*: 454—461 (1932).
7. W. E. C. Wacker, *Magnesium and Man*, Harvard University Press, Cambridge, MA, 1980, pp. 11—51.

8. W. E. C. Wacker, B. L. Vallee, in *Mineral Metabolism: An Advanced Treatise,* (C. L. Comar, F. Bronner, eds.), Vol. 2., Pt. A, Academic Press, New York, 1964, pp. 483–521.

9. J. W. Jones, T. P. Kohman, Synchrocyclotron production and properties of Mg^{28}, *Phys. Rev., 90*: 495–496 (1953).

10. R. K. Sheline, N. R. Johnson, New long lived Mg^{28} isotope, *Phys. Rev., 89*: 520–521 (1953).

11. R. Schwartz, Stable ^{26}Mg for dietary magnesium availability measurements, in *Stable Isotopes in Nutrition* (J. R. Turnland, P. E. Johnson, eds.), ACS Symposium Series No. 258, American Chemical Society, Washington, D.C., 1984, pp. 77–89.

12. R. Schwartz, H. Spencer, R. A. Wentworth, Measurement of magnesium absorption in man using stable ^{26}Mg as a tracer, *Clin. Chim. Acta, 87*: 265–273 (1978).

13. J. K. Aikawa, *The Role of Magnesium in Biologic Processes,* C. C. Thomas, Springfield IL, 1963.

14. M. Schofield, Magnesium: Epsom to Hartlepool. Quoted by J. K. Aikawa, 1963, in *The Role of Magnesium in Biologic Processes,* C. C. Thomas, Springfield IL, 1950, pp. 3–14.

15. F. Jolyet, M. Cahours, Sur l' action physiologique des sulfates de potasse, de soude at de magnesie en injection dans le sang, *Arch. Physiol. Norm. Path., 2*: 113–120 (1869).

16. A. H. Curtis, Magnesium sulphate solution as an aid in anesthesia *J. Am. Med. Assoc.,* 1477, 1492 (1921).

17. S. J. Meltzer, J. Auer, Physiological and pharmacological studies of magnesium salts I. General anesthesia by subutaneous injections, *Am. J. Physiol. Path., 14*: 366–388 (1905).

18. S. J. Meltzer, J. Auer, Physiological and pharmacological studies of magnesium salts II. The toxicity of intravenous injections. In particular the effects upon the centers of the medulla oblongata *Am. J. Physiol., 15*: 387–405 (1905).

19. J. A. Blake, The use of magnesium sulphate in the production of anesthesia and in the treatment of tetanus, *Surg. Gynecol. Obstet., 2*: 541–550 (1906).

20. E. M. Lazard, A preliminary report on the intracenous use of magnesium sulphate in puerperal eclampsia, *Am. J. Obstet. Gynecol., 9*: 178–188 (1925).

21. M. S. Seelig, *Magnesium Deficiency in the Pathogenesis of Disease,* Plenum Med. Book Co., New York, 1980, pp. 347–370.

22. H. D. Kruse, E. R. Orent, E. V. McCollum, Studies on magnesium deficiency in animals. 1. Symptomatology resulting from magnesium deprivation, *J. Biol Chem., 96*: 519–539 (1982).

23. T. B. Osborne, L. B. Mendel, The inorganic elements in nutrition, *J. Biol. Chem., 34*: 131–139 (1918).

24. J. Leroy, Necessite du magnesium sur la croissance de la souris, *Compt. Rond. Seance Soc. Biol., 94*: 431–433 (1926).

25. E. V. Tufts, D. M. Greenberg, The biochemistry of magnesium deficiency I. Chemical changes resulting from magensium deprivation, *J. Biol. Chem.*, *122*: 693–714 (1938).

26. E. V. Tufts, D. M. Greenberg, The biochemistry of magnesium deficiency II. The minimum magnesium requirement for growth, gestation and lactation, and the effect of the dietary calcium level thereon, *J. Biol. Chem.*, *122*: 715–726 (1938).

27. J. K. Aikawa, *The Relationship of Magnesium to Disease in Domestic Animals and in Humans*, C. C. Thomas, Springfield, IL, 1971.

28. J. Duckworth, W. Godden, G. M. Warnock, The effect of acute magnesium deficiency on bone formation in rats, *Biochem. J.*, *34*: 97–108 (1940).

29. F. W. Heaton, The parathyroid glands and magnesium metabolism in the rat, *Clin. Sci.*, *28*: 543–553 (1965).

30. E. Watchorn, R. A. McCance, Subacute magnesium deficiency in rats, *Biochem. J.*, *31*: 1379–1389 (1937).

31. H. A. Heggtveit, L. Herman, R. K. Mishra, Cardiac necrosis and calcification in experimental magnesium deficiency. A light and electron microscopic study, *Am. J. Pathol.*, *45*: 757–782 (1964).

32. E. R. Orent, H. D. Kruse, E. V. McCollum, Studies on magnesium deficiency in animals. VI. Chemical changes in the bone with associated blood changes resulting from magnesium deprivation, *J. Biol. Chem.*, *106*: 573–593 (1934).

33. C. Lida, K. Fuwa, W. E. C. Wacker, A general method for magnesium analysis in biological materials by atomic absorption spectroscopy, *Anal. Biochem.*, *18*: 18–26 (1967).

34. J. K. Aikawa, Gastrointestinal absorption of Mg^{28} in rabbits, *Proc. Soc. Exp. Biol. Med.*, *100*: 293–295 (1959).

35. J. K. Aikawa, E. L. Rhoades, G. S. Gordon, Urinary and fecal excretion of orally administered Mg^{28}, *Proc. Soc. Exp. Biol. Med.*, *98*: 29–31 (1958).

36. J. K. Aikawa, G. S. Gordon, E. L. Rhoades, Magnesium metabolism in human beings: Studies with Mg^{28}, *J. Appl. Physiol.*, *15*: 503–507 (1960).

37. L. V. Avioli, M. Berman, Mg^{28} kinetics in man, *J. Physiol.*, *226*: 653–674 (1966).

38. A. D. Care, D. B. Ross, Gastro-intestinal absorption of ^{28}Mg in sheep, *Proc. Nutr. Soc.*, *21*: ix (1962).

39. A. C. Field, Studies on magnesium in ruminant nutrition. Distribution of Mg^{28} in the gastro-intestinal tract and tissues of sheep, *Br. J. Nutr.*, *15*: 349–359 (1961).

40. A. C. Field, B. S. W. Smith, Effect of magnesium deficiency on the uptake of Mg^{28} by the tissues in mature rats, *Br. J. Nutr.*, *18*: 103–112 (1964).

41. L. A. Graham, J. J. Caesar, A. S. V. Burgen, Gastroinestinal absorption and excretion of Mg^{28} in man, *Metabolism*, *9*: 646–659 (1960).

42. D. M. McAleese, M. C. Bell, R. M. Forbes, Mg^{28} studies in lambs. *J. Nutr.*, *74*: 505–514.

43. L. Silver, J. S. Robertson, L. K. Dahl, Magnesium turnover in the human studied with Mg^{28}, *J. Clin. Invest.*, *39*: 420–425 (1960).

44. A. Dimich, J. E. Rizek, S. Wallach, W. Silver, Magnesium transport in patients with thyroid disease, *J. Clin. Endocrin. Metab.*, *26*: 1081–1092 (1966).

45. V. E. Currie, F. W. Lengemann, R. A. Wentworth, R. Schwartz, Stable ^{26}Mg as an in vivo tracer in investigation of magnesium utilization, *Int. J. Nucl. Med. Biol.*, *2*: 159–164 (1975).

46. S. Chandra, G. H. Morrison, Imaging elemental distribution and ion transport in cultured cells with ion microscopy, *Science*, *228*: 1543–1544 (1985).

47. G. O. Ramseyer, J. T. Brenna, G. H. Morrison, R. Schwartz, Elemental isotopic abundance determination of Magnesium in biological materials by secondary ion mass spectrometry, *Anal. Chem.*, *56*: 402–407 (1984).

48. M. Walser, Magnesium metabolism, *Rev. Physiol. Biochem. Exp. Pharmacol.*, *59*: 185–296 (1967).

49. R. Schwartz, N. A. Woodcock, J. D. Blakeley, F. L. Wang, E. A. Khairallah, Effect of magnesium deficiency in growing rats on synthesis of liver proteins and serum albumin, *J. Nutr.*, *100*: 123–128 (1970).

50. J. L. Greger, R. Schwartz, Cellular changes in the exocrine pancreas of rats fed two levels of magnesium and protein, *J. Nutr.*, *104*: 1610–1617 (1974).

51. H. F. Chou, R. Schwartz, L. Krook, R. H. Wasserman, Intestinal calcium absorption and bone morphology in magnesium deficient chicks, *Cornell Vet.*, *69*: 88–103 (1979).

52. J. L. Greger, R. Schwartz, Subcellular changes in protein metabolism and magnesium content of the pancreas in rats fed two levels of dietary protein and magnesium, *J. Nutr.*, *104*: 1618–1629 (1974).

53. R. L. Griswold, N. Pace, The Intracellular distribution of metal ions in rat liver, *Exp. Cell. Res.*, *11*: 362–367 (1956).

54. R. E. Thiers, B. L. Vallee, Distribution of metals in subcellular fractions of rat liver, *J. Biol. Chem.*, *226*: 911–920 (1957).

55. E. Urban, H. P. Schedl, Net movements of magnesium and calcium in the rat small intestine in vivo, *Proc. Soc. Exp. Biol. Med.*, *132*: 1110–1113 (1969).

56. H. Rasmussen, D. P. B. Goodman, N. Friedmann, J. E. Allen, K. Kurokawa, Ionic control of metabolism, in *Handbook of Physiology* (G. D. Aurbach, ed.), vol. 7, American Physiological Society, Washington, D.C., 1976, pp. 226–264.

57. F. W. Heaton, The kidney and magnesium homeostasis, *Ann. NY Acad. Sci.*, *162*: 775–785 (1969).

58. M. E. Shils, Experimental human magnesium depletion, *Am. J. Clin. Nutr.*, *15*: 133–143 (1964).

59. M. E. Shils, Magnesium, in *Present Knowledge of Nutrition*, 5th Edition, Nutr. Rev. Nutrition Foundation, Inc. Washington, D.C., 1984, pp. 422–438.

60. A. C. Alfrey, N. L. Miller, R. Trow, Effect of age and magnesium depletion on bone magnesium pools in rats, *J. Clin. Invest.*, *54*: 1074–1081 (1974).

61. I. MacIntyre, S. Hannah, C. C. Booth, A. E. Read, Intracellular magnesium deficiency in man, *Clin. Sci.*, *20*: 297–305 (1961).

62. D. B. Lyon, Studies on the Solubility of Ca, Mg, Zn and Cu in cereal products, *Am. J. Clin. Nutr.*, *39*: 190–195 (1984).

63. D. R. Williams, *The Metals of Life: The Solution Chemistry of Metal Ions in Biological Systems*, Van Nostrand Reinhold Co. New York, 1971, pp. 8–22.

64. R. Schwartz, D. L. Grunes, R. A. Wentworth, E. H. Wien, Magnesium absorption from leafy vegetables intrinsically labelled with the stable isotope ^{26}Mg, *J. Nutr.*, *110*: 1365–1371 (1980).

65. R. Schwartz, H. Spencer, J. E. J. Walsh, Magnesium absorption in human subjects from leafy vegetables intrinsically labeled with ^{26}Mg, *Am. J. Clin. Nutr.*, *39*: 571–576 (1984).

66. R. A. McCance, E. M. Widdowson, H. Lehmann, The effect of protein intake on the absorption of calcium and magnesium, *Biochem. J.*, *36*: 686–691 (1942).

67. J. L. Kelsay, K. M. Behall, E. S. Prather, Effect of fiber from fruits and vegetables on metabolic responses of human subjects, *Am. J. Clin. Nutr.*, *32*: 1876–1880 (1979).

68. R. A. McCance, E. M. Widdowson, The fate of calcium and magnesium after intravenous administration to normal persons, *Biochem. J.*, *33*: 523–529 (1939).

69. B. E. C. Nordin, (Ed). *Calcium, Phosphate and Magnesium Metabolism*, Churchill Livingstone, New York, pp. 97–112.

70. J. G. Reinhold, G. B. Faradji, P. Abadi, F. Ismail-Beigi, Decreased absorption of calcium, magnesium, zinc and phosphorus by humans due to increased fiber and phosphorus consumption as wheat bread, *J. Nutr.*, *106*: 493–503 (1976).

71. P. Roth, E. Werner, Intestinal absorption of magnesium in man, *Int. J. Appl. Rad. Isotopes*, *30*: 523–526 (1979).

72. M. O. J. Olson, Metal ion effects on nuclear protein phosphorylation, in *Metal Ions in Genetic Information Transfer* (G. L. Eichhorn, L. G. Marzilli, Eds.), Elsevier/North-Holland, New York, 1981 pp. 167–191.

73. P. Brannan, P. Vergne-Marini,, C. Y. C. Pak, A. R. Hull, J. S. Fordtran, Magnesium absorption in the human small intestine. Results in normal subjects, patients with chronic renal

disease, and patients with absorptive hypercalciuria, *J. Clin. Invest.*, *57*: 1412–1418 (1976).

74. R. Meneely, L. Lepper, F. K. Gishan, Intestinal maturation: in vivo magnesium transport, *Pediat. Res.*, *16*: 295–298 (1982).

75. J. G. Chutkow, Sites of magnesium absorption and excretion in the intestinal tract of the rat, *J. Lab. Clin. Med.*, *63*: 71–79 (1964).

76. B. G. Danielson, G. Johansson, B. Jung, S. Ljunghall, H. Lundquist, P. Malmborg, Gastrointestinal magnesium absorption. Kinetic studies with [28]Mg and a simple method for determination of fractional absorption, *Min. Electr. Metab.*, *2*: 116–123 (1979).

77. T. A. M. Aldor, E. W. Moore, Magnesium absorption by everted sacs of rat intestine and colon, *Gastroenterology*, *59*: 745–753 (1970).

78. J. Z. Hendrix, N. W. Alcock, R. M. Archibald, Competition between calcium, strontium and magnesium for absorption in the isolated rat intestine, *Clin. Chem.*, *9*: 734–744 (1963).

79. D. B. Ross, Influence of sodium on the transport of magnesium across the intestinal wall of the rat in vitro, *Nature*, *189*: 840–841 (1961).

80. D. B. Ross, In vitro studies on the transport of magnesium across the intestinal wall of the rat, *J. Physiol*, *160*: 417–428 (1962).

80a. D. B. Ross, E. D. Care, The movement of [28]Mg across the cell wall of guinea pig small intestine in vitro, *Biochem. J.*, *82*: 21 P (1962).

81. J. Behar, Effect of calcium on mangesium absorption, *Am. J. Physiol.*, *229*: 1590–1595 (1975).

82. J. Behar, Magnesium absorption by the rat ileum and colon, *Am. J. Physiol*, *227*: 334–340 (1974).

83. A. D. Care, A. Th. Van't Clooster, In vitro transport of magnesium and other cations across the wall of the gastrointestinal tract of sheep, *J. Physiol*, *177*: 174–191 (1965).

84. C. Schmulen, M. Lerman, C. Y. C. Pak, J. Zerwekh, P. Vergne-Marini, S. Morawski, J. S. Fordtran, Effect of 1,25-dihydroxy-vitamin D_3 therapy on intestinal absorption of magnesium in patients with chronic renal disease, *Am. J. Physiol.*, *238*: G349–352 (1980).

85. D. R. Wilz, R. W. Gray, J. H. Dominguez, J. Lemann Jr., Plasma 1,25 $(OH)_2$ vitamin D concentrations and net intestinal calcium, phosphate and magnesium absorption in humans, *Am. J. Clin. Nutr.*, *32*: 2052–2060 (1979).

86. P. J. Milla, P. J. Aggett, O. H. Wolff, J. T. Harries, Studies in primary hypomagnesemia evidence for defective carrier-mediated small intestinal transport of magnesium, *Gut*, *20*: 1028–1033 (1979).

87. A. R. Portis Jr., H. W. Heldt, Light dependent changes of the Mg^{2+} concentration in the stroma in relation to the Mg^{2+} depen-

dency of CO_2 fixation in intact chloroplasts, *Biochim. Biophys. Acta, 449*: 434–446 (1976).

88. M. E. Shils, Magnesium, calcium and parathyroid hormone interactions, *Ann. N.Y. Acad. Sci., 355*: 165–180 (1980).

89. M. G. Brunette, M. D. Crochet, A microinjection study of nephron permeability to calcium and magnesium, *Am. J. Physiol., 221*: 1442–1448 (1975).

90. M. G. Brunette, N. Vignault, S. Carriere, Micropuncture study of magnesium transport along the nephron in the young rat, *Am. J. Physiol., 227*: 891–896 (1974).

91. F. Morel, N. Roinel, C. LeCrimelle, Electron probe analysis of tubular fluid composition, *Nephron, 6*: 350–364 (1969).

92. G. A. Quamme, Effect of intraluminal furosemide on calcium and magnesium transport in the rat nephron, *Am. J. Physiol., 241*: F340–347 (1981).

93. G. A. Quamme, J. H. Dirks, Intraluminal and contraluminal magnesium on magnesium and calcium transfer in the rat nephron, *Am. J. Physiol., 238*: 198 (1980).

94. G. A. Quamme, J. H. Dirks, Magnesium transport in the nephron, *Am. J. Physiol., 239*: F393–401 (1980).

95. G. A. Quamme, C. M. Smith, Magnesium transport in the proximal straight tubule of the rabbit, *Am. J. Physiol., 246*: F544–550 (1984).

96. N. L. M. Wong, J. H. Dirks, G. A. Quamme, Tubular reabsorptive capacity for magnesium in the dog kidney, *Am. J. Physiol., 244*: F78–83 (1983).

97. Q. M. Yi, K. P. Wong, The effects of magnesium ions on the hydroynamic shape, conformation, and stability of the ribosomal 235 RNA from *E. Coli, Biochem. Biophys. Res. Comm., 104*: 733–729 (1982).

98. J. E. Jones, R. Schwartz, L. Krook, Calcium homeostasis and bone pathology in magnesium-dificient rats, *Calc. Tiss. Int., 31*: 231–238 (1980).

99. S. Breibart, J. S. Lee, A. McCoord, G. Forbes, Relation of age to radiomagnesium in bone, *Proc. Soc. Exp. Biol. Med., 105*: 361–363 (1960).

100. F. W. Neumann, F. J. Mulryan, Synthetic hydroxyapatite crystals. IV. Magnesium incorporation, *Calc. Tiss. Res., 7*: 133–138 (1971).

101. L. Martindale, F. W. Heaton, The relationship between skeletal and extracellular fluid magnesium in vitro, *Biochem. J., 97*: 440–443 (1965).

102. A. C. Alfrey, N. L. Miller, D. Butkus, Evaluation of body magnesium stores, *J. Lab. Clin. Med., 84*: 153–162 (1974).

103. D. Mauzerall, Why chlorophyll?, *Ann. NY Acad. Sci., 206*: 483–494 (1973).

104. D. E. Metzler, *Biochemistry. The Chemical Reactions of Living Cells*, Academic Press, New York, 1977, p. 387.

105. G. Girault, J. M. Galmiche, C. Lemaire, Binding and exchange of nucleotides on the chloroplast coupling factor CF1: The role of magnesium, *Eur. J. Biochem.*, *128*: 405–411 (1982).

106. B. M. Henkin, K. Sauer, Magnesium effects on chloroplast photosystem II fluorescence and photochemistry, *Photochem. Photobiol.*, *26*: 277–286 (1977).

107. G. H. Krause, Light-induced movement of magnesium ions in intake chloroplasts, *Biochim. Biophys. Acta*, *460*: 500–510 (1977).

108. W. E. C. Wacker, J. K. Aikawa, C. K. Davis, D. J. Horvath, W. L. Lindsay, in *Geochemistry and the Environment*, (W. Hertz, Ed.), vol. 2. National Academy of Science, Washington, D.C., 1977.

109. G. L. Eichhorn, The effects of metal ions on the structure and function of nucleic acids, in *Metal Ions in Genetic Information Transfer*, (G. L. Eichhorn, L. G. Marzilli, eds.), Elsevier/North-Holland, New York, 1981, pp. 1–46.

110. M. Grunberg-Manago, G. Hui Bon Hoa, P. Douzow, A. Wishnia, Cation control of equilibrium and rate processes in initiation of protein synthesis, in *Metal Ions in Genetic Information Transfer*, (G. L. Eichhorn, L. G. Marzilli, Eds.), Elsevier/North-Holland, New York, 1981, pp. 193–232.

111. L. A. Loeb, A. S. Mildvan, The role of metal ions in the fidelity of DNA and RNA synthesis, in *Metal Ions in Genetic Information Transfer* (G. L. Eichhorn, L. G. Marzilli, Eds.), Elsevier/North-Holland, New York, 1981, pp. 125–142.

112. A. Masini, D. Ceccarelli-Stanzani, U. Muscatelli, Effect of the external Magnesium ion concentration on functional steady-states of isolated rat liver mitochondria, *IRCS Med. Sci: Biochem. Cell Membrane Biol. Metab. Nutr.*, *9*: 769–770 (1981).

113. A. S. Mildvan, L. A. Loeb, The role of metal ions in the mechanisms of DNA and RNA polymerases, in *Metal Ions in Genetic Information Transfer*, (G. L. Eichhorn, L. G. Marzilli, eds.), Elsevier/North-Holland, New York, 1981, pp. 103–123.

114. G. L. Eichhorn, Complexes of polynnucleotides in nucleic acids, in *Inorganic Biochemistry*, G. L. Eichhorn, ed.), Elsevier Scientific Publishing Co., New York, 1973, pp. 1191–1209.

115. E. Gerlo, W. Freist, J. Charlier, Arginyl-tRNA synthetase from *Escherichia coli* K_{12}: Specificity with regard to ATP analogs and their magnesium complexes, *Hoppe-Seyler's Zeitschr. Physiol. Chem.*, *363*: 365–373 (1982).

116. K. W. Rhee, R. O. Potts, C. C. Wang, M. J. Fournier, N. C. Ford Jr., Effects of magnesium and ionic strength on the diffusion and charge properties of several single tRNA species, *Nucl. Acids Res.*, *9*: 2411–2420 (1981).

117. B. L. Black, J. M. McDonald, L. Jarett, Characterization of

Mg^{2+}- and (Ca^{2+} + Mg^{2+})- ATPase activity in adipocyte endoplasmic reticulum, *Arch. Biochem. Biophys.*, *199*: 92–102 (1980).

118. P. Marche, S. Koutouzov, P. Meyer, Metabolism of phosphoinositides in the rat erythrocyte membrane: A reappraisal of the effect of magnesium on the ^{32}P incorporation into polyphosphoinositides, *Biochim. Biophys. Acta*, *710*: 332–340 (1982).

119. T. Kornberg, M. L. Gefter, Purification and DNA synthesis in cell-free extracts: Properties of DNA polymerase II, *Proc. Natl. Acad. Sci. USA*, *68*: 761–764 (1971).

120. F. Ramirez, J. F. Marecek, Coordination of magnesium with adenosine 5'-diphosphate and triphosphate, *Biochim. Biophys. Acta*, *589*: 21–29 (1980).

121. K. K. Shukla, H. M. Levy, F. Ramirez, J. J. Marecek, B. McKeever, S. S. Margossian, Oxygen-exchange studies on the pathways for magnesium adenosine 5'-triphosphate hydrolysis by actomyosin, *Biochemistry*, *22*: 4822–4830 (1983).

122. K. K. Shukla, H. M. Levy, F. Ramirez, J. F. Marecek, Evidence from oxygen exchange studies that the two heads of myosin are functionally different, *J. Biol. Chem.*, *259*: 5423–5429 (1984).

123. R. H. Smith, Calcium and magnesium metabolism in calves. IV. Bone composition in magnesium deficiency and control of plasma magnesium, *Biochem. J.*, *71*: 609–614 (1959).

124. P. J. England, R. H. Denton, P. J. Randle, The influence of magnesium ions and other bivalent metal ions on the aconitase equilibrium and its bearing on the binding of magnesium ions by citrate in rat heart, *Biochem. J.*, *105*: 32c–33c (1967).

125. I. Rose, The state of magnesium in cells as estimated from the adenylate kinase equilibrium, *Proc. Natl. Acad. Sci.*, *61*: 1079–1086 (1968).

126. D. Veloso, R. W. Gwynn, M. Oskersson, R. L. Veech, The concentrations of free and bound magnesium in rat tissues, *J. Biol. Chem.*, *248*: 4811–4819 (1973).

127. H. Rubin, Central role for magnesium in coordinate control of metabolism and growth in animal cells, *Proc. Natl. Acad. Sci. USA*, *72*: 3551–3555 (1975).

128. H. Rubin, T. Koide, Mutual potentiation by magnesium and calcium of growth in animal cells, *Proc. Natl. Acad. Sci. USA*, *75*: 4379–4383 (1976).

129. J. H. Duffus, L. J. Patterson, The cell cycle in the fission yeast schizosaccharomyces pombe: Changes in activity of magnesium-dependent ATPase and in total internal magnesium in relation to cell division, *Z. Allg. Mikrobiol.*, *14*: 727–729 (1974).

130. G. H. Walker, J. H. Duffus, Magnesium as the fundamental regulator of the cell cycle, *Magnesium*, *2*: 1–16 (1983).

131. E. H. Back, R. D. Montgomery, E. E. Ward, Neurological manifestations of magnesium deficiency in infantile gastroenteritis and malnutrition, *Arch. Dis. Child.*, *37*: 106–109 (1962).

132. B. A. Barnes, O. Cope, T. Harrison, Magnesium conservation in the human being on a low magnesium diet, *J. Clin. Invest.*, *37*: 430–440 (1958).

133. J. G. Chutkow, J. D. Grabow, Clinical and chemical correlations in magnesium-deprivation encephalopathy of young rats, *Am. J. Physiol.*, *223*: 1407–1414 (1972).

134. P. Bois, Effect of magnesium deficiency on mast cells and urinary histamine in rats, *Br. J. Exp. Pathol.*, *44*: 151–155 (1963).

134b. H. Gitelman, L. Welt, Magnesium deficiency, *Ann. Rev. Med.*, *20*: 233–242 (1969).

135. J. A. F. Rook, Spontaneous and induced magnesium deficiency in ruminants, *Ann. NY Acad. Sci.*, *162*: 727–731 (1969).

136. B. Sjollema, L. Seekles, The magnesium content of the blood, particularly in tetany, *Klin. Wochenschr.*, *11*: 989–990 (1932).

137. F. L. Lakshmanan, R. B. Rao, W. W. Kim, J. L. Kelsay, Magnesium intakes, balances and blood levels of adults consuming self-selected diets, *Am. J. Clin. Nutr.*, *40*: 1380–1389 (1984).

138. B. L. Vallee, W. E. C. Wacker, D. D. Werner, The magnesium deficiency tetany syndrome in man, *New Engl. J. Med.*, *262*: 155–161 (1960).

139. J. G. Chutkow, S. Meyers, Chemical changes in the cerebrospinal fluid and brain in magnesium deficiency, *Neurology*, *18*: 963–974 (1968).

140. R. Heiperts, K. Eickhoff, K. H. Karters, Magnesium and inorganic phosphate content in CSF related to blood brain barrier function in neurological disease, *J. Neurol. Sci.*, *40*: 87–96 (1979).

141. D. J. Reed, M. H. Ten, The role of the cat choroid plexus in regulating cerebrospinal fluid magnesium, *J. Physiol.*, *281*: 477–485 (1978).

142. E. Watchorn, R. A. McCance, Inorganic constituents of cerebrospinal fluid. II. The ultrafiltration of calcium and magnesium from human sera, *Biochem. J.*, *26*: 54–74 (1932).

143. W. W. Oppelt, I. MacIntyre, D. P. Rall, Magnesium exchange between blood and cerebrospinal fluid, *Am. J. Physiol.*, *205*: 959–962 (1963).

144. P. F. Baker, Regulation of intracellular Ca and Mg in squid axons, *Fed. Proc.*, *35*: 2589–2595 (1976).

145. J. J. Boullin, The action of extracellular cations on the release of sympathetic transmittor from peripheral nerves, *J. Physiol. (London)*, *189*: 85–99 (1967).

146. N. W. Alcock, M. E. Shils, Comparison of magnesium deficiency in the rat and mouse, *Proc. Soc. Exp. Biol. Med.*, *146*: 137–141 (1974).

147. J. G. Chutkow, Studies on the metabolism of magnesium in the magnesium deficient rat, *J. Lab. Clin. Med.*, *65*: 912–926 (1965).

148. J. E. Welsh, R. Schwartz, L. Krook, Bone pathology and parathyroid gland activity in hypocalcemic magnesium deficient chicks, *J. Nutr.*, *111*: 514–524 (1981).

149. N. D. Grace, B. L. Odell, Relation of polysome structure to ribonuclease and ribonuclease inhibitor activities in the livers of magnesium deficient guinea pigs, *Can. J. Biochem.*, *48*: 21–26 (1970).

150. B. H. Altura, B. T. Altura, A. Grebrewold, H. Ising, T. Gunther, Magnesium deficiency and hypertension: correlation between magnesium deficient diets and microcirculatory changes in situ, *Science*, *238*: 1315–1317 (1984).

151. R. P. Breitenbach, W. A. Gounerman, W. L. Erfling, C. S. Anast, Dietary magnesium, calcium homeostasis and parathyroid gland activity of chickens, *Am. J. Physiol.*, *225*: 12–17 (1973).

152. J. Levi, S. G. Massry, J. W. Boburn, F. Leach, C. R. Kleeman, Hypocalcemia in magnesium depleted dogs: Evidence for reduced responsiveness to parathyroid hormone and relative failure of parathyroid gland function, *Metabolism*, *23*: 323–335 (1974).

153. R. Didier, E. Gueux, Y. Rassiguier, Magnesium deficiency in the Japanese quail, *Comp. Biochem. Physiol.*, *79*: 223–227 (1984).

154. M. J. Dunn, Magnesium depletion in the Rhesus monkey: Induction of magnesium dependent hypocalcemia, *Clin. Sci.*, *41*: 333–344 (1971).

155. W. Menaker, Influence of protein intake on magnesium requirement during protein synthesis, *Proc. Soc. Exp. Biol. Med.*, *85*: 149–151 (1954).

156. J. J. Vitale, P. L. White, M. Nakamura, D. M. Hegsted, N. Zamcheck, E. E. Hellerstein, Interrelationships between experimental hypercholesteremia, magnesium requirement, and experimental atherosclerosis, *J. Lab. Clin. Med.*, *53*: 433–441 (1957).

157. S. L. Krauter, R. Schwartz, Blood and mast cell histamine levels in magnesium-deficient rats, *J. Nutr.*, *110*: 851–858, *Chem.*, *96*: 519–539 (1980).

158. G. M. Haas, G. H. Laing, R. M. Galt, Induction of acute lymphomaleukemia in rats by deprivation of magnesium was prevented by dietary liver powder procured from normal but not from magnesium deficient rats, *Magnesium*, *1*: 49–56 (1982).

159. B. J. Hunt, L. F. Belanger, Localized multiform sub-periosteal hyperplasia and generalized osteomyelosclerosis in magnesium deficient rats, *Calc. Tiss. Res.*, *9*: 17–27 (1972).

160. J. H. Mirra, N. A. Alcock, M. E. Shils, P. Tannenbaum, Effects of calcium and magnesium deficiencies on rat skeletal development and parathyroid gland area, *Magnesium*, *1*: 16–33 (1982).

161. E. D. Eanes, S. L. Rattner, The effect of magnesium on apatite

formation in seeded sypersaturated solutions at pH 7.4, *J. Dent. Res.*, *60*: 1719–1723 (1981).

162. T. P. Feenstra, J. Hop, P. L. deBruyn, The influence of small amounts of magnesium on the formation of calcium phosphates in moderately supersaturated solutions, *J. Colloid Interface Sci.*, *83*: 583–588 (1981).

163. C. Y. C. Pak, E. C. Diller, Ionic interactions with bone mineral. V. Effect of Mg^{2+}, Citrate $^{3-}$, F^{-}, and SO_4^{2-} on the solubility, dissolution and growth of bone mineral, *Calc. Tiss. Res.*, *4*: 69–77 (1969).

164. J. J. Freitag, K. J. Martin, M. B. Conrades, E. Bellorin-Font, S. Teitelbaum, S. Klahr, Evidence for skeletal resistance to parathyroid hormone in magnesium deficiency. Studies in isolated perfused bone, *J. Clin. Invest.*, *64*: 1238–1244 (1979).

165. C. J. Johnson, D. P. Peterson, E. K. Smith, Myocardial tissue concentrations of magnesium and potassium in men dying suddenly from Ischemic heart disease, *Am. J. Clin. Nutr.*, *32*: 967–970 (1970).

166. S. Ralston, I. T. Boyle, R. A. Cowan, G. P. Crean, A. Jenkins, W. S. Thomson, PTH and vitamin D responses during treatment of hypomagnesemic Hypoparathyroidism, *Acta Endocrinol.*, *103*: 535–538 (1983).

167. J. F. Habener, J. T. Potts Jr., Relative effectiveness of magnesium and calcium on the secretion and biosynthesis of parathyroid hormone in vitro, *Endocrinology*, *98*: 197–202 (1976).

168. P. Mennes, R. Rosenbaum, K. Martin, E. Slatopolski, Hypomagnesemia and impaired parathyroid hormone secretion in chronic renal disease, *Ann. Int. Med.*, *88*: 206–209 (1978).

169. S. R. Wilkinson, J. A. Stuedemann, Tetany hazard of grass as affected by fertilization with nitrogen, potassium, or poultry litter and methods of grass tetany prevention, in *Grass Tetany* (V. V. Rendig, D. L. Grunes, Eds.), American Society for Agronomy, Madison, WI, 1979, pp. 93–121.

170. R. H. Smith, A. B. McAllan, Binding of magnesium and calcium in the contents of the small intestine of the calf, *Br. J. Nutr.*, *29*: 703–718 (1966).

171. R. Burau, P. R. Stout, Trans-aconitic acid in range grasses in early spring, *Science*, *150*: 766–767 (1965).

172. J. B. Russell, Enrichment and isolation of rumen bacteria that reduce trans-aconitic acid to tricarballylic acid, *Appl. Environ. Microbiol.*, *49*: 120–126 (1985).

173. J. Devane, M. P. Ryan, Diuretics and magnesium excretion, *Magnesium-Bulletin*, *3*: 122–125 (1981).

174. P. Fourman, D. B. Morgan, Chronic magnesium deficiency, *Proc. Nutr. Soc.*, *21*: 34–41 (1962).

175. P. Lim, E. Jacob, Magnesium status of alcoholic patients, *Metabolism*, *21*: 1045–1051 (1972).

176. R. D. Montgomery, Magnesium balance studies in marasmic kwashiorkor, *J. Pediat.*, *59*: 119–123 (1961).
177. R. G. King, S. W. Stanbury, Magnesium metabolism in primary hyperparathyroidism, *Clin. Sci.*, *39*: 281–303 (1970).
178. M. G. Fitzgerald, P. Fourman, An experimental study of magnesium deficiency in man, *Clin. Sci.*, *15*: 635–647 (1956).
179. P. A. Byrne, J. L. Caddell, The magnesium load test: II. Correlation of clinical and laboratory data in neonates, *Clin. Pediatr.*, *14*: 460–465 (1975).
180. J. L. Caddell, R. Suskind, H. Sillup, R. E. Olson, II. Parenteral magnesium load evaluation of malnourished Thai children, *J. Pediatr.*, *83*: 129–135 (1973).
181. J. L. Caddell, N. Ratananon, P. Trangratapit, Parenteral magnesium load test in postpartum Thai women, *Am. J. Clin. Nutr.*, *26*: 612–615 (1973).
182. J. L. Caddell, F. L. Saier, C. A. Thomason, Parenteral magnesium load test in postpartum American women, *Am. J. Clin. Nutr.*, *28*: 1099–1104 (1975).
183. J. L. Caddell, P. A. Byrne, R. A. Triska, A. E. McElfresh, The magnesium load test: III. Correlation of clinical and laboratory data in infants from one to six months of age, *Clin. Pediatr.*, *14*: 478–484 (1975).
184. J. E. Jones, R. Manolo, E. B. Flink, Magnesium requirements in adults, *Am. J. Clin. Nutr.*, *20*: 632–635 (1967).
185. M. S. Seelig, The requirement of magnesium by the normal adult, *Am. J. Clin. Nutr.*, *14*: 342–390 (1964).
186. H. I. Chu, S. H. Lui, H. C. Hsu, H. C. Choa, S. H. Cheu, Calcium, phosphorus, nitrogen, and magnesium metabolism in normal young chinese adults, *Chinese Med.*, *59*: 1–33 (1941).
187. H. Collumbine, V. Basnayake, J. Lemottee, T. W. Wickramanayake, Mineral metabolism on rice diets, *Br. J. Nutr.*, *4*: 101–111 (1950).
188. A. R. P. Walker, F. W. Fox, J. T. Irving, Studies in human mineral metabolism, *Biochem. J.*, *42*: 452–462 (1948).
189. C. F. Consolazio, L. O. Matoush, R. A. Nelson, R. S. Harding, J. E. Canham, Excretion of sodium, potassium, magnesium and iron in human sweat and the relation of each to balance and requirements, *J. Nutr.*, *79*: 407–415 (1963).
190. R. Whang, J. Aikawa, T. Hameter, Routine serum magnesium determination: An unrecognized need. *Abst. 2nd. Int. Symposium on Magnesium*, Montreal, quoted by W. E. C. Wacker, *Magnesium and Man*, Harvard University Press, Cambridge, MA, 1980.
191. J. R. Marier, Quantitative factors regarding magnesium status in the modern-day-world, *Magnesium*, *1*: 3–15 (1982).

192. A. J. Johannesson, L. G. Raisz, Effects of low medium magnesium concentration on bone resorption in response to parathyroid hormone and 1,25-dihydroxy vitamin D in organ culture, *Endocrinology*, *113*, 2294–2298 (1983).
193. W. Menaker, I. S. Kleiner, Effect of deficiency of magnesium and other minerals on protein synthesis, *Proc. Soc. Exp. Biol. Med.*, *81*: 377–378 (1952).
194. E. M. Widdowson, J. W. T. Dickerson, Chemical composition of the body. in *Mineral Metabolism*, (C. L. Comar, F. Bronner, Eds.), Vol. 2, Pt. A, Academic Press, New York, 1964, pp. 1–247.
195. N. L. M. Wong, G. A. Quamme, J. H. Dirks, Effect of chlorthiazide on renal calcium and magnesium handling in the hamster, *Can. J. Physiol. Pharmacol.*, *60*: 1160–1165 (1982).
196. A. Walsh, The application of atomic absorption spectra to chemical analysis, *Spectrochim. Acta*, *7*: 108–117 (1955).

BIBLIOGRAPHY

G. L. Eichhorn, L. G. Marzilli, Eds., Elsevier/North-Holland, New York, 1981, pp. 167–191

Quoted by W. E. C. Wacker, *Magnesium and Man*, Harvard University Press, Cambridge, MA, 1980, pp. 11–51.

B. A. Hemsworth, J. F. Mitchell, The characteristics of acetyl choline release in the auditory cortex, *Br. J. Pharmacol.*, *36*: 161–170 (1969).

R. A. McCance, E. M. Widdowson, Mineral metabolism of healthy adults on white and brown bread dietaries, *J. Physiol.*, *101*: 44–85 (1942).

R. J. McCollister, E. B. Flink, M. D. Lewis, Urinary excretion of magnesium in man following the ingestion of ethanol, *Am. J. Clin. Nutr.*, *12*: 415–420 (1963).

E. R. Miller, D. E. Ullrey, C. L. Zutaut, B. V. Baltzer, D. A. Schmidt, J. A. Hoefer, R. W. Luecke, Magnesium requirement of the baby pig, *J. Nutr.*, *85*: 13–20 (1965).

L. Paunier, I. C. Radde, S. W. Kooh, D. Fraser, Primary hypomagnesemia with secondary hypocalcemia in an infant, *Pediatrics*, *41*: 385–402 (1965).

R. P. Ruben, The role of calcium in the release of neurotransmittor substances and hormones, *Pharmacol. Rev.*, *22*: 389–428 (1970).

C. R. Reddy, J. W. Coburn, D. L. Hartenbower, R. M. Friedler, A. S. Brickman S. Massry, J. Jowsey, Studies on mechanisms of hypocalcemia of magnesium depletion, *J. Clin. Invest.*, *52*: 3000–3010 (1973).

R. Schwartz, C. C. Giesecke, Mass spectrometry of a bolatile Mg chelate in the measurement of stable ^{26}Mg when used as a tracer, *Clin. Chim. Acta*, *97*: 1–8 (1979).

R. H. Smith, Importance of magnesium in the control of plasma calcium in the calf, *Nature*, *191*: 181–182 (1961).

F. L. Stuzman, D. S. Amatuzio, A study of serum and spinal fluid calcium and magnesium in humans, *Arch. Biochem.*, *39*: 271–275 (1952).

S. Wallach, J. E. Rizek, A. Dimick, N. Prasad, W. Silver, Magnesium transport in normal and uremic patients, *J. Clin. Endocrinol. Metab.*, *26*: 1069–1980 (1966).

5
Iron

DONALD B. THOMPSON / The Pennsylvania State University,
University Park, Pennsylvania

INTRODUCTION

Iron is the most abundant element on earth. In the earth's crust it
is second among metals only to aluminum. Some common igneous rocks
contain 5% iron; in soil the average concentration is 3.8% (1). On the
other hand, ocean water contains relatively little iron, between a few
tenths and 3 μg/L. The average residence time of this iron is 140
years (2), at the end of which time it is precipitated as sediment.
In our aerobic worl Fe(III) is the stable redox state, and insoluble
ferric oxide is the stable form (3). As a consequence, deficiency in
the midst of abundance is possible. A fully iron-replete 70-kg adult
male contains only 3—5 gm of iron, for an average concentration of
about 60 ppm.

Perhaps due to the difficulty in assimilating iron from food, the
human body has adopted an unusual strategy to maintain optimal iron
status. Body losses of absorbed iron are very small. Unlike many
nutrients, homeostasis seems to be regulated primarily by absorption,
not by excretion, even though the capacity for absorption is not
great. This unusual scheme may be a consequence of the solution
behavior of iron, which necessitates an intricate system to maneuver
even small quantities of iron through the body. There are two implications of the strategy of regulation by absorption: if deficiency
occurs it may be difficult to quickly overcome by diet alone, and if
overload occurs the ability to excrete the excess is limited. The two
problems can be either "tired blood" or "rusty livers" (4).

This chapter will present aspects of iron metabolism, the requirement for iron, food iron content, iron chemistry, processing effects on iron in food, and iron and immunity. Primary emphasis will be given to iron absorption, particularly to the lumenal behavior of nonheme iron.

IRON METABOLISM

Quantitative Distribution

Included in the 3—5 gm iron in the 70-kg adult male are nonfunctional iron stores, usually in the range of 500—1000 mg. Iron is stored as ferritin (about two-thirds of the total) or hemosiderin (5). Ferritin appears to be ubiquitous in cells (6). This 24-subunit protein is assembled in a hollow spherical structure, the size of which appears ideal to include polymerized iron in Fe(III) form. Hemosiderin has a more amorphous structure. It may represent a degraded form of ferritin (7). Most storage iron occurs in the liver, spleen, and bone (7). Consequently, determination of this storage iron is an invasive procedure. More recently it has been shown that minute quantities of ferritin are present in serum, and that this ferritin correlates well with iron stores determined invasively (5) or by iron absorption determination (8). Consequently, iron stores are now more conveniently estimated by serum ferritin determination.

Of the remaining functional iron the great majority is in the form of hemoglobin. The iron in the porphyrin complex of hemoglobin is able to associate reversibly with a sixth ligand, the oxygen molecule. This protein occurs in circulating red blood cells and serves to transport oxygen from the lungs to actively respiring tissues, where the released oxygen functions as the final electron acceptor in the electron transport chain.

A similar association of iron and oxygen occurs in myoglobin, a single peptide homologue of hemoglobin. This protein functions to store oxygen delivered to the tissues by hemoglobin. In the human the total body myoglobin iron is approximately one-eighth that of hemoglobin, or about 8% of the total body iron (7). Levels of this protein may be quite high in muscle tissue of diving animals, such as whales.

In both hemoglobin and myglobin the iron is maintained as Fe(II) in the living animal. In these proteins it is the ligand-binding behavior rather than the redox behavior that is functionally important.

About 3% of the functional iron is in the form of heme enzymes, including catalase and peroxidase (9). In these enzymes iron functions not as a stable binder of ligands but as a facile redox reactant. It appears that Fe(IV) and possibly even Fe(V) oxidation states may be involved in the reactions (10). Other heme enzymes, such as the cytochromes, also depend upon the redox behavior of iron, but in these cases the couple is of the more tame Fe(II)/Fe(III) variety. In the various cytochromes, the $E^{o'}$ values are modified by the surrounding protein to provide an appropriate electrochemical gradient for efficient electron transfer and ATP generation.

Another 3% of the functional iron is in the form of nonheme enzymes (9), including the iron-sulfur proteins and those enzymes that utilize iron as a cofactor. As with the cytochromes, the iron-sulfur proteins are involved in electron transport, with the redox couple based upon the iron present. Enzymes utilizing iron as a cofactor include xanthine oxidase, aconitase, and several of the amino acid hydroxylases.

The vast majority of iron in blood is associated with hemoglobin. The iron is maintained within this molecule within the erythrocyte until the cell is catabolized. A small amount of iron (about 3 mg) is present in the serum bound to a different protein, transferrin (11). This 80,000 MW glycoprotein is comprised of two peptide chains, each of which has one iron-binding site, which binds iron tenaciously. The binding constant has been estimated in the vicinity of 10^{20} M^{-1} or greater at physiological pH (12,13,14). As pointed out by Aisen and Brown (12), if $K = 10^{24}$ M^{-1}, one liter of plasma would contain less than one free ferric ion. It is in this form that iron is moved through the blood. Normally transferrin is about 35% saturated with iron at the specific iron-binding sites. Higher transferrin saturation may predispose to infection, whereas lower transferrin saturation may indicate iron deficiency.

Kinetics

Iron in hemoglobin remains within the red blood cell as part of the protein until the aged cell and its components are dismantled by the reticuloendothelial system. The estimated life span of these cells is 120 days. The iron released associates with transferrin and is delivered to the bone marrow, where nascent red blood cells incorporate the iron into new hemoglobin molecules. The numbers involved are impressive: about 2.5×10^6 erythrocytes enter the circulation each second, each containing about 280×10^6 molecules of hemoglobin (15). The efficiency of the recycling process is an important determinant of iron nutritional needs. In an adult male, 95% of the 18 mg iron needed for hemoglobin synthesis is obtained from degraded cells (16); therefore little additional iron must be obtained through the diet.

Balance

The adult male loses about 1 mg of iron per day (17,18), about two-thirds from the gut as sloughed epithelial cells and digestive secretions (unabsorbed dietary iron is not included here). The other one-third is lost primarily in urine and from the skin (18). This loss dictates the amount that must be obtained from the diet. Due to the high proportion of body iron in blood, loss of blood results in significant additional loss of iron. Consequently, women of childbearing age will lose almost twice the iron per day (1.8 mg) as men. Conversely, gain of blood iron through transfusions to remedy certain blood abnormalities may result in a significant excess of iron, leading to iron toxicity. It has recently been shown that use of oral contraceptives can lead to increased iron stores (19), possibly due to decreased menstrual blood loss.

Iron Deficiency

When dietary iron level or bioavailability becomes insufficient to meet the needs of a normal individual or one whose iron needs are elevated due to blood loss or other reasons, the body responds first by drawing upon iron stores. As stores become progressively depleted over time, the efficiency of dietary iron absorption increases (20,21). When this adjustment is not sufficient to remedy the situation, serum ferritin may drop below 12 mg/L, and essential iron in the tissues will decrease (20). Transferrin saturation may then drop to below 15%. Eventually, hemoglobin concentration will drop sufficiently that anemia may be diagnosed.

Beard and Finch (22) have recently pointed out an important distinction between the clinical and the public health definitions of iron deficiency. Clinical iron deficiency will be manifest as the anemia of iron deficiency. Anemia is defined not by reference to an arbitrary hemoglobin value, but by a hemoglobin increase as a result of iron therapy. The above authors emphasize that "anemia is not suitable as an initial procedure for the recognition of iron deficiency," because it is possible either for a physiologically adequate individual hemoglobin value to be low without deficiency or for an individual hemoglobin value to appear normal for a population in spite of that individual's deficiency.

Severe cases of iron deficiency would be identified by the accompanying anemia, whereas some less severe cases would be missed if hemoglobin were the only parameter determined (23). Beard and Finch (22) suggest determination of transferrin saturation, free erythrocyte protoporphyrin, or serum ferritin to evaluate iron deficiency in a population. As is obvious from the above considerations, estimates of iron deficiency in a population will depend upon what type of testing was performed. For this reason, it is difficult to estimate the magnitude of iron deficiency worldwide. Florentino and Guirriec (24) have discussed the prevalence of nutritional anemia in detail for several individual developing countries. Several authors point to iron deficiency as the most common nutritional deficiency worldwide (16) affecting as many as one billion individuals.

The functional implications of iron deficiency have recently been reviewed (25). These include decreased immunocompetence and increased infection, a decrease in physical work capacity, less efficient thermoregulation, and diminished mental development. Immunocompetence and infection will be discussed further. The decrease in physical work capacity may be understood not only by decreased delivery of oxygen due to anemia but also by suboptimal cellular function, particularly in the mitochondrial respiratory chain. These respiratory enzyme deficiencies may also be related to problems with thermoregulation, but additional factors are increased catecholamine levels and impaired conversion of T4 to T3 in the thyroid (25). It would appear that mental development and behavior are adversely influenced in deficiency, but such conclusions are difficult to draw with certainty at this point.

Iron Overload and Toxicity

Toxic effects of dietary iron are uncommon, at least partially because it is rare to encounter food with sufficient and sufficiently bioavailable iron to cause toxicity. On the other hand, it has been estimated that about 2000 cases per year of iron poisoning are observed in the United States, consisting primarily of children ingesting iron supplements (17). Other researchers have estimated 500 hospitalizations and several deaths per year (26). Death may occur within 4–6 hours due to severe necrotizing gastroenteritis (9).

Iron overload from dietary sources has been observed among Bantu tribesmen consuming beer brewed in iron kettles. The "Bantu siderosis" appears to result from this maize-sorghum beverage, which can contain 40 mg Fe/L. Moreover, this iron is present in a highly available form (27).

Prolonged treatment of certain anemias by transfusion, as noted above, may result in excessive body iron. Treatment with various iron chelating agents to remove iron has been attempted to minimize the consequences of the primary treatment (28).

Idiopathic hemochromatosis allows excess absorption of iron from foods and results in tissue damage due to iron deposition. Concern for individuals with this genetic disorder was one of the reasons an increased level of enrichment iron in bread, proposed by the FDA in 1970, was finally rejected in 1977 (29).

IRON REQUIREMENT

The amount of iron lost from the body of a healthy individual forms the basis for the amount of iron needed in the diet. In the 1980 RDAs, the average availability of food iron is assumed to be 10%. Thus the amount of iron required, multiplied by a factor of 10, forms the estimate of the RDA (17) as expressed in table form. The accompanying text recognizes the uncertainty in this assumption of 10% absorption, and it is suggested that an estimate of absorbable iron may be more meaningful (17), as proposed by Monsen et al. (30). For each meal total iron, heme iron, nonheme iron, ascorbic acid, and the amount of meat, poultry, and/or fish must be known. On this basis absorbable iron may be estimated.

The requirement varies according to age and sex. For the adult male 10 mg is the RDA based on losses estimated at 1 mg/day. For the female of reproductive age, the losses are estimated at 1.8 mg/day; the greater losses are the result of losses of 1 mg/day (over a month) due to menstrual bleeding. Thus the RDA is 18 mg. Based on the figure of 6 mg iron/1000 kcal in the average Western diet, this total may be difficult if not impossible for a woman to achieve without caloric excess. Furthermore, 10% of women lose 1.4 mg iron per day through menstruation, implying that their actual requirement would be 2.2 mg iron/day (11) and that dietary iron needed would be 22 mg. Periods of special concern regarding iron intake are late infancy (6 months to 2—3 years) during which the infant is increasingly dependent on dietary iron and for whom iron stores have dwindled (16), adolescence, particularly during the growth spurt (16), and pregnancy, especially the latter half of the pregnancy (16).

It should be emphasized that, based on the strong affect of diet on iron absorption, all numerical estimates of iron needed must be interpreted as guidelines (16).

IRON CONTENT OF FOODS

Native Iron

The recent emphasis on iron bioavailability detracts from the more mundane considerations of the amount of iron in foods. Nevertheless it should be borne in mind that, while iron content of an individual food is important, the iron content of the meal itself may be of overriding importance. Iron bioavailability, expressed on a percentage basis, gains practical meaning only when meal iron content is known.

From the perspective of its behavior in food systems and during digestion and absorption, iron may be classified as heme or nonheme, depending upon whether it is associated with porphyrin in the food. As will be discussed below, the proportion in each form is an important determinant of absorption. Heme iron occurs primarily in muscle foods. In the intact animal, the majority of the heme iron is in the form of hemoglobin; however, after bleeding the remaining heme iron is primarily present as myoglobin. Of total iron the proportion of heme iron in raw pork, lamb, and beef was 49, 57, and 62%, respectively (31). These values were based on 10, 16, and 26 µg total iron per gm tissue, respectively.

Table 1. Iron Content of Foods

Food	mg iron/100 kcal	mg iron/100 gm edible portion of food
Low Iron Content (<0.7 mg iron/100 kcal)		
Apples, raw	0.6	0.3
Bologna, frankfurters	0.6	1.8
Bread, white, unenriched	0.3	0.7
Egg whites	0.2	0.1
Lamb, loin chop, broiled	0.4	1.3
Milk	trace	trace
Potato chips	0.3	1.8
Rice, brown, cooked	0.4	0.5
Rice, white, unenriched, cooked	0.2	0.2
Sugar, white, granulated	0.1	0.1
Wheat flour, white, unenriched	0.2	0.8
Medium Iron Content (0.7–1.9 mg iron/100 kcal)		
Beef, ground, cooked	1.1	3.2
Beef, round, broiled	1.9	3.7
Bread, white, enriched	0.9	2.5
Bread, whole wheat	0.9	2.3
Chicken, dark, cooked	1.0	1.7
Chicken, white, cooked	0.8	1.3
Eggs, whole	1.4	2.3
Raisins, uncooked	1.2	3.5
Sugar, brown	0.9	3.4
Wheat flour, white, enriched	0.8	2.9
Wheat flour, whole grain	1.0	3.3
High Iron Content (>2.0 mg iron/100 kcal)		
Asparagus, cooked	3.0	0.6
Beans, green, cooked	2.4	0.6
Broccoli, cooked	3.1	0.8
Lettuce, romaine	7.8	1.4
Liver, calves, fried	5.4	14.2
Spinach, fresh, cooked	9.6	2.2
Wheat bran, commercially milled	7.0	14.9

Source: Ref. 2.

The iron in plant-derived foods is almost exclusively in nonheme form. Iron content of a variety of foods has been collected and grouped into foods of low, medium, and high iron content, normalized to an equal calorie basis (2). A selection of these values is shown in Table 1.

Several authors have observed that most Western diets contain 6–7 mg iron/1000 kcal, and it is on this basis that many women have difficulty achieving the RDA for iron at their optimal caloric intake.

Recently the results of the FDA's total diet study (1982–1984) appeared. Levels of dietary iron were below the RDA for various groups: infants (6–11 mos.), 79.3% of RDA; children (2 yr), 56.0%; girls (14–16 yr), 59.4%; and women (25–30 yr), 57.8% (32).

The iron content of foods may be influenced by food refinement and food fortification. As shown in Table 1, whole wheat flour contains about 33 ppm iron, whereas white flour contains only about 8 ppm. The difference is largely due to removal of bran and germ, at 150 and 94 ppm iron, respectively. Enriched white flour has about the same iron content as whole wheat, at 29 ppm.

Contamination Iron

The effect of food preparation on iron content is not shown in Table 1. Iron gained in preparation might be thought of as contamination iron. It might result from dirt, the water supply, industrial processing equipment, or home cooking utensils.

The amount of contamination iron in a food is difficult to determine. In one attempt, an in vitro approach was used (33), in which contamination iron was estimated as the proportion of the total iron that did not exchange with a tracer dose of ^{59}Fe. On this basis a variety of Asian meals, prepared after careful washing, contained between 4 and 50% of the total iron as contamination iron (33).

An earlier application of the same procedure (34) suggests that little contamination iron is present in a variety of Western meals. A hamburger meal, pizza, soup and bread, spaghetti in meat sauce, and two vegetarian meals varied from 93 to 107% exchangeable iron. Intentional contamination in the form of soil was variably exchangeable, casting doubt as to the reliability of this method to estimate contamination iron. Iron in red soil from Brazil appears inert with respect to exchange with the label, whereas iron in clays from Thailand appears about one-third exchangeable (34).

Iron in the water supply will vary considerably as to the water source. "Average" river water has been shown to contain 670 µg/L (0.67 ppm) (2). Analysis would be done after removal of particulates >0.45 µm in diameter. At usual pH's, the iron would be in the form of colloidal suspensions of ferric hydroxide. Between 0.5 and 10 ppm iron may occur under certain conditions in ground water with very low dissolved oxygen (2). At neutral pH and measurable dissolved oxygen, content would be much lower, in the ppb range.

The standard for the U.S. Public Health Service for drinking water is 0.3 ppm iron maximum, a standard which is generally achieved in the United States. Water with greater than 0.3 ppm iron will stain plumbing and laundry (2). It has been estimated that iron from drinking water might contribute as much as 10% of the daily intake in extreme cases (35).

The maximum contribution of contamination iron to the iron content of soy protein isolate processed by acid precipitation using water with 0.3 ppm iron has been estimated by this author to be 19 ppm. This estimate assumes that all iron from the water ends up in the product, very much an extreme-case assumption since the yield of solids may only be 35% from the original defatted soy (36). The actual proportion of soy protein isolate iron contributed by contamination is not known. Sources other than from the water used are of potential importance. Food-processing equipment in general may contribute to iron content of soy and other products.

Equipment used to prepare food has been shown to contribute considerable iron to food under certain conditions. The best-documented situation involves the previously mentioned iron overload which results from drinking beer brewed in iron containers. Derman et al. (27) mimicked the production of such beer and showed that soluble iron (that which would end up in the beer) could be as high as 89 mg/L.

Moore (37) compared iron contents of selected foods cooked in a cast-iron "Dutch oven" versus the same foods cooked in glass. Iron content increased for all foods, but dramatically so for spaghetti sauce and apple butter, which increased from 3.0 and 0.5 mg/100 gm to 87.5 and 52.5 mg/100 gm, respectively. Unfortunately, experimental details, particularly the condition and treatment of the cast-iron vessel, were not stated. A subsequent critique of this and other work (38) concluded conservatively that experiments had shown only that under certain conditions food iron levels could be increased by preparation in cast-iron cookware. Furthermore, this article, written in the midst of the controversy surrounding a proposed increase in bread enrichment levels, suggests that the often-cited shift away from cast-iron cookware may be less than overwhelming, based on national sales figures. Until very recently the issue of increased iron content during preparation in cast-iron vessels had been accepted uncritically, despite a lack of throughly documented investigation. A study describing both cooking conditions as well as conditioning and treatment of the cookware (cast-iron versus glass) has clearly demonstrated the magnitude of the iron content increase for a variety of foods (39). Reminiscent of Moore's (37) data, spaghetti sauce iron content was higher, but not as dramatically — 5.8 versus 0.7 mg/100 gm — for the sauce cooked in cast iron. Applesauce iron levels also were higher, 7.4 versus 0.3 mg/100 gm. Most foods tested contained more iron when cooked in cast iron, but the final levels were lower than Moore

(37) found. Brittin and Nossaman (39) suggest that a poorly conditioned pan and cover could be important determinants that might lead to higher iron content. Even with a well-conditioned vessel, the iron obtained from the pan can be significant. Such foods as fried egg, bacon, or fried chicken have their original iron contents doubled (39). The bioavailability of this iron may be very high, as recently observed (40,41).

Conceptually, fortification iron might be thought of as "intentional contamination." The bioavailability of this iron is variable and may be strongly affected by processing.

IRON CHEMISTRY

Solubility

In biological systems iron exists primarily in two redox states, Fe(II) and Fe(III) (10). Salts of both forms may be at least somewhat soluble at low pH, but as pH increases Fe^{+3} solubility decreases due to its strong tendency to hydrolyze to $Fe(OH)_3$ and polymerize (42). The tendency of Fe^{+2} ions to hydrolysis is much less strong: the K_{sp} of $FeOH_2$ and $FeOH_3$ is about 10^{-15} and 10^{-37}, respectively. The concentration of free, aqueous Fe^{+3} at neutral pH probably does not exceed 10^{-16} M, whereas 10^{-1} M Fe^{+2} may exist in free, aqueous form (4). Much has been made of this differing solubility behavior. However, it must be borne in mind that this behavior is observed in the absence of complex formation with other ligands (43). The concentration of iron maintained in soluble form will be stongly influenced by other ligands present.

Aisen (44) has pointed out that although the solubility of $Fe(OH)_2$ is far higher than $Fe(OH)_3$ at neutral pH, in most biological fluids near pH 7 at atmospheric oxygen tension, Fe^{+2} will be oxidized to Fe^{+3}. Thus the practical importance of Fe^{+2} solubility may be limited. The oxidation of Fe^{+2} may be encouraged by molecules which interact with Fe^{+3}. Fe^{+3} is the smaller of the two ions, interacting strongly with "hard" ligands such as oxygen, which may readily substitute for the water molecules in the hydration sphere of aquo-Fe^{+3} (43,45). The ability of Fe^{+3} to form soluble complexes may dominate the solubility behavior of iron in foods.

Even if sufficient free aqueous Fe^{+3} were present to form $Fe(OH)_3$ at neutral pH, the polymeric product would not necessarily precipitate immediately (46). Even highly hydrolyzed Fe(III) solutions contain a wide variety of colloidal particles of ill-defined and rapidly varying composition (43). Actual precipitation may take days, months, or even years (46).

Based on the above considerations, reference to K_{sp} of $Fe(OH)_2$ and $Fe(OH)_3$ may have little relevance to the solution behavior of

Fe^{+2} and Fe^{+3} in the complex food environment or the even more complex environment during digestion of the food.

Redox State

Review of Concepts

The spontaneous nature of a reaction is given by ΔG, the change in free energy. This quantity is described as

$$\Delta G = \Delta G^\circ + RT \ln Q \tag{1}$$

in which the free energy change for a reaction is shown to be a function of the standard free energy change (ΔG°), the temperature, and the reaction quotient. The change in free energy also represents the maximum electrochemical work that may be done:

$$\Delta G = - n F E \tag{2}$$

E in this equation represents the electrochemical potential. The Nernst equation may be derived from equations (1) and (2).

$$E = E^\circ - \frac{RT}{nF} \ln Q \tag{3}$$

E is the measured potential and E° is the potential when all reactants and products are in their standard states, i.e., at unit activity (or, less precisely, at 1 M concentration). For a particular reaction E° may be determined by reference to a table of standard reduction potentials for half-cell reactions. For the overall reaction the appropriate half-cells must be combined algebraically. For example, the autoxidation of Fe^{+2} by molecular oxygen is given by

$$4\ Fe^{+2}\ (aq) + O_2 + 4\ H^+ \rightleftarrows 4\ Fe^{+3}\ (aq) + 2\ H_2O \tag{4}$$

which may be seen to be a combination of two half-cells:

$$O_2 + 4\ H^+ + 4\ e^- \rightleftarrows 2\ H_2O \qquad E^\circ = +1.229 \tag{5}$$

$$Fe^{+3}\ (aq) + e^- \rightleftarrows Fe^{+2}\ (aq) \qquad E^\circ = +0.771 \tag{6}$$

The standard reduction potential for the reaction may be calculated to be +1.229 - (+0.771) = +0.458 volts. Whan all reactants and products are in the standard states $Q = 1$ in reaction (3), and therefore $E = E^\circ$. Under any other conditions the actual E will result from E° as modified by the term including Q. The figure +0.458 volts for the E° of reaction (4) implies the direction which will be spontaneous when all reactants and products originate in the standard state. Thus

Fe^{+3} will be produced at the expense of Fe^{+2}. The extent to which the reaction may proceed (if it proceeds at a measurable rate) will depend upon Q. In reaction (4) it is important to note that 1 M H^+ implies pH 0.

It should be recalled that the adjective *spontaneous* does not imply that the reaction will occur on a practically important time scale. Without a suitable mechanism a spontaneous reaction may be postponed indefinitely. In biological systems enzymes are often needed to provide suitable mechanisms for redox reactions.

Practical Considerations Regarding Application of Redox Potentials in Food Systems

Aisen (44) has pointed out that the mechanism for autoxidation is actually quite complex, with the following reactions postulated:

$$Fe^{+2} + OH^- \rightleftharpoons (FeOH)^+$$

$$O_2 + OH^- \rightleftharpoons (O_2OH)^- \qquad (7)$$

$$(FeOH)^+ + (O_2OH)^- \rightleftharpoons Fe(OH_2)^+ + O_2^-.$$

This sequence of events is inferred from the following rate law:

$$\frac{d\ Fe(II)}{dt} = K\ [Fe^{+2}]\ [O_2]\ [OH^-]^2 \qquad (8)$$

Because the standard state of H^+ in reaction (4) implies pH 0, one may reason that as pH increases there will be a greater tendency to maintain iron as Fe^{+2}. Such reasoning flies in the face of experience, which says that Fe^{+2} solutions rapidly autoxidize at neutral pH but not in acid. The resolution to this apparent paradox is provided by the mechanism of autoxidation, as described by Aisen (44). The second-order dependence on $[OH^-]$ implies a strong *kinetic* pH dependence. At pH 0, $[OH^-] = 10^{-14}$ M; thus an appropriate mechanism would seem to be absent.

A further consideration regarding reaction (4) is that it applies only to iron ions in free aqueous form. As noted above, Fe^{+3} in aqueous form is present at only vanishingly low concentration at neutral pH, with hydrolysis to polymers occurring when the concentration is exceeded. Thus reaction (4) will proceed to completion at neutral pH because a suitable mechanism is available and because a secondary reaction (hydrolysis) rapidly removes the Fe^{+3} (aq) product.

The effect of pH on the calculated $E°$ for reaction (4) may be quantitated. Biologists have found it convienient to use tables of $E°'$, the standard half-cell potentials corrected to pH 7. When a reaction involves H^+, $E°$ values are adjusted by 0.42 volts. $E°'$ for equation (4) would be calculated as +0.81 - (+0.771) = +0.04 volts. A similar correction is also necessary when a half-cell reaction involves OH^-, since $[H^+]$ and $[OH^-]$ are related through k_w.

The above discussion shows the potential confusion that may result when the assumption is made that the redox state of iron may be predicted from tables of standard redox potentials. It should be remembered that such predictions are based on thermodynamic considerations, when in fact the actual reaction may be dominated by kinetics. Furthermore, these predictions assume a very simple reaction medium, something far different from the complicated collection of molecules in food. The sparing solubility of Fe^{+3} (aq) and the strong tendency of this ion to form complexes suggests that the complexation and hydrolysis behavior of this ion will strongly influence the redox state of iron in a food.

Interactions with Ligands

Redox Considerations

Interaction with ligands is capable of influencing the amount of Fe(II) and Fe(III) in soluble form. Such interaction is also capable of altering $E°'$ for the Fe(III)/Fe(II) redox couple. One of the reasons that iron is so important in biological systems is that the potential of this redox couple is sensitive to the environment of the iron ion. Much of electron transport from NADH + H^+ to H_2O is possible due to the range of redox potentials for iron complexed with sulfur or porphyrin in protein, as illustrated in Table 2. Complexation with other more simple ligands also affects the redox potential (see Table 2).

A phenanthroline-iron complex will be stable as Fe(II) when the redox potential is below +1.12, whereas the CN^- complexation will tend to be stable as Fe(III) when the redox potential is above +0.36 (see Table 2). Thus the redox state of the iron will depend upon the redox potential of the system as well as the nature of the ligands available for complexation (10). The thermodynamic binding strength of the complexes with the various available ligands may determine which ligands exert the strongest influence in a complex food mixture. On the other hand, complexation kinetics may be more important in determing which ligands are associated with the iron. The redox state of the iron in a food is therefore only partially determined by the redox potential of the complex food system.

Binding Strength and Kinetics

When an iron complex forms it may be more or less soluble than the aqueous ions. If the complex is insoluble its availability to an organism will probably be low; if it is soluble its availability to an organism *may* be great.

Table 2. Redox Potentials for Selected Fe(III)/Fe(II) Half-Cells

half-cell [Fe(III)/Fe(II)]	$E^{\circ\prime}$
NADH dehydrogenase	-0.30, +0.03
succinate dehydrogenase	0.00
cytochromes	
b_k	+0.030
c_1	+0.225
c	+0.235
a	+0.210
a_3	+0.385
ferredoxin (Chromatium)	-0.49
Fe phenanthroline^{+3}/Fe phenanthroline^{+2}	+1.12
$Fe(CN)_6^{-3}$ / $Fe(CN)_6^{-4}$	+0.36

Source: Refs. 3 and 230.

Binding strength may be expressed by the association constant, K_a, defined as follows:

Binding molecule + metal ion $\rightleftarrows$ complex

$$K_a = \frac{[\text{complex}]}{[\text{molecule}] \, [\text{ion}]}$$

At equilibrium the concentrations of the three species are determined or calculated. K_a is a thermodynamic concept, predicated upon achievement of equilibrium. The rate of attainment of equilibrium is ignored. For Fe^{+3}, rate considerations may be critical at neutral pH due to the low concentration of Fe^{+3} (aq). One would predict the spontaneous transfer of Fe(III) bound to one complex to a different complex if K_a for the second complex were greater; nevertheless, for this transfer to actually occur, a mechanism would be needed.

It has been argued that for an iron complex to be available to an organism, the binding must be strong enough to protect the iron from precipitation but not so strong that the iron is not accessible for absorption. Even if the binding strength were in some appropriate range, some means of release would still be needed.

Transfer of iron from one complex to a stronger complex has been studied. When iron is received by transferrin, a molecule with $K_a = 10^{20}$ M^{-1}, the rate of exchange appears unrelated to the strength of the binding in the weaker donating complex (47,48). Stereochemical considerations have been postulated to influence the mechanism of exchange, thus explaining the tremendous variation in the rate of exchange.

The kinetic "stability" constant, the half-time of iron exchange between two iron complexes, has recently been studied for complexes of importance in food systems (49). It is well known that thermodynamic association constants are influenced by pH. Iron release from transferrin following receptor-mediated endocytosis is thought to be encouraged by a drop in pH (50). It is likely that kinetic stability constants are also influenced by pH. This behavior might be very important during the process of neutralization of chyme in the small intestine.

IRON ABSORPTION

Overview

Iron absorption has been reviewed by several authors (9,18,42,51—56). The numerous influences on iron absorption fit generally under one of two headings: chemical effects involving the contents of the gastrointestinal tract, and effects related to the physiological behavior of the organism ingesting the iron. Put another way, absorption may be thought of as the sum of intestinal lumen-related effects and intestinal mucosal cell-related effects.

The behavior of iron during digestion and absorption differs for heme and nonheme iron. Heme iron-containing proteins are digested to release the iron-porphyrin heme moiety. It has been suggested that the brush border may contain receptors for heme (57). Heme is thought to be absorbed intact, after which the porphyrin ring is degraded and the iron presumably becomes indistinguishable from absorbed nonheme iron (58,59). The cellular details of heme uptake and degradation have been elucidated by electron microscopy (60).

Nonheme iron makes up the greatest proportion of dietary iron, especially in undeveloped countries, whose populations experience the highest prevalence of iron deficiency. As a result, the following review applies primarily to nonheme iron unless otherwise stated.

Physiology of Absorption

Sites of Absorption

Iron is absorbed primarily in the small intestine. Within this region iron absorption capacity is greatest proximally. This capacity has been demonstrated experimentally under controlled lumenal conditions, ruling out pH differences as an explanation for decreased absorption in the distal direction (51).

Iron Uptake and Transfer

To be absorbed, iron must be taken up by the intestinal mucosal cell and subsequently be moved across the cell, across the serosal membrane, and finally to the blood, where it is picked up by transferrin and presumably becomes indistinguishable from previously absorbed iron. It is important to realize that absorption is the sum of uptake and transfer (61). Due to the rapid turnover of the mucosal cells, uptake does not imply absorption. If the mucosal iron is not transferred before the cell is sloughed into the lumen, this iron is no longer part of the organism; rather it contributes to the lumenal contents.

Homeostasis

Unlike most nutrients, for which homeostasis is achieved by regulation of excretion, iron homeostasis appears to be regulated primarily by absorption, as originally proposed by McCance and Widdowson (62). From a physiological perspective, iron absorption is influenced both by the dose level and by the iron status of the organism. In general, a lower proportion of iron is absorbed as the dose increases. Furthermore, as iron status worsens, the percentage absorption of a physiological dose trends to increase. By these means the organism attempts to regulate its iron content within a narrow range.

In the original "mucosal block" theory of homeostasis (63), iron absorption was prevented after a mucosal iron acceptor became saturated with iron. Granick (64) contributed to this theory by suggesting that ferritin was the mucosal iron acceptor, proposing that absorption ceased when the protein became saturated. Conrad and Crosby (65) suggested that the amount of ferritin incorporated into the developing mucosal cells somehow dictated the subsequent ability of those cells to allow absorption during their brief lifetimes. The concept of acceptor saturation was inverted: a "block" occurred only when the acceptor remained unsaturated and able to sequester additional iron (66). Higher body stores of iron would dictate greater apoferritin deposition, which in turn would lead to lower iron absorption. As Brozovic (54) has pointed out, this newer approach has merit but is not entirely satisfactory.

In studies with rats, Savin and Cook (67) found evidence to indicate that iron absorption is dictated by two mucosal iron-binding proteins: mucosal ferritin and mucosal transferrin. Ferritin serves a similar function as in the mucosal block theory of Crosby (66), whereas mucosal transferrin has an opposite effect; it encourages transport across the mucosal cell to the serosal membrane. In their work, the *ratio* of transferrin:ferritin was predictive of iron absorption by the ligated rat intestine (67). Other workers also have suggested that these two proteins work in concert to maintain iron homeostasis (68).

In distinction to this line of thought, others have focused on the brush border of the mucosal cell to explain how absorption could be altered for maintenance of homeostasis (69,70). This line of thought is consonant with the original mucosal block concept of Hahn (63). Cox and Peters (71,72) show evidence that iron deficiency may induce iron carriers in the brush border of human tissue, and these investigators suggest that initial entry into the cell may be a major regulatory step. In subsequent work with rabbits (73), important control of iron absorption was ascribed to an inducible active process allowing iron to enter the mucosal cell. In other work rabbit microvillus membranes were shown to possess high-affinity glycoprotein receptors for iron, the activity of which reflected homeostatic changes (74).

Huebers and associates have studied the influence of mucosal transferrin for some time (75), although originally it was referred to simply as a nonferritin iron-binding protein. In addition to the mucosal transferrin within the cell, they also studied an "elutable factor" of high molecular weight on the lumenal surface (76). Most recently these workers studied iron absorption from ^{125}I-labeled plasma transferrin in rat intestinal segments. Based on intact transferrin uptake by the mucosal cells, and subsequent reappearance of ^{125}I-labeled apotransferrin in the lumen, they suggest that mucosal transferrin may act as a shuttle for iron entry into the cell (77).

The above homeostatic mechanisms are predicated on the generally accepted idea, stated above as fact, that iron homeostasis is regulated by absorption and not by excretion (62). In challenge to this entire line of thought is a recently proposed, well-argued hypothesis (78). These authors suggest that homeostasis is maintained by both absorption and excretion; the macrophages of the lamina propria of the intestinal villi play a central role even at normal iron intake. Under normal circumstances, this model may be viewed as an extension of the mucosal block concept of Crosby (66), as the macrophage may be viewed as an additional hurdle (after mucosal ferritin) to be cleared before iron would reach the capillary. Iron not destined for the blood would be shunted from these macrophages to goblet cells for excretion into the intestinal lumen. Linder and Munro (53) had pointed out the possible excretory role of macrophages, but concluded that this means of excretion is only important in pathological states. The hypothesis of Refsum and Schreiner (78) is not entirely revolutionary even for normal individuals. As Brozovic (54) has reviewed, others have previously suggested an excretory role for macrophages in iron homeostasis (66,79). This hypothesis might provide a theoretical framework to explain the observation by Bjorn-Rasmussen et al. (80) that a portion of a test dose of iron retained in the gastrointestinal tract of humans after two weeks is lost over an additional two-week period. It has generally been assumed that iron remaining two weeks after the test dose would be considered absorbed.

Refsum and Schreiner (78) also suggested a novel way for transferrin saturation to influence mucosal cells at the time of their formation in the crypt region. If the developing mucosal cell had a need for a fixed quantity of iron, varying amounts of transferrin would be withdrawn by it from the blood, according to the degree of transferrin saturation. These authors suggest that the amount of apotransferrin left in the cell after iron extraction could help determine the way iron is handled by that cell in its brief lifetime (78). This concept is consistent with that suggested by Savin and Cook (67). Refsum and Schreiner (78) further hypothesize that the transferrin would have been incorporated by pinocytosis in combination with a transferrin receptor, which might move to the lumenal surface where it could play a functional role in absorption. This aspect of the hypothesis is reminiscent of the suggestions of Cox and co-workers and also of Huebers and co-workers, as discussed above.

The mechanism by which iron homeostasis is maintained is undoubtedly complex. It has been extensively studied without any clear resolution. A complex, multifaceted process is likely. In most situations the body is somehow able to communicate its iron needs to the cells involved in absorption so that they may respond appropriately.

Effects of Diet on Absorption From a Single Test Meal

The iron status of an animal will reflect its integrated absorption over a period of time relative to its needs integrated over the same period. Thus a diet low in iron, low in iron bioavailability, or both, may lead to nutritional iron deficiency. As a result the organism would tend to absorb more of a given iron dose than if it were iron-replete. In this obvious way the diet fed over a period of time can influence iron absorbed from an individual meal.

The effect of iron levels of meals consumed before a test meal may be the result of a more subtle consideration than the integrated long-term effect on iron status. Fairweather-Tait and Wright (81) showed that iron concentration of the diet for the three-day period before the meal is exponentially related to iron retention from a labeled meal fed to a rat. Furthermore, the iron content of even one meal during this period can profoundly affect iron retention. In a subsequent study either high or low iron content of a single meal within two days before the test dose could significantly affect iron retention from the test dose (82). Topham et al. (83) showed that short-term exposure to a low iron diet could affect the ratio of mucosal ferritin:mucosal transferrin, the distribution of administered ^{59}Fe in these two proteins, and subsequent absorption. The implication of these studies is that the iron status of the intestinal mucosal cells at the time of the test dose may be more important than the iron status of the entire animal.

In some situations it appears that absorption from a single meal may be influenced by the diet in a way unrelated to iron status or previous meal iron content. For example, iron absorption after a fast was compared in infants fed either breast milk or cow's milk formula. The breast-fed infants had higher serum ferritin, indicating better iron status. Nevertheless, they absorbed twice as much of an identical dose of inorganic iron than did those infants fed the cow's milk formula (84). It appears that the breast milk conditioned the intestine in some way to effect better iron absorption. One may only speculate how strong this conditioning might actually be because the breast-fed infants absorbed more of the dose in spite of their better iron status. Had iron status been equal for the two groups the difference might have been even greater.

Studies with rats have shown a similar phenomenon. Oxalic acid fed prior to a test dose of iron has been shown to increase absorption of iron from the test dose, which itself contained no oxalic acid (85). Studies with soy protein isolate- and casein-based diets show different iron retention even when test meals are identical (86). Less iron was retained by rats fed the soy protein isolate-based diet before and after the test dose, despite the apparently lower iron status (as indicated by hemoglobin concentration) of the animals fed this diet. Subsequent work showed that the effect of diet on iron retention could be partitioned into an effect of the diet before the dose and an effect of the diet after the iron dose (87).

An explanation of the effect(s) of diet on iron absorption is not presently available. The adaptive intestinal behavior might include changes in the rate of mucosal cell turnover or some specific intracellular conditioning (88).

Lumenal Considerations

Application of Iron Chemistry to Digestion

As stated above, iron is maximally absorbed in the proximal small intestine. To understand food iron behavior in this region, one must consider the effects of all treatments to that point. Industrial treatments as well as home preparation may influence the chemistry of iron in the food prior to ingestion. (These influences will be discussed later.) The ultimate food processor, the digestive tract, will have profound influences on the chemical behavior of the iron in food.

After a brief period of chewing in the mouth, food is deposited in the stomach. The influence of the brief oral residence time is probably insignifinificant with respect to iron chemistry except that food particles are reduced in size. When the food reaches the stomach, the effect on iron of a rapid drop in pH can be profound. As stated earlier, solubility of the free hydrated iron ions can be high enough to be important at low pH. In addition, many species which bind iron strongly at neutral pH bind much less strongly at low pH (89). The initial pepsin digestion products provide an increase in potential protein-derived iron-binding materials, although the actual amount of binding at this low pH is probably small, and that iron binding that does occur is probably weak. As mentioned earlier, the redox potential of the system may also be related to pH.

As portions of chyme are released into the proximal small intestine, the material encounters pancreatic secretions, which contain additional digestive enzymes, including proteases, lipases, and amylases. At the same time bicarbonate ion serves to neutralize each portion of the material fairly rapidly. Although the activities of these pancreatic enzymes are highest around neutral pH, they will become increasingly effective as pH increases. As a consequence, the food iron in each portion of chyme is present in a rapidly changing environment. As the digesta works its way down the small intestine, its chances of being taken up by the mucosal cells decrease, as a result of both the higher pH and the decreased ability of mucosal cells of the jejunum and ileum to take up iron. The critical period with respect to iron absorption is that period during which the environment changes from that of the stomach to the small intestine. Throughout the small intestine absorption of small molecules also occurs as they become available. Thus from the perspective of a particular duodenal mucosal cell, the iron in its proximity passes by in a state of ferment in a number of respects. Predictions based on thermodynamic stability considerations will likely be fortuitous, if correct, because the thought of equilibrium during this process is absurd.

This complex picture notwithstanding, various researchers have sought to understand the chemical behavior of iron during the lumenal aspects of digestion. Various in vitro procedures have been developed to allow controlled study of iron behavior during digestion. More recent approaches have been reviewed by Van Campen (90) and Miller and Schricker (91).

Narasinga Rao and Prabhavathi (92) subjected food to a pepsin treatment at pH 1.35, adjusted the pH to 7.5, filtered, and determined "ionizable" iron. As they define the latter term, it refers to ferrous iron reacting with α, α-dipyridyl after the supernatant was exposed to the reducing agent. Iron in soluble form but complexed with protein, persumably as Fe(III), might not be included as "ionizable," but it would be included in the total ("soluble") iron in the filtrate. These workers correlated percent ionizable iron values with percent absorption by human adult males with normal hematological status.

Miller et al. (93) found that the above method was not readily applicable to meals from semisynthetic ingredients and proposed a different method. The major differences from the previous method are that pH is adjusted to neutrality by dialysis of $NaHCO_3$ from inside a dialysis bag into the digesta, and that iron availability is determined by the amount of soluble iron able to dialyze into the same dialysis bag (6000–8000 MW cutoff). These same workers have published a detailed comparison of their in vitro method with a series of human tests involving absorption of labeled iron from semisynthetic meals (94). On the basis of a good correlation with this series of human studies the method has been widely used.

Two studies from Bates' laboratory (89,95) utilize in vitro methodology in order to understand the details of iron behavior in the intestinal lumen. In the first study, successive incubations at pH 2 and pH 6 were employed in the absence of digestive enzymes. Iron in cooked pinto beans was found in three populations: 1) spontaneously soluble, 2) not spontaneously soluble but potentially mobilized by reducing or chelating agents, and 3) bound firmly to insoluble material, even in the presence of reducing or chelating agents. An important observation in this study is that ascorbic acid was able to produce its solubilization effect at pH 2 in 20 min or less (89), presumably by a reductive mechanism. Another observation of analytical interest is that bathophenanthroline sulfonate, the reagent used to detect Fe^{+2}, when present throughout the incubation procedure, was able to increase the soluble iron as well as to transform it entirely into the Fe^{+2} state even in the absence of the dithionite added to reduce Fe^{+3} to Fe^{+2}. The change in iron solubility due to strong binding to this reagent would indicate that the soluble Fe^{+3}, as well as some insoluble iron, is in equilibrium with the soluble Fe^{+2}.

Based on titration of iron salts and iron ascorbate complexes with NaOH, Gorman and Clydesdale (96) concluded that under some conditions ascorbic acid may chelate Fe^{+3} faster than it reduces these ions to Fe^{+2}. In the presence of bathophenanthroline, the Fe^{+2}-bathophenanthroline complex formed even when the titration results indicated that the original complex was as Fe^{+3}-ascorbate. Gorman and Clydesdale (96) suggested that bathophenanthroline could strongly bind to any small amoung of Fe^{+2} (aq) present, indirectly affecting the binding equilibrium for the Fe^{+3}-ascorbate complex to form more free Fe^{+3}, which would then itself be reduced and complexed to bathophenanthroline. These authors suggested that this mechanism might explain how Kojima et al. (89) found iron in Fe^{+2} form by bathophenanthroline analysis even in the presence of various chelating agents. As noted above, Kojima et al. (89) observed that bathophenanthroline itself was capable of reducing iron when used as a chelator and not as an analytical reagent.

More recently Reddy et al. (95), in a similar study from Bates' laboratory, employed digestive enzymes and a different Fe^{+2}-specific reagent, ferrozine. In this work the concept of "redox flux" is introduced, meaning the capacity for formation of Fe^{+2} from super-natant Fe^{+3} by an appropriate chelating or reducing agent in the presence of a strong Fe^{+2}-binding molecule, ferrozine. They were able to demonstrate that phosvitin, an iron-binding protein in egg yolk, had a high redox flux in the presence of ascorbic acid, but that little or no redox flux could be observed for iron associated with pyrophosphate in the presence of ascorbate (95).

An important aspect of this study is that it confirms that, under some conditions, ascorbate may be an effective chelator of Fe^{+3}, keeping this redox state in (or drawing it into) soluble form; furthermore, in the presence of a suitable Fe^{+2}-binding molecule, ascorbate's reducing ability may provide a mechanism for ready delivery of this same iron in the form of Fe^{+2}, accounting for the observed "redox flux." In this way the steady state form of soluble iron might be as Fe^{+3}, but with a high potential to *become* Fe^{+2}.

Previous workers have recognized the possible importance of the kinetics in the transfer of iron from one strong chelator to another (47,48). In these studies it was conceded that equilibrium considerations regarding iron binding are dictated by relative thermodynamic stability constants, but shown that kinetics of iron transfer in the direction of equilibrium appeared to be related to steric factors regarding the two complexes involved. Forth and Rummel (42) recognized the potential importance of this concept and coined the hybrid term "kinetic stability constant" to describe the half-time for iron exchange between two complexes. More recently, Gorman and Clydesdale (49, 96) used the rate of transfer of iron from various complexing agents to apotransferrin as an indicator of potential release of lumenal iron. They recognized that the initial period of exchange might be of greatest importance with respect to iron behavior under lumenal conditions (49). This model of transfer of iron to apotransferrin might be especially appropriate if transferrin actually does act as a shuttle to bring iron into the intestinal mucosal cell, as suggested by Huebers et al. (77).

May et al. (97) discussed the influence of low molecular weight Fe(III) complexes as a means of mediating transfer of iron between two strongly iron-binding proteins. As these authors pointed out, considerable transfer between strongly iron-binding proteins may occur in the presence of low molecular weight compounds even when the low molecular weight complexes are at the limit of analytical detection. Citrate-mediated exchange of Fe(III) has been studied for transferrin. This concept again emphasizes the importance of kinetic considerations rather than relying upon thermodynamic stability constants. One implication of this concept is that low molecular weight complexes formed during digestion may be important beyond the magnitude implied by their concentration at any given time.

Iron salts appear to be best absorbed when taken without food; under such conditions Fe^{+2} salts are better utilized than Fe^{+3} salts.

In the presence of a meal, utilization of ingested iron is lessened.
Even orange juice, with its high content of ascorbic and citric acids,
depresses iron absorption relative to iron salts alone (98). The impact
of food iron valence on iron absorption is not so easy to determine.
It is possible that solubility considerations might actually favor higher
bioavailability from Fe(III) due to its ability to form soluble complexes
with ligands in food. Even if absorption were only allowed for Fe^{+2},
it might be chemically advantageous to keep the iron solubilized as
Fe(III) until the appropriate time. Such reasoning might explain how
cysteine improved iron absorption from ferric chloride to a level above
that observed from ferrous sulfate without cysteine (99).

It should be noted that the above mentioned depressing effect of
orange juice on absorption of iron from iron salts given alone does
not contradict the observation that *food iron* absorption may be
enhanced by ascorbic acid alone or in orange juice.

The Pool Concept

The reason for the classification of food iron as heme iron or nonheme
iron is that the behavior of iron in these two forms is different and
the behavior of each is independent of the other. Hallberg and Bjorn-
Rasmussen (100) showed that heme and nonheme iron could be independently labeled (with ^{55}Fe and ^{59}Fe) to allow calculation of total absorption from a meal containing both types of iron. This observation implied that these two forms of iron did not interact, behaving as two
distinct pools.

At about the same time, Cook et al. (101) studied nonheme iron
absorption with both ^{55}Fe and ^{59}Fe. These workers biosynthetically
labeled various plants using ^{55}Fe, thus labeling the iron intrinsic to
the plant as it grew. Then, after processing a portion of the ^{55}Fe-
radiolabeled plant for incorporation into a meal, they added a tracer
dose of ^{59}Fe as a soluble inorganic salt. Foods labeled in two ways
—intrinsically and extrinsically— were fed to human subjects. When
the extrinsic iron was mixed throughly with the food before ingestion,
the ratio of the proportion of extrinsic to intrinsic labels absorbed
was only slightly above one, indicating essentially complete exchange
of the two radiolabels.

Hallberg (102) discussed the implications of the concept of a common
nonheme iron pool in a particular meal. A most important implication
is that nonheme iron absorption from a food or from a meal may be
studied by the absorption of a tracer dose of iron appropriately added.
The concept of complete exchange has been validated in a number
of subsequent studies and has been extended to the rat model (103).

Complete (or nearly complete) exchange of iron in the nonheme
pool has been empirically demonstrated under a range of conditions,
with only a few exceptions noted (104—107). What accounts for this
observed complete exchange?

Hallberg first attempted to answer this question (102). He acknowledged that some of the native iron is probably either partly soluble and/or partly complexed. For isotopic exchange to occur, he reasoned that some portion of the pool must be "active." It is easy to imagine complete isotopic exchange among soluble, uncomplexed (or loosely complexed) iron. It is more difficult to understand how the iron in this active pool is able to interact with the remaining iron so that exchange is complete throughout. Furthermore, it must be borne in mind that this exchange is rapid. As Consaul and Lee (106) point out, very little is known about how the exchange occurs, or even if it occurs before or after ingestion.

The paradox of rapid, complete exchange of poorly bioavailable iron has not been addressed. The food iron must be bound in a sufficiently labile manner to allow exchange, yet be sufficiently strongly bound to hinder uptake from the lumen. Hallberg (102) suggests that the size of the "active pool" may determine bioavailability. A different explanation may be more satisfying. Although considerable iron in a meal may remain insoluble at low pH, the strength of the binding of this iron to food components is decreased, due to the increased tendency of various iron-binding ligands to exist in protonated form at low pH. Furthermore, solubility of the hydrated cations Fe^{+2} and Fe^{+3} increases dramatically as pH drops. Even iron in insoluble, complexed form might both dissociate and reform its complex more rapidly, thus allowing complete, rapid isotope exchange at gastric pH.

This complete exchange need not be discordant with low bioavailability, because the chemical behavior of iron during the process of neutralization will change drastically from that at low pH. This behavior will be strongly influenced by the amount and nature of the various ligands either originally present or recently produced. Many products of digestion may potentially function as iron-binding ligands. Digested protein in the form of peptides and amino acids may have a very important influence on iron bioavailability from a meal. The types of peptides formed as well as the rate of their formation may swing the balance in one direction or the other. For example, certain peptides produced rapidly might increase bioavailability, whereas those same peptides produced slowly would have little effect. In the latter case, if the peptide were produced after the iron had interacted with the previously available ligands as the pH increased, an insurmountable kinetic barrier might result.

Under some circumstances incomplete isotopic exchange with nonheme iron has been observed. Insoluble forms of radiolabeled fortification iron were absorbed less readily than radiolabeled wheat iron in bread (107). As mentioned earlier, a method to estimate the amount of contamination iron is based upon the extent to which exchange is incomplete (33), although the varying extent of exchange for different forms of contamination iron makes this method subject to error. In this respect contamination iron and fortification iron are similar, in that the completeness of isotopic exchange depends upon the chemical species. This difference may well be related to the solubility and/or to the rate of dissociation/reassociation for these materials at low pH.

This problem with varying exchangeability of different forms of contamination iron is particularly troubling, because there is no good way to distinguish native from contamination iron. For example, if a food contained 20% of its iron as contamination iron refractory to isotope exchange (and completely unavailable), an extrinsic radiolabel would only label 80% of the nonheme iron. The effective iron dose would also only be 80% of what the experimenter would assume based on iron content; consequently the proportion of the label absorbed would not apply to the total iron analytically present, but the investigator would not know it. The estimate of bioavailability would be incorrectly high for all the iron physically present. In fact, whenever exchange is incomplete, an incorrectly high estimate may result. Conceptually it might be safer to assume a three-pool model for iron: heme iron, nonheme iron that is exchangeable with extrinsic iron, and nonheme iron that is not exchangeable with extrinsic iron. It would be important to recognize that conditions in food processing might alter the relative proportions of iron in these three pools (108–112).

Van Campen (90), Smith (105), and Consaul and Lee (106) have all warned against incautious application of the extrinsic label to the nonheme iron pool. In spite of the above concerns, the use of this methodology has contributed tremendously to our understanding of iron bioavailability (104).

Bioavailability of Nonheme Iron

Implicit in the two-pool model for iron in a meal is the concept that all the nonheme iron in the meal is equally subject to the same environment following the point at which complete exchange occurs. Thus bioavailability of the iron in one food in the meal is influenced by all the other foods in the meal, even those which originally may have contained little or no iron.

Bioavailability of nonheme iron has been reviewed by numerous authors (19,18,113–116); in fact an entire book on the topic has appeared (117). Because this review literature is readily accessible and the primary literature is immense, the present review will focus on aspects of bioavailability considerations. Reference will be made to recent reviews, as these may allow the reader rapid entry into the literature.

Enhancers of Nonheme Iron Bioavailability. The two most important enhancers of food iron bioavailability are ascorbic acid and animal tissue (113,104). As described earlier, it has been suggested that "absorbable" iron may only be calculated if the ascorbic acid level and the amount of animal tissue is known (17).

The interaction of ascorbic acid and iron absorption has been extensively reviewed by Lynch and Cook (118) and by Monsen (119). Lynch and Cook (118) combined the results from a number of investigations to show a strong correlation between the ascorbic acid:iron ratio and percent iron absorption. The enhancing effect of ascorbic acid has been ascribed to its ability to chelate Fe^{+3} (120) as well as to its ability to either maintain iron in the Fe^{+2} state (104) or to reduce Fe^{+3} to Fe^{+2} before the pH increases (51). Kojima et al. (89) point out that under some circumstances citrate more effectively chelates Fe^{+3} than does ascorbate. Nevertheless, ascorbate is more effective in enhancing iron absorption. In addition to chelation, the reducing capacity of ascorbate may be important according to the hypothesis of Reddy et al. (95) in that, in addition to solubilizing Fe^{+3}, the ascorbate may allow a high "redox flux," the conversion of Fe^{+3} to Fe^{+2}.

Conrad and Schade (120) have shown that ascorbate can apparently interact with iron as pH is lowered to 2, and that when pH is again increased, the complex is stable even at higher pHs at which complexation was previously impossible. This result would seem to be an example of the importance of kinetics throughout the period of changing pH in the digestive tract. In fact, this example might serve as a model to understand the effects of other ligands in food on iron absorption. Only after complete exchange occurs at low pH and the pH begins to increase again do all the ligands previously unassociated with iron get the chance to chelate the meal iron. Wherever each ion ends up may well not be the most thermodynamically stable state, especially as conditions continue to change throughout the neutralization and digestive process. Ascorbate is able to chelate iron more effectively beginning at lower pH (pH 5 and below) than citrate, which chelates iron more effectively after pH has increased further (to around pH 6) (89). The significance of this chelation behavior may lie in the tendenency of ascorbate to bind *first* during the period of pH neutralization. Like an unfair game of musical chairs, the ascorbate may get to sit down before the music stops, whereupon the other ligands must scramble among themselves.

Hungerford and Linder (121) claimed that at least some of an iron-ascorbate complex may be absorbed intact into the mucosal cell, whereupon it dissociates. A small portion of the iron in NaFe(III)EDTA has also been shown to be absorbed as the intact complex, but most of this absorbed complex appears in the urine (122,123). The rest of the iron in this complex appears to dissociate in the lumen, however.

From a practical standpoint the amount of ascorbic acid present in one-third of a glass of orange juice (25 mg) can lead to a significant increase in iron absorption (104). A glass of orange juice with a hamburger, string bean, and mashed potato meal nearly doubled nonheme iron absorption (124).

Animal tissue in the form of red meat, poultry, and fish contains both heme and nonheme iron. The heme iron tends to be more highly available and less influenced by enhancers and inhibitors affecting nonheme iron. The nonheme iron from animal tissue also tends to be of higher bioavailability due to an enhancer of nonheme iron absorption. This enhancement of nonheme iron absorption is especially important when the remaining nonheme iron in the meal is considered. It has been estimated that 1 gm of meat is approximately equivalent to 1 mg ascorbic acid in its effect on nonheme iron bioavailability (30).

The identity of the "factor" in meat which enhances iron bioavailability is not known. Animal protein in general does not enhance absorption, as observed for egg white (125) and cow's milk (124). Martinez-Torrez et al. (99) showed that cysteine ingested with a meal in a gelatin capsule was capable of enhancing iron absorption to the same extent as fish. When mixed with the food, the effect of cysteine was lost. More recently, this same laboratory has presented evidence that unoxidized cysteine-containing peptides released during digestion of beef could enhance nonheme iron absorption (126). This work suggests that the meat factor may be these peptides.

Iron availability from human milk has been shown to be high. When fed as a single extrinsically labeled test meal, the 21% absorption of iron suggested the presence of some enhancing substance (127). What this substance might be is not known.

Fructose appears to enhance iron absorption under certain conditions (29). Ferric fructose was shown to be more available to the guinea pig than $FeSO_4$ when administered after a fast (128). This same group showed that a hydrolytic ferric fructose polymer could form which was quite reactive (129), dissociating in excess sugar or excess chelating agents (4). It is unclear whether such a mechanism might be important during digestion of a food. One advantage of these complexes would be elimination of metallic taste (130). Whether fructose in this form should be considered an enhancer of absorption or part of a highly available iron fortificant is open to interpretation.

In their studies of iron absorption of maize and sorghum beer, Derman et al. (27) studied the effect of the lactic acid produced during fermentation. They attributed the high iron bioavailability from this beer at least partially to the lactic acid content. (It may be of interest to note that Lactobacillus delbruckii was employed as a pure culture in this study, and it would be expected that bacterial siderophores would likely not have been present as a result (131).)

This same group studied the absorption of iron from a variety of vegetables (132). They showed that those associated with moderate or good bioavailability contained one or more of malic, citric, or ascorbic acids. Iron absorption from a rice meal was enhanced to about the same extent by each of the following organic acids alone: citric acid, ascorbic acid, L-malic acid, and tartaric acid. Unfortunately, other components of the vegetables studied were capable of strong inhibition, as described below. Nevertheless, this work suggested that ascorbic acid may not stand alone in its ability to enhance iron absorption. Each of these acids would have the potential to maintain the Fe^{+3} state in soluble form as pH increases, as has been shown for ascorbic acid. Results of this study suggest that reducing ability may be secondary to the ability to form soluble iron complexes.

Inhibitors of Nonheme Iron Bioavailability. Compounds inhibitory to iron absorption could fall into two classes: molecules capable of binding iron so strongly that they will not release it, regardless of whether the complex were soluble; and molecules capable of binding iron such that the complex itself becomes (or remains) insoluble during the critical period in the proximal small intestine. General classes of compounds which may inhibit nonheme iron absorption include phenolics, certain proteins, polysaccharide components of dietary fiber, and phosphates. Other materials include wheat bran, EDTA, and possibly oxalates, among others.

Phenolic Compounds. Phenolic compounds are widespread in nature, occurring as phenolic acids; flavones and flavonols; chalcones; anthocyanins; and as combined forms (133). These combined forms would include tannins, both hydrolyzable and condensed. Lignins might also be considered combined forms. Certain bacterial siderophores are efficient binders of ferric ions due to mono- or dihydroxybenzoic acid moieties (134). The strength of metal complexes will vary with specific phenolic compounds. Ribereau-Gayon (133) has pointed out that horticulturalists have long turned hydrangeas a deep blue by adding iron or aluminum salts, which complex with dihydroxylic anthocyanins. Nevertheless, these metals have no effect on roses, because this species also contains flavonols, which bind iron more strongly than the anthocyanins.

Disler et al. (135) first reported the inhibitory effect of tea. In a subsequent report (136), they clearly demonstrated that when tannins were extracted before preparing the tea infusion, the inhibitory effect was lost. When the extracted tannins were added back, the effect was similar to tea brewed normally. These workers show that 80% of the original tea leaf tannins (which are 10% by weight of the dried leaves) may be present in brewed tea. Rossander et al. (137) and Hallberg and Rossander (124) have verified the inhibitory effect of tea.

Disler et al. (136) warned that a variety of tannins are present in vegetable foods. Several of these same investigators were involved in a more recent study of iron absorption from vegetables. An inverse correlation was observed between polyphenol content and iron absorption (132). Foods high in polyphenolics included spinach, lentils, and beet greens. Other human studies have implied low iron bioavailability for spinach; nevertheless, in rats bioavailability was equal to $FeCl_3$ (85). Narasinga Rao and Prabhavathi (138) suggested that high tannin levels in Indian diets may be responsible for low nonheme iron bioavailability.

Coffee has recently been shown to be inhibitory to nonheme iron absorption, although not quite to the same extent as tea (124,139). The mechanism for this effect is not as clear-cut as for tea. Based on the in vitro work of Kojima et al. (89), it would appear that decreased solubility of meal iron is not affected by coffee as it is by tea. The presence of soluble iron-phenolic complexes in coffee has been suggested (113).

Roy and Mukherjee (140) reviewed the influence of food tannins on iron metabolism. Intestinal lesions due to high doses of the tannic acid of tea and coffee have also been reported (141).

Selected Proteins. Cook and Monsen (142) observed that not all animal-derived proteins enhanced iron absorption as do meat, poultry, and fish flesh. Neither whole milk, whole egg, nor cheese had any enhancing effect when substituted for egg white in a purified diet. When the proportion of egg white in the basic diet was further studied (125), an inverse relationship between absorption and dietary level of egg white protein content was observed. An explanation for this observation remains to be made.

Peters et al. (143) reported low bioavailability of egg yolk iron in human work as well as in vitro. They saw an equally strong effect with the phosphovitellin fraction alone. The specific protein in question is more commonly referred to as phosvitin; it is the principal phosphoprotein of the yolk, accounting for 7% of the yolk protein (144). It contains about 0.4% iron, accounting for most if not all the iron present (145). Pertinent features of this protein have recently been reviewed extensively (146). Perhaps most important chemically is that 50% or more of the amino acid residues are phosphoserine, accounting for the high iron-binding capacity (146). Sato et al. (147) showed that the phosvitin-iron complex could promote iron precipitation in the intestine. Subsequently work from this laboratory suggested that phosphopeptides produced during digestion could strongly bind iron (148). Reddy et al. (95) showed that the iron so bound may be readily solubilized by ascorbic acid.

Legume proteins are generally acknowledged to adversely affect iron bioavailability (149). Following work of Cook et al. (150), in which a strong inhibitory effect of soy was observed in humans, considerable effort has been expended to understand this effect in humans (151−156) and in animals (86,87,157−160). It would appear that legumes as a class may show poor iron bioavailability themselves and also decrease overall bioavailability in a meal (154). In vitro studies suggest that the inhibitory effect may be related to the protein component of soy products (161). Kane and Miller (162) used in vitro methodology to suggest that the affinity of indigested or partially digested protein for iron, compared to the affinity of low molecular weight digestion products for this mineral, may determine iron absorption. Schnepf and Satterlee (163) described iron trapped within large peptide aggregates after in vitro digestion of soy. The iron was released only upon dissociation of the aggregates. The effect of dietary protein on iron bioavailability has been recently reviewed (164).

Polysaccharide Components of Dietary Fiber. Kelsay (165) has reviewed studies of iron balance in humans as influenced by foods with varying levels and types of dietary fiber. No consistent effect was observed, probably because other dietary components may have been even more important. In most studies iron balance was not affected by fiber. Ali et al. (166) also note conflicting results regarding the influence of dietary fiber.

It would seem likely that the wide variety of materials included in the term "dietary fiber" would interact differently with minerals. In the perfused intestinal segment of the dog, Fernandez and Phillips (167) showed that lignin and psyllium mucilage (Mucilose, Winthrop Laboratories) were potent inhibitors of iron absorption. Pectin was less potent, and cellulose was without effect. These same workers showed that in vitro binding of iron to the insoluble fraction correlated well with the observation using dogs (168). Pectin but not cellulose has been shown to hinder iron absorption in patients with idiopathic hemochromatosis (169).

Several investigators have studied binding of iron to specific polysaccharides (170–173). For sodium alginate, the complexes which formed appeared to be disrupted by an in vitro digestion treatment (171,173).

It is important to recognize that properties of a given type of polysaccharide can vary according to the precise chemical description. For example, a wide variety of commercial pectins are sold, varying in source, degree of purity, methoxyl content, and distribution of methoxyl groups. Clydesdale (110) has described the discrepancies in the literature relating to iron binding by cellulose. He notes reasonable consistency when the same types of cellulose are studied at the same pH, but much different behavior among the different forms of cellulose.

Ranhotra et al. (174) studied the effect of added cellulose in breads and saw no reduction of iron absorption relative to the control bread-based diet. Wheat bran does seem to reduce iron absorption, but whether the fiber fraction is responsible is doubtful (175). Using an in vitro method, Leigh and Miller (176) concluded that the effect of an isolated bran fiber fraction was highly dependent on the other ligands present in the meal.

The chemical binding of iron to food fiber sources has been studied for rice hemicellulose (177), a variety of brans and hulls (178), and wheat and corn brans (179).

Phosphate-containing Compounds. The high phosphoserine content of phosvitin appears to be responsible for the inhibitory effect of egg yolk. The lack of enhancement of iron absorption with cow's milk and cheese (142) may be related to the phosphate content of casein (143, 180). Monsen and Cook (181) showed that when physiological levels of both calcium and phosphate (as $CaHPO_4$ or as $CaCl_2$ and K_2HPO_4) were present simultaneously, iron absorption was reduced. When one or the other was present, no effect was observed. The implication of this work was that iron apparently associated with the phosphate in the insoluble calcium phosphate. In human work, Snedecker et al. (182) reported trends that support Monsen and Cook (181).

Greger (183) has reviewed the effect of phosphorus-containing compounds on iron absorption, concluding that the results have been inconsistent. She points out that the different forms of inorganic phosphate used in foods may behave differently. Based on in vitro work, Subba Rao and Narasinga Rao (184) suggest that certain poly-phosphates may actually enhance iron absorption.

Mahoney and Hendricks (185) showed that a variety of phosphate compounds decreased iron absorption by the rat, with pyrophosphate and tripolyphosphate having a stronger effect than orthophosphate alone. Zemel and Bidari (186) also suggest that polyphosphates may have a stronger inhibitory effect on iron utilization by the rat than orthophosphates. In vitro studies suggest that tripolyphosphate may exert a stronger inhibitory effect than hexametaphosphate (187). In humans, hexametaphosphate apparently interacted less strongly with added calcium than orthophosphate, with less effect on iron absorption as a result (188).

Phytate, or inositol hexaphosphate, is capable of strong binding with Fe^{+3}; in fact, this behavior at acid pH is the basis for many of the analytical procedures for phytate (189). The compound which precipitates in the analysis is approximately tetraferric phytate. Nevertheless, the iron of monoferric phytate has been shown to be highly available (190). Only when purified sodium phytate is added to meals is iron absorption decreased (113).

Beard (191) showed that increasing the native level of phytate in soy products does not alter the inhibition of iron absorption by these products. Simpson et al. (192) showed that removal of phytate from wheat bran had no effect on the inhibitory effect of the bran. Although Simpson et al. (192) saw no increase in iron absorption after enzymatic hydrolysis of phytate in wheat bran, Hallberg et al. (192a) have recently described phytate as the main cause of wheat bran inhibition of iron absorption. The effect of phytate on iron bioavailability has been reviewed (193).

Other Inhibitory Factors. Wheat bran has been shown to inhibit iron absorption by some means other than phytate. At the same time, Fairweather-Tait (194) found that the fiber itself was not responsible either. Anderson et al. (195) agreed that the bran fiber itself had no direct impact on iron absorption. Simpson et al. (192) suggested that a soluble, phosphate-rich fraction of dephytinized bran may be responsible for the inhibitory effect of the bran.

Under certain conditions EDTA may be an inhibitory factor, as described by Cook and Monsen (196). When the EDTA:iron molar ratio was 1:1, absorption was decreased 28%; a ratio of 2:1 decreased iron absorption by 50%. Nevertheless, NaFe(III)EDTA has been proposed as an appropriate fortificant, and its use as such has been recently reviewed (123). At lower molar ratios the adverse effect is not seen. It has been pointed out that if 5 mg iron as NaFe(III) EDTA were added to 15 mg dietary iron, the dietary ratio would only be 0.25. These authors express concern regarding the impact of current levels of EDTA as a food additive based on this ratio (196).

Chemically, the example of EDTA is of interest because, although the binding equilibrium would be constant regardless of initial concentrations of iron and EDTA, the amount of iron free to interact with other ligands would be small under conditions of a molar excess of EDTA. At low molar ratios the effect may be to help maintain Fe^{+3} in soluble form. Of further interest chemically is the fact that tea, but not wheat bran, is able to inhibit iron absorption from this complex (122). Thus EDTA appears to be protective with respect to certain inhibitory ligands.

Just as EDTA may enhance or inhibit nonheme iron absorption, depending upon the ratio of chelator to iron, so soy isolate and other nonenhancing proteins have been suggested as potential vehicles for iron fortification after binding additional iron (197). Good iron bioavailability is suggested by iron release after in vitro digestion (198). The removal of native iron from soy protein isolate is much more difficult, as recently described (199).

Practical Aspects of Bioavailability. Bioavailability of iron is generally considered to be the proportion of the mineral present that may be utilized by the animal. Consequently it is often expressed as a percentage with respect to some total amount. It is important to keep in mind that an individual needs an *amount* of iron. Consequently a meal of low iron bioavailability but high iron content might be quite appropriate. For example, in a study of iron bioavailability from an egg-containing breakfast, bioavailability was decreased but the total iron absorbed was increased when the egg was present. The basis for the percentage changed due to the high iron content of the eggs (137). Studies in which total iron content is strictly controlled are scientifically more satisfying but may be less relevant practically.

At the not inconsequential risk of confusing the reader, another example is presented. Initial work showed the inhibitory effect of soy products on nonheme iron absorption (150). Absorption of iron was subsequently studied in a hamburger meal in which half of the meat protein was replaced by soy products. Decreasing the proportion of meat led to less "meat factor," an enhancer of nonheme iron absorption. Furthermore, the added soy products were inhibitory to nonheme iron absorption. Nevertheless, the *amount* of nonheme iron absorbed was the same, due to the high iron content of the soy. Calculation of absorption of heme iron allowed estimation of total iron absorption, which was lower due to the decreased amount of heme iron (152). If soy were *added* to an intact hamburger meal, total iron absorbed would increase despite lower bioavailability of nonheme iron (152).

An additional twist has been provided by Lynch et al. (200). Rather than calculate heme iron absorption (152), these investigators measured it as well as nonheme iron absorption from a dual-labeled ground beef meal. Surprisingly, they found that the soy products apparently enhanced the absorption of heme iron (200). Consequently, the effect of substituting soy products for meat is less strong on two counts: the iron content of soy products is high, and an enhancing effect on heme iron absorption accompanies the inhibitory effect of nonheme iron absorption. A reduction in total iron absorption still occurs, but it is not as strong as was originally thought (150). Long-term human studies are consistent with this conclusion (153).

This discussion of the practical aspects of bioavailability is not meant to imply that bioavailability is unimportant. In fact several authors have suggested that increased bioavailability, rather than increased iron content, is the most efficient way to improve iron nutrition (see for example Ref. 149). Nevertheless, bioavailability information may lead to erroneous practical interpretations if not properly interpreted.

Bioavailability may be related to the physical as well as to the chemical behavior of a meal. For example, a meal of a hamburger, french fries, and milkshake was either served normally or after "homogenization" in a blender. Slower gastric emptying and increased heme and nonheme iron absorption were observed for the treated meal (201).

Physical behavior of a meal may also be important in the presence of hydrocolloid gums and soluble dietary fiber. Effects on mineral absorption might result from reduction of diffusibility due to increased viscosity of lumenal contents or from shortened intestinal transit time, as suggested by Harmuth-Hoene and Schelenz (202).

EFFECTS OF PROCESSING ON THE STATE OF IRON IN FOODS

The complex process of digestion can have subtle effects on the bio-availability of meal iron. The digestive process has as its raw material the variety of foods present in the meal. It is obvious that the

predigestion state of the iron, as well as that of factors either immediately or potentially capable of enhancing or inhibiting iron absorption,
will have an influence on the bioavailability of iron from the meal. Iron
will originally occur in some native form in a particular food. Iron
compounds added as fortificants will have a well-defined chemical
state when added. During processing and storage various changes
may occur: the state of the native food iron may be altered; fortification iron may interact with food constituents; contamination iron may
be introduced and itself interact with food constituents; ligands may
be produced or destroyed; and changes in pH or redox potential may
occur. The great variety of potential changes makes the study of
the effects of food processing on iron bioavailability difficult. Ideally,
one would want to use human studies to gauge the impact of processing treatments. Because of the multivariate nature of the problem,
this approach is not feasible. Instead, in vitro studies and rat-
feeding studies have been primarily employed to study processing
effects. Iron chemistry and bioavailability in food processing has
been reviewed (109).

Lee and Clydesdale (203) described a method of determining the
"iron profile" in a food. Elemental iron is separated using a magnetic
stir bar. Remaining insoluble iron is measured after centrifugation
(204), leaving total soluble iron. "Free" ionic iron in the supernatant
is determined by the reaction with bathophenanthroline in the presence
of hydroxylamine HCl, a strong reducing agent. Soluble iron that
does not react with the bathophenanthroline is assumed to be in a
strongly bound complex. Ionic iron as Fe^{+2} is determined by the
reaction with bathophenanthroline in the absence of the reducing
agent; the remainder of the ionic iron is assumed to be as Fe^{+3}. This
methodology led to the important understanding of the changes undergone by fortificant iron after processing and storage (203,204). The
behavior of various food fortificants during processing has been
reviewed by the same authors (29). "Iron profile" methodology has
been applied in the same laboratory to study the effect of drying
processing (204), thermal processing of spinach (205), ascorbic acid
added to soy protein (206), solubilization of cereal iron by milk (207),
organic acids added to soy-extended meat (208), and organic acids
added to fortified corn-soy-milk blend with different heat treatments
(209).

A different group has used the "iron profile" approach to study
the effect of spray-drying soy beverage (111) and the effect of various ingredients in the breadmaking process (210). The effect of
retorting on bioavailability of fortification iron added to a milk product
has also been examined (211).

Determination of the "iron profile" after processing suggests
strongly that certain processing conditions and ingredient additions
may enhance iron bioavailability. Certainly this approach shows that
the state of the iron present may be quite responsive to subsequent
treatment.

Although the "iron profile" method is not the same as studying bioavailability, it does give an indication of potential changes in bioavailability. An in vitro digestion procedure, such as that of Miller et al. (93), allows a better estimate of bioavailability, still not as good as a human study, but much more convenient and less costly. Shricker and Miller (108) showed that baking into bread did not alter relative bioavailabilities of iron fortification compounds, and that meal composition was more important than the type of fortification iron. In this same laboratory a wide variety of processed soy products were studied (34 samples total). It was concluded that iron bioavailability was lowest when the degree of protein refinement was highest (161). In a different study the in vitro methodology was used to suggest that the type of breakfast cereal processing is related to iron bioavailability from meals containing the cereal (212).

Several authors have suggested that ferric hydroxide polymerization will result in poorly available iron. The actual situation may be more complicated, as shown by Berner et al. (46) using in vitro and rat methodology. These workers suggested that polymerization prior to ingestion probably does not account for decreased bioavailability from a meal. Subsequent study of these polymers introduced into ligated intestinal segments showed lower availability from these polymers (213), suggesting that depolymerization may occur prior to the intestine, probably in the stomach. The conclusion remains that preformed polymeric iron may not be significant in depressing iron absorption.

Work with animals suggests that processing treatments are in many cases beneficial in terms of iron bioavailability, especially of fortification iron. Theuer et al. (214) showed by rat hemoglobin repletion that in sterilized infant formula, bioavailability of most iron fortificants tested was improved after processing. The sterilization process itself was shown to be responsible for the improvement (112).

Several authors have examined the possibility that meat curing may affect iron bioavailability to rats. Mahoney et al. (215) determined that the level of nitrite used in curing of beef affected iron bioavailability, whereas Lee et al. (216) observed no effect of either nitrite or erythorbate on iron absorption. The effects of nitrite-cured meats on iron bioavailability has been reviewed by Lee and Greger (217). A human metabolic balance study suggested a possible (but nonsignificant) increased iron bioavailability from nitrite-cured sausage (217).

Food preparation may alter the heme:nonheme iron ratio in meat (31). This change would lead to lower bioavailability. Prolonged warming of food (75° C for 4 hr) caused a not unexpected decrease (up to 80%) in ascorbic acid content. Although the chemical state of iron was not determined, it would be anticipated that iron bioavailability from a meal would be decreased based on the loss of this enhancer of nonheme iron absorption (218). Losses of ascorbic acid during heat processing would also have an impact on iron nutriture. It has been suggested that ascorbic acid inclusion in diets would be more effective than addition of iron with regard to increasing iron absorption from the diet (149). Unfortunately, the heat-lability of unprotected ascorbic acid precludes its effective use. Fe^{+3} itself may lead to increased oxidation of ascorbic acid on storage, leading to the need to encapsulate one or both reactants (219). Stabilized ethyl cellulose-coated ascorbic acid has been added to corn-soy-milk (CSM) and wheat-soy blend (WSB), with much greater stability (220). Because ferrous sulfate can catalyze fat oxidation and lead to off-flavors, encapsulation has been employed, with no apparent effect on bioavailability (221).

The types of iron fortificants have recently been reviewed in the same book: elemental sources (222), nonelemental sources (221), and experimental sources (123) are all described, taking into account bioavailability considerations, organoleptic effects, and relative cost.

IRON AND IMMUNITY

Various authors have recently reviewed aspects of iron and the immune response (223–226). Iron deficiency can lead to a less efficient immune system, resulting in increased mortality upon bacterial challenge. On the other hand, one strategy of the efficient immune system of an iron-replete individual may be to withhold iron in response to infection, thus reducing access of the infectious agent to this essential nutrient. Thus transferrin saturation may be lower in response to the infection.

Weinberg (224) has described the distinction between acquired immunity and natural immunity. Acquired immunity involves both the humoral and cell-mediated systems of responding to challenge. Natural immunity may be viewed as a generally effective set of responses by the organism. Included in the concept of natural immunity are the alternate complement pathway, lysozyme, β-lysins, interferon, and iron withholding (224). Based on this classification, the iron withholding response may be clearly differentiated from the phenomenon of acquired immunity.

Iron withholding by an organism may be an important function of various transferrin proteins which specifically bind iron, and which are localized at sites of potential microbial invasion. Serum transferrin was first identified based on its biological ability to inhibit microbial growth (227). A more general-purpose example is lactoferrin, which has been found in tears, saliva, gastrointestinal fluid, hepatic bile, seminal fluid, and both cervical and bronchial mucus (144).

Weinberg (224) discussed the means by which the host may mobilize its iron-withholding mechanisms in response to invasion. Transferrin saturation (and thus serum iron) decreases, as does intestinal absorption. The virulence of a microbial species is in some cases related to the efficiency with which it can acquire iron (131), and decreased transferrin saturation makes this acquisition less efficient. It has been pointed out that the term "anemia of infection" may have led to the inappropriate therapeutic administration of iron, wheareas the term "hypoferremic response" would contraindicate this inappropriate treatment (224). This normal response to infection may be an important confounding influence on studies of iron absorption and iron status if not properly considered.

The decreased efficiency of acquired immunity in iron *deficiency* is not to be confused with the temporary "hypoferremic response." The intricacies of the various effects of iron deficiency on the acquired immunity system have been reviewed (223,225).

A potentially interesting aspect of iron and nutritional immunity has recently been suggested (228,229). Iron binding by bifidobacteria has been studied as a mechanism to partly explain the nutritional immunity conferred by breast milk, which encourages growth of these organisms. Bottle-fed infants have lower counts of this organism, which may partially explain their lessened ability to inhibit the growth of pathogens in the intestinal lumen (229).

REFERENCES

1. H. J. M. Bowen, *Trace Elements in Biochemistry*, Academic Press, London, 1966.
2. Anonymous, *Iron*, Subcommittee on Iron, Committee on Medical and Biologic Effects of Environmental Pollutants, University Park Press, Baltimore, 1979.
3. T. G. Spiro, Chemistry and biochemistry of iron, in *Proteins of Iron Metabolism* (E. B. Brown, P. Aisen, J. Fielding, and R. R. Crichton, eds.), Grune and Straton, New York, 1977.
4. P. Saltman, J. Hegenauer, and J. Christopher, Tired blood and rusty livers, *Annals of Clin. Lab. Sci.*, 6(2): 167–176 (1976).
5. J. D. Cook, and B. S. Skikne, Serum ferritin: a possible model for the assessment of nutrient stores, *Am. J. Clin. Nutr.*, 35: 1180–1185 (1982).
6. R. R. Crichton, Iron uptake and utilization by mammalian cells. II. Intracellular iron utilization, *T.I.B.S.*, 9(6): 283–286 (1984).
7. A. Jacobs, Iron metabolism in the bone marrow, in *Metals in Bone* (N. D. Priest, ed.), MTP Press, Ltd., Lancaster, England, 1985.

8. J. D. Cook, D. A. Lipschitz, L. E. M. Miles, and C. A. Finch, Serum ferritin as a measure of iron stores in normal subjects, *Am. J. Clin. Nutr.*, *27*: 681–687 (1974).

9. S. R. Lynch, Iron, in *Absorption and Malabsorption of Mineral Nutrients* (N. W. Solomons and I. H. Rosenberg, eds.), Alan R. Liss, Inc., New York, 1984.

10. C. A. Reed, Oxidation states, redox potentials and spin states, in *The Biological Chemistry of Iron* (H. B. Dunford, D. Dolphin, K. N. Raymond and L. Sieker, eds.), D. Reidel Publishing Co., Dordrecht, Holland, 1982.

11. L. Hallberg, Iron, in *Present Knowledge in Nutrition*, 5th ed., The Nutrition Foundation, Washington, D.C., 1984.

12. P. Aisen, and E. B. Brown, The iron-binding function of transferrin in iron metabolism, *Seminars in Hematology*, *14*: 31–53 (1977).

13. J.-N. Octave, Y.-J. Schneider, A. Tronet, and R. R. Crichton, Iron uptake and utilization by mammalian cells. I. Cellular uptake of transferrin and iron, *T.I.B.S.*, *8*(6): 217–220 (1983).

14. N. D. Chasteen, Transferrin: a perspective, in *Iron Binding Proteins without Cofactors or Sulfur Clusters* (E. C. Theil, G. C. Eichorn, and L. G. Marzilli, eds.), Elsevier, New York, 1983.

15. R. L. Pike and M. S. Brown, *Nutrition. An Integrated Approach*, 3rd ed., John Wiley & Sons, New York, 1984.

16. R. R. Dallman and M. A. Siimes, *Iron deficiency in infancy and childhood*, The Nutrition Foundation, New York, 1979.

17. National Research Council, *Recommended Dietary Allowances*, 9th ed., National Academy of Sciences, Washington, D. C., 1980.

18. D. Narins, Absorption of nonheme iron, in *Biochemistry of Nonheme Iron* (A. Bezkorovainy, ed.), Plenum Press, New York, 1980.

19. E. P. Frassinelli-Gunderson, S. Margen and J. R. Brown, Iron stores in users of oral contraceptive agents, *Am. J. Clin. Nutr.*, *41*: 703–712 (1985).

20. C. A. Finch and J. D. Cook, Iron deficiency, *Am. J. Clin. Nutr.*, *39*: 471–477 (1984).

21. W. H. Crosby and M. A. O'Neil-Cutting, A small-dose iron tolerance test as an indicator of mild iron deficiency, *JAMA*, *251*: 1986–1987 (1984).

22. J. L. Beard and C. A. Finch, Iron deficiency, in *Iron Fortification of Foods* (F. M. Clydesdale and K. L. Wiemer, eds.), Academic Press, New York, 1985.

23. P. R. Dallman and J. D. Reeves, Laboratory diagnosis of iron deficiency, in *Iron Nutrition in Infancy and Childhood* (A. Steckel, ed.), Raven Press, New York, 1984.

24. R. F. Florentino and R. M. Guirriec, Prevalence of nutritional anemia in infancy and childhood with emphasis on developing countries, in *Iron Nutrition in Infancy and Childhood* (A. Steckel, ed.), Raven Press, New York, 1984.

25. D. Vyas and R. K. Chandra, Functional implications of iron deficiency, in *Iron Nutrition in Infancy and Childhood* (A. Steckel, ed.), Raven Press, New York, 1984.

26. V. R. Gordeuk, G. M. Brittenham, C. E. McLaren, M. A. Hughes, and L. J. Keating, Carbonyl iron therapy for iron deficiency anemia, *Blood, 67*: 745–752 (1986).

27. D. P. Derman, T. H. Bothwell, J. D. Torrance, W. R. Bezwoda, A. P. MacPhail, M. C. Kew, M. H. Sayers, P. B. Disler, and R. W. Charlton, Iron absorption from maize (Zea mays) and sorghum (Sorghum vulgare) beer, *Br. J. Nutr., 43*: 271–279 (1980).

28. C. G. Pitt and A. E. Martell, The design of chelating agents for the treatment of iron overload, in *Inorganic Chemistry in Biology and Medicine* (A. E. Martell, ed.), American Chemical Society, Washington, D. C., 1980.

29. K. Lee and F. M. Clydesdale, Iron sources used in food fortification and their changes due to food processing, *CRC Critical Rev. Food Sci. Nutr., 11*: 117–153 (1979).

30. E. R. Monsen, L. Hallberg, M. Layrisse, D. M. Hegsted, J. D. Cook, W. Mertz, and C. A. Finch, Estimation of available dietary iron, *Am. J. Clin. Nutr., 31*: 134–141 (1978).

31. B. R. Schricker, D. D. Miller, and J. R. Stouffer, Measurement and content of nonheme and total iron in muscle, *J. Food Sci., 47*: 740–743 (1982).

32. J. A. T. Pennington, B. E. Young, D. B. Wilson, R. D. Johnson and J. E. Vanderveen, Mineral content of foods and total diets: The selected minerals in foods survey, 1982 to 1984, *J. Am. Dietetic Assoc., 86*: 876–891 (1986).

33. L. Hallberg, E. Bjorn-Rasmussen, L. Rossander, R. Suwanik, R. Pleehachinda, and M. Tuntawiroon, Iron absorption from some Asian meals containing contamination iron, *Am. J. Clin. Nutr., 37*: 272–277 (1983).

34. L. Hallberg and E. Bjorn-Rasmussen, Measurement of iron absorption from meals contaminated with iron, *Am. J. Clin. Nutr., 34*: 2801–2815 (1981).

35. B. C. J. Zoeteman and F. J. J. Brinkman, Human intake of minerals from drinking water in European communities, in *Hardness of Drinking Water and Public Health*, Pergamon Press, London, 1975.

36. D. B. Thompson, The effect of soy products in the diet on retention of non-heme iron from radiolabeled test meals fed to marginally iron-deficient young rats, Ph. D. dissertation, University of Illinois at Urbana-Champaign, 1984.

37. C. V. Moore, Iron nutrition and requirements, *Series Haemato-logica*, *6*: 1—14 (1965).
38. G. S. Sharon, Of (iron) pots and pans. *Nutr. Today*, March/April, 34—35 (1972).
39. H. C. Brittin and C. E. Nossaman, Iron content of food cooked in iron utensils, *J. Am. Dietetic Assoc.*, *86*: 897—901 (1986).
40. A. Rosanoff and B. M. Kennedy, Bioavailability of iron produced by the corrosion of steel in apples, *J. Food Sci.*, *47*: 609—613 (1982).
41. F. E. Martinez and H. Vannuchi, Bioavailability of iron added to the diet by cooking food in an iron pot, *Nutr. Res.*, *6*:421—428 (1986).
42. W. Forth and W. Rummel, Iron absorption, *Physiol. Rev.*, *53*: 724—792 (1973).
43. T. G. Spiro and P. Saltman, Polynuclear complexes of iron and their biological implications, *Structure and Bonding*, *6*: 116—156 (1969).
44. P. Aisen, Some physicochemical aspects of iron metabolism, in *Iron Metabolism*, CIBA Foundation Symposium 51 (new series), Elsevier, Amsterdam, 1977.
45. T. G. Spiro and P. Saltman, Inorganic chemistry, in *Iron Bio-chemistry and Medicine* (A. Jacobs, and M. Worwood, eds.), Academic Press, New York, 1974.
46. L. A. Berner, D. D. Miller and D. Van Campen, Availability to rats of iron in ferric hydroxide polymers, *J. Nutr.*, *115*: 1042—1049 (1985).
47. G. W. Bates, C. Billups and P. Saltman, The kinetics and mechanism of iron(III) exchange between chelates and transferrin. I. The complexes of citrate and nitrilotriacetic acid, *J. Biol. Chem.*, *242*: 2810—2815 (1967).
48. G. W. Bates, C. Billups and P. Saltman, The kinetics and mechanism of iron(III) exchange between chelates and transferrin. II. The presentation and removal with ethylenediaminetetracetate, *J. Biol. Chem.*, *242*: 2816—2821 (1967).
49. J. E. Gorman and F. M. Clydesdale, Thermodynamic and kinetic stability constants of selected carboxylic acids and iron, *J. Food Sci.*, *49*: 500—503 (1984).
50. A. Dautry-Varsat, A. Ciechanover and H. F. Lodish, pH and the recycling of transferrin during receptor-mediated endocytosis, *Proc. Nat. Acad. Sci. USA*, *80*: 2258—2262 (1983).
51. R. W. Charlton and T. H. Bothwell, Iron absorption, *Ann. Rev. Med.*, *34*: 55—68 (1983).
52. E. J. Underwood, *Trace Elements in Human and Animal Nutrition*, Academic Press, New York, 1977.
53. M. C. Linder, H. N. Munro, The mechanism of iron absorption and its regulation, *Fed. Proc.*, *36*: 2017—2023 (1977).

54. B. Brozovic, Absorption of iron, in *Intestinal Absorption in Man* (J. McColl and G. E. Sladen, eds), Academic Press, New York, 1975.

55. D. Van Campen, Regulation of iron absorption. *Fed. Proc.*, *33*: 100–105 (1974).

56. A. Turnbull, Iron absorption, in *Iron in Biochemistry and Medicine* (A. Jacobs, and M. Worwood, eds.), Academic Press, London, 1974.

57. R. Tenhunen, R. Grasbeck, I. Kuovonen, and M. Lundberg, An intestinal receptor for heme: its partial characterization, *Int. J. Biochem.*, *12*: 713–716 (1980).

58. L. R. Weintraub, M. B. Weinstein, H. J. Huser, and S. Rafal, Absorption of hemoglobin iron: the role of a heme-splitting substance in the intestinal mucosa, *J. Clin. Invest.*, *47*: 531–539 (1968).

59. M. S. Wheby, G. E. Suttle, and K. T. Ford, Intestinal absorption of hemoglobin iron, *Gastroent.*, *58*: 647–654 (1970).

60. J. C. Wyllie, and N. Kauffman, An electron microscopic study of heme uptake by rat duodenum, *Lab. Invest.*, *47*: 471–476 (1982).

61. J. G. Manis, and D. Schachter, Active transport of iron by intestine: features of the two-step mechanism, *Am. J. Physiol.*, *203*: 73–80 (1962).

62. R. A. McCance, and E. M. Widdowson, Absorption and excretion of iron, *Lancet*, *2*: 680–684 (1937).

63. P. F. Hahn, W. F. Bale, J. F. Ross, W. M. Balfour, and G. H. Whipple, Radioactive iron absorption by gastrointestinal tract. Influence of anemia, anoxia and antecedent feeding. Distribution in growing dogs, *J. Exper. Med.*, *78*: 169–188 (1943).

64. S. Granick, Ferritin. IX. Increase of the protein apoferritin in the gastrointestinal mucosa as a direct response to iron feeding. The function of ferritin in the regulation of iron absorption, *J. Biol. Chem.*, *164*: 737–746 (1946).

65. M. E. Conrad, and W. H. Crosby, Intestinal mucosal mechanisms controlling iron absorption, *Blood*, *22*: 406–415 (1963).

66. W. H. Crosby, The control of iron balance by the intestinal mucosa, *Blood*, *22*: 441–449 (1963).

67. M. A. Savin, and J. D. Cook, Mucosal iron transport by rat intestine, *Blood*, *56*: 1029–1035 (1980).

68. F. A. El-Shobaki and W. Rummel, Mucosal transferrin and ferritin factors in the regulation of iron absorption, *Res. Exp. Med.*, *171*: 243–253 (1977).

69. N. J. Greenberger, S. P. Balcerzak and G. A. Ackerman, Iron uptake by isolated intestinal brush borders. Changes induced by alterations in iron stores, *J. Lab. Clin. Med.*, *73*: 711–721 (1969).

70. C. L. Kimber, T. Mukherjee and D. J. Deller, *In vitro* iron

attachment to the intestinal brush border: effect of iron stores and other environmental factors, *Am. J. Dig. Dis.*, *18*: 781–791 (1973).

71. T. M. Cox and T. J. Peters, The kinetics of iron uptake *in vitro* by human duodenal mucosa: studies in normal subjects, *J. Physiol.*, *289*: 469–478 (1979).

72. T. M. Cox and T. J. Peters, Cellular mechanisms in the regulation of iron absorption by the human intestine: studies in patients with iron deficiency before and after treatment, *Br. J. Haematol.*, *44*: 75–86 (1980).

73. T. M. Cox and M. W. O'Donnell, Studies on the control of iron uptake by rabbit small intestine, *Br. J. Nutr.*, *47*: 251–258 (1982).

74. M. W. O'Donnell and T. M. Cox, Microvillar iron-binding glycoproteins isolated from the rabbit small intestine, *Biochem. J.*, *202*: 107–115 (1982).

75. H. Huebers, E. Huebers, W. Forth and W. Rummel, Binding of iron to a non-ferritin protein in the mucosal cells of normal and iron-deficient rats during absorption, *Life Sciences*, *10* (part I): 1141–1148 (1971).

76. H. Huebers and W. Rummel, Iron binding proteins: mediators in iron absorption, in *Intestinal Permeation* (M. Kramer and F. Lauterbach, eds.), Proceedings of the Fourth Workshop Conference Hoechst, Schloss Reinsenberg, Excerpta Medica, Amsterdam, 1975.

77. H. A. Huebers, E. Huebers, E. Csiba, W. Rummel and C. A. Finch, The significance of transferrin for intestinal iron absorption, *Blood*, *61*: 283–290 (1983).

78. S. B. Refsum and B. B. Schreiner, Regulation of iron balance by absorption and excretion. A critical review and a new hypothesis, *Scand. J. Gastroent.*, *19*: 867–874 (1984).

79. G. Astaldi, G. Meardi and T. Lisino, The iron content of jejunal mucosa obtained by Crosby's biopsy in hemochromatosis and hemosiderosis, *Blood*, *28*: 70–82 (1966).

80. E. Bjorn-Rasmussen, J. Carneskog and A. Cederblad, Losses of ingested iron temporarily retained in the gastrointestinal tract, *Scand. J. Haematol.*, *25*: 124–126 (1980).

81. S. J. Fairweather-Tait and A. J. A. Wright, The influence of previous iron intake on the estimation of bioavailability of Fe from a test meal given to rats, *Br. J. Nutr.*, *51*: 185–191 (1984).

82. S. J. Fairweather-Tait, T. E. Swindell and A. J. A. Wright, Further studies in rats on the influence of previous iron intake on the estimation of bioavailability of Fe, *Br. J. Nutr.*, *54*: 79–86 (1985).

83. R. W. Topham, S. A. Joslin and J. S. Prince Jr., The effect of short-term exposure to low-iron diets on the mucosal processing of ionic iron, *Biochem. Biophys. Res. Commun.*, *133*: 1092–1097 (1985).

84. U. M. Saarinen, M. A. Siimes and P. R. Dallman, Iron absorption in infants: High bioavailability of breast milk iron as indicated by the extrinsic tag method of iron absorption and by the concentration of serum ferritin, *J. Pediatrics, 91*: 36–39 (1977).

85. D. Van Campen and R. M. Welch, Availability to rats of iron from spinach: effects of oxalic acid, *J. Nutr., 110*: 1618–1621 (1980).

86. D. B. Thompson and J. W. Erdman Jr., The effect of soy protein isolate in the diet on retention by the rat of iron from radiolabeled test meals, *J. Nutr., 114*: 307–311 (1984).

87. D. B. Thompson and J. W. Erdman Jr., The effect of diet on retention by the rat of iron from a radiolabeled casein test meal, *J. Nutr., 115*: 319–326 (1985).

88. Anonymous, Effect of meals on iron absorption, *Nutr. Rev., 44* (1): 22 (1986).

89. N. Kojima, D. Wallace and G. W. Bates, The effect of chemical agents, beverages, and spinach on the *in vitro* solubilization of iron from cooked pinto beans, *Am. J. Clin. Nutr., 34*: 1392–1407 (1981).

90. D. Van Campen, Iron bioavailability techniques: an overview, *Food Tech., 37*(10): 127–132 (1983).

91. D. D. Miller and B. R. Schricker, In vitro estimation of food iron bioavailability, in *Nutritional Bioavailability of Iron*, (C. Kies, ed.), American Chemical Society, Washington, D. C., 1982.

92. B. S. Narrasinga Rao and T. Prabhavathi, An *in vitro* method for predicting the bioavailability of iron from foods, *Am. J. Clin. Nutr., 31*: 169–175 (1978).

93. D. D. Miller, B. R. Schricker, R. R. Rasmussen and D. Van Campen, An *in vitro* method for estimation of iron availability from meals, *Am. J. Clin. Nutr., 34*: 2248–2256 (1981).

94. B. R. Schricker, D. D. Miller, R. R. Rasmussen and D. Van Campen, A comparison of *in vivo* and *in vitro* methods for determining availability of iron from meals, *Am. J. Clin. Nutr., 34*: 2257–2263 (1981).

95. M. B. Reddy, M. V. Chidambaram, J. Fonseca and G. W. Bates, Potential role of *in vitro* iron bioavailability studies in combatting iron deficiency: A study of the effects of phosvitin on iron mobilization from pinto beans, *Clin. Physiol. Biochem., 4*: 78–86 (1986).

96. J. E. Gorman and F. M. Clydesdale, The behavior and stability of iron-ascorbate complexes in solution, *J. Food Sci., 48*: 1217–1220, 1225 (1983).

97. P. M. May, D. R. Williams and P. W. Linder, Biological significance of low molecular weight iron(III) complexes, in *Metal Ions in Biological Systems*, vol. 7 (H. Sigel, ed.), Marcel Dekker, New York, 1978.

98. J. Schultz and N. J. Smith, A quantitative study of the absorption of food iron in infants and children, *Am. J. Dis. Child.*, *95*: 109–119 (1958).

99. C. Martinez-Torres, E. Romano and M. Layrisse, Effect of cysteine on iron absorption in man, *Am. J. Clin. Nutr.*, *34*: 322–327 (1981).

100. L. Hallberg and E. Bjorn-Rasmussen, Determination of iron absorption from whole diet. A new two-pool model using two radioiron isotopes given as haem and non-haem iron, *Scand. J. Haematol.*, *9*: 193–197 (1972).

101. J. D. Cook, M. Layrisse, C. Martinez-Torres, R. Walker, E. Monsen and C. A. Finch, Food iron absorption measured by an extrinsic tag, *J. Clin. Invest.*, *51*: 805–815 (1972).

102. L. Hallberg, The pool concept in food iron absorption and some of its implications, *Proc. Nutr. Soc.*, *33*: 285–291 (1974).

103. E. R. Monsen, Validation of an extrinsic iron label in monitoring absorption of nonheme food iron in normal and iron-deficient rats, *J. Nutr.*, *104*: 1490–1495 (1974).

104. L. Hallberg, Bioavailability of dietary iron in man, *Ann. Rev. Nutr.*, *1*: 123–147 (1981).

105. K. T. Smith, Effects of chemical environment on iron bioavailability measurements, *Food Tech.*, *39* (10): 115–120 (1983).

106. J. R. Consaul and K. Lee, Extrinsic tagging in iron bioavailability research: a critical review, *J. Agric. Food Chem.*, *31*: 684–689 (1983).

107. J. D. Cook, V. Minnich, C. V. Moore, A. Ramussen, W. B. Bradley and C. A. Finch, Absorption of fortification iron in bread, *Am. J. Clin. Nutr.*, *26*: 861–872 (1973).

108. B. R. Schricker and D. D. Miller, *In vitro* estimation of relative iron availability in breads and meals containing different forms of fortification iron, *J. Food Sci.*, *47*: 723–727 (1982).

109. K. Lee, Iron chemistry and bioavailability in food processing, in *Nutritional Bioavailability of Iron* (C. Kies, ed.), American Chemical Society, Washington, D.C., 1982.

110. F. M. Clydesdale, Physicochemical determinants of iron bioavailability, *Food Tech.*, *37* (10): 133–144 (1983).

111. R. S. Kadan and G. M. Ziegler Jr., Effects of ingredients on iron distribution in spray-dried experimental soy beverage, *Cereal Chem.*, *61*: 5–8 (1984).

112. R. C. Theuer, Fortification of infant formula, in *Iron Fortification of Foods* (F. M. Clydesdale and K. L. Wiemer, eds.), Academic Press, New York, 1985.

113. T. A. Morck and J. D. Cook, Factors affecting the bioavailability of dietary iron, *Cereal Foods World*, *26*: 667–672 (1981).

114. E. R. Morris, An overview of current information on bioavailability of dietary iron to humans, *Fed. Proc.*, *42*: 1716–1720 (1983).

115. J. D. Cook, Determinants of nonheme iron absorption in man, *Food Tech.*, *37* (10): 124–126 (1983).
116. R. M. Forbes and J. W. Erdman Jr., Bioavailability of trace mineral elements, *Ann. Rev. Nutr.*, *3*: 213–231 (1983).
117. C. Kies, ed., *Nutritional Bioavailability of Iron*, American Chemical Society, Washington, D. C., 1982.
118. S. R. Lynch and J. D. Cook, Interaction of vitamin C and iron, *Ann. N. Y. Acad. Sci.*, *355*: 32–43 (1980).
119. E. R. Monsen, Ascorbic acid: an enhancing factor in iron absorption, in *Nutritional Bioavailability of Iron*, (C. Kies, ed.), American Chemical Society, Washington, D. C., 1982.
120. M. E. Conrad and S. G. Schade, Ascorbic acid chelates in iron absorption: a role for hydrochloric acid and bile, *Gastroent.*, *55*: 35–45 (1968).
121. D. M. Hungerford and M. C. Linder, Aspects of the effect of vitamin C on iron absorption, in *The Biochemistry and Physiology of Iron* (P. Saltman and J. Hegenauer, eds.), Elsevier North Holland, Inc., New York, 1982.
122. A. P. MacPhail, T. H. Bothwell, J. D. Torrance, D. P. Derman, W. R. Bezwoda, R. W. Charlton and F. Mayet, Factors affecting the absorption of iron from Fe(III)EDTA, *Br. J. Nutr.*, *45*: 215–227 (1981).
123. P. MacPhail, R. Charlton, J. H. Bothwell and W. Bezwoda, Experimental fortificants, in *Iron Fortification of Foods*, (F. M. Clydesdale and K. L. Wiemer, eds.), Academic Press, New York, 1985.
124. L. Hallberg and L. Rossander, Effect of different drinks on the absorption of non-heme iron from composite meals, *Human Nutr.: Applied Nutr.*, *36A*: 116–123 (1982).
125. E. R. Monsen and J. D. Cook, Food iron absorption in human subjects. V. Effects of the major dietary constituents of a semi-synthetic meal, *Am. J. Clin. Nutr.*, *32*: 804–808 (1979).
126. P. G. Taylor, C. Martinez-Torres, E. L. Ramano, and M. Layrisse, The effect of cysteine-containing peptides released during meat digestion on iron absorption in humans, *Am. J. Clin. Nutr.*, *43*: 68–71 (1986).
127. J. A. McMillan, S. A. Landaw, and F. Oski, Iron sufficiency in breast-fed infants and the availability of iron from human milk. *Pediatrics*, *58*: 686–691 (1976).
128. G. W. Bates, J. Boyer, J. C. Hegenauer, and P. Saltman, Facilitation of iron absorption by ferric fructose, *Am. J. Clin. Nutr.*, *25*: 983–986 (1972).
129. G. Bates, J. Hegenauer, J. Renner, and P. Saltman, Complex formation, polymerization and autoreduction in the ferric fructose system, *Bioinorganic Chem.*, *2*: 311–327 (1973).
130. H. Cross, T. Pepper, M. W. Kearsley, and G. G. Birch, Mineral complexing properties of food carbohydrates, *Starch*, *37*: 132–135 (1985).

131. J. B. Neilands, Iron absorption and transport in microorganisms, Ann. Rev. Nutr., 1: 27—46 (1981).

132. M. Gillooly, T. H. Bothwell, J. D. Torrance, A. P. MacPhail, D. P. Derman, W. R. Bezwoda, W. Mills, and R. W. Charlton, The effects of organic acids, phytates and polyphenols on the absorption of iron from vegetables, Br. J. Nutr., 49: 331—342 (1983).

133. P. Ribereau-Gayon, Plant Phenolics, Hafner Publishing Co., New York, 1972.

134. J. B. Neilands, Microbial iron compounds, Ann. Rev. Biochem., 50: 715—731 (1981).

135. P. B. Disler, S. R. Lynch, R. W. Charlton, J. D. Torrance, and T. H. Bothwell, The effect of tea on iron absorption, Gut, 16: 193—200 (1975).

136. P. B. Disler, S. R. Lynch, J. D. Torrance, M. H. Sayers, T. H. Bothwell, and R. W. Charlton, The mechanism of the inhibition of iron absorption by tea, S. Afr. J. Med. Sci., 40: 109—116 (1975).

137. L. Rossander, L. Hallberg, and E. Bjorn-Rasmussen, Absorption of iron from breakfast meals, Am. J. Clin. Nutr., 32: 2484—2489 (1979).

138. B. S. Narasinga Rao, and T. Prabhavathi, Tannin content of foods commonly consumed in India and its influence on ionisable iron, J. Sci. Food Agric., 33: 89—96 (1982).

139. T. A. Morck, S. R. Lynch, and J. D. Cook, Inhibition of food iron absorption by coffee, Am. J. Clin. Nutr., 37: 416—420 (1983).

140. S. N. Roy, and S. Mukherjee, Influence of food tannins on certain aspects of iron metabolism: Part I - Absorption and excretion in normal and anemic rats, Indian J. Biochem. Biophys. 16: 93—98 (1979).

141. N. C. Panda, B. K. Sahu, S. K. Panda, and A. G. Rao, Damage done to intestine, liver and kidneys by tannic acid of tea and coffee, Ind. J. Nutr. Dietet., 18: 97—103 (1981).

142. J. D. Cook and E. R. Monsen, Food iron absorption in human subjects III. Comparison of the effect of animal proteins on nonheme iron absorption, Am. J. Clin. Nutr., 29: 859—867 (1976).

143. T. Peters, L. Apt, and J. F. Ross, Effect of phosphates upon iron absorption studied in normal human subjects and in an experimental model using dialysis, Gastroent., 61: 315—322 (1971).

144. A. Bezkorovainy, Biochemistry of Nonheme Iron, Plenum Press, New York, 1980.

145. G. Taborsky, Iron in egg yolk, in The Biochemistry and Physiology of Iron (P. Saltman and J. Hegenauer, eds.), Elsevier North Holland, Inc., New York, 1982.

146. G. Taborsky, Phosvitin, in *Iron Binding Proteins Without Cofactors on Sulfur Clusters* (E. C. Theil, G. C. Eichorn, and L. G. Marzilli, eds.) Elsevier, New York, 1983.

147. R. Sato, Y.-S. Lee, T. Noguchi, and H. Naito, Iron solubility in the small intestine of rats fed egg yolk protein, *Nutr. Rep. Int.*, *30*: 1319–1326 (1984).

148. R. Sato, T. Noguchi, and H. Naito, The formation and iron-binding property of phosphopeptides in the small intestinal contents of rats fed egg yolk diet, *Nutr. Rep. Int.*, *31*: 245–252 (1985).

149. T. H. Bothwell, F. M. Clydesdale, J. D. Cook, P. R. Dallman, L. Hallberg, D. Van Campen, and W. J. Wolf, *The effects of cereals and legumes on iron availability*, The Nutrition Foundation, New York, 1982.

150. J. D. Cook, T. A. Morck, and S. R. Lynch, The inhibitory effect of soy products on nonheme iron absorption in man, *Am. J. Clin. Nutr.*, *34*: 2622–2629 (1981).

151. T. A. Morck, S. R. Lynch, and J. D. Cook, Reduction of the soy-induced inhibition of nonheme iron absorption, *Am. J. Clin. Nutr.*, *36*: 219–228 (1982).

152. L. Hallberg and L. Rossander, Effect of soy protein on nonheme iron absorption in man, *Am. J. Clin. Nutr.*, *36*: 514–520 (1982).

153. C. E. Bodwell, Effects of soy protein on iron and zinc utilization in humans, *Cereal Foods World*, *28*: 342–348 (1983).

154. S. R. Lynch, J. L. Beard, S. A. Dassenko, and J. D. Cook, Iron absorption from legumes in humans, *Am. J. Clin. Nutr.*, *40*: 42–47 (1984).

155. L. Hallberg and L. Rossander, Improvement of iron nutrition in developing countries: comparison of adding meat, soy protein, ascorbic acid, citric acid, and ferrous sulfate on iron absorption from a sample Latin American-type of meal, *Am. J. Clin. Nutr.*, *39*: 577–583 (1984).

156. M. Gillooly, J. D. Torrance, T. H. Bothwell, A. P. MacPhail, D. Derman, W. Mills, and F. Mayet, The relative effect of ascorbic acid on iron absorption from soy-based and milk-based infant formulas, *Am. J. Clin. Nutr.*, *40*: 522–527 (1984).

157. B. R. Schricker, D. D. Miller, and D. Van Campen, Effects of iron status and soy protein on iron absorption by rats, *J. Nutr.*, *113*: 996–1001 (1983).

158. C. R. Ranger and R. J. Neale, Iron availability from soy, meat and soy/meat samples in anemic rats with and without prevention of caprophagy, *J. Food Sci.*, *49*: 585–589 (1984).

159. M. F. Picciano, K. E. Weingartner, and J. W. Erdman Jr., Relative bioavailability of dietary iron from three processed soy products, *J. Food Sci.*, *49*: 1558–1561 (1984).

160. J. L. Greger and J. Mulvaney, Absorption and tissue distribution of zinc, iron, and copper by rats fed diets containing lactalbumin, soy, and supplemental sulfur-containing amino acids, *J. Nutr.*, *115*: 200–210 (1985).

161. B. R. Schricker, D. D. Miller and D. Van Campen, *In vitro* estimation of iron availability in meals containing soy products, *J. Nutr.*, *112*: 1696–1705 (1982).

162. A. P. Kane and D. D. Miller, *In vitro* estimation of the effects of selected proteins on iron bioavailability, *Am. J. Clin. Nutr.*, *39*: 393–401 (1984).

163. M. I. Schnepf and L. D. Satterlee, Partial characterization of an iron soy protein complex, *Nutr. Rep. Int.*, *31*: 371–380 (1985).

164. L. A. Berner and D. D. Miller, Effects of dietary proteins on iron bioavailability—a review, *Food Chem.*, *18*: 47–69 (1985).

165. J. L. Kelsay, Effect of diet fiber level on bowel function and trace mineral balances of human subjects, *Cereal Chem.*, *58*: 2–5 (1981).

166. R. Ali, H. Staub, G. Coccodrilli, and L. Schanbacher, Nutritional significence of dietary fiber: effect on nutrient bioavailability and selected gastrointestinal functions, *J. Agric. Food Chem.*, *29*: 465–472 (1981).

167. R. Fernandez and S. F. Phillips, Components of fiber impair iron absorption in the dog, *Am. J. Clin. Nutr.*, *35*: 107–112 (1982).

168. R. Fernandez and S. F. Phillips, Components of fiber bind iron *in vitro*, *Am. J. Clin. Nutr.*, *35*: 100–106 (1982).

169. L. Monnier, C. Colette, L. Aguirre, and J. Mirouze, Evidence and mechanism for pectin-reduced intestinal inorganic iron absorption in idiopathic hemochromatosis, *Am. J. Clin. Nutr.*, *33*: 1225–1232 (1980).

170. A. L. Camire and F. M. Clydesdale, Effect of pH and heat treatment on the binding of calcium, magnesium, zinc, and iron to wheat bran and fractions of dietary fiber, *J. Food Sci.*, *46*: 548–551 (1981).

171. L. A. Berner and L. F. Hood, Iron binding by sodium alginate, *J. Food Sci.*, *48*: 755–758 (1983).

172. S. R. Platt and F. M. Clydesdale, Binding of iron by cellulose, lignin, sodium phytate and beta-glucan, alone and in combination, under simulated gastrointestinal pH conditions, *J. Food Sci.*, *49*: 531–535 (1984).

173. M. B. Zemel and P. C. Zemel, Effects of food gums on zinc and iron solubility following *in vitro* digestion, *J. Food Sci.*, *50*: 547–550 (1985).

174. G. S. Ranhotra, C. Lee, and J. A. Gilroth, Bioavailability of iron in high-cellulose bread, *Cereal Chem.*, *56*: 156–158 (1979).

175. W. Van Dokkum, A. Wesstra, and F. A. Schippers, Physiological effects of fibre-rich types of bread, *Br. J. Nutr.*, 47: 451–460 (1982).

176. M. J. Leigh and D. D. Miller, Effects of pH and chelating agents on iron binding by dietary fiber: implications for iron availability, *Am. J. Clin. Nutr.*, 38: 202–213 (1983).

177. R. R. Mod, R. L. Ory, N. M. Morris, and F. L. Normand, Chemical properties and interactions of rice hemicellulose with trace minerals *in vitro*, *J. Agric. Food Chem.*, 29: 449–454 (1981).

178. S. A. Thompson and C. W. Weber, Influence of pH on the binding of copper, zinc, and iron in six fiber sources, *J. Food Sci.*, 44: 752–754 (1979).

179. J. G. Reinhold, S. S. Garcia, and P. Garzon, Binding of iron by fiber of wheat and maize, *Am. J. Clin. Nutr.*, 34: 1384–1391 (1981).

180. J. Hegenauer, P. Saltman, D. Ludwig, L. Ripley, and A. Ley, Iron-supplemented cow milk. Identification and spectral properties of iron bound to casein micelles, *J. Agric. Food Chem.*, 27: 1294–1301 (1979).

181. E. R. Monsen and J. D. Cook, Food iron absorption in human subjects IV. The effects of calcium and phosphate salts on the absorption of nonheme iron, *Am. J. Clin. Nutr.*, 29: 1142–1148 (1976).

182. S. M. Snedeker, S. A. Smith, and J. L. Greger, Effect of dietary calcium and phosphorus levels on the utilization of iron, copper and zinc by adult males, *J. Nutr.*, 112: 136–143 (1982).

183. J. L. Greger, Effects of phosphorus-containing compounds on iron and zinc utilization, in *Nutritional Bioavailability of Iron* (C. Kies, ed.), American Chemical Society, Washington, D. C., 1982.

184. K. Subba Rao and B. S. Narasinga Rao, Studies on iron chelating agents for use as absorption promoters—Effect of inorganic polyphosphates on dietary ionizable iron, *Nutr. Rep. Int.*, 27: 1209–1219 (1983).

185. A. W. Mahoney and D. G. Hendricks, Some effects of different phosphate compounds on iron and calcium absorption, *J. Food Sci.*, 43: 1473–1476 (1978).

186. M. B. Zemel and M. T. Bidari, Zinc, iron and copper availability as affected by orthophosphates, polyphosphates and calcium, *J. Food Sci.*, 48: 567–573 (1983).

187. M. B. Zemel, In vitro evaluation of the effects of orthotripoly-, and hexametaphosphate on zinc, iron and calcium bioavailability, *J. Food Sci.*, 49: 1562–1565 (1984).

188. N. J. S. Bour, B. A. Soullier, and M. B. Zemel, Effect of level and form of phosphorus and level of Ca intake on zinc, iron and copper bioavailability in man, *Nutr. Res.*, 4: 371–379 (1984).

189. D. B. Thompson and J. W. Erdman Jr., Phytic acid determination in soybeans, *J. Food Sci.*, *47*: 513—517 (1982).

190. E. R. Morris and R. Ellis, Isolation of monoferric phytate from wheat bran and its biological value as an iron source to the rat, *J. Nutr.*, *106*: 753—760 (1976).

191. J. Beard, personal communication, 1986.

192. K. M. Simpson, E. R. Morris, and J. D. Cook, The inhibitory effect of bran on iron absorption in man, *Am. J. Clin. Nutr.*, *34*: 1469—1478 (1981).

192a. L. Hallberg, L. Rossander, and A. B. Skanberg, Phytates and the inhibitory effect of bran on iron absorption in man, *Am. J. Clin. Nutr.*, *45*: 988—996 (1987).

193. E. R. Morris and R. Ellis, Phytate, wheat bran, and bioavailability of dietary iron, in *Nutritional Bioavailability of Iron*, (C. Kies, ed.), American Chemical Society, Washington, D. C., 1982.

194. S. J. Fairweather-Tait, The effect of different levels of wheat bran on iron absorption in rats from bread containing similar amounts of phytate, *Br. J. Nutr.*, *47*: 243—249 (1982).

195. H. Anderson, B. Navert, S. A. Bingham, H. N. Englyst, and J. H. Cummings, The effect of breads containing similar amounts of phytate but different amounts of wheat bran on calcium, zinc and iron balance in man, *Br. J. Nutr.*, *50*: 503—510 (1983).

196. J. D. Cook and E. R. Monsen, Food iron absorption in man II. The effect of EDTA on absorption of dietary non-heme iron, *Am. J. Clin. Nutr.*, *29*: 614—620 (1976).

197. K. J. Nelson and N. N. Potter, Iron binding by wheat gluten, soy isolate, zinc, albumin, and casein, *J. Food Sci.*, *44*: 104—107 (1979).

198. K. J. Nelson and N. N. Potter, Iron availability from wheat gluten, soy isolate, and casein complexes, *J. Food Sci.*, *45*: 52—55 (1980).

199. C. V. Morr and A. Seo, Comparison of methods to remove endogenous iron from soy protein isolates, *J. Food Sci.*, *51*: 994—996 (1986).

200. S. R. Lynch, S. A. Dassenko, T. A. Morck, J. L. Beard, and J. D. Cook, Soy protein products and heme iron absorption in humans, *Am. J. Clin. Nutr.*, *41*: 13—20 (1985).

201. B. S. Skikne, S. R. Lynch, R. G. Robinson, J. A. Spicer, and J. D. Cook, The effect of food consistency on iron absorption, *Am. J. Gastroent.*, *78*: 607—610 (1983).

202. A.-E. Harmuth-Hoene and R. Schelenz, Effect of dietary fiber on mineral absorption in growing rats, *J. Nutr.*, *110*: 1774—1784 (1980).

203. K. Lee and F. M. Clydesdale, Quantitative determination of the elemental, ferrous, ferric, soluble and complexed iron in foods, *J. Food Sci.*, *44*: 549—554 (1979).

204. K. Lee and F. M. Clydesdale, Chemical changes of iron in food and drying processes, *J. Food Sci.*, *45*: 711–715 (1980).
205. K. Lee and F. M. Clydesdale, Effect of thermal processing on endogenous and added iron in canned spinach, *J. Food Sci.*, *46*: 1064–1073 (1981).
206. S. W. Rizk and F. M. Clydesdale, Effect of iron sources and ascorbic acid on the chemical profile of iron in a soy protein isolate, *J. Food Sci.*, *48*: 1431–1435 (1983).
207. F. M. Clydesdale and D. B. Nadeau, Solubilization of iron in cereals by milk and milk fractions, *Cereal Chem.*, *61*: 330–335 (1984).
208. S. W. Rizk and F. M. Clydesdale, Effect of organic acids in the *in vitro* Solubilization of iron from a soy-extended meat patty, *J. Food Sci.*, *50*: 577–581 (1985).
209. S. W. Rizk and F. M. Clydesdale, Effects of baking and boiling on the ability of selected organic acids to solubilize iron from a corn-soy-milk food blend fortified with exogenous iron sources, *J. Food Sci.*, *50*: 1088–1091 (1985).
210. R. S. Kadan and G. M. Ziegler Jr., Effects of ingredients on iron solubility and chemical state in experimental breads, *Cereal Chem.*, *63*: 47–51 (1986).
211. R. A. Clemens and K. C. Mercurio, Effects of processing on the bioavailability and chemistry of iron powders in a liquid milk-based product, *J. Food Sci.*, *46*: 930–935 (1981).
212. B. L. Carlson and D. D. Miller, Effects of product formulation, processing, and meal composition on *in vitro* estimated iron availability from cereal-containing breakfast meals, *J. Food Sci.*, *48*: 1211–1216 (1983).
213. L. A. Berner, D. D. Miller, and D. Van Campen, Absorption of iron from ferric hydroxide polymers introduced into ligated rat duodenal segments, *J. Nutr.*, *116*: 259–264 (1986).
214. R. C. Theuer, K. S. Kemmerer, W. H. Martin, B. L. Zoumas, and H. P. Sarett, Effect of processing on availability of iron salts in liquid infant formula products, *J. Agric. Food Chem.*, *19*: 555–558 (1971).
215. A. W. Mahoney, D. G. Hendrickes, T. A. Gillett, D. R. Buck, and C. G. Miller, Effect of sodium nitrite on the bioavailability of meat iron for the anemic rat, *J. Nutr.*, *109*: 2182–2189 (1979).
216. K. Lee, B. L. Chinn, J. L. Greger, K. L. Graham, J. E. Shimaoka, and J. C. Liebert, Bioavailability of iron to rats from nitrite and erythorbate cured processed meats, *J. Agric. Food Chem.*, *32*: 856–860 (1984).
217. K. Lee and J. L. Greger, Bioavailability and chemistry of iron from nitrite-cured meats, *Food Tech.*, *37*(10): 139–144 (1983).
218. L. Hallberg, L. Rossander, H. Persson, and E. Svahn, Deleterious effects of prolonged warming of meals on ascorbic acid content and iron absorption, *Am. J. Clin. Nutr.*, *36*: 846–850 (1982).

219. S. R. Tannenbaum, V. R. Young, and M. C. Archer, Vitamins and minerals, in *Food Chemistry*, 2nd ed. (O. R. Fennema, ed), Marcel Dekker, New York, 1985.

220. F. Barrett and P. Ranum, Wheat and blended cereal foods, in *Iron Fortification of Foods* (F. M. Clydesdale and K. L. Wiemer, eds.), Academic Press, New York, 1985.

221. R. F. Hurrell, Nonelemental sources, in *Iron Fortification of Foods* (F. M. Clydesdale and K. L. Wiemer, eds.), Academic Press, New York, 1985.

222. J. Patrick, Elemental sources, in *Iron Fortification in Foods* (F. M. Clydesdale and K. L. Wiemer, eds.), Academic Press, New York, 1985.

223. R. K. Chandra, B. Au, G. Woodford, and P. Hyam, Iron status, immune response and susceptibility to infection, in *Iron Metabolism*, Ciba Foundation Symposium 51 (new series), Elsevier, Amsterdam, 1977.

224. E. D. Weinberg, Iron withholding: a defense against infection and neoplasia, *Physiol. Rev.*, *64*: 65–102 (1984).

225. A. Sherman, Iron, infection and immunity, in *Malnutrition, Disease Resistance, and Immune Function* (R. R. Watson, ed.), Marcel Dekker, New York, 1984.

226. E. D. Weinberg, Iron, infection and neoplasia, *Clin. Physiol. Biochem.*, *4*: 50–60 (1986).

227. A. L. Schade and L. Caroline, An iron-binding component in human blood plasma, *Science*, *104*: 340–341 (1946).

228. N. Topouzian and A. Bezkorovainy, Iron uptake by *Bifidobacterium bifidum* var. *Pennsylvanicus*: the effect of sulfhydryl reagents and metal chelators, *IRCS Med. Sci.*, *14*: 275–276 (1986).

229. A. Bezkorovainy, N. Topouzian, and R. Miller-Catchpole, Mechanisms of ferric and ferrous iron uptake by *Bifidobacterium bifidum* var. *Pennsylvanicus*, *Clin. Physiol. Biochem.*, *4*: 150–158 (1986).

230. A. L. Lehninger, *Biochemistry*, 2nd ed., Worth Publishers, New York, 1977.

6
Zinc

KENNETH T. SMITH / The Procter & Gamble Company, Cincinnati, Ohio

INTRODUCTION

The study of trace elements in the nutritional sciences has progressed rapidly in the past 20 years. Interestingly, the term "trace element" was originally used to denote a quantity of mineral so small that it could not be accurately measured. As our instrumentation and analytical capabilities have continually improved, so too has our knowledge and understanding of trace element nutrition. While some elements have surpassed others with respect to the progress made, factors such as ease of identification, number of measurable biochemical parameters, unique versus ubiquitous distribution, toxicity profiles, and deficiency symptoms have all contributed to the complex and as yet incomplete evolutionary emergence of our understanding of each trace element in question. Zinc is certainly one element that has undergone a tremendous explosion in information during this relatively short time.

DISTRIBUTION AND ESSENTIALITY

Zinc is present in microorganisms, plants, and animals alike. Perhaps such a wide distribution originally made it seem unlikely that alterations in zinc metabolism could lead to significant biochemical or, for that matter, clinical manifestations of zinc deficiency. Zinc ranks among the top 25 abundant elements in the outer layer of the earth. It exists as a wide range of combinations of five stable isotopes and six radioisotopes. Most familiar to the biochemist, nutritionist, food scientist, and clinician would be the more common radioisotope ^{65}Zn

and the lesser used radioisotopes [67]Zn and [69m]Zn. In quantitative terms, the actual concentration of zinc in soil (though variable) is often reported to be in the 50 ppm range, while sea water and fresh water have been reported to contain on average 8 ppb and 64 ppb, respectively (1,47). Of course, man-made contributions, which primarily arise from manufacturing, processing, or waste, can contribute to these values to a varying degree. For more detailed information, the reader is referred to a recent report from the National Research Council (2). Most important to the chemist and biochemist is the fact that zinc is a divalent ion and, as such, can form a wide variety of inorganic salts and organic zinc complexes of varying stability. It is the latter form of zinc, particularly its association in a variety of zinc proteins and the resultant deficiency or toxicity associated with nutritional states, that has received the greatest share of attention in zinc biochemistry to date.

It was over 120 years ago that Raulin, a pupil of Pasteur, identified zinc as an essential nutrient for the growth of *Aspergillus niger* (3). This particular observation was confirmed 40 years later by Bertrand and Javillier (4). In the first part of this century, Birckner (5) made the observation that egg yolk, cow's milk, and human milk all contain zinc. Moreover, as a direct result of these observations, Birckner also concluded that this particular element was of nutritive value. It is somewhat amazing that one can find in the 1980s detailed investigations that are still building upon this original observation of the unique presence of zinc in milk (32–35). In fact, we still know precious little about the forms and functions of zinc in this most essential of biological fluids. From the demonstration of the presence of zinc in tissues, fluids, and animals, it was but a relatively short time until it was confirmed as an essential element for the maintenance of plant life (6). In contrast, a considerably longer time frame was required to demonstrate the true essentiality of this nutrient in animals. Of course, a major contributing factor was the ability, or lack thereof, to be able to produce a suitable zinc-deficient diet. Contamination, by ingredients or process, was a major obstacle that continually plagued early researchers and still serves to confuse modern research efforts, even in today's more controlled and sophisticated environment. Nevertheless, in 1934, Todd and co-workers (7) succeeded in the production and feeding of a truly zinc-deficient diet, which demonstrated that zinc is, in fact, an essential nutrient for the rat. Twenty years after this observation, Tucker and Salmon (8) showed that in swine, parakeratosis was the direct consequence of zinc deficiency. These particular findings have relevance even today, since they provided the impetus to fortifying feeds and feedstuffs with supplemental zinc, a practice which persists even among some foods consumed by humans. Therefore, these landmark studies provided the basis for demonstrating the essentiality of zinc as manifested among a variety of animal species. With respect to man, zinc's essentiality as determined by zinc deficiency was first suggested by Prasad and his colleagues (9) in 1960 and 1961 (10). Their particular description of individuals in the Middle East with a syndrome of dwarfism, hypogonadism, and severe iron deficiency has become a landmark observation in the zinc literature. Their reports and experiences with iron and zinc supplementation regimens represent the first report of a zinc deficiency condition in man and its subsequent response to supplementation. There have been other reports since this original observation but none with as much an impact on zinc's essential and basic role in human nutrition.

Although these original observations dealt with true zinc deficiency more recently, another set of observations attest to zinc's continued importance in nutrition. These observations are the relatively recent reports of Hambidge et al. (11), who have shown low hair zinc concentrations in children among middle class families in the United States. Similar to the original observations of Prasad, these children had relatively poor growth, poor appetite, and impaired taste perception. Of prime importance was the fact that appropriate zinc supplementation resulted in an overall improvement of their condition. Most important is the potential impact of this more subtle, marginal zinc deficiency in humans.

Since the discovery of zinc as an essential nutrient and its subsequent investigation by a wide variety of disciplines many excellent and detailed reviews have been written (12,13,16,38,47,105). The reader is referred to any of these reviews for more detailed description of the specific subtopics covered. In this regard, the recent work of Hambidge (13) represents the most complete and comprehensive review of zinc. Since these reviews provide an excellent background for zinc biochemistry, it is not the intent of this work to repeat such facets of zinc biology. Rather than providing a detailed review of the complete spectrum of zinc, the intent of this overview is to provide a perspective and a few of the key features of zinc metabolism that have developed and are pertinent to foods and nutrition over the past several years.

BIOCHEMISTRY AND ZINC METABOLISM— ZINC ABSORPTION

The prototypical "adult 70-kg man" contains an average of 1.5–2.5 grams of zinc. This value is significant since of the trace elements in nutrition, zinc is second only to iron in terms of its total pool in the human body. With respect to its distribution and concentration in tissues, the occurrence of zinc is remarkably consistent. For example, on average, the zinc concentration in virtually all of the tissues where it is found ranges from 10 to 100 ppm (wet weight).

Moreover, reports from numerous investigators suggest that there is very little species difference in the distribution and concentration of zinc (12,13,47). Some tissues are sensitive to changes in dietary zinc intake, for example, bone, blood and liver; while other tissues are refractory to alterations in dietary zinc levels, i.e., lung, heart and muscle. Zinc can be found in all of the visceral organs: brain, muscle, bones, teeth, skin, eyes, gonads, blood, milk, saliva, pancreatic secretions, biliary secretions, and cerebrospinal fluid. In conjunction with its ubiquitous distribution among tissues, zinc's role as an essential cofactor in an ever-growing list of enzymes and enzymatic functions when integrated into intermediary metabolism results in the fact that tissues and organs are quite susceptible, though somewhat variable in degree, to zinc deficiency. As a general rule, the tissues that undergo the highest metabolic rates (i.e., mucosa, intestinal epithelial cells, skin, etc.) appear to be the most sensitive to zinc deficiency. Once again, for a detailed and comprehensive review of the distribution and functionality of zinc in a variety of tissues, the reader is referred to the work of Hambidge, Casey, and Krebs (13).

ZINC ABSORPTION

Though the mechanistic aspects of the zinc absorption process have been studied extensively by a number of laboratories, the exact details of this process remain uncertain. There are, however, some generalities that can be made. For example, zinc absorption is approximately 30% effieient, thus, with an intake of 15mg/day (the current RDA), the average net absorption, barring other abnormalities, would be roughly 5 mg. Absorption, for the most part, takes place in the small intestine. Althought the exact site of maximum absorption as defined either by kinetic rate or quantity has not been clearly defined, it seems likely that the entire intestine may in fact play a role in the absorption process (14). Zinc absorption is homeostatically regulated. Under conditions of zinc deficiency, an increased percentage of dietary zinc will be absorbed across intestinal cells, while under states of zinc excess, the efficiency of zinc absorption is reduced. The exact control steps involved in this process are still being actively investigated. Nevertheless, it has been a consistent finding that an intestinal metalloprotein, i.e., metallothionien, is directly or indirectly involved in the regulatory processes. More recently, a second brush-border-membrane protein has been identified as being involved in the zinc absorption process (15). However, the exact role and nature of this newly found protein remains to be elucidated.

The actual process of zinc absorption appears to be regulated by a carrier-mediated mechanism (16). Consistent with these observations is the evidence (17,18), whereby zinc was shown in two different laboratories to undergo two distinct phases during the absorption process. The uptake phase or accumulation of zinc in the intestinal mucosal cells was followed by a transfer or slower release phase, whereby zinc was released across the basolateral membrane to the plasma. Even these findings, however, are not universally accepted, since others (19) have suggested alternative mechanisms. Pregnancy and lactation are known to represent states of increased zinc need, and consistent with this is the observation that zinc absorption is increased during pregnancy and lactation (20).

The nature of the form of zinc carried across the intestine is also an area of uncertainty. Protein and amino acids have been shown to quantitatively increase zinc absorbed (21—23) from the diet. Yet these effects are still somewhat controversial, with some investigators showing positive effects (24), while others report neutral interactions of zinc with protein and amino acid mixtures (25). Similarly, phytate and fiber have been reported to either decrease zinc absorption (26,27) or exert a minimal to negligible effect (28,29). In the area of mineral interactions, to be covered in detail later, conflicting data also exist (30).

Perhaps one of the most controversial areas in the field of zinc absorption/nutrition has been the nature of the zinc-binding ligands in human and bovine milk. In part, the interest in this area developed from observations on the value of human milk zinc in the treatment of acrodermatitis enteropathica (31). Since this original observation, a list of potential candidates for the elusive zinc-binding ligand in human milk has evolved (32—34). Each candidate has had its potential assessed by a number of laboratories, only to have the final results remain inconclusive. Most recently, some laboratories have proposed a complete and definitive identification of this ligand (35), but it appears that final consensus has yet to be reached. In examining the relative complex nature of the zinc absorption process, coupled with the potential for interactions of zinc with ligands and/or other metals during absorption, it may well be that the ligand(s) are multifactorial in nature and influenced by a variety of conditions, including the experimental treatments involved (36). Such a scenario would include such factors as the negative influence of components in cow's milk (i.e., calcium), present in human milk at much lower concentration, which could result in a shift in the extent of mineral-mineral interaction (37). While the use of radioisotopes has undoubtedly advanced our knowledge in the area of zinc absorption, their use has also led to some potential artifactual results that are very difficult to interpret. Only further research will resolve this question in sufficient detail to bring the role of zinc-binding ligands to full definition.

ZINC IN PLASMA

Newly absorbed zinc is translocated across epethelial cells in a manner which has not been fully elucidated. During this process the absorbed mineral interacts with a variety of cellular constituents, perhaps the best known of which is the previously mentioned zinc-binding protein metallothionein (38). This inducible protein, which is capable of binding mercury, gold, copper, cadmium, and other divalent ions as well, is the subject for a more detailed discussion later. Nevertheless, despite its interaction with this potent zinc-binding protein, newly absorbed zinc eventually makes its way across the enterocyte and into the plasma compartment.

Once in plasma, an active reservoir for zinc distribution/concentration, zinc is normally associated with the following blood constituents. Erythrocytes are the highest in zinc content and contain approximately 75–80% of the metal. This is predominantly in the form of carbonic anhydrase (39). The plasma has approximately an additional 20–25% of the quantity of zinc in the blood (40,41), with the remainder among the leucocytes and platelets (42). The plasma component is clearly the most active, and zinc is associated here primarily with the plasma transport protein albumin. Serum albumin has been identified as possessing a large number of binding sites for metals (43,44), and although the specificity of zinc binding remains under investigation, it has been convincingly demonstrated that albumin is the acceptor protein for newly absorbed zinc (45,46). In fact, serum zinc levels tend to correlate fairly well with serum albumin levels (47). Moreover, the hereditary disease condition in humans, i.e., hyperzincemia, has been identified as the consequence of abnormally high amounts of zinc bound to albumin (48).

There is a small component of zinc associated with low molecular weight ligands. These ligands, most often the amino acids cysteine and histidine, may play a role in zinc transport. In fact, recent evidence suggests that these and other amino acids are, in fact, capable of influencing the zinc absorption process in the rat (49).

ZINC DISTRIBUTION AND FUNCTIONS

Once in the plasma, zinc is available for distribution to other tissues or organs. This includes the redirection or redistribution of zinc from plasma to intestinal epithelial cells for eventual excretion. This particular topic will be covered in the following section. Once again, the mechanism whereby zinc in the blood moves from plasma to cells has been the subject of investigation by numerous laboratories. This process is beyond the scope of the current description, yet suffice it to say that the current thinking on this complex and con-

troversial topic is covered in detail in the excellent recent review of
Cousins (38). With respect to some general principles on the tissue
distribution of zinc, the liver, spleen, intestine, pancreas, thyroid,
kidney, pituitary, testes, adrenals, and ovaries have all been shown
to exhibit rapid uptake and turnover. In contrast, muscle, brain,
and erythocytes demonstrate a slower turnover, with hair and bone
exhibiting the slowest accumulation of zinc, probably due to their
inherent slow rate of growth and turnover. This latter fact, i.e.,
the slow accumulation of bone zinc, has been used by some investigators as a potential bioassay for dietary zinc bioavailability (50).
Essentially, the newly absorbed zinc is deposited in the matrix of
newly forming bone tissue and is thus a reflection of the dietary
zinc status of the animal. Some investigators have had more success
with this assay (50,51) than others (52). At the same time, the
physiological significance of bone zinc has never been fully established.
Nor has it been adequately established that the deposited zinc is
available for recycling or redistribution by the animal during periods
of mild to severe zinc defiency.

In terms of the function of zinc in intermediary metabolism, a
very complex picture emerges. For example, zinc has been shown to
have roles in enzyme funtions (53), structural integrity (54), and
regulation of hormonal actions under normal conditions and during
infectious episodes (55). This multiplicity in functionality is such
that today there are over 200 zinc enzymes or proteins that have
been identified and isolated to one degree or another. This growth
has occurred in the last 40--50 years!

Zinc has well-established roles in such basic enzymatic functions
and DNA and RNA polymerases. Moreover, it has been suggested
that zinc may actually stabilize these nucleotides (DNA, RNA) as
well (56,57). Similarly, membrane stability has been identified as a
role in which zinc plays an important participatory role (54,58).
Thymidine kinase is yet another critical enzyme where zinc is involved
biochemically, and such basic involvement in nuclear biochemical and
regulatory events may partially explain why rapidly replicating tissues and cell types, i.e., intestinal and mucosal epithelia, may be
particularly sensitive to zinc deficiency. In fact, after only one to
two weeks on zinc-deficient diets, rats have been observed to have
severely reduced cellular integrity of the gastrointestinal tract. The
integument of the small intestine becomes thin and almost transparent
with a greatly reduced zinc content (59).

Yet other functions of zinc are involvement in glucose, lipid,
and protein metabolism; cellular differentiation during embryogenesis;
and hormonal interactions (calmodulin, growth hormone, somatamedin,
gonadatropin, prolactin, thyroid-stimulating hormone, and corticosteroids) (12,13). From this list, it is not difficult to imagine that

zinc plays a major role in growth and cellular function as well. Zinc deficiency is clearly a key determinant of growth and maturation and as such, zinc defiency is known to retard growth and development.

In fact, in studies that have often been described as landmark findings, Walravens and Hambidge (1976) and Golden and Golden (1981) found that such effects are not limited to severe states of deficiency (60,61). Infants and adolescents, when fed diets marginal in zinc content, failed to grow at the same rate as their zinc-adequate counterparts. Zinc supplements produced a significant positive effect in the measurements of linear growth. Consistent with these findings are others that have related zinc to taste perception, food consumption, and food efficiency. Zinc's role in reproduction is also firmly established (62–64). More recently, newer roles for zinc have been suggested in the field of immunocompetence. Such activities as T-cell function, cell-mediated cytotoxicity, neutrophile function, and natural killer cell function have all been described in association with zinc to some extent in the literature (65--68). Finally, the classical signs of zinc defiency, such as alopecia and dermatitis, have firmly established zinc as essential in the process of keratogenesis and wound healing.

In all, there seems to be no single biochemical fuction where zinc has not been shown to at least partially impact on the primary or secondary nature of the process. Integration of this rapidly expanding area of trace element nutrition into the processes of biochemistry will no doubt continue to be the challenge for the 1990s and beyond.

ZINC EXCRETION

Zinc is lost from the body predominatly via the gastrointestinal tract, with losses in the urine (69–71) and sweat (72–74) as relative minor components. Zinc in the feces is a reflection of dietary zinc, pancreatic zinc, zinc from intestinal secretions, and sloughed intestinal mucosal cells. Hence one of the difficulties in the use of radioisotopes or stable isotopes of zinc as tracers in the study of the absorption/distribution process is the fact that the specific activity of the isotope may in fact be influenced by this endogenous secretory compartment. Estimates have been made that as much as 10–50% of the zinc in the feces may, in fact, be endogenous in origin. Nevertheless, our current understanding of zinc absorption is largely based on isotopic studies. In essence, these specific activity considerations must be kept in mind and may, in fact, help to resolve some of the previously controversial data that have perplexed many investigators.

ZINC IN THE FOOD SUPPLY

Sources of Zinc

Categorization of the zinc contents of foods is difficult. In general, meat (in particular organ meats), eggs, shellfish, and some vege- tables are high in zinc. In contrast, milk, fruits, and other vege- tables are reported to be lower in zinc content. On top of this per- haps inherent variation in foods is also the fact that zinc in the soil can vary and further contribute to differences observed in zinc concen- trations around the world (75). Overall, however, there is a similar concentration of zinc found in foods from a wide variety of sources.

Processing and cooking have little effect on zinc in most foods. Thus, total dietary zinc intakes are more influenced by food choices than anything else. A rough correlate exists between protein intake and zinc intake, but even this relationship is not foolproof. Currently, the best estimate for zinc intake across a wide range of countries and dietary patterns is 10—14 mg zinc per day. This value approaches the current RDA for zinc in the United States, but at the same time it must be recognized that marginal deficiency states might/do exist as previously described by the group in Denver (60). Obviously, with the wide distribution and multifunctional roles in biochemical processes, there are active research programs attempting to identify the factor(s) associated with such marginal deficiencies. Though some indices are promising, to date there is no definitive and con- clusive results to firmly establish a single marker as the primary and most sensitive biochemical lesion to indicate zinc status.

ZINC—MINERAL INTERACTIONS

One of the more recent and exciting areas of trace mineral nutrition is that of mineral—mineral interactions. Although this general area was observed and commented on nearly 30 years ago (76), it really has not received adequate attention until the past two to five years. The premise of this area has its basis in the fact that mineral elements share some common chemical structures and similarities, and these may in fact result in competitive chemical interactions. These inter- actions could potentially occur within a food or food system, at the site of absorption, or at some postabsorptive yet metabolically active site. The competition may be between two nutritive elements, a nutritive and nonnutritive element, or between two toxic elements. Thus, such interactions may be deleterious or potentially advan- tageous, depending on the elements and/or interactions in question. Since trace elements by definition are required at very low levels by organisms, a small reduction in absorptive capacity may in fact be viewed as a particularly significant occurrence. Similarly, interactions could occur between elements required at much larger concentrations, i.e., Ca, Mg, and a second trace element, i.e., iron, zinc, and copper, such that the quantitative ratios of these elements to each other serve to further distort or amplify the interaction. Such a condition may aggravate a deficiency or amplify a toxicity depending on the minerals in question. Similar to the early days of biological investigations when physiologists were identifying the correct balance of salts in solution to enable them to work with the isolation of tissues and organs, modern investigators in trace element nutrition are faced with defining the conditions around the balance of trace nutrients and their effects on nutrition and metabolism.

As stated by W. G. Hoekstra in 1964 (77), "Because of the involved nature of mineral imbalance, it is not surprising to find conflicts in the literature." Such is clearly the case when observations on zinc interactions and interrelationships are considered. Let us take as an example the interaction between zinc and dietary phytate. This particular relationship, i.e., the interaction of phytate and zinc has been identified for nearly three decades (78,79). In the original work of Forbes and O'Dell, the first suggestion of a phytic acid−zinc interaction was made. Both Prasad (61) and Reinhold (80), in observations about 10 years apart, speculated that phytate was in part responsible for the effects noted in the Middle East. Yet, over the course of time, the literature is confusing with respect to the degree and even existence of this effect. For example, Ellis et al. (81) have recently shown that dietary phytate levels up to 1.14 g/day had no effect on zinc absorption in man. In this study, it is interesting to consider the argument presented by C. F. Mills that implicates a complicating factor of calcium in relation to these observations (82). This observation was previously identified by Morris and Ellis (83,84), who showed that calcium in diets that contained phytate had a dramatic effect on the zinc availability in the diet. Basically, the fact that high calcium diets are more detrimental to the phytate− zinc interaction may well be explained by an additional component or synergistic effect of calcium on this interaction. At the same time, calcium may exert a direct negative effect by interacting with the absorption of zinc (85,86), yet even this area is also not without conflicting observations (87).

Perhaps one of the more often studied and cited interrelationships of zinc with minerals is the zinc−copper antagonism. In the late 1960s, Van Campen showed a Zn−Cu as well as a Cu−Zn effect in vitro (88). Similarly, the absorption of these two elements has been shown to be interrelated in a number of animal studies (89,90). For example, copper absorption has been shown to be increased during states of zinc deficiency (91), and the copper concentration in organs and milk from zinc-deficient cows has been observed to be significantly elevated (92). In contrast, the effect of copper on zinc absorption in humans was recently found to be negligible (113). The mechanism of such interaction is at present centered around the intestinal binding protein metallothionein (93,94). In fact, in a pilot series of experiments using radiozinc, radiocopper, and radiocadmium, the absorption of these three minerals could be influenced by the concentration and sequence in which these radioisotopes were placed in the lumen of the isolated vascularly perfused rat intestine (59). This observation was indicative of a complex series of interactions that may occur during the absorption process. Both the chemical similarities and biochemical (binding) influences could determine the final absorption values. Clearly, much work is needed to further elucidate this intriguing interaction potential.

Among, humans, similar observations can be described. For example, copper-deficiency symptoms have been observed to be aggravated by zinc supplementation. Long-term treatment of patients with sickle cell anemia with zinc supplements (150 mg, 6x daily) resulted in the appearance of copper deficiency (95). Subsequent administration of additional copper allowed a return to more normal conditions. This particular phenomenon has been observed by others as well (96). In fact, elevated zinc intakes have been used in the treatment of the copper accumulation disease (Wilson's disease) by limiting the amounts of dietary copper available for absorption due to interactions (97). Tin has also been shown to exert a negative effect on zinc absorption in humans (114,115).

Along similar lines, but apparently to a lesser extent, zinc and iron have been identified as interacting also. In fact, Solomons et al. have recently raised concerns over nutritional multimineral supplements and infant formula, which may contain inappropriate zinc:iron ratios that would exacerbate this interrelationship (98).

From this brief review, it is clear that mineral interaction may well play a more determinant role in nutrition than previously anticipated. Such interactions can be potentially useful in therapies, diet recommendations, supplements and food fortication regimens. Yet whenever their potential exists, they must be taken into account when mineral nutrition experiments are designed, conducted or interpreted.

ZINC DEFICIENCY AND TOXICITY

As with many nutrients, zinc exhibits a profile of intakes which progress through the stages of deficiency—adequacy—toxicity. Although the levels of this nutrient in which the three stages exist may vary somewhat, there are a number of generalizations that can be made. Of course, clear-cut cases of either deficiency or toxicity as those produced in experimental animals are fairly easy to distinguish, whereas the marginal deficiency or marginal toxicity states are much

more difficult to define. In general, zinc is not very toxic to animals
or man. In fact, deficiencies with respect to intake, absorption,
and metabolism appear to be a far greater concern than toxicity.

ZINC DEFICIENCY

As previously mentioned, zinc deficiency in humans is a relatively
new finding. Even the observations on true zinc-deficient states in
animals date back only 50 years. Common to the syndrome of zinc
deficiency are a number of dominant features: retarded growth, loss of
of appetite, hypogonadism, dermatitis, alopecia, and resistance to
infection. Not only is the sense of taste impaired, but smell and
sight (night blindness) have also been identified as a result of zinc
deficiency (99–101). Similarly, wound healing, ulcerations (both
external and intestinal), and parakeratosis have also been related to
zinc deficiency (102–104). In fact, the hereditary disorder acrodermatitis enteropathica (104) is characterized by a number of these
factors, including severe skin abnormalities. Infants suffering from
acrodermatitis usually respond quite well to zinc therapy. Of course,
the classic reversal of growth and hypogonadism with zinc therapy
are well documented in the cases reported from the Middle East.
Delayed wound healing and zinc's role in the immunological area are
two examples of additional roles that zinc may play in the abnormal
metabolism associated with zinc deficiency.

All of these symptoms are perhaps not so surprising when one
considers, as mentioned earlier, that zinc is an essential cofactor in
the enzymatic profiles of over 200 proteins from a wide variety of
species (12). These enzymes have roles in the synthesis and degradation of all major cellular constituents. In fact, protein, carbohydrate,
lipids, and nucleic acids all have representative enzymatic functional-
ities, which include zinc metalloenzymes in their various metabolic
pathways. Some examples of these are collagenases, synthetases,
aldolases, ligases, carboxylases, creatinases, anhydrases, peptidases,
and dehydratases.

ZINC TOXICITY

Zinc is a homeostatically regulated metal nutrient (105). Thus, for
the most part, the regulation of zinc via the absorptive processes
previously described controls the level present in an organism and
renders true toxicity to be a rare occurrence. The primary mode of
exposure in relation to zinc toxicity is via the lungs. In a response
known as metal fume fever, the inhalation of zinc metal (normally
zinc oxide) results in a series of symptoms, including weakness,
fever, pain, and hyperventilation (106,107). With todays standards

for health and safety in the workplace, the occurrence of this syndrome is virtually nonexistent. It is interesting to note, however, that the symptoms normally subside within a day or two. The mechanism of this particular toxic response is not well understood at all.

With respect to oral exposure, much less is known. Although there have been reports of zinc toxicity where galvanized containers were used to store acid foods, these have generally been isolated or anecdotal in nature. In some cases, a rise in serum zinc values well above normal have been documented. In fact, there is one extreme case of an individual who was exposed to dialysis fluid contaminated by storage in a galvanized tank (108). In this particular case, the serum zinc rose to over 400 μg/dl, and yet no toxic symptoms were observed. In general most zinc salts are gastrointestinal irritants, and thus the immediate reflex response of nausea and vomiting may serve to self-limit exposure via the oral route.

An additional factor that may aid in the regulation of zinc toxicity is the presence of the inducible protein metallothionein in the intestinal mucosal cells. Metallothionein is a metal-binding protein, which has the ability to bind a variety of metals, including copper, zinc, cadmum, mercury, and gold (109,110). This protein has been found to occur in nearly all species examined to date and in a wide variety of tissues as well (111). It is a relatively small protein of approximately 10,000 daltons, and 25—30% of the amino acids are the sulfur-containing amino acid cysteine. This protein has the ability to bind 5—7 gram atoms of metal per mole of protein with different binding affinities. Interestingly, one of the first characterizations of the protein was performed with cadmium (109,112), and much research has addressed its relationship to cadmium toxicity. More recently, this protein has been identified as playing a pivotal role in the zinc absorptive and metabolic processes (105). Such a role may well serve to protect organisms from repeated exposure to high levels of metal due to its inducible nature.

In a number of laboratories, the dietary factors responsible for and involved in the induction of this unique metalloprotein have been studied. For a detailed review of these studies, the reader is referred to recent articles (16, 105), which describe the salient features of this protein and its relation to the overall absorptive processes. In a brief overview of this process, however, the following can be noted. Changes in dietary zinc (or other metal capable of inducing metallothiomein) result in an increase in this protein at the liver, kidney, and intestinal level. The response is somewhat rapid, on the order of 24—48 hours, and the newly synthesized metallothinein in the intestinal cell is capable of binding any excess metal present. The metal is thus sequestered in the enterocyte and is destined for excretion as the mucosal cell migrates to the tip of the villi for eventual desquamation. Although simplified, this process can be seen to be an important mechanism to limit both cells and the organism to potential toxic levels of a variety of metals. Although the exact function(s) of this protein are still being defined by many laboratories, the role of this protein in detoxification and regulation is certain to remain a pivotal point in its overall function.

In general then, the overall low toxicity, regulated absorption and excretion and self-limiting nature of the ingestion of this metal all combine to form a relatively safe profile for this nutrient metal. Except for the ingestion via inhalation, little concern remains for the toxic potential of zinc. In fact, this profile has been the prime factor in the lack of concern over public exposure. On the other hand, from a nutrient interaction perspective, zinc may well exert negative effects on other trace minerals in the diet.

CONCLUSIONS

The tremendous growth that we have witnessed in the area of zinc nutrition over the past 20 years is directly attributable to a consistent pattern of scientific breakthroughs, many of which owe their beginnings to improved analytical capabilities. The wide distribution and participation of this trace element in virtually all physiologic and biochemical functions places this mineral at the very cornerstone of biology. Its proteinaceous participation and its regulatory roles can account for a seemingly endless crucial involvement in metabolism. From cell replication to cellular communication, zinc is a common thread. These wide metabolic manifestations have been cause of a constant growth in knowledge and at the same time a good deal of confusion about zinc. Due to the nature of its pervasive involvement in a variety of interconnecting and interdependent processes, it is sometimes difficult to completely separate cause from effect. Nevertheless, the field continues to grow and evolve.

Due in part to the diversity of zinc involvement in biochemistry and medicine as well as the complex nature of its actions at the cellular and subcellular level, a general overview of the features of this element as they relate to nutrition has been provided. Once again, the popularity of this element has led to many recent and excellent detailed reviews and/or texts on specific subtopics of interest to a variety of readers (12,13,16,23,30,38,47,62,69,82,104). One thing is certain—zinc will continue to be the cornerstone of many biological processes upon which the foundations of life exist. From birth to death and during all intermediate stages, zinc is truly a metal of life.

REFERENCES

1. B. L. Vallee. Biochemistry, physiology, and pathology of zinc. *Physiol. Rev.*, *39:* 443–490 (1959).
2. National Research Council, Subcommittee on Zinc, "Zinc" University Park Press, Baltimore (1979).
3. J. Raulin. Etudes cliniques sur la vegetation. *Ann. Sci. Natur. Botan. Biol. Vegetale*, *11:* 93–299 (1869).
4. C. Bertrand and M. Javillier. Influence du zinc et du manganese sur la composition minerale de l'*Aspergillus niger*. *C. R. Acad. Sci.*, *152:* 1337–1340 (1911).
5. V. Birckner. The zinc content of some food products. *J. Biol. Chem.*, *38:* 191–203 (1919).
6. A. Sommer and C. B. Lipman. Evidence on the indispensable nature of zinc and boron for higher green plants. *Plant Physiol.*, *1:* 231–249 (1926).
7. W. R. Todd, C. A. Elvehjem, and E. B. Hart. Zinc in the nutrition of the rat. *Amer. J. Physiol.*, *107:* 146–156 (1934).
8. H. F. Tucker and W. D. Salmon. Parakeratosis or zinc deficiency disease in the pig. *Proc. Soc. Exp. Biol. Med.*, *88:* 613–616 (1955).
9. J. A. Halstead and A. S. Prasad. Syndrome of iron deficiency anemia, hepatosplenomegaly, hypogonadism, dwarfism and geophagia. *Trans. Amer. Clin. Climat. Assoc.*, *72:* 130–149 (1960).
10. A. S. Prasad, J. A. Halstead, and M. Nadimi. Syndrome of iron deficiency anemia, hepatosplenomegaly, hypogonadism, and geophagia. *Am. J. Med.*, *31:* 532–536 (1961).
11. P. A. Walravens, N. F. Krebs, and M. Hambidge. Linear growth of low income preschool children receiving a zinc supplement. *Am. J. Clin. Nutr.*, *38:* 195 (1983).
12. E. J. Underwood. *Trace Elements in Human and Animal Nutrition*, 4th edition. Academic Press, New York (1977).
13. M. Hambidge. In *Trace Elements in Human and Animal Nutrition*, 5th edition, edited by W. Mertz. Academic Press, New York (1986).
14. D. L. Antonson, A. J. Barak, and J. A. Vanderhoff. Determination of the site of zinc absorption in rat small intestine. *J. Nutr.*, *109:* 142–147 (1979).
15. M. P. Menard and R. J. Cousins. Zinc transport by brush border membrane vesicles from rat intestine. *J. Nutr.*, *113:* 1434–1442 (1983).
16. N. W. Solomons and R. J. Cousins. Zinc, In *Absorption and Malabsorption of Mineral Nutrients*, edited by Solomons and Rosenberg, A. R. Liss, New York, pp. 125–197 (1984).

17. K. T. Smith and R. J. Cousins. Quantitative aspects of zinc absorption by isolated, vascularly perfused rat intestine. *J. Nutr.*, *110:* 316–323 (1980).

18. N. T. Davies. Studies on the absorption of zinc by rat intestine. *Br. J. Nutr.*, *43:* 189–203 (1980).

19. M. J. Jackson, D. A. Jones, and R. H. T. Edwards. Zinc absorption in the rat. *Br. J. Nutr.*, *46:* 15–27 (1981).

20. N. T. Davies and R. B. Williams. The effect of pregnancy and lactation on the absorption of zinc and lysine by the rat duodenum in situ. *Br. J. Nutr.*, *38:* 417–423 (1977).

21. E. Giroux and N. J. Prakash. Influence of zinc ligand mixtures on serum zinc levels in rats. *J. Pharm. Sci.*, *66:* 391–392 (1977).

22. R. A. Wapnir, D. E. Khani, M. A. Bayne, and F. Lifshitz. Absorption of zinc by the rat ileum, effects of histidine and other low molecular weight ligands. *J. Nutr.*, *113:* 1346–1354 (1983).

23. N. W. Solomons. Biological availability of zinc in humans. *Am. J. Clin. Nutr.*, *35:* 1048–1075 (1982).

24. R. A. Wapnir and L. Stiel. Zinc intestinal absorption in rats: Specificity of amino acids as ligands. *J. Nutr.*, *116:* 2171–2179 (1986).

25. P. Oestreicher and R. J. Cousins. Influence of intraluminal constituents on zinc absorption by the isolated vascularly perfused rat intestine. *J. Nutr.*, *112:* 1978–1982 (1982).

26. J. G. Reinhold, K. Nasr, A. Lahimagarzadeh, and H. Hedagati. Effects of purified phytate and phytate rich bread upon metabolism in man. *Lancet, 1:* 283–288 (1973).

27. J. L. Kelsay, R. Jacobs, and E. S. Prather. Effect of fiber from fruits and vegetables or metabolic responses of human subjects. III. Zinc, copper, and phosphorous balances. *Am. J. Clin. Nutr.*, *32:* 2307 (1979).

28. B. E. Guthrie and M. F. Robinson. Zinc balance studies during wheat bran supplementation. *Fed. Proc.*, *37:* 254 (1978).

29. N. T. Davies, V. Hristic, and A. A. Flett. Phytate rather than fiber in bran as the major determinant of zinc availability to rats. *Nutr. Rep. Intl.*, *15:* 207–214 (1977).

30. T. Hazell. Minerals in foods: dietary sources, chemical forms, interactions, bioavailability. *Wld. Rev. Nutr. Diet.*, *46:* 1–123 (1985).

31. E. J. Moynahan. Acrodermatitis enterophathica: a lethal inherited zinc deficiency disorder. *Lancet, 2:* 399–400 (1974).

32. G. W. Evans and P. E. Johnson. Characterization and quantitation of a zinc-binding ligand in human milk. *Pediatr. Res.*, *14:* 876–880 (1980).

33. B. Lonnerdal, A. G. Stanislowski, and L. S. Hurley. Isolation of a low molecular weight zinc binding ligand from human milk. *J. Inorg. Biochem.*, *12:* 71—78 (1980).

34. L. S. Hurley and B. Lonnerdal. Picolinic acid as a zinc binding ligand in human milk: an unconvincing case. *Pediatr. Res.*, *15:* 166—167 (1981).

35. B. Lonnerdal, B. Hoffman, and L. S. Hurley. Zinc and copper binding proteins in human milk. *Am. J. Clin. Nutr.*, *35:* 1170—1176 (1982).

36. R. J. Cousins and K. T. Smith. Zinc binding properties of bovine and human milk in vitro: influence of changes in zinc content. *Am. J. Clin. Nutr.*, *33:* 1083—1087 (1980).

37. K. T. Smith and J. T. Rotruck. *Trace Mineral and Calcium Interactions in TEMA-VI.* Plenum Press (in press) (1987).

38. R. J. Cousins. Absorption, transport, and hepatic metabolism of copper and zinc: Special reference to metallothionein and ceruloplasmin. *Physiol. Rev.*, *65:* 238—309 (1985).

39. B. L. Vallee, H. D. Lewis, M. D. Altschule, and J. G. Gibson. The relationship between carbonic anhydrase activity and zinc content of erythrocytes in normal, in anemia and other pathologic conditions. *Blood*, *4:* 467—476 (1949).

40. A. S. Prasad and D. Oberleas. Binding of zinc to amino acids and serum proteins in vitro. *J. Lab. Clin. Med.*, *76:* 416—425 (1970).

41. B. J. Scott and A. R. Bradwell. Identification of the serum binding proteins for iron, zinc, cadmium, nickel, and calcium. *Clin. Chem.*, *29:* 629—633 (1983).

42. E. Dennes, R. Tupper, and A. Wormall. Studies on zinc in blood. *Biochem. J.*, *82:* 466—476 (1962).

43. F. R. N. Gurd and D. S. Goodman. Preparation and properties of serum and plasma proteins. XXXII. The interaction of human serum albumin with zinc ions. *J. Am. Chem. Soc.*, *74:* 670—675 (1952).

44. F. Friedberg. Albumin as the major metal transport agent in blood. *FEBS Letters*, *59:* 140 (1975).

45. J. K. Chesters and M. Will. Zinc transport proteins in plasma. *Br. J. Nutr.*, *46:* 111—118 (1981).

46. K. T. Smith, M. L. Failla, and R. J. Cousins. Identification of albumin as the plasma carrier for zinc absorption by the perfused rat intestine. *Biochem. J.*, *184:* 627—633 (1979).

47. J. A. Halsted, J. C. Smith, Jr., and M. I. Irwin. A conspectus of research on zinc requirements of man. *J. Nutr.*, *104:* 347—378 (1974).

48. M. L. Failla, M. Van de Veerdonk, W. T. Morgand, and J. C. Smith, Jr. Characterization of zinc binding proteins of plasma in familial hyperzincemia. *J. Lab. Clin. Med.*, *100:* 943—952 (1982).

49. R. A. Wapnir, J. Garcia-Aranda, D. Mevorach, and F. Lifshitz. Differential absorption of zinc and low molecular weight ligands in the rat gut in protein energy malnutrition. *J. Nutr.* (1985).
50. E. J. Murry and H. H. Messer. Turnover of bone during normal and accelerated bone loss in rats. *J. Nutr.*, *111:* 1641 (1981).
51. N. R. Calhoun, E. G. McDaniel, M. P. Howard, and J. C. Smith, Jr. Loss of zinc from bone during deficiency state. *Nutr. Rep. Int.*, *17:* 299 (1978).
52. B. F. Harland, M. R. Spivey-Fox, and B. F. Freg. Protection against zinc deficiency by prior excess dietary zinc in young Japanese quail. *J. Nutr.*, *105:* 1509 (1975).
53. B. L. Vallee. In *Zinc Enzymes*, edited by T. G. Spiro. Wiley, New York (1983).
54. M. Chvapil. Effect of zinc on cells and biomembranes. *Med. Clin. North Am.* *60:* 799 (1976).
55. B. Sugarman. Zinc and infection. *Rev. of Inf. Disease, 5 (1):* 137–147 (1983).
56. W. E. C. Wacker and B. L. Vallee. Nucleic acids and metals. *J. Biol. Chem.*, *234:* 3257 (1959).
57. Y. A. Shin and G. L. Eichhorn. Interactions of metal ions with polynucleotides and related compounds. XI. The reversible unwinding and rewinding of deoxyribonucleic acid by zinc (II) ions through temperature manipulation. *Biochemistry*, *7:* 1026 (1968).
58. W. J. Bettger and B. L. O'Dell. A critical physiological role of zinc in the structure and function of biomembranes. *Life Sci.*, *28:* 1425–1438 (1981).
59. K. T. Smith. Some aspects of the mechanism and regulation of zinc absorption in the isolated, vascularly perfused rat intestine. Ph. D. thesis, Rutgers University (1979).
60. P. A. Walravens and K. M. Hambidge. Growth of infants fed a zinc supplemented formula. *Am. J. Clin. Nutr.*, *29:* 1114 (1976).
61. M. H. N. Golden and B. E. Golden. Effect of zinc supplementation on the dietary intake, rate of weight gain, and energy cost of tissue deposition in children recovering from severe malnutrition. *Am. J. Clin. Nutr.*, *34:* 900 (1981).
62. J. K. Chesters. Biochemical functions of zinc in animals. *World Rev. Nutr. Diet.*, *32:* 135 (1978).
63. L. S. Hurley and H. Swenerton. Congenital malformations resulting from zinc deficiency in rats. *Proc. Soc. Exp. Biol. Med.*, *123:* 692 (1966).
64. I. E. Dreosti. In *Neurobiology of the Trace Elements*, edited by Smith. Humana Press, Clifton, NJ (1983).
65. J. K. Chesters. The role of zinc ions in the transformation of lymphocytes by phytohaemagglutinin. *Biochem. J.*, *130:* 133 (1972).

66. R. K. Chandra and B. An. Single nutrient deficiency and cell-mediated immune responses. *Am. J. Clin. Nutr., 33:* 736 (1980).
67. F. J. Cummings, J. I. Fardy and M. H. Brigs. Trace elements in human milk. *Obstet. Gynecol., 62:* 506 (1983).
68. F. Ferry and M. Donner. In vitro modulation of murine natural Killer cytotoxicity by zinc. *Scand. J. Immunol. 19 (5):* 435–445 (1984).
69. R. E. Burch, H. K. J. Hahn, and J. F. Sullivan. Newer aspects of the roles of zinc, manganese, and copper in human nutrition. *Clin. Chem., 21:* 501 (1975).
70. R. A. McCance and E. M. Widdowson. The absorption and excretion of zinc. *Biochem. J., 36:* 692–696 (1942).
71. G. S. Fell, A. Fleck, D. P. Cuthbertson, K. Queen, C. Morrison, and S. C. Husain. Urinary zinc levels as an indication of muscle metabolism. *Lancet, 1:* 280–282 (1973).
72. M. T. Bauer and J. C. King. Tissue zinc levels and zinc excretion during experimental zinc depletion in young men. *Am. J. Clin. Nutr., 39:* 556 (1984).
73. K. K. Schraer and D. H. Calloway. Zinc balance in pregnant teenagers. *Annals Nutr. Metab., 17:* 205 (1974).
74. D. B. Milne, W. K. Canfield, J. R. Mahalko, and H. H. Sandstead. Effect of oral folic acid supplments on zinc, copper, and iron absorption and excretion. *Am. J. Clin. Nutr., 39:* 535 (1984).
75. J. F. Hodgson, W. H. Allaway, and R. B. Lackman. Regional plant chemistry as a reflection of environment. In *Environmental Geochemistry in Health and Disease.* The Geological Society of America (1971).
76. R. M. Forbes. Nutritional interactions of zinc and calcium. *Federation Proc., 19:* 643 (1960).
77. W. G. Hoekstra. Recent observations on mineral interrelationships. *Fed. Proc., 23:* 1068–1076 (1964).
78. R. M. Forbes. Nutritional interactions of zinc and calcium. *Fed. Proc.* 643–647 (1960).
79. B. L. O'Dell and J. E. Savage. Effect of phytate on zinc availability. *Proc. Soc. Exp. Biol. Med., 103:* 304–309 (1960).
80. J. G. Reinhold. High phytate content of rural Iranian bread: a possible cause of human zinc deficiency. *Am. J. Clin. Nutr., 24:* 1204–1206 (1971).
81. R. Ellis, E. R. Morris, A. D. Hill, J. C. Smith, Jr. Phytate: zinc molar ration of regular, ovo-lacto-vegetarian, and soy meat substitute hospital diets. *Fed. Proc., 40:* 938 (1981).
82. C. F. Mills. Dietary interactions involving the trace elements. *Ann. Rev. Nutr., 5:* 173–193 (1985).
83. E. R. Morris and R. Ellis. Effect of dietary phytate/zinc molar ratio or growth and bone zinc response of rats fed semi-purified diets. *J. Nutr., 110:* 1037–1045 (1980).

84. E. R. Morris and R. Ellis. Bioavailability to rats of iron and zinc in wheat bran. Response to low-phytate/zinc molar ratio. *J. Nutr.*, *110:* 2000—2010 (1980).

85. D. A. Heth and W. G. Hoekstra. Zinc-65 absorption and turnover in rats 1. A procedure to determine zinc-65 absorption and the antagonistic effect of calcium in a practical diet. *J. Nutr. Nutr.*, *85:* 367—374 (1965).

86. R. M. Forbes and M. Yoke. Zinc requirement and balance studies with the rat. *J. Nutr.*, *70:* 53—57 (1960).

87. B. Dawson-Hughes, F. H. Seligson, and V. A. Hughes. Effects of calcium carbonate and hydroxyapatite on zinc and iron retention in postmenopausal women. *Am. J. Clin. Nutr.*, *44:* 83—88 (1986).

88. D. R. Van Campen. Copper interference with the intestinal absorption of zinc-65 in rats. *J. Nutr.*, *97:* 104 (1969).

89. G. W. Evans, C. I. Grace, and C. Hahn. The effect of copper and cadmium on ^{65}Zn absorption in zinc-deficient and zinc-supplemented rats. *Bioinorg. Chem.*, *3:* 115—120 (1974).

90. P. W. F. Fischer, A. Giroux, and M. L'Abbe. The effect of dietary zinc on intestinal copper absorption. *Am. J. Clin. Nutr.*, *34:* 1670—1675 (1981).

91. A. C. Magee and G. Matrone. Studies on growth, copper metabolism and iron metabolism of rats fed high levels of zinc. *J. Nutr.*, *72:* 233—242 (1960).

92. M. Kirchgessner, F. J. Schwarz, E. Grassman, and H. Steinhart. In *Copper in the Environment, II,* edited by Nigo, Wiley, New York (1979).

93. P. Oestreicher and R. J. Cousins. Copper and zinc absorption in the rat: mechanism of mutual antagonism. *J. Nutr.*, *115:* 159—166 (1985).

94. A. C. Hall, B. W. Youngs, and I. Bremner. Intestinal metallothionein and the mutual antagonism between copper and zinc in the rat. *J. Inorg. Biochem.*, *11:* 57—66 (1979).

95. A. S. Prasad, G. J. Brewer, E. B. Schoamaker, and P. Rabbani. Hypocupremia induced zinc therapy in adults. *J. Am. Med. Assoc.*, *240:* 2166—2168 (1978).

96. K. G. Porter, D. McMaster, M. E. Elmes, and A. M. Love. Anemia and low serum copper during zinc therapy. *Lancet, 2:* 774 (1977).

97. G. J. Brewer, G. M. Hill, A. S. Prasad, Z. T. Cossack, and P. Rabbani. Oral zinc therapy for Wilson's disease. *Ann. Intern. Med.*, *99:* 314—320 (1983).

98. N. W. Solomons, R. A. Jacob, and O, Pineda. Studies on the bioavailability of zinc in man. *J. Lab. Clin. Med.*, *94:* 335—343 (1979).

99. R. K. Chesters and J. Quarternan. Effects of zinc deficiency on food intake and feeding patterns of rats. *Br. J. Nutr.,* *24:* 1061 (1970).

100. S. A. Morrison, R. M. Russell, E. A. Carney, and E. V. Oakes. Zinc deficiency as a cause of abnormal dark adaptation in cirrhosis. *Am. J. Clin. Nutr.,* *31:* 276−281 (1978).

101. R. I. Henkin, P. J. Schechter, R. Hoye, and C. F. T. Mattein. Idiopathic hypogeusia with dysgeusia, hyposmia. A new syndrome. *J.A.M.A.,* *217:* 434−440 (1971).

102. P. K. Lewis, W. G. Hoekstra, R. H. Grummer, and P. H. Phillips. The effect of certain nutritional factors including calcium, phosphorous and zinc on parakeratosis in swine. *J. Animal Sci.,* *15:* 741 (1956).

103. J. A. Halsted and J. C. Smith, Jr. Plasma zinc in health and disease. *Lancet, 1:* 322−324 (1970).

104. D. L. Larson. In *Clinical Applications of Zinc Metabolism,* edited by W. J. Porres. Thomas, Springfield, IL (1974).

105. R. J. Cousins. Regulatory aspects of zinc metabolism in liver and intestine. *Nutrition Reviews, 37:* 97−103 (1979).

106. J. P. Papp. Metal fume fever. *Postgrad. Med., 43:* 160−163 (1968).

107. L. C. Rohrs. Metal fume fever from inhaling zinc oxide. *Arch. Ind. Hyg. Occup. Med., 16:* 42−47 (1957).

108. E. D. M. Gallery, J. Blomfield, and S. R. Dixon. Acute zinc toxicity in hemodialysis. *Br. Med. J., 4:* 331 (1972).

109. J. H. Kagi and M. Nordberg. *Metallothionein.* Birkhauser, Basel (1978).

110. J. H. Kagi and B. L. Vallee. Metallothionein: a cadmium-and zinc containing protein from equine renal cortex. *J. Biol. Chem., 235:* 3460−3465 (1960).

111. S. Klausen, J. H. Kagi, and K. J. Wilson. Characterization of isoprotein patterns in tissue extracts and isolated samples of metallothionein by reverse-phase HPLC. *Biochem. J., 209:* 71−80 (1983).

112. Y. Kojima, C. Berger, B. L. Vallee, and J. H. Kagi. Amino acid sequence of equine renal metallothionein I. *Proc. Nat. Acad. Sci. U.S.A., 73:* 3413−3417 (1976).

113. L. S. Valberg, P. R. Flanagan, and M. J. Chamberlain. Effects of iron, tin, and copper on zinc absorption in humans. *Am. J. Clin. Nutr., 40:* 536−541 (1984).

114. J. L. Greger and M. A. Johnson. Effects of dietary tin on zinc, copper, and iron utilization by rats. *Food Cosmet. Toxicol., 34:* 475−482 (1981).

115. M. A. Johnson, M. J. Baier, and J. L. Greger. Effect of dietary tin on zinc, copper, iron, manganese and magnesium metabolism in adult males. *Am. J. Clin. Nutr., 35:* 1332−1338 (1982).

7
Chromium

RICHARD A. ANDERSON / United States Department of Agriculture,
Beltsville Human Nutrition Research Center, Beltsville, Maryland

INTRODUCTION

There is increasing evidence that a number of common diseases, including cardiovascular diseases and maturity-onset or type II diabetes,
are related to the foods eaten throughout life. For example, cholesterol or high-fat foods are associated with cardiovascular diseases, and
sugar and refined carbohydrates have been associated with maturity-
onset diabetes. However, decreased intake of these foods may only
be part of the solution, and an increased intake of trace elements,
especially chromium, may also be important in lessening the chances
of developing cardiovascular diseases and/or maturity-onset diabetes.
Fortunately, dietary improvements, including decreasing dietary intake
of fat, cholesterol, sugar, and other refined carbohydrates, would
be associated with the increased intake of more nutritional foods, including fruits and vegetables, that are usually lower in fat, cholesterol,
and refined carbohydrates, and also often good sources of chromium.
Studies involving humans as well as experimental animals suggest very
strongly an association of dietary Cr intake with signs and symptoms
associated with cardiovascular diseases and maturity-onset diabetes but
the problems are also amplified by the observation that dietary intake
of Cr even by affluent, apparently healthy individuals is suboptimal.

SIGNS AND SYMPTOMS OF CHROMIUM DEFICIENCY

Glucose Intolerance

Many of the signs and symptoms of chromium deficiency (see Table 1) are not specific for Cr but may also be attributed to other nutritional or nonnutritional problems. Glucose intolerance is an example of a malady that is not specific for Cr, but if Cr deficiency is present glucose intolerance is almost always also present. Glucose intolerance is not specific for Cr deficiency, but Cr deficiency is specific for glucose intolerance. There have not been any studies involving human subjects that have reported signs and symptoms of Cr deficiency that have not included glucose intolerance. Overt signs of diabetes, including glucose intolerance, unexpected weight loss, and impaired nerve conduction, that were refractory to insulin were reversed by supplemental chromium in a female patient receiving total parenteral nutrition (1). Following Cr supplementation, the 45 units of exogenous insulin that were administered prior to supplemental Cr were no longer needed, and diabetic symptoms returned to normal. Similar results have been reported for another patient receiving total parenteral nutrition (2).

Improvements in glucose tolerance following Cr supplementation are not limited to patients on total parenteral nutrition. Children with protein calorie malnutrition (3,4), the elderly (5), insulin and non insulin-requiring diabetics (6–8), and hypoglycemics (9) have all been shown to respond to supplemental Cr. Improvements in glucose intolerance following Cr supplementation are related to previous dietary Cr intake. For example, malnourished children in Turkey, Jordan, and Nigeria consuming a diet low in Cr responded to supplemental Cr with improvements in glucose tolerance (3,4), while malnourished children from Egypt consuming a diet high in Cr did not respond to supplemental Cr (10).

Anderson et al. (12) reported that the response to supplemental Cr was dependent upon the degree of glucose intolerance of the subjects. Study design was double-blind crossover with each test period lasting three months. Seventy-six subjects were divided into three groups based upon their blood glucose values 90 min after a glucose load (1 gm glucose/kg body wt). Glucose tolerance of 18 of 20 subjects with marginally elevated blood glucose (90 min glucose greater than 100 mg/dl) was improved following Cr supplementation. Glucose tolerance of subjects with optimal glucose tolerance (90-min glucose less than 100 mg/dl but greater than fasting) was nearly identical before and after Cr supplementation. These subjects had good glucose tolerance and presumably good Cr status and therefore should not and did not respond to supplemental Cr. Blood glucose of subjects with low blood sugar (90-min glucose values less than fasting) also improved following Cr supplementation. In a separate study (9), hypoglycemic glucose values of eight hypoglycemic female patients improved significantly following Cr supplementation. Hypoglycemic symptoms, including blurred vision, sweating, shaking, and extreme fatigue, also improved following Cr supplementation. Improvements in blood glucose and symptoms were accompanied by improvements in insulin binding and insulin receptor number (9). The effects of supplemental Cr on hyperglycemics, normal, and hypoglycemic subjects is summarized in Fig. 1. Chromium tends to normalize blood sugar by decreasing blood sugar of subjects with elevated glucose values, having no effect on subjects with near-optimal glucose tolerance and increasing that of subjects with low blood sugar. These effects are likely due to increased insulin efficiency.

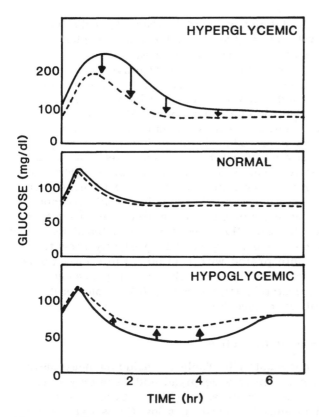

Fig. 1. Effects of supplemental Cr on hyperglycemic, normal, and hypoglycemic subjects. Solid line is before and dashed line is following Cr supplementation.

Chromium functions in glucose metabolism primarily through its action of insulin potentiation; much higher amounts of insulin are required at suboptimal levels of chromium both in vivo and in vitro. Chromium potentiates insulin but does not replace insulin. Overall insulin function can be increased by increasing insulin concentration but also by increasing biologically active forms of Cr (13). Since some of the secondary complications linked to diabetes, including the increased incidence of cardiovascular diseases, have been associated with elevated insulin (14), it is advantageous to keep circulating insulin levels well within the normal range. Chromium functions not only in increasing the activity of insulin, but with this increased activity less insulin is required. Therefore, if the dietary intake of Cr is sufficient, lower levels of insulin are usually present.

Cardiovascular Diseases

Signs and symptoms of Cr deficiency listed in Table 1, including elevated serum cholesterol and triglycerides, increased incidence of aortic plaques, increased aortic intimal plaque areas, as well as elevated circulating insulin, would all tend to be negative factors in the prevention of cardiovascular diseases. Animal studies indicate that with increasing age, rats fed a low-Cr diet have increased serum cholesterol, aortic lipids, and plaque formations (15). Plaque formations can not only be prevented due to sufficient Cr, but in rabbits aortic intimal plaque areas were reduced due to injection of exogenous Cr (16). Individuals who died of coronary artery disease were reported to have significantly lower concentrations of Cr in aortic tissues than subjects dying from accidents, although the Cr concentrations of the other tissues were similar (17). Subjects with coronary artery disease have also been reported to have lower serum Cr than subjects free of disease (18,19). However, tissue and blood Cr levels reported in these studies are much higher than presently accepted values and data must be interpreted with caution. Chromium supplementation of human subjects has also been shown to improve blood lipids, including total cholesterol, triglycerides, HDL-cholesterol, and total cholesterol:HDL ratio (see Ref. 11).

Other Signs and Symptoms of Cr Deficiency

Other signs of Cr deficiency listed in Table 1, including neuropathy and encephalopathy, do not appear widespread and have only been observed in patients on total parenteral nutrition. Corneal lesions appear to be restricted to animals raised on a low-Cr as well as low-protein diet. The effects of Cr on fertility and sperm count have only been studied in rats. Low-Cr male rats, with glucose and lipid parameters similar to the Cr-supplemented control rats, displayed

Table 1. Signs and Symptoms of Chromium Deficiency

Functions	Animals
Impaired glucose tolerance	Human, rat, mouse, squirrel monkey, guinea pig
Elevated circulating insulin	Human, rat
Glycosuria	Human, rat
Fasting hyperglycemia	Human, rat, mouse
Impaired growth	Rat, mouse, turkey
Hypoglycemia	Human
Elevated serum cholesterol and triglycerides	Human, rat, mouse
Increased incidence of aortic plaques	Rabbit, rat, mouse
Increased aortic intimal plaque areas	Rabbit
Neuropathy	Human
Encephalopathy	Human
Corneal lesions	Rat, squirrel monkey
Decreased fertility and sperm count	Rat
Decreased longevity	Rat, mouse
Decreased insulin binding	Human
Decreased insulin receptor number	Human

Source: Adapted from Ref. 40.

marked decreases in sperm count and fertility (20). Decreases in fertility and sperm count of male rats were evident prior to detectable changes in glucose and lipid metabolism.

DIETARY CR INTAKE

Reported Cr intakes have decreased in past decades due primarily to increased awareness of Cr contamination, recent use of standard reference materials, and improved analytical instrumentation and procedures. Several studies have reported daily dietary Cr intakes in excess of 100 or even 200 µg, and some studies have reported even higher values (see Ref. 21 for original references). However, recent well-controlled studies suggest that dietary Cr intake is approximately 50 µg or less in Finland, Canada, England, and the United States (21–24). Values for dietary Cr intake for Canada (23) were slightly above 50 µg, while the daily dietary Cr intake for the other three countries was approximately 30 µg. (Fifty micrograms would be equal to the minimum suggested intake as suggested by the National Academy of Sciences, who have recommended a suggested safe and adequate intake of 50–200 µg (25).)

Many of the earlier studies that reported erroneously high levels for dietary Cr intake also determined the intake of several other elements, and unless conditions for sample collection and analysis are optimized for Cr, values for Cr would be expected to be too high. For example, if conditions are not optimized for Cr and stainless steel blender blades are used during the homogenization of the samples, erroneously high values for dietary Cr intake would be observed. Chromium leaching from stainless steel, which is approximately 18% Cr, into foods and beverages is well documented (26–28).

Appreciable losses of chromium occur in the refining and processing of certain foods. The recovery of Cr in white flour was only 35 to 44% of that of the parent wheat products (29). Chromium content of unrefined sugar was only 60% of the parent molasses, and brown and white sugar contained only 24 and 8%, respectively, of the Cr content of the unrefined product (30). The high intake of refined sugar in typical American diets (about 120 gm/day/person) not only contributes marginal levels of dietary Cr, but refined carbohydrates stimulate losses of endogenous Cr (31) (see section on urinary Cr excretion).

There are presently no reliable data bases from which to calculate dietary Cr intake from diet records. Therefore, dietary Cr intake must be measured. The most reliable technique is the duplicate plate technique, where, whatever a subject eats or drinks, a duplicate portion prepared in the same manner is added to a container for analysis. The dietary Cr intake, using the duplicate plate technique, of 10 male and 22 female subjects consuming self-selected diets for seven consecutive days is shown in Fig. 2. S.S.A.I. denotes the minimum suggested safe and adequate intake of 50 µg as established by the National Academy of Sciences (25). More than 90% of the daily diets contained less than the minimum suggested safe and adequate intake, and when the seven-day average Cr intake was determined, none of the subjects consumed the minimum suggested intake. Mean daily Cr intake was 33 ± 3 µg for the male subjects and 25 ± 1 µg for the females. These subjects were normal affluent subjects that should have an average or better knowledge of good nutritional practices.

Institutional diets prepared by nutritionists appear to be slightly higher in Cr but still do not meet the minimum suggested safe and adequate intake (26,31). Mean Cr intake of institutional diets was

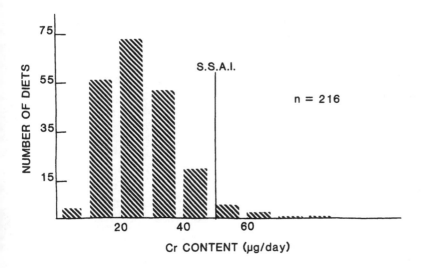

Fig. 2. Frequency distribution of dietary Cr intake. Diets from 10 males and 22 females were collected for 7 consecutive days and analyzed for Cr.

reported to be approximately 16 µg per 1000 calories (31). Therefore, more than 3000 calories would need to be consumed of a nutritious, well-balanced diet to meet the minimum suggested intake and more than 12,000 calories per day to achieve the upper limit of 200 µg of Cr.

SOURCES OF DIETARY CR

There have not been any comprehensive studies to determine the Cr content of individual foods eaten in the United States. However, Finnish workers completed a large study involving the nutrient content, including Cr, of numerous foods from each of the basic food groups. Foods high in Cr are listed in Table 2. Even foods that are high in Cr are only slightly higher than a number of other foods, and there are no known foods eaten that are outstanding sources of dietary Cr. The approximate daily dietary intake of Cr from the various food groups is listed in Table 3. The distribution is similar among fruits, vegetables, dairy products, beverages, and meat with lesser amounts from cereal products and negligible dietary Cr from fish and seafood. Beverages, including milk, account for one-third of the daily intake of Cr. Normal cow's milk is reported to contain 5 to 15 ng of Cr per ml (22) and cow's colostrum five-fold higher levels. Breast milk has been reported to contain similar levels, but

Table 2. Foods High in Chromium

Mushrooms

Brewer's yeast

Black pepper

Prunes

Raisins

Nuts

Asparagus

Beer

Wine

Source: Ref. 40.

Table 3. Daily Intake of Chromium from Various Food Groups

Food Group	Average daily intake (μg)	Comments
Cereal products	3.7	55% from wheat
Meat	5.2	55% from pork; 25% from beef
Fish and seafoods	0.6	
Fruits, vegetables, nuts and mushrooms	6.8	70% from fruits and berries
Dairy products, eggs and margarine	6.2	85% from milk
Beverages, confectionaries, sugar and condiments	6.6	45% from beer wine and soft drinks
Total	29.1	

Source: Ref. 41.

Table 4. Correlation of Chromium Intake with Other Nutrients

Parameter	Correlation coefficient
Potassium	0.66**
Energy	0.61**
Fat	0.60**
Saturated fat	0.59**
Sodium	0.58**
Oleic acid	0.57*
Phosphorus	0.51*
Vitamin B6	0.51*
Copper	0.51*
Ash	0.50*
Protein	0.47*
Total carbohydrate	0.47*
Linoleic acid	0.44
Cholesterol	0.42
Pantothenic	0.37
Calcium	0.36
Ascorbic acid	0.36
Vitamin B12	0.35
Riboflavin	0.34
Starch	0.27
Pentose	0.26
Thiamin	0.26
Iron	0.25
Folic acid	0.25
Lactose	0.25
Maltose	0.24
Vitamin A	0.22
Sucrose	0.18

Table 4. (continued)

Parameter	Correlation coefficient
Niacin	0.18
Fiber	0.15
Zinc	0.12
Reducing sugars	0.10
Cellulose	0.07
Fructose	0.05
Sorbitol	0.05
Hemicellulose	0.01
Lignin	0.01
Galactose	0.10
Glucose	-0.09
Pectin	-0.09

**Significant correlation at $P < 0.001$
*Significant correlation at $P < 0.01 - P \leqslant 0.001$.
Source: Ref. 21.

recent reports indicate that breast milk contains approximately 0.3 to 0.4 ng of Cr per ml (32). Some brands of beer are very good sources of Cr, and a single serving (12 oz) may contain approximately 20 μg or approximately two-thirds of the normal dietary Cr intake (28). However, some brands of beer contain much lower concentrations of Cr, and the mean Cr intake per serving of beer is approximately 7% of the minimum suggested safe and adequate intake (28).

Diets high in Cr also appear to be high in potassium, energy, fat, and sodium (Table 4). Chromium content also was correlated with oleic acid, phosphorus, vitamin B_6, copper, protein, and total carbohydrate. Copper is the only trace element that correlated with chromium. Gibson and Scythes (23) also reported a correlation between dietary intake of chromium and copper.

CHROMIUM ABSORPTION

Chromium compounds are poorly absorbed in animals and humans, with mean Cr absorption of 0.4 to 3 percent or less (11,21,33). There is some evidence that natural complexes in the diet are absorbed better than simple Cr salts (33). Chromium absorption is rapid, with substantial absorption of radioactively labeled Cr within two hours of Cr ingestion (11).

Chromium absorption by human subjects consuming normal self-chosen diets is inversely related to dietary intake at levels found in the diet (Fig. 3). At a daily intake of 10 µg, Cr absorption is approximately 2%, and with increasing intake to 40 µg, Cr absorption decreases to 0.5%. At dietary intakes above 40 µg, Cr absorption appears constant at approximately 0.4%. Increasing Cr intake from normal basal levels of 30 to 40 µg where absorption would be approximately 0.4% due to supplemental inorganic Cr to an intake of approximately 240 µg did not alter Cr absorption (5). Chromium absorption from high-Cr foods contributing more than 40 µg per day needs to be established. The site and mechanism of Cr absorption also need to be elucidated.

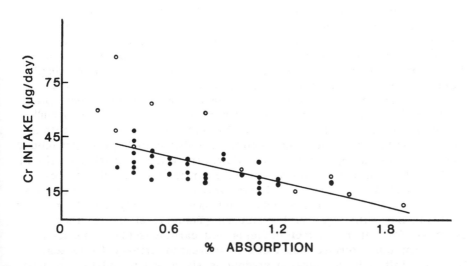

Fig. 3. Chromium absorption of adult subjects at varying Cr intakes. Solid symbols denote 7-day average values and open symbols daily values. (From Ref 21).

CHROMIUM EXCRETION AND EFFECTS OF STRESS

Essentially all absorbed Cr is excreted in the urine with only small amounts lost in hair, perspiration, and bile (35,36). The reported values for urinary Cr excretion have decreased more than 700-fold in the past two decades due to improved methods of sampling, prevention of contamination, and improved instrumentation (34,37,38). Presently accepted values for urinary Cr excretion from several countries are less than 0.5 μg/day for normal free-living subjects (11,34). These lower values for urinary Cr excretion are compatible with presently accepted intake and absorption values. For example, assuming a dietary intake of approximately 40 μg (see section on Cr intake) and an absorption of 0.5% (see section of Cr absorption), approximately 0.2 μg would be absorbed and excreted, which is consistent with the presently accepted values for urinary Cr excretion.

Elevated Cr levels are present during various forms of stress, such as elevated intake of glucose or refined carbohydrates, strenuous exercise, and physical trauma. A single exposure to elevated glucose intake, such as a glucose tolerance test, leads to marked increases in urinary Cr concentration 90 min following the challenge compared to basal levels (34). These increased acute losses of Cr following a glucose load are sustained when dietary intake of refined sugars is increased. The increase in urinary Cr excretion following consumption of increased refined sugar diet is shown in Fig. 4. Thirty-seven subjects, 19 men and 18 women, consumed for 12 weeks reference diets formulated by nutritionists to contain optimal levels of protein, fat, carbohydrate, and other nutrients; the following six weeks, subjects consumed high-sugar diets. The reference diets contained 35% of total calories from complex carbohydrates and 15% from simple sugars, and the high-sugar diets contained 15% complex and 35% simple carbohydrates. Compared to the reference diets, consumption of the high-sugar diets increased urinary Cr losses from 10 to 300% for 27 of 37 subjects (Fig. 4; 31). In that study dietary Cr intake was constant, since nutritious foods, in addition to high-sugar foods, were eaten to supply dietary Cr. However, under normal conditions, foods high in simple sugars would be eaten in place of other more nutritious foods. Therefore, dietary intake of Cr would be decreased and urinary Cr losses increased, which may ultimately lead to marginal Cr deficiency with signs and symptoms (Table 1) similar to those observed for maturity-onset diabetes and cardiovascular diseases.

Strenuous exercise also leads to increased urinary Cr losses. In nine adult male runners, a strenuous six-mile run led to a nearly fivefold increase in urinary Cr excretion two hours after running, and total urinary Cr excretion on the day of running was more than double the urinary Cr losses observed on a nonrun day (38).

Physical trauma severe enough to require treatment at a shock trauma unit also led to increased Cr losses. Urinary Cr excretion of seven severely traumatized male patients was more than 50-fold above normal basal levels the first 25 hours following trauma and was still elevated 72 hours following injury. However, the effects of trauma alone are difficult to assess due to the contaminating Cr present in the fluids used in the treatment of the injuries (39).

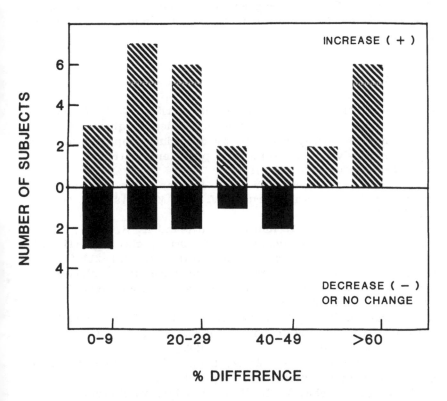

Fig. 4. Percent differences in chromium excretion between reference and high sugar diet periods. Subjects with increased urinary losses while consuming the high sugar diets are denoted by the hatched bars and subjects with the same or lower Cr excretion are denoted by the solid bars. (From Ref. 31.)

SUMMARY

In summary, dietary Cr intake of most normal individuals appears to be suboptimal. Signs of marginal Cr deficiency are similar to those observed for maturity-onset diabetes and (or) cardiovascular diseases. Improvements in glucose intolerance and related variables as well as improvements in risk factors associated with cardiovascular diseases following Cr supplementation are well documented. The distribution of Cr among various food groups is similar, and beverages, including milk, account for approximately one-third of the daily intake of Cr. Stress, including strenuous exercise, high-sugar diets, and physical trauma lead to increased Cr losses. Nutritional problems related to marginal Cr status appear widespread, and considerably more research is needed to elucidate the dietary role of Cr in human health and disease.

REFERENCES

1. K. N. Jeejeebhoy, R. C. Chu, E. B. Marliss, G. R. Greenberg, and A. Bruce-Robertson, Chromium deficiency, glucose intolerance, and neuropathy reversed by chromium supplementation in a patient receiving long-term total parenteral nutrition, *Am. J. Clin. Nutr.*, 30: 531–538 (1977).

2. H. Freund, S. Atamian, and J. E. Fischer, Chromium deficiency during total parenteral nutrition, *J. Am. Med. Assoc.*, 241: 496–498 (1979).

3. L. L. Hopkins, Jr., O. Ransome-Kuti, and A. S. Majaj, Improvement of impaired carbohydrate metabolism by chromium (III) in malnourished infants, *Am. J. Clin. Nutr.*, 21: 203–211 (1968).

4. C. T. Gurson and G. Saner, Effect of chromium on glucose utilization in marasmic protein-calorie malnutrition, *Am. J. Clin. Nutr.*, 24: 1313–1316 (1971).

5. R. A. Levine, D. H. P. Streeten, and R. J. Doisy, Effects of chromium supplementation on the glucose tolerance of elderly subjects, *Metabolism*, 17: 114–125 (1968).

6. W. H. Glinnsmann and W. Mertz, Effect of trivalent chromium on glucose tolerance. *Metabolism*, 15: 510–520 (1966).

7. R. Nath, J. Minocha, V. Lyall, S. Sunder, V. Kumar, S. Kapoor, and K. L. Dhar, Assessment of chromium metabolism in maturity onset and juvenile diabetes using chromium-51 and therapeutic response of chromium administration on plasma lipids, glucose tolerance and insulin levels, in *Chromium in Nutrition and Metabolism* (D. Shapcott and J. Hubert, eds.), Elsevier/ North Holland, Amsterdam, 1979, pp. 213–222.

8. R. T. Mossop, Effects of chromium (III) on fasting glucose,

cholesterol, and cholesterol HDL levels in diabetics, *Cent. Afr. J. Med.*, *29*: 80–83 (1983).

9. R. A. Anderson, M. M. Polansky, N. A. Bryden, S. J. Bhathena, and J. J. Canary, Effects of supplemental chromium of patients with symptoms of reactive hypoglycemia, *Metabolism*, *36*: 351–355 (1987).

10. J. P. Carter, A. Kattab, K. Abd-El-Hadi, J. Davis, A. El Gholmy, and V. N. Patwardhan, Chromium (III) in hypoglycemia and in impaired glucose utilization in kwashiorkor, *Am. J. Clin. Nutr.*, *21*: 195–202 (1968).

11. J. S. Borel and R. A. Anderson, Chromium, in *Biochemistry of the Essential Ultratrace Elements* (E. Frieden, ed.), Plenum Publishing Corporation, New York, 1984, pp. 175–199.

12. R. A. Anderson, M. M. Polansky, N. A. Bryden, E. E. Roginski, W. Mertz, and W. Glinsmann, Chromium supplementation of human subjects: effects on glucose, insulin and lipid variables, *Metabolism*, *32*: 894–899 (1983).

13. R. A. Anderson, J. H. Brantner, and M. M. Polansky, An improved assay for biologically active chromium, *J. Agric. Food Chem.*, *26*: 1219–1221 (1978).

14. R. W. Stout, The relationship of abnormal circulating insulin levels to atherosclerosis, *Atherosclerosis*, *27*: 1–13 (1977).

15. H. A. Schroeder and J. J. Balassa, Influence of chromium, cadmium, and lead on rat aortic lipids and circulating cholesterol, *Am. J. Physiol*, *209*: 433–437 (1965).

16. A. S. Abraham, M. Sonnenblick, M. Eini, O. Shemash, and A. P. Batt, The effect of chromium on established atherosclerotic plaques in rabbits, *Am. J. Clin. Nutr.*, *33*: 2294–2298 (1980).

17. H. A. Schroeder, A. P. Nason, and I. H. Tipton, Chromium deficiency as a factor in atherosclerosis, *J. Chronic Dis.*, *23*: 123–142 (1970).

18. H. A. I. Newman, R. F. Leighton, R. R. Lanese, and N. A. Freedland, Serum chromium and angiographically determined coronary artery disease, *Clin. Chem.*, *24*: 541–544 (1978).

19. M. Simonoff, Y. Llabador, C. Hamon, A. M. Peers, and G. N. Simonoff, Low plasma chromium in patients with coronary artery and heart diseases, *Biol. Trace Element Res.*, *6*: 431–439 (1984).

20. R. A. Anderson and M. M. Polansky, Dietary chromium deficiency: effect on sperm count and fertility in rats, *Biol. Trace Element Res.*, *3*: 1–5 (1981).

21. R. A. Anderson and A. S. Kozlovsky, Chromium intake, absorption and excretion of subjects consuming self-selected diets, *Am. J. Clin. Nutr.*, *41*: 1177–1183 (1985).

22. P. Koivistoinen, Mineral element composition of Finnish foods: N, K, Ca, Mg, P, S, Fe, Cu, Mn, Zn, Mo, Co, Ni, Cr, F, Se, Si, Rb, Al, B, Br, Hg, As, Cd, Pb, and Ash, *Acta Agric. Scand.*, *Suppl. 22*, Stockholm, 1980.

23. R. S. Gibson and C. A. Scythes, Chromium, selenium, and other trace element intakes of a selected sample of Canadian premenopausal women, *J. Biol. Trace Element Res.*, *6*: 105–116 (1984).

24. W. Bunker, M. S. Lawson, H. T. Delves, B. E. Clayton, The uptake and excretion of chromium by the elderly, *Am. J. Clin. Nutr.*, *39*: 717–802 (1984).

25. National Research Council, *Recommended Dietary Allowances*, National Academy of Sciences, Washington, D.C., 1980.

26. J. T. Kumpulainen, W. R. Wolf, C. Veillon, and W. Mertz, Determination of chromium in selected United States diets, *J. Agric. Food Chem.*, *27*: 490–494 (1979).

27. E. G. Offenbacher and F. X. Pi-Sunyer, Temperature and pH effects on the release of chromium from stainless steel into water and fruit juices, *J. Agric. Food Chem.*, *31*: 89–92 (1983).

28. R. A. Anderson and N. A. Bryden, Concentration, insulin potentiation and absorption of chromium in beer, *J. Agric. Food Chem.*, *31*: 308–311 (1983).

29. E. G. Zook, E. F. Greene, and E. R. Morris, Nutrient composition of selected wheats and wheat products. VI. Distribution of manganese, copper, nickel, zinc, magnesium, lead, tin, cadmium, chromium, and selenium as determined by atomic absorption spectroscopy and colorimetry, *Cereal Chem.*, *47*: 720–727 (1970).

30. R. Masironi, W. Wolf, and W. Mertz, Chromium in refined and unrefined sugars—possible nutritional implications in the etiology of cardiovascular diseases, *Bull. WHO*, *49*: 322–324 (1973).

31. A. S. Kozlovsky, P. B. Moser, S. Reiser, and R. A. Anderson, Effects of diets high in simple sugars on urinary chromium losses, *Metabolism 35*: 515–518 (1986).

32. C. E. Casey and K. M. Hambidge, Chromium in human milk from American mothers, *52*: 73–77 (1984).

33. W. Mertz, Chromium occurrence and function in biological systems, *Physiol Rev.*, *49*: 163–239 (1969).

34. R. A. Anderson, M. M. Polansky, N. A. Bryden, K. Y. Patterson, C. Veillon, and W. H. Glinsmann, Effect of chromium supplementation on urinary Cr excretion of human subjects and correlation of Cr excretion with selected clinical parameters, *J. Nutr.*, *113*: 276–281 (1983).

35. L. L. Hopkins, Jr., Distribution in the rat of physiological amounts of injected Cr^{51} (III) with time, *Am. J. Physiol.*, *209*: 731–735 (1965).

36. R. J. Doisy, D. H. P. Streeten, M. L. Souma, M. E. Kalafer, S. L. Rekant, and T. G. Dalakos, Metabolism of chromium 51 in human subjects, in *Newer Trace Elements in Nutrition* (W. Mertz and W. E. Cornatzer, eds.), Dekker, New York, 1971, pp. 115–168.

37. C. Veillon, K. Y. Patterson, and N. A. Bryden, Direct determination of chromium in human urine by electrothermal atomic absorption spectrometry, *Anal. Chim. Acta.*, *136*: 233—241 (1982).

38. R. A. Anderson, M. M. Polansky, N. A. Bryden, E. E. Roginski, K. Y. Patterson, and D. C. Reamer, Effects of exercise (running) on serum glucose, insulin, glucagon, and chromium excretion, *Diabetes*, *31*: 212—216 (1982).

39. J. Borel, T. C. Majerus, M. M. Polansky, P. B. Moser, and R. A. Anderson, Chromium intake and urinary chromium excretion of trauma patients, *J. Biol. Trace Element Res.*, *6*: 317—326 (1984).

40. R. A. Anderson, Chromium supplementation: effects on glucose tolerance and lipid metabolism, in *Trace Elements in Health and Disease* (H. Bostrom and N. Ljungstedt, eds.), Norstedts Tryckeri, Stockholm, Sweden, 1985, pp. 110—124.

41. R. A. Anderson, Chromium requirements and needs in the elderly, in *Handbook of Nutrition in the Aged* (R. R. Watson, ed.), CRC Press Inc., Boca Raton, FL, 1985, pp. 137—144.

8
Iodine

JEAN A. T. PENNINGTON / Center for Food Safety and Applied Nutrition, Food and Drug Administration, Washington, D.C.

INTRODUCTION

Properties and Classification of Iodine

Iodine, a nonmetallic, nearly black crystalline solid which sublimes at room temperature to a deep violet vapor, was termed "substance X" by the French chemist Bernard Courtois in 1811. In 1813 the English chemist Sir Humphrey Davy recognized substance X as an element analogous to chlorine and suggested the name iodine from the Greek "ioeides," meaning violet-colored.

Iodine, fluorine, chlorine, bromine, and astatine are the halogen elements that constitute group VIIA of the Periodic Table. Because of their great reactivity, halogens are not found free (uncombined) in nature. The common oxidation states of iodine are -1 (iodides), +5 (iodates), and +7 (periodates). Iodine combines readily with most metals and some nonmetals to form iodides. The iodide ion is a strong reducing agent which readily gives up electrons. The only naturally occurring isotope of iodine is ^{127}I; the radioactive isotopes ^{131}I and ^{125}I are commonly used in metabolic research and medical diagnosis.

Commercial Sources of Iodine

As a trace element in the earth's crust, iodine at 0.5 ppm is not concentrated enough to form independent minerals. The major commercial sources of iodine are deposits of saltpeter (potassium and sodium nitrate) from Chile in which iodine is present as solid iodates, especially calcium iodate; brine extracted from oil wells, which contains mainly sodium iodine (NaI); and seaweed extracts. Some species of seaweed are able to concentrate iodine manyfold from seawater.

249

Commercial Uses of Iodine

Iodine is an essential dietary element which may be added to table salt and dietary supplements to prevent iodine deficiency in humans and to animal feed and salt licks to prevent iodine deficiency in animals. Reddish-brown solutions of iodine in water or alcohol are widely used as disinfectants and antiseptics to kill bacteria, viruses, and fungi. Two-percent and five-percent (Lugol's solution) aqueous solutions of iodine and dilute solutions (usually two percent) of iodine in alcohol (tincture of iodine) have been used as topical antiseptics for cuts and wounds. Iodophors are used in the food industry to clean, sanitize, and disinfect food service equipment. Calcium and potassium iodates may be used as dough conditioners to make bread and other bakery products. Iodine compounds are used in analytical chemistry, photography, and metallurgy. Silver iodide in its crystalline form is used to seed clouds for artificial rain making.

Iodine isotopes and other iodine compounds are used to diagnose a variety of diseases, to assess the function of various organs and systems, and to trace the course of compounds in metabolism. Radioactive iodine isotopes are useful for diagnosing thyroid disorders such as hyperthyroidism, hypothyroidism, and thyroid cancer by determining the size, shape, and activity of the thyroid gland. Radioactive iodine, which is ingested in aqueous solution, is absorbed, circulated, metabolized, and concentrated in the thyroid gland like other iodides. The magnitude of iodide uptake and binding (measured by placing a counter close to the thyroid gland) can then be quantified in patients suspected of having thyroid disease. Thyroid clearance of radioactive iodine after 1 to 2 hours and uptake over 24 to 48 hours are convenient measures of thyroid function. Radioactive iodine isotopes are also used as tracers for medical diagnosis and metabolic research for conditions other than thyroid concerns, such as diagnosis of kidney and liver disorders, screening for pulmonary emboli, measurement of blood/plasma volume, measurement of cardiac output, and location of tumors.

The minute amounts of radioactive iodine used for diagnostic purposes have no known significant effect on thyroid function, but when large doses of radioactive iodine are administered, thyroid cells may be damaged. Radioactive iodine concentrates in the thyroid and emits beta particles, which can destroy a portion of the thyroid cells. An appropriate dose can restore a hyperactive gland to normal. This treatment is usually reserved for adults and those who are poor surgical risks. It is not used during pregnancy as it may destroy the fetal thyroid. Radioactive iodine should not be used during lactation as it is secreted into the breast milk and may damage the infant thyroid and increase the risk of thyroidal cancer developing in later life (1). Radioactive iodine can also be used as therapy for certain types of thyroid cancer, although surgery and external radiation are often more appropriate.

Radiopaque contrast media (gas, barium sulfate, or organic iodine compounds) may be introduced into tissues to make detailed examination of the structures feasible. The contrast media allow for greater difference in density between adjacent tissues. Iodine atoms absorb more radiation than surrounding tissues, thus causing denser shadows. Iodinated contrast compounds are widely used for arteriograms, urograms, intravenous pyelograms, infusion computer tomographic scans, and oral cholecystograms. Thus, by use of iodine contrast media, the biliary tract, the urinary tract, renal function, the vascular system, the chambers of the heart, the spinal canal, and other body systems can be visualized.

IODINE PHYSIOLOGY

Body Distribution and Functions of Iodine

Iodine is present in all body tissue and secretions, but the total amount in the body is quite small. Estimates of total body iodine range from 10 to 50 mg for an adult. It is estimated that 70 to 90% of the total body iodine is concentrated in thyroglobulin (an iodinated glycoprotein) of the thyroid gland. The thyroid gland is located just below the larynx at the base of neck and consists of two lobes joined by a bridge that crosses the trachea.

Iodine is required for synthesis of the thyroid hormones, thyroxine (T4) and triiodothyronine (T3), which are iodinated molecules of the essential amino acid tyrosine. T4 contains four iodine atoms, and T3 contains three iodine atoms. Synthesis of thyroid hormones is the only known role for iodine in animal bodies, and there is no evidence of an iodine requirement for plants or microorganisms. Only vertebrates and closely related protochordates have the ability to synthesize significant amounts of biologically active thyroid hormones.

The thyroid hormones regulate a wide variety of physiological processes in body tissues. Fundamental to many of these processes is their effect on the rate of cellular oxidation and hence on calorigenesis, thermoregulation, and intermediary metabolism. Thyroid hormones exert a calorigenic effect (increased oxygen consumption) on heart, liver, and kidney cells, but not on brain cells. T4 and T3 are necessary for protein synthesis. They promote nitrogen retention, glycogenolysis, intestinal absorption of glucose and galactose, lipolysis, and uptake of glucose by adipocytes. They also sensitize adipose tissue to actions of other lipolytic hormones and decrease serum cholesterol.

Iodine Absorption and Excretion

Iodine occurs in foods mainly as inorganic iodide, which is readily and completely absorbed from the gastrointestinal tract. Other forms of iodine in foods are reduced to iodide before absorption. Iodide is rapidly distributed throughout the body via the circulatory system. Approximately 30% is removed by the thyroid for hormonal synthesis, and the remainder is excreted by the kidneys. Because iodine intake in excess of requirement is excreted primarily through the urine, urinary iodine is a good measure of iodine status. Minor amounts of iodine are also excreted in feces, sweat, and expired air. Large amounts of iodine can be lost with lactation, as it is readily secreted into milk.

Iodine Metabolism

Inorganic iodide is actively transported from the blood into thyroid cells. This process, which is stimulated by thyroid-stimulating hormone (TSH or thyrotropin) from the anterior lobe of the pituitary gland is referred to as the "iodide pump," the "iodide-concentrating mechanism," or the "iodide-trapping mechanism." The normal ratio of thyroid to plasma iodide is 25 to 1. Once inside the thyroid cells, iodide is oxidized to iodine by the enzyme iodide peroxidase.

Thyroid tissue is organized into follicules in which thyroid cells surround a central cavity. The follicules contain a gelatinous pink material, the colloid, in which thyroid hormones are stored bound to a glycoprotein called thyroglobulin. Synthesis of thyroid hormones occurs at the border between the cells and the colloid by successive iodination and condensation of tyrosine residues attached to the thyroglobulin in the colloid. The oxidized (active) iodine ions react with the tyrosine components of thyroglobulin to form 3-monoiodotyrosine and 3,5-diiodotyrosine. These two molecules then couple to form the active hormones T4 and T3. T4 (3,4,3',5'-tetraiodothyronine) and T3 (3,5,3'-triiodothyronine) attached to thyroglobulin in the thyroid follicules are stored as the main component of the thyroid colloid.

Proteolytic enzymes (from the lysozomes) release T3 and T4 from thyroglobulin. These hormones move into capillaries by simple diffusion and are released into the blood. In the blood, 99% of T4 is bound to globulin and albumin proteins; T3 is bound to a lesser extent. These bound hormones constitute the protein-bound iodine of the plasma. The hormones must be removed from the proteins to function. As the free hormones are removed from the circulation by entering body cells, they are replaced by T4 and T3 from the bound pool. Iodine is removed from the hormones primarily in the liver and kidney. The liberated iodine may be recaptured and reutilized by the thyroid gland.

Synthesis and secretion of T4 and T3 are under control of TSH. TSH stimulates almost all aspects of thyroid function, including iodide transport, oxidation to iodine, and iodine binding to tyrosine. Synthesis of thyroid hormones is regulated by the level of circulating free T4 and T3 as a negative feedback mechanism operating by direct action on the TSH-secreting cells of the pituitary gland and by indirect action on the hypothalamus and its thyrotropin-releasing factor. During fasting, serum T3 concentrations decrease while serum T4 concentrations do not change (due to reduced peripheral conversion of T4 to T3), and TSH secretion is inhibited (2,3).

Disturbances in Iodine Metabolism

Some individuals are unable to form T4 and T3 because of defects (which are probably enzymatic defects) involving iodine metabolism. These defects, which can result in thyroid enlargement (goiter), consist of the inability of the thyroid to trap iodide, to iodinate tyrosine (organify iodine), and to couple iodotyrosines to form T4 and T3. Hashimoto's thyroiditis is a disease in which there is a disturbance in the thyroidal uptake of iodine. This disease leads to goiter and hypothyroidism, but is successfully treated with thyroid therapy. Toyoshima et al. (4) reported five cases in which there was an absence of the iodide concentrating mechanism. All subjects had goiter and responded to potassium iodide therapy. Case reports of organification defects resulting in goiter and/or hypothyroidism or cretinism have been reported (5,6). Elte (7) estimated that about 14% of patients with goiter have organification defects. Such patients must be treated with thyroid hormones.

Iodine Requirements

To ensure an adequate supply of thyroid hormones, the thyroid must trap about 60 μg of iodine per day (8). The daily iodine requirement for prevention of goiter in adults is 50—75 μg, or approximately 1 μg/kg of body weight (9). To provide a margin of safety, an allowance of 150 μg is recommended for adolescents and adults (10). The recommended allowances are 40—50 μg/day for infants and 70—120 μg/day for children 1 to 10 years old (10). Additional allowances of 25 and 50 μg/day for pregnant and lactating women, respectively, are recommended to cover the demands of the fetus and provide the extra iodine secreted into breast milk (10). The recommendations for infants and, by extrapolation, for the children's age groups, are based on the assumption that an infant receives 30 μg or more of iodine per day in milk from an adequately fed mother (10).

Food products designed to provide complete nutritional maintenance of individuals (e.g., infant formula, total parenteral nutrition, and other special medical diets) should contain iodine levels that meet the Recommended Dietary Allowances (RDA). Odne et al. (11) developed an intravenous trace element solution containing 56 µg iodine/ml to meet patient iodine requirements. The nutrient specifications established by the Food and Drug Administration (FDA) for infant formula indicate a minimum of 5 µg iodine and a maximum 75 µg per 100 calories (12).

IODINE DEFICIENCY

Endemic Goiter

Dietary iodine deficiency results in decreased plasma levels of T4 and T3 and a compensatory increase in TSH secretion. TSH causes the thyroid to increase iodine uptake in an attempt to increase the output of thyroid hormone. Overactivity of the iodine-trapping mechanism is usually associated with thyroid hypertrophy as the gland proliferates its epithelium, producing more cells and increasing cell size. Prolonged TSH action causes a marked enlargement of the gland. The enlargement of the thyroid gland due to iodine deficiency is called endemic goiter.

Iodine intakes consistently lower than 50 µg/day usually result in goiter; however, not all people with iodine deficiency develop goiters. Women and particularly adolescent girls seem to be especially at risk. Because goiter is a compensatory mechanism, most goiterous individuals are clinically euthyroid. With goiter, the T3 secretion of the thyroid gland increases and the deiodination of T4 and T3 increases (13). This increase in T3 is an adaptive response of the thyroid to iodine deficiency and is responsible for maintaining euthyroidism. Affected individuals may continue in general good health despite the goiters. In females, endemic goiters generally persist throughout life, while in males they tend to decrease following adolescence.

With large goiters, obstructive complications may result from compression of the trachea, esophagus, and blood vessels of the neck. Goiters may become so large as to interfere with speaking and breathing. Also, goiters are associated with an increased risk of other thyroid diseases and malignant growth (14). There is an association between endemic goiter and thyroid follicular carcinoma, possibly resulting from prolonged exposure of the thyroid to increased TSH activity.

Endemic goiter, which was recognized as a deficiency disease in the late nineteenth century, is found in areas where soil and water and hence plant and animal materials are low in iodine. The distribution of iodine over the earth's surface is uneven and may reflect glacial effects and/or weathering of the soil. Aston and Brazier (15) reported that most areas of endemic goiter contained relatively young glacial soils and that iodine in soils was independent of parent rock types. Iodine is retained and enriched during vegetative recycling processes (15).

Endemic goiter is found particularly in mountainous regions and areas far from salt water; the incidence of endemic goiter is lowest along sea coasts. Currently this condition is confined to developing countries and typically occurs in mountainous areas such as the Andes, Himalayas, and the mountain chain extending through Southeast Asia and Oceania. Matovinovic (14) describes endemic goiter in various countries including Yugoslavia, Ecuador, Peru, Lebanon, Pakistan, Southeast Asia, India, Africa, New Guinea, and Eastern Mediterranean countries.

Iodine malabsorption, as a consequence of protein-calorie malnutrition, may be a contributory factor to endemic goiter (16—18). Ingenbleek and Beckers (16) reported that thyroidal iodine clearance and radioiodine uptake decreased with protein-calorie malnutrition, and that iodine metabolism was related to morphological changes of the intestinal wall induced by protein-calorie malnutrition.

Goitrogens

The development of endemic goiter due to iodine deficiency may be exacerbated by the ingestion of substances that impair iodine uptake by the thyroid or impair incorporation of iodine into tyrosine. Such substances are called goitrogens. Goitrogens that prevent the binding of iodide to thyroglobulin within the thyroid include thiouracil and other drugs with a thiocarbamide grouping. Goitrogens that inhibit the trapping of iodine in the thyroid cells include thioglucosides that yield thiocyanate and isothiocyanates on hydrolysis. Thioglucosides are found in vegetables of the genus Brassica and family Crucifera. These vegetables include cabbage, cauliflower, broccoli, brussels sprouts, kale, kohlrabi, turnips, and rutabaga. Other foods (not of the Brassica genus) that yield thiocyanate on hydrolysis include nuts, cassava, maize, bamboo shoots, sweet potatoes, and lima beans. A goitrogenic factor present in millet may contribute to the incidence of goiter in Western Sudan, other parts of Africa, and India (18,19).

Goitrogens fed to female mammals may affect the iodine content of milk, or they may be secreted into milk and affect the thyroid function of the offspring. Rapeseed meal, which contains high levels of glucosinolates, when fed to dairy cows reduced the iodine content of milk (20). Oyelola et al. (21) reported that the content of thiocyanate (presumably from cassava) in the breast milk of Nigerian women may be high enough to affect thyroid function in nursing infants in the presence of insufficient iodine. This may result in infant goiter. Because thiouracil is actively transported into breast milk, mothers who require this drug should not breast-feed as their infants may develop goiter (22).

Goitrogenic substances in food are of particular significance when usual iodine intake is low or borderline. An adequate dietary iodine intake can usually overcome the goitrogenic effects of thiocyanates derived from foods. Also, the enzymes which activate progoitrogens to goitrogens may be destroyed by cooking (8). Dietary iodine cannot, however, prevent goiter caused by thiouracil and related drugs.

Hypothyroidism

With severe and prolonged iodine deficiency, the effects of a deficient supply of thyroid hormones may occur. This condition, which is referred to as hypothyroidism or myxedema, is characterized by reduced metabolic rate, diminished vigor, decreased basal body temperature, and cold intolerance. There is weight gain in spite of decreased appetite. The skin becomes dry, rough, and thickened, and the hair becomes coarse. Loss of hair is also a characteristic symptom. A jellylike substance, mucin, accumulates in body tissue. Mucin accumulation under the skin makes the facial features puffy and coarse. Mucin weakens muscular function and thickens the vocal cords, producing a hoarse voice. Other symptoms include easy bruisability and drooping of upper eyelids. Constriction of blood arterioles causes increased resistance to blood flow and prolonged circulation time. Body water and extracellular fluid volume increase, causing edema (nonpitting myxedema). Myxedematous disease of the heart is one of the conditions leading to heart failure in hypothyroidism.

Metabolic changes resulting from hypothyroidism include positive nitrogen balance due to a slowing of protein degradation; slowed glucose absorption due to delayed insulin secretion; a flat glucose tolerance curve; decreased lipid metabolism; accumulation of triglycerides and cholesterol in the blood; and decreased conversion of carotene to vitamin A, which gives the skin a yellowish color. Neurological complications of hypothyroidism include mental sluggishness, psychomotor slowing, various psychiatric symptoms with paranoid overtones, hearing loss, and disturbances of gait, coordination, and rapid movements. Additional symptoms that may appear in juvenile hypothyroidism include delayed eruption and deformation of teeth, increased dental caries, excessive body and facial hair, brittle scalp hair, increased muscle size, growth retardation, and decreased intellectual performance. These complications of adult and juvenile hypothyroidism due to iodine deficiency are reversible with iodine therapy or supplementation.

Endemic Cretinism

A deficiency of iodine during pregnancy, infancy, or early childhood
may cause endemic cretinism in an infant or child because of lack of
thyroid hormones during critical periods of development. Cretinism
is associated with endemic goiter but is not always found where goiter
is endemic. The factor that determines whether cretinism accompanies
endemic goiter is the severity of iodine deficiency. Cretinism is manifest in areas where urinary iodine excretion is less than 20 µg/day.
This disease is now confined to developing countries. Cretinism is
the most severe form of iodine deficiency, and the effects, which
become apparent during the first 12 months of life, are not reversible.
     Two types of endemic cretinism (neurological and myxedematous)
have been described. Neurological endemic cretinism, which predominates in most regions, is characterized by mental and physical
retardation and deaf-mutism. Clinical hypothyroidism is usually
absent. This type of cretinism is the result of maternal iodine deficiency that adversely affects the fetus before the thyroid is functional.
Myxedematous endemic cretinism is characterized by congenital hypothyroidism. This form predominates in some Central African countries
and may be due to dietary goitrogens acting in utero and/or during
early childhood on a thyroid already compromised by iodine deficiency.
     Symptoms of cretinism which appear during the first three to
five months of life include delayed union of skull bones; torpid behavior; slow nursing; spells of choking; cyanosis (blue skin) during
feeding; excessive sleepiness; hoarseness; constipation; pallid, cold,
gray-mottled or yellow-tinged skin; tongue enlargement; umbilical
herniation; flabby musculature; low pulse rate; and low body temperature. Cretinism which develops in the later months of the first
year may show any of these symptoms as well as delay or reversal of
sitting, crawling, walking, talking, and head lifting. Delays in teeth-
ing, failure to grow in length at a normal rate, excessive body and
facial hair, increased neck or hip fat, and low pulse pressure may
also be present.
     The full picture of endemic cretinism is characterized by mental
retardation, retarded growth, deafness, deaf-mutism, delayed/arrested
psychomotor development, and various neurological abnormalities. The
skin is pale, grey, thickened, and lacking in mobility; the hair is
dry and scanty. Other symptoms include poor appetite, breathing
difficulties, low body temperature, cold intolerance, decreased sweating, and decreased vigor and activity. Cretins have shortened stature
with infantile body proportions (short legs and arms) and depression
of the bridge of the nose.

Iodine Supplementation

Addition of iodine to table salt is recognized as an economical, convenient, and effective means of preventing goiter in areas of endemic

goiter. In the United States, an iodized salt plan, which was first adopted by the Michigan Department of Health in 1924 and thereafter became available throughout the United States, eradicated the problems of endemic goiter in this country. Iodized salt is available in many other countries including Argentina, Czechoslovakia, France, New Zealand, Switzerland, Yugoslavia, and Canada. In some countries (e.g., Canada), iodization of salt is mandatory. In the United States, the FDA proposed and enacted regulations that require label statements to specify whether table salt is iodized or noniodized. This allows the consumer to purchase the desired type of salt. Slightly more than half of the table salt consumed in the United States is iodized (10). The level of iodine in iodized salt in the United States provides 76 µg of iodine per gram of table salt (208 µg of iodine per ½ teaspoon of iodized salt). Other countries have different (usually lower) levels of iodine in iodized salt.

Many endemic goiter areas are in poorly developed countries with limited access to medical care. Intramuscular injection of iodine is the recommended method of prophylaxis in such areas (e.g., New Guinea, Zaire) where distribution of iodized salt is not practical. These injections release iodine slowly over 1 to 3 years.

FOOD SOURCES OF IODINE

Analytical Methodology

The concentration of iodine in most foods is low; accurate determination of iodine in foods requires sensitive analytical methods and freedom from reagent contamination. The traditional method used for iodine analysis is a chemical one (the Sandell and Kolthoff method), which is based on the catalytic effect of iodine on the reduction of ceric sulfate by arsenious acid. The rate of reduction is followed by the disappearance of the yellow color of the ceric ion. Other methods used to determine iodine in food include the selective ion electrode method, neutron activation analysis, gas chromatography, differential pulse polarography, and X-ray fluorescence spectrometry.

Factors Affecting the Iodine Content of Foods

The iodine content of foods is affected by several factors. The iodine content of drinking water reflects the iodine content of the rocks and soils with which it has had contact. The iodine content of plant foods and feeds reflects the iodide content of soil, irrigation water, and fertilizers. Foods of marine origin (salt water fish and seaweed) have higher natural concentrations of iodine than other foods because they concentrate iodine from seawater. The iodine content of meat, poultry, eggs, and dairy products may reflect the iodine content of drinking water, forage, and feeds; use of iodine feed supplements; use of iodine-containing salt licks; use of iodine-containing veterinary drugs; and contact of the foods with iodine-containing disinfectants and sanitizers (commonly used in the diary industry).

The concentration of iodine in processed foods is affected by the use of iodine-containing food additives including calcium iodate, potassium iodate, cuprous iodide, potassium iodide, erythrosine, and iodophors. The use and maximum limits (where appropriate) of these food additives are specified in the Code of Federal Regulations (23). Calcium and potassium iodate are dough strengtheners; cuprous and potassium iodide may be added to table salt. Iodophors are sanitizing agents, and erythrosine (FD&C Red No. 3) is a red food dye. Red No. 3 is 58% iodine on a weight basis; however, the bioavailability of iodine from this compound is probably only 2 to 5% (24,25).

Iodine Content of Total Diet Study Foods

The iodine content of foods included in the FDA's Total Diet Study for 1982—1984 (26) are summarized by food groups in Table 1. Beverages, fats and sauces, nuts, fruits, fruit juices/drinks, and sweets provided an average of 3 µg/serving or less. Notable exceptions were foods which may have contained the red food coloring erythrosine (e.g., strawberry gelatin dessert and canned fruit cocktail). Vegetables provided an average of 6 µg/serving; breakfast/luncheon meats provided 7 µg/serving; soups contained 12 µg/serving. Grain-based desserts provided 13 µg/serving, and cheese contained 14 µg/serving. Although milk is usually high in iodine, most iodine partitions into whey during cheese processing, leaving little in the finished cheese (27).

Ready-to-eat breakfast cereals provided an average of 19 µg/serving (excluding fruit-flavored presweetened cereals which probably contained erythrosine). Grain products provided 20 µg/serving. Meat and poultry provided 20 µg/serving; eggs provided 32 µg/serving; cooked grains provided 45 µg/serving; and mixed dishes provided 56 µg/serving. Legumes, fluid milk and cottage cheese, fish, and dairy-based desserts provided 58, 60, 72, and 72 µg/serving, respectively.

The average iodine content of infant foods was 2 µg/serving for fruits and juices and for cereals, 3 µg/oz for infant formula, 6 µg/jar for vegetables, 7 µg/jar for mixed dishes, 26 µg/jar for pudding/custard, and 33 µg/jar for meats and poultry. Other authors providing detailed information on the iodine content of specific foods are Kidd et al. (28); Dunsmore (29); Wenlock et al. (30); Varo et al. (31); Bruhn et al. (32); Allegrini et al. (33); and Dellavalle and Barbano (27).

Table 1. Iodine Content of Total Diet Study Foods

	Serving portion, household measure (g)[a]		Iodine, µg/serving (n)
BEVERAGES			
cola, soda, canned	12 fl oz	(370g)	0 (8)
coffee, decaffeinated, from instant	8 fl oz	(240g)	0 (8)
tea, from bag	8 fl oz	(240g)	0 (7)
whisky	1½ fl oz	(42g)	0 (8)
water	8 fl oz	(240g)	0 (7)
coffee, from instant	8 fl oz	(240g)	0 (8)
beer, canned	12 fl oz	(356g)	0 (8)
wine, table	4 fl oz	(118g)	2 (8)
lemon-lime soda, canned	12 fl oz	(368g)	4 (8)
cola, diet soda, canned	12 fl oz	(355g)	4 (8)
cherry drink, from powder	8 fl oz	(262g)	5 (8)
average			1/serving
FATS & SAUCES			
margarine	1 T	(14g)	0 (7)
italian dressing	1 T	(15g)	0 (8)
corn oil	1 T	(14g)	0 (8)
mayonnaise	1 T	(14g)	1 (8)
butter	1 T	(14g)	1 (8)
gravy, from mix	1 T	(16g)	1 (8)
cream substitute, powdered	1 T	(6g)	1 (7)
white sauce, medium, homemade	1 T	(16g)	3 (8)
cream, half & half	1 T	(15g)	4 (8)
average			1/serving
NUTS			
pecans	1 oz	(28g)	0 (7)
peanut butter	2 T	(32g)	2 (8)
peanuts, roasted	1 oz	(28)	3 (8)
average			2/serving
FRUITS			
banana	1 medium	(114g)	0 (7)
watermelon	1 cup	(160g)	0 (8)
applesauce, canned	½ cup	(122g)	0 (8)
grapes	1 cup	(160g)	0 (8)
plums, purple	2 medium	(132g)	0 (8)
grapefruit	½ medium	(118g)	0 (8)
cherries, sweet	15 medium	(102g)	0 (8)

Table 1. (continued)

	Serving portion, household measure (g)[a]	Iodine, µg/serving (n)
FRUITS (continued)		
apple	1 medium (138g)	1 (8)
pineapple, juice pack, canned	½ cup (125g)	1 (8)
orange	1 medium (131g)	1 (8)
avocado	½ medium (100g)	1 (8)
pear	1 medium (166g)	2 (8)
peach	1 medium (87g)	3 (8)
raisins, dried	½ cup (73g)	4 (8)
peach, heavy syrup, canned	½ cup (128g)	4 (8)
pear, heavy syrup, canned	2 halves (158g)	5 (8)
prunes, dried	3 prunes (25g)	5 (7)
cantaloupe	1 cup (160g)	6 (8)
strawberries	1 cup (149g)	13 (8)
fruit cocktail, heavy syrup, canned[b]	½ cup (128g)	50 (8)
average without fruit cocktail		2/serving
FRUIT JUICES/DRINKS		
pineapple juice, canned	8 fl oz (250g)	0 (8)
lemonade, from frozen concentrate	8 fl oz (248g)	0 (8)
grapefruit juice, from frozen concentrate	8 fl oz (247g)	2 (8)
orange juice, from frozen concentrate	8 fl oz (249g)	2 (8)
orange drink, canned	8 fl oz (248g)	2 (8)
grape juice, canned	8 fl oz (251g)	3 (8)
prune juice, bottled	8 fl oz (256g)	8 (8)
apple juice, canned	8 fl oz (248g)	10 (8)
average		3/serving
SWEETS		
sugar	1 T (12g)	0 (4)
pancake syrup	1 T (20g)	1 (6)
catsup	1 T (15g)	1 (8)
chocolate powder for milk	3 hp t (22g)	1 (7)
grape jelly	1 T (18g)	2 (5)
honey	1 T (21g)	3 (8)
milk chocolate candy	1 oz (28g)	9 (8)
caramels	1 oz (28g)	9 (7)

Table 1. (continued)

	Serving portion, household measure (g)[a]		Iodine, μg/serving (n)
SWEETS (continued)			
gelatin dessert, strawberry[b]	½ cup	(120g)	43 (8)
average without strawberry gelatin dessert			3/serving
VEGETABLES			
collards, fresh/frozen, boiled	½ cup	(95g)	0 (8)
lettuce, raw	5 leaves	(100g)	0 (8)
cabbage, boiled	2/3 cup	(100g)	0 (8)
broccoli, fresh/frozen, boiled	2/3 cup	(104g)	0 (8)
asparagus, fresh/frozen, boiled	6 spears	(90g)	0 (8)
cauliflower, fresh/frozen, boiled	2/3 cup	(83g)	0 (8)
tomato, raw	1 medium	(123g)	0 (7)
tomato juice, canned	8 fl oz	(243g)	0 (8)
cucumber, pared, raw	½ cup slices	(52g)	0 (7)
winter squash, fresh/frozen, boiled	½ cup	(102g)	0 (8)
carrot, raw	1 medium	(72g)	0 (8)
beets, canned	½ cup	(85g)	0 (8)
sweetpotato, baked w/skin	1 medium	(114g)	0 (8)
sweet pepper, green, raw	½ cup chopped	(50g)	1 (8)
corn, boiled	½ cup	(82g)	1 (7)
coleslaw, homemade	1 cup	(120g)	1 (8)
sauerkraut, canned	½ cup	(118g)	1 (8)
celery, raw	½ cup diced	(60g)	1 (8)
tomato sauce, canned	½ cup	(122g)	1 (8)
summer squash, fresh/frozen, boiled	½ cup slices	(90g)	1 (8)
mixed vegetables, canned	½ cup	(82g)	1 (8)
mushrooms, canned	½ cup pieces	(78g)	1 (7)
radish, raw	½ cup slices	(58g)	1 (8)
onion, raw	½ cup chopped	(80g)	2 (8)
spinach, fresh/frozen, boiled	½ cup	(90g)	2 (8)
green beans, fresh/frozen, boiled	1 cup	(125g)	3 (8)
spinach, canned	½ cup	(107g)	3 (8)
pickle, dill	1 medium	(65g)[c]	3 (8)
tomatoes, canned	½ cup	(120g)	4 (8)
green beans, canned	1 cup	(136g)	4 (8)
potato, boiled w/o peel	1 medium	(135g)	5 (7)
peas, green, frozen, boiled	2/3 cup	(107g)	5 (8)
potato chips	20 chips	(40g)	5 (8)
corn, canned	½ cup	(82g)	6 (8)

Table 1. (continued)

	Serving portion, household measure (g)[a]		Iodine, µg/serving (n)
VEGETABLES (continued)			
corn, cream style, canned	½ cup	(128g)	10 (7)
peas, green, canned	2/3 cup	(113g)	14 (8)
sweetpotato, candied, homemade	½ cup	(100g)	15 (8)
french fries	20 pieces	(100g)	26 (8)
potatoes, mashed, from flakes	½ cup	(100g)	28 (6)
potato, baked w/peel	½ potato	(101g)[d]	29 (7)
potatoes, scalloped, homemade	½ cup	(122g)	34 (8)
onion rings, breaded, fried, frozen, heated	7 rings	(70g)	46 (8)
average			6/serving
BREAKFAST/LUNCHEON MEATS			
bacon, pork, fried	3 slices	(19g)	2 (7)
frankfurter, canned	1 frank	(57g)	7 (8)
salami	1 oz	(28g)	8 (8)
bologna	1 oz	(28g)	10 (8)
average			7/serving
SOUPS			
chicken noodle, canned reconstituted with water	1 cup	(241g)	5 (7)
beef bouillon, canned reconstituted with water	1 cup	(241g)	10 (8)
tomato, canned, reconstituted with whole milk	1 cup	(248g)	17 (7)
vegetable beef, canned reconstituted with water	1 cup	(244g)	17 (8)
average			12/serving
GRAIN-BASED DESSERTS			
yellow cake with white icing, from mix	1/12 cake	(92g)	5 (5)
apple pie, frozen	1/6 pie	(92g)	7 (6)
chocolate cake with chocolate icing, commercial	1/6 cake	(85g)	9 (8)
danish/sweet roll	4¼" diameter	(65g)	11 (6)
doughnut, cake type	3¼" diameter	(42g)	14 (8)
coffeecake, commercial	1/8 of 12 oz cake	(42g)	14 (8)
cookies, sandwich type	2 cookies	(20g)	16 (8)

Table 1. (continued)

	Serving portion, household measure (g)[a]		Iodine, µg/serving (n)
GRAIN-BASED DESSERTS (continued)			
cookies, chocolate chip	2 cookies	(21g)	29 (8)
average			13/serving
CHEESE			
american cheese	1 oz	(28g)	9 (7)
cheddar cheese	1 oz	(28g)	19 (8)
average			14/serving
READY-TO-EAT CEREALS			
granola	½ cup	(61g)	8 (7)
raisin bran	1 cup	(56g)	11 (7)
shredded wheat	1 biscuit	(24g)	13 (8)
corn flakes	1 cup	(25g)	15 (6)
oat ring	1 cup	(25g)	24 (8)
crisped rice	1 cup	(28g)	45 (8)
fruit flavored, presweetened[b]	1 cup	(30g)	2,226 (6)
average without fruit flavored cereal			19/serving
GRAIN PRODUCTS			
noodles, cooked	½ cup	(80g)	3 (8)
popcorn, popped	2 cups	(18g)	10 (8)
corn chips	1 oz	(28g)	12 (5)
biscuit	1 medium	(28g)	13 (8)
roll, white	1 roll	(28g)	13 (8)
bread, rye	1 slice	(25g)	16 (8)
macaroni, cooked	½ cup	(70g)	17 (7)
tortilla, flour	1 medium	(30g)	17 (7)
saltines	10 crackers	(28g)	22 (6)
bread, white	1 slice	(23g)	23 (8)
pancake	6" diameter	(73g)	26 (6)
bread, whole wheat	1 slice	(23g)	26 (8)
cornbread	2½"x2½"x1 3/8"	(55g)	32 (7)
muffin, blueberry/plain	1 muffin	(40g)	46 (8)
average			20/serving
MEAT AND POULTRY			
pork chop, cooked	3½ oz	(100g)	7 (8)
ham, cured, cooked	3½ oz	(100g)	11 (8)
pork roast, cooked	3½ oz	(100g)	11 (8)
lamb chop, cooked	3½ oz	(100g)	14 (8)
beef chuck roast, cooked	3½ oz	(100g)	15 (7)

Table 1. (continued)

	Serving portion, household measure (g)[a]		Iodine, µg/serving (n)
MEAT AND POULTRY (continued)			
beef loin/sirloin, cooked	3½ oz	(100g)	15 (8)
veal cutlet, cooked	3½ oz	(100g)	16 (8)
beef, ground, cooked	3½ oz	(100g)	19 (8)
pork sausage, cooked	3½ oz	(100g)	19 (8)
turkey breast, roasted	3½ oz	(100g)	19 (7)
chicken, roasted	3½ oz	(100g)	21 (8)
chicken, fried	3½ oz	(100g)	32 (8)
beef round steak, stewed	3½ oz	(100g)	35 (8)
liver, fried	3½ oz	(100g)	42 (8)
meatloaf, homemade	3½ oz	(100g)	123 (8)
average without meatloaf			20/serving
EGGS			
eggs, soft boiled	1 medium egg (50g)		20 (7)
eggs, fried	1 medium egg (46g)		33 (8)
eggs, scrambled	1 medium egg (64g)		42 (8)
average			32/serving
GRAINS			
oatmeal, cooked	½ cup	(117g)	22 (8)
farina, cooked	½ cup	(116g)	24 (8)
rice, white, cooked	½ cup	(102g)	47 (6)
corn grits, cooked	½ cup	(121g)	86 (8)
average			45/serving
MIXED DISHES			
chili con carne, canned	1 cup	(255g)	8 (6)
chicken noodle casserole, homemade	1 cup	(240g)	19 (7)
pizza, cheese, frozen, heated	1/3 of 10" diameter pizza (114g)		23 (5)
1/4 lb hamburger, fast food	1 hamburger (174g)		30 (7)
spaghetti & meatsauce, homemade	1 cup	(248g)	45 (8)
fried chicken dinner w/ potatoes, cornbread & vegetable, frozen, heated	1 dinner	(325g)	49 (7)
lasagne, homemade	1 serving	(170g)	54 (8)
beef & vegetable stew, homemade	1 cup	(245g)	56 (8)
chicken potpie, frozen, heated	1 potpie	(227g)	64 (7)

Table 1. (continued)

	Serving portion, household measure (g)[a]		Iodine, μg/serving (n)
MIXED DISHES (continued)			
pork chow mein, homemade	1 cup	(250g)	65 (8)
macaroni & cheese, from box mix	1 cup	(200g)	82 (8)
spaghetti in tomato sauce, canned	1 cup	(250g)	180 (8)
average			56/serving
LEGUMES			
pork & beans, canned	½ cup	(126g)	6 (8)
lima beans, mature, boiled	½ cup	(94g)	30 (8)
pinto beans, boiled	½ cup	(85g)	47 (8)
kidney beans, boiled	½ cup	(92g)	63 (8)
navy beans, boiled	½ cup	(91g)	78 (8)
cowpeas, boiled	½ cup	(86g)	79 (8)
lima beans, immature, boiled	½ cup	(91g)	104 (8)
average			58/serving
FLUID MILK & COTTAGE CHEESE			
milk, evaporated	½ cup	(126g)	40 (7)
yogurt, strawberry	1 cup	(227g)	43 (8)
cottage cheese	1 cup	(225g)	54 (8)
buttermilk	1 cup	(245g)	56 (8)
milk, whole	1 cup	(244g)	61 (8)
milk, skim	1 cup	(246g)	64 (8)
milk, lowfat	1 cup	(244g)	66 (8)
yogurt, lowfat	1 cup	(227g)	73 (8)
milk, chocolate, lowfat	1 cup	(250g)	83 (8)
average			60/serving
FISH			
tuna, canned in oil	3½ oz	(100g)	28 (8)
shrimp, breaded & fried	3½ oz	(100g)	29 (7)
fish sticks, frozen, heated	4½ sticks	(100g)	55 (7)
cod/haddock, cooked	3½ oz	(100g)	175 (8)
average			72/serving
DAIRY-BASED DESSERTS			
ice cream sandwich	1 bar	(62g)	29 (8)
ice milk, vanilla	1 cup	(131g)	45 (8)
pudding, chocolate, from mix	½ cup	(130g)	46 (7)

Table 1. (continued)

	Serving portion, household measure (g)[a]		Iodine, μg/serving (n)
DAIRY-BASED DESSERTS (continued)			
Pumpkin pie, frozen	1/6 pie	(152g)	61 (8)
Ice cream, chocolate	1 cup	(133g)	94 (8)
Milkshake, chocolate, fast food	1 average	(283g)	158 (7)
Average			72/serving
INFANT FRUITS/JUICES/FRUIT DESSERTS			
Prunes/plums	1 jar	(220g)	0 (8)
Banana & pineapple	1 jar	(220g)	0 (8)
Fruit dessert	1 jar	(220g)	0 (8)
Dutch apple/apple betty	1 jar	(220g)	2 (8)
Applesauce	1 jar	(213g)	2 (8)
Peaches	1 jar	(220g)	2 (8)
Orange juice	1 jar	(130g)	3 (8)
Pears	1 jar	(213g)	4 (8)
Apple juice	1 jar	(130g)	8 (8)
Average			2/jar
INFANT CEREALS			
Oatmeal with applesauce & banana	1 jar	(135g)	1 (8)
Mixed cereal prepared with whole milk	1 oz	(28g)	2 (7)
Average			2/serving
INFANT FORMULAS			
Milk-based with iron	1 fl oz	(30g)	3 (8)
Milk-based without iron	1 fl oz	(30g)	3 (8)
Average			3/fl oz
INFANT VEGETABLES			
Sweet potato/yellow squash	1 jar	(220g)	0 (8)
Carrots	1 jar	(213g)	2 (8)
Green beans	1 jar	(206g)	2 (8)
Peas	1 jar	(206g)	2 (8)
Mixed/garden vegetables	1 jar	(213g)	4 (8)
Corn, creamed	1 jar	(213g)	6 (8)
Spinach, creamed	1 jar	(213g)	23 (8)
Average			6/jar

Table 1. (continued)

	Serving portion, household measure (g)[a]		Iodine, µg/serving (n)
INFANT MIXED DISHES			
vegetables with beef	1 jar	(213g)	0 (8)
high meat ham & vegetables	1 jar	(128g)	1 (7)
vegetables with bacon/ham	1 jar	(213g)	2 (8)
tomatoes, beef & macaroni	1 jar	(213g)	6 (8)
vegetables with turkey/ chicken	1 jar	(213g)	9 (8)
high meat beef & vegetables	1 jar	(128g)	9 (6)
turkey & rice	1 jar	(213g)	11 (7)
chicken & noodles	1 jar	(213g)	19 (8)
high meat chicken/turkey & vegetables	1 jar	(128g)	64 (7)
average without high meat chicken/turkey & vegetables			7/jar
INFANT MILK-BASED DESSERT			
pudding/custard	1 jar	(220g)	26 (8)
INFANT MEATS & POULTRY			
pork	1 jar	(99g)	17 (8)
beef	1 jar	(99g)	21 (8)
chicken/turkey	1 jar	(99g)	60 (8)
average			33/jar

[a]Serving portions and gram weights are adopted from revised Agriculture Handbook No. 8 (1976-86), Agriculture Handbook No. 456 (1975), and the USDA Provisional Table on Fast Foods (1984).

[b]High levels of iodine may be due to the presence of erythrosine (FD&C Red No. 3).

[c]3 3/4" long & 1½" in diameter.

[d]½ of a potato that is 4 3/4" long & 2 1/3" in diameter.

Iodine Content of Cow Milk

The two major sources of iodine in cow milk are the iodine added to cattle feed in the form of ethylenediamine dihydroiodide (EDDI) and the iodine derived from iodophors, iodine-containing sanitizing agents widely used in the dairy industry. Table 2 summarizes data from various investigators on the iodine content of cow milk, and Table 3 presents information on the effects of the use, nonuse, and overuse of EDDI and iodophors on the iodine content of milk. The wide

Table 2. Iodine Content of Cow Milk, Human Milk, and Infant Formula $(\mu g/L)^a$

Mean	Range	Reference
Cow milk		
706	320–1170	Dunsmore, 1976 (New South Wales) (29)
	95–540	Bakker, 1977 (34)
25		Costa et al., 1977 (Italy) (35)
422		Bruhn and Franke, 1978 (36)
203		Conrad and Hemken, 1978 (37)
646	40–4840	Hemken, 1979 (38)
400		Bergerioux and Boisvert, 1979 (Canada) (39)
	40–337	Sheldrake et al., 1980 (Australia) (40)
620		Lacroix and Wong, 1980 (41)
182		Moxon and Dixon, 1980 (42)
117		Wheeler et al., 1980 (New South Wales) (43)
125	32–214	Ohno, 1980 (Japan) (44)
370		Hillman and Curtis, 1980 (45)
509		Fischer and L'Abbe, 1981 (46)
361	52–3609	Curtis and Hamming, 1982 (47)
753		Wheeler, Fleet, and Ashley, 1983 (Australia) (48)
525	448–693	Lawrence, Chadha, and Conacher, 1983 (49)
466		Ruegsegger, Kuehn, and Schultz, 1983 (50)
	247–304	Larson et al., 1983 (51)
768		Franke, Bruhn, and Osland, 1983 (52)
99		Fardy and McOrist, 1984 (53)
92	61–159	Olson et al., 1984 (54)
152		Etling and Gehin-Fouque, 1984 (55)
254		Pennington et al., 1986 (26)
		Bruhn et al., 1983 (average/year) (32)
372	60–4006	1977
400	67–3015	1978
587	27–2169	1979
489	51–2101	1980
264	23–4173	1981
Human milk		
150		Bergerioux and Boisvert, 1979 (39)
		5 women in Montreal, Canada
149	22–310	Bruhn and Franke, 1983 (56)
		16 women in San Jose, California
64		Fardy and McOrist, 1984 (53)
		3 samples in Australia
178	29–490	Gushurst et al., 1984 (57)
		37 women in Durham, North Carolina
82		Etling and Gehin-Fouque, 1984 (55)
		68 women in Paris, France

Table 2. (continued)

Mean	Range	Reference
Milk-based infant formula		
134		Dunsmore, 1976 (new South Wales) (29)
376		Miles, 1978 (58)
300		Bergerioux and Boisvert, 1979 (Canada) (39)
	59—177	Fischer and L'Abbe, 1981 (46)
107		Pennington et al., 1986 (26)
Soy-based infant formula		
629		Miles, 1978 (58)
86	83—89	Fisher and L'Abbe, 1981 (46)

[a]Values reported as μg/kg in original references were converted to μg/L on the basis of 1.031 kg/L for cow milk, 1.048 kg/L for human milk, and 1.015 kg/L for infant formula.

ranges in the iodine content of milk shown in Table 2 probably result from the variable use of EDDI in feed and the variable use of iodo-phors as sanitizing agents.

The mammary gland does not limit the amount of iodine secreted in milk as it does with many other elements; the iodine content of milk is highly correlated with intake. EDDI, which contains about 80% iodine, may be added to cattle feed at a level of about 4—16 mg per cow per day to prevent goiter, but considerably more than this amount may be added in the belief that the extra will prevent infectious pododermatitis (foot rot), soft tissue lumpy jaw, and bronchitis, and/or improve reproductive efficiency. EDDI may also be given as an expectorant in the treatment of bovine respiratory infections. There is no scientific evidence to support the various purported uses of EDDI other than to prevent iodine deficiency. As Table 3 indicates, use of EDDI in normal concentrations increases the iodine content of milk (52), and excess EDDI fed to cattle results in excessive iodine levels in milk (45,50,54).

In addition to increasing the iodine content of milk, use of EDDI in cow feed at high levels may result in iodine toxicosis or iodism (1). Olson et al. (54) reported that daily iodine intake of over 68 to 600 mg/day by cows was associated with iodine intoxication. Signs of toxicity were reversed by lowering supplemental intake to less than 12 mg/day. Hillman and Curtis (45) reported that herds fed excessive dietary iodide (164 mg/day as EDDI) displayed signs of iodism including nasal and lacrimal discharge, coughing, bronchopneumonia, hair loss, dermatitis, hyperglycemia, hypocholesterolemia, and a neutrophilic-lymphopenic shift in blood leukocytes. Hillman and Curtis (45) concluded that dietary iodide should be limited to nutritional

Table 3. Effect of Use of EDDI and Iodophors on the Iodine Content of Milk (μg/L)[a]

| EDDI | | | |
None	Normal	Excess	Reference
171	768[b]		Franke, Bruhn, and Osland, 1983 (52)
	92[c] (61–159)	190–1250[d]	Olson et al., 1984 (54)
	466[e]	2503–6225[f]	Ruegsegger, Kuehn, and Schultz, 1983 (50)
	370[g]	2160[h]	Hillman and Curtis, 1980 (45)

| Iodophor | | |
None	With	Reference
(12–49)		Binnerts, 1979 (Netherlands) (59)
184	203	Conrad and Hemken, 1978 (37)
268	753	Wheeler, Fleet, and Ashley, 1983 (Australia) (48)
37 (1–60)	706 (320–1170)	Dunsmore, 1976 (New South Wales) (29)
22	40–337[i]	Sheldrake et al., 1980 (Australia) (40)

[a]Values reported as μg/kg in original references were converted to μg/L on the basis of 1.031 kg/L.
[b]4 ppm EDDI.
[c]12 mg I/day from EDDI.
[d]68–600 mg I/day from EDDI.
[e]Bulk tank milk samples; amount of EDDI not known.
[f]1 g I/day from EDDI for 2 weeks to cows with mastitis.
[g]16 mg I/day from EDDI.
[h]164 mg I/day from EDDI.
[i]Varied depending on iodine concentration of iodophor (1000 or 5000 mg I/L) and whether teats were washed, scrubbed, and dried; washed and dried; or not washed or dried after iodophor use.

requirements, and that prolonged prophylatic or therapeutic use should be avoided; Hemken et al. (60) cautioned that herds fed EDDI should receive the product as recommended by the manufacturer.

Teat dips containing iodophors have been reported to significantly reduce various bacterial infections (61,62) and are used to control mastitis (udder inflammation). In addition to their use in teat dips and udder salves, iodophors may be used on the farm and in processing plants to clean equipment such as milking machines, milk vats, transport tankers, and milk tanks. The iodine concentration of the milk is affected by the iodine concentration, viscosity, and method of application of teat dips (40,63,64). Conrad and Hemken (37) reported that the primary mode of increased iodine from iodophors appeared to be absorption through the skin and entry into the milk during the milk synthesis process, rather than by contamination from the teat surface.

Dunsmore and Nuzum (65) reported that iodophor udder washes increased iodine in milk by 34 µg/L and iodine-bearing udder salves increased iodine in milk by 54 µg/L. Hemken et al. (60) reported that teat dips and udder washes, if used as recommended, should not contribute more than 100–150 µg/L of iodine. Indiscriminate use of iodophors can, however, add significant amounts of iodine to milk. Table 3 indicates iodine increases of 20 to over 600 µg/L resulting from use of iodophors. Iodine-containing medications given to cows may also affect the iodine content of milk. Intrauterine infusion of Lugol's solution to treat bovine endometritis increased milk iodine from a range of 90–300 µg/L to 1700–6100 µg/L, and the levels remained elevated for four days (66).

The iodine content of milk may have changed and be changing over time. Hemken et al. (60) reported that values of iodine in cow milk published before 1970 were generally less than 100 µg/L, but those published after 1970 have been higher. The values summarized in Table 2 (reported between 1976 and 1985) are generally in excess of 100 µg/L. In late 1980, members of the California dairy industry voluntarily decided to eliminate iodine supplementation of dairy feeds, an action which resulted in lower iodine levels in milk in 1981 (mean 264 µg/L) than in the previous four years (means 372–587 µg/L) (32) (see Table 2).

Iodine Content of Human Milk and Infant Formula

Table 2 also presents information on the iodine content of human milk and infant formula as summarized from published papers. The mean iodine content of the human milk samples ranged from 64 to 178 µg/L. Gushurst et al. (57) found a significant correlation between the iodine level of breast milk and dietary iodine as estimated by a food frequency questionnaire. Iodized salt intake was significantly related to the iodine content of the breast milk. By regulation, infant formula must contain a minimum of 5 to a maximum of 75 µg of iodine per 100 calories (12). Since most infant formulas contain about 676 calories/L, this would amount to 34 to 507 µg iodine/L. Most of the iodine values for infant formulas summarized in Table 2 are within these limits.

Iodine Content of Bread

In the late 1960s and early 1970s some commercial baking companies were manufacturing bread made by a continuous mix process that used iodate dough conditioners. Bread made by this process contained about 500 µg/100 gm or about 150 µg/slice (67), whereas bread made by the batch process contained very little iodine. During the late 1960s and early 1970s, the high-iodine bread may have accounted for 43% of all white bread (67). In 1980, the NAS (10) reported that the practice of using iodate in bread was on the decline, as most major baking companies voluntarily changed to other conditioning agents since becoming aware of the concern about high iodine intakes. The white bread in the Total Diet Study (samples collected in 1982–1984) contained 23 µg/slice (see Table 1).

Other Sources of Iodine

Other potential sources of iodine include iodized salt, iodized water, dietary supplements, and medications. Although fresh water is generally low (1–18 µg/L) in iodine, some small towns in remote areas may use iodine (instead of chlorine) to purify drinking water. Effective bacteriological control is obtained at 1 mg/L water (68,69). Such levels could add 1000–2000 µg to daily iodine intake. The Environmental Protection Agency recommends against iodization of drinking water for communities except on a temporary basis.

Dietary supplements may be another source of iodine. Many multivitamins with minerals supply 100–150 µg of iodine per capsule; however, if they are colored with erythrosine, they will contain more than the label indicates. One brand of supplement contained 375 µg/tablet (70). Iodine supplements and seaweed extracts are also available for purchase.

Iodine may also be present in medications or diagnostic materials including oral drugs, topical medications, and iodinated contrast media. Elevated plasma and/or urine iodine levels noted in infants exposed to iodine-containing topical medications (71–73) and various adverse metabolic effects resulting from iodine-containing topical medications (74–79) indicated that iodine can be absorbed through the skin.

IODINE STATUS IN THE UNITED STATES

Measurement of Iodine Status

For large-scale studies the most practical measures of iodine status are observation for goiter and measurement of urinary iodine excretion. If large numbers of a population have thyroid enlargements and excrete less than 50 µg of iodine per day, endemic goiter due to iodine deficiency would be suspected. For smaller studies or for individual patients, iodine status may also be assessed by the presence of goiter and urinary iodine excretion plus other measures, such as thyroidal iodine clearance, radioactive iodine uptake, levels of serum T4 and T3, levels of plasma inorganic iodide, and dietary iodine intake estimates. The combination of low serum T4, low plasma inorganic iodide, high thyroidal uptake of radioactive iodine, and decreased urinary excretion of iodine are characteristic of iodine deficiency. Likewise high serum T4, high plasma inorganic iodine, low thyroidal uptake of radioactive iodine, and high urinary excretion of iodine are characteristics of excessive intake. In some cases, measurement of TSH level might be appropriate to rule out secondary hypothyroidism caused by pituitary deficiency.

Goiter Incidence and Urinary Excretion of Iodine

In the early decades of the twentieth century, endemic goiter was found throughout the Appalachian Mountain range, in the Rocky Mountain area, and in northern states from the Great Lakes to the Pacific Northwest. Endemic goite was the principal medical reason for rejecting recruits for the United States Army in World War I (80). The high incidence of goiter found among World War I draftees stimulated the use of iodized salt as prophylaxis in large population groups
    Measurements of iodine status in the late 1960s and early 1970s indicated that iodine deficiency was no longer a public health problem in the United States. There was little evidence of goiter, and that which was found was not associated with low levels of urinary iodine excretion. The Ten State Survey, conducted by the Department of Health, Education, and Welfare in 1968−1970, provided information on iodine excretion and goiter prevalence. Examinations for thyroid size in 35,999 persons indicated an overall goiter prevalence of 3.1%, with the highest prevalence in adolescent girls (5.0%) and women (5.7%) (81). Normal to high iodine excretion levels were noted in both goitrous and nongoitrous subjects, and a higher prevalence of goiter was found among persons excreting high levels of iodine. Goitrogens and thyroid abnormalities accounted for less than 8% of the goiters; other goiters were not explained.
    Of 2371 preschool children (ages 1−6 years) examined in 1968−1970 by Owen et al. (82), only five (0.2%) had goiter. Urinary iodine excretion of the children reflected use of iodized salt. Goiter examinations performed in 1971−1972 on 7785 children 9−16 years of age in Michigan, Kentucky, and Georgia revealed a goiter prevalence of 6.8% (83). Most children with goiter had palpably but not visibly enlarged thyroids and showed no evidence of clinical or biochemical thyroid abnormality. Children with goiter and those in areas with high goiter prevalence tended to have higher rather than lower iodine excretion, indicating that goiter in these children could not be related to iodine deficiency.

Thyroidal Uptake of Radioactive Iodine

Also by the late 1960s and early 1970s, physicians noted that the thyroid glands of patients were taking up less radioactive iodine, suggesting that their thyroids were already saturated with this element. The high dietary intake of iodine in the United States altered the interpretation of tests measuring radioactive iodine uptake by the thyroid gland, as normal values for uptake and retention of radioiodine were lower than previously.

Pittman et al. (67) reported that thyroidal radioiodine uptake of euthyroid subjects decreased from 28.6% in 1959 to 15.9% in 1967–1968. The subjects in 1967–1968 also displayed elevated urinary iodine excretion and elevated plasma inorganic iodide, and depressed thyroidal iodide clearance. Sostre (84) noted that radioactive iodine uptake decreased between 1955 and the late 1960s, but stabilized from the late 1960s to 1976. Data on 1575 euthyroid subjects indicated no change in uptake between 1967 and 1976. Wong and Schultz (85) reported that radioactive iodine uptake in the Minneapolis area averaged 25% in 1957, 12% in 1971, and 21% in 1975. Culp and Huskison (86) reported that 24-hour radioactive iodine uptakes decreased from the late 1960s (15–45%) to the early 1970s (9–32%) because of increased iodine in the diet.

Dietary Intake of Iodine

Pittman et al. (67) and Wong and Schultz (85) identified bread made with iodate dough conditioners as the major source of iodine affecting radioactive iodine uptake. Kidd et al. (28), who obtained dietary histories from 754 children 9–16 years old coupled with urinary iodine excretion, indicated that milk, bread made with iodine-containing dough conditioners, and iodized salt were significant sources of iodine.

In 1974, FDA included iodine in the Total Diet Study to determine iodine intakes of specific age-sex groups and to identify the major sources of iodine in their diets. Iodine intakes were estimated for 15 to 20-year-old males, 2-year-old children, and 6-month-old infants between 1974 and 1981/82 (87), and for eight age-sex groups in 1982–1984 (26). The results of these Total Diet Studies, along with results for dietary studies from other countries, are shown in Table 4.

Between 1974 and 1981/82 iodine levels of the Total Diet Study for 15- to 20-year-old males were 213–551% RDA, those for 6-month-old infants were 264–1152% RDA, and those for 2-year-old children were 330–1040% RDA (87). Discretionary salt, which was included only for the 15- to 20-year-old male diet, was not iodized. For the 15- to 20-year-old males, the major contributor to iodine intake was dairy products (45.5%), followed by grain and cereal products (23.6%), sugars and adjuncts (14.1%), and meat, fish, and poultry (10.7%).

Table 4. Iodine Content of Daily Diets (μg/day)

Iodine		Country	Reference
60–70		Denmark	Dige-Peterson, 1977 (88)
312[a]		Sweden	Abdulla et al., 1981 (89)
78[a]		Sweden—vegans	Abdulla et al., 1981 (89)
340		Finland	Varo et al., 1982 (31)
323		England	Wenlock et al., 1982 (30)
210		Netherlands	Van Dokkum et al., 1982 (90)
533	(274–842)	U.S. hospital diets	Pittman, Dailey, and Beschi,1969 (67)
677	(595–713)	U.S. hospital diets	Pittman, Dailey, and Beschi,1969 (67)
		U.S. FDA Total Diet Studies	
359[b]	(182–576)	6 mo infants	Pennington et al., 1984 (87)
435[b]	(231–728)	2 yr children	Pennington et al., 1984 (87)
527[b]	(319–827)	15–20 yr males	Pennington et al., 1984 (87)
200[c]		6–11 mo infants	Pennington et al., 1986 (26)
460[c]		2 yr children	Pennington et al., 1986 (26)
420[c]		14–16 yr females	Pennington et al., 1986 (26)
710[c]		14–16 yr males	Pennington et al., 1986 (26)
270[c]		25–30 yr females	Pennington et al., 1986 (26)
520[c]		25–30 yr males	Pennington et al., 1986 (26)
250[c]		60–65 yr females	Pennington et al., 1986 (26)
340[c]		60–65 yr males	Pennington et al., 1986 (26)

[a]Assuming a caloric intake of 2000 calories/day; values were expressed per 1000 calories in original reference.
[b]Averages for six years of data between 1974 and 1981/82.
[c]Averages for two years of data, 1982–1984.

The fruit and vegetable group and the beverage group each contributed less than 3% of total iodine. For the 2-year-old diet, major contributors to iodine intake were milk (44.1%), grain and cereal products (24.4%), other dairy products (11.5%), meat, fish, and poultry (8.5%), and sugars and adjuncts (5.6%). Other food groups contributed less than 3% of total iodine. For the infant diet, the main source of iodine was milk (67.4%), followed by other dairy products (14.6%), grain and cereal products (6.0%), and meat, fish and poultry (4.9%). Other commodity groups contributed less than 3% of the total iodine intake.

Table 5. Contribution of Food Groups to Iodine Intake: Selected Minerals in Food Survey, 1982–1984[a] (26)

	6-11 mo.	2 yr	14-16yr female	14-16yr male	25-30yr female	25-30yr male	60-65yr female	60-65yr male
	(% of daily iodine intake)							
grains and grain products	18	59	46	52	34	47	34	32
Milk and milk products	62	21	24	22	22	17	18	19
Meat, fish, poultry and eggs	9	6	9	8	18	14	20	22
Mixed dishes and soups	4	4	8	7	10	8	8	7
Vegetables and fruits (including juices and fruit drinks)	3	3	5	5	8	7	9	9
Desserts	4	6	8	7	8	6	10	11
Other beverages[b]	0	0	1	0	2	1	1	1

[a]Contributions of nuts, gravy, sauce, fats, condiments, and sweeteners to iodine intake were less than 1%.
[b]Coffee, tea, soda, alcoholic beverages, and water.

The average daily intakes of iodine for the 1982–1984 Total Diets exceeded RDAs for all eight age-sex groups: 400% RDA for 6–11-month infants; 657% RDA for 2-year-old children; 280% RDA for teenage girls; 473% RDA for teenage boys; 180% RDA for adult females; 347% RDA for adult males; 167% RDA for older females; and 227% RDA for older males. As shown in Table 5, the major sources of iodine in these diets were grains and grain products and milk and milk products. These two food groups provided 51 to 80% of the total iodine intake. The high levels of iodine from the grain group were due primarily to the inclusion of the fruit-flavored, presweetened breakfast cereal. This food item represents all presweetened cereals in the Total Diet Study food list. (Each of the 234 foods in the Total Diet Study represents an aggregate of similar foods; one item in each aggregate was selected to represent the group.) Because fruit-flavored cereals often have added colors (e.g., erythrosine) to correspond to the fruit flavors, estimates of iodine from this group may be overestimated. Iodine intakes continue to be monitored annually throught the FDA's Total Diet Study.

Iodine intakes from other countries (Table 4) appear to be on the order of 200–340 µg/day (30,31,89,90), except those for Denmark (88) and vegans in Sweden (89), where intakes were 60–80 µg/day.

Results from measures of goiter incidence, urinary iodine excretion, radioactive iodine uptake, and estimated dietary levels of iodine indicate that iodine intakes in the United States are in excess of that needed for adequate nutrition. The present high iodine intake does not currently pose a public health concern; however, any additional increase in the iodine content of the food supply should be viewed with caution (10). NAS (10) recommended that the many adventitious sources of iodine, such as iodophors, alginates (food additives derived from seaweeds), food dyes, and dough conditioners, should be replaced by compounds containing less or no iodine. Because of the high levels of iodine in the United States food supply, the FDA issued regulations setting limits for the amounts of certain iodine compounds that could be used as direct and indirect food additives (23). There are limits set for the amount of potassium iodate and calcium iodate that can be used as dough conditioners and for the amount of cuprous iodide and potassium iodide added to table salt. Since 1977 FDA has approved no new iodine compounds for use in foods for food processing. FDA has urged industry to eliminate indiscriminate use of iodophors, and has cautioned the dairy industry and feed manufacturers about the overuse of EDDI in cattle feed.

IODINE SENSITIVITY

Iodine in the form of drugs (oral medications or iodinated contrast media) may have side effects in sensitive individuals. Huang and Peterson (91) suggest that such reactions result when iodine forms a complex with plasma proteins, creating antigens. Case histories of parotitis (enlargement of the salivary glands) (92–94), ioderma (95), and anaphylactoid reactions (96) are reported with use of iodinated contrast media. Pretreatment with corticosteroids and antihistamines (96) and desensitization (by administration of increasing doses of iodinated contrast media) (97,98) have been used to reduce the frequency and severity of subsequent reactions in sensitive patients.

Case histories of sensitivity to iodine-containing oral medications have been reported (91,99–102). The reported symptoms include ioderma of skin and conjunctiva, urticaria, angioderma, fever, coryza, and pulmonary edema. Even though iodine can cause acniform eruptions in sensitive persons, iodine intake does not appear to affect the incidence or severity of acne vulgaris (103,104).

IODINE TOXICITY

Iodine intakes of between 50 and 1,000 μg/day by adults can be considered safe (10); however, intakes greater than 2000 μg per day are considered excessive and potentially harmful (8). Intakes of this magnitude from the food supply are unlikely, but they could result from iodine-containing medications. Prolonged exposure to high levels of iodine carries considerable risk that certain thyroid diseases such as goiter, hypothyroidism, or hyperthyroidism (thyrotoxicosis) will develop.

Iodine-induced Goiter/Hypothyroidism

Large dietary or therapeutic intakes of iodine may inhibit organic iodine formation (prevent the binding of iodine to tyrosine in the thyroid). The resulting depression of circulating T4 and T3 induces an increase in TSH, resulting in iodide goiter (the Wolff-Chaikoff effect) in certain susceptible individuals. This effect is usually transient, and subjects treated with iodide usually escape from this inhibition after several days. Those who do not escape develop goiters and become hypothyroid. Individuals with Hashimoto's thyroiditis seem to be particularly susceptible to iodine-induced goiter.

Tai et al. (105) reported a goiter incidence of 7.3% and enlarged thyroid incidence of 28.3% in the inhabitants of a Chinese village who drank deep-well water with a high iodine content. Villagers drinking water with normal iodine concentrations had incidences of goiter and enlarged thyroids of 1.5% and 8.7%, respectively. Freund et al. (68) used iodine as a means of disinfecting the water supply of a prison community. At a concentration of 1 mg iodine per liter of water, 2 of 15 inmates had impaired organification of thyroidal iodide.

Iodine-containing compounds have been widely used as expectorants for patients with asthma or chronic bronchitis. Case reports of individuals developing hypothyroidism while taking iodine-containing expectorants for asthma/bronchitis are reported by Herxheimer (106), Penfold et al. (107), and Korsager and Kristensen (108). An infant with cystic fibrosis developed hypothyroidism without goiter following short-term iodide therapy (109). Because of the demonstrated complications of iodide therapy, and lack of documentation of its efficacy as a mucolytic or expectorant, the authors expressed doubt that such therapy has a role in the treatment of cystic fibrosis (109).

Iodine can cross the placenta and interfere with synthesis of thyroid hormone in the fetus. Administration of iodine to a pregnant woman is especially risky since the fetal thyroid is less able to escape the inhibitory effects of iodine on the concentrating mechanism. Iodine-induced goiters with or without hypothyroidism may occur in newborn infants of mothers who have taken pharmacologic doses of

iodine during pregnancy. The goiter usually regresses spontaneously after several months, but deaths due to a compression of the
trachea have occurred. Verma and Dhar (110) report two infant
deaths from iodine-induced neonatal congenital goiter with hypothyroidism. In both cases, the mothers had consumed large doses of
iodine during pregnancy. Penfold et al. (107) report the birth of
a goitrous infant to an asthmatic mother who had taken an iodine-
containing expectorant. Melvin et al. (111) report an infant with
congenital goiter, hypothyroidism, respiratory distress, and cardiac
failure due to maternal treatment of asthma with an iodine-containing
drug. Takasu et al. (112) reported transient neonatal hypothyroidism
in the infant of a mother who was receiving treatment for primary
hypothyroidism due to Hashimoto's thyroiditis.

Chabrolle and Rossier (75,76) and Prager and Gardner (77)
present cases of hypothyroidism resulting from topical application of
iodine-containing medications.

Iodine-Induced Hyperthyroidism (Thyrotoxicosis)

Excessive intake of iodine may cause overstimulation of the thyroid
gland, which produces excess hormone and causes hyperthyroidism.
Symptoms of hyperthyroidism include nervousness, tremor, heat intolerance, goiter, bulging eyes, tachycardia, fatigue, weakness, increased appetite, and weight loss. Hyperthyroidism has been reported
from use of iodine-containing expectorants (79,113—115); an iodine-
containing antiarrhythmic drug (116); Lugol's solution for goiter (117);
hydroxyquinoline for amoebiasis (117); iodinated contrast media (79,
118); a multivitamin preparation containing 375 μg iodine per dose
(70); topical medications (74,78,79); and an iodine-containing vaginal
solution (119).

Patients with euthyroid Graves Disease may become hyperthyroid
when given large doses of iodine or iodinated contrast media (120).
Increased dietary iodine may be responsible for the decreased effectiveness of oral antithyroid drugs. The remission rate with antithyroid drugs has declined from the 1950s and early 1960s; however,
in areas where iodine intakes are low (less than 250 μg/day) remission rates of 50—60% are still seen (120).

Thyrotoxicosis may occur in endemic goitrous areas following
prophylactic programs (salt iodination or iodized oil injection).
This syndrome, which is called Jod-Basedow, occurs in persons chronically exposed to iodine deficiency, who respond to iodine repletion
by secreting an excessive amount of thyroid hormone. It most often
occurs in the middle-aged and older persons of a community supplemented with iodine. Lewis (121) noted the increased prevalence of
thyrotoxicosis in the elderly in Tasmania, Australia, which paralleled
the increase of iodine availability. Livadas et al. (122) reported 16
cases of toxic adenoma of the thyroid aggravated by small doses of

potassium iodide. Baker and Phillips (123) found that the incidence of thyrotoxicosis in 12 towns in England and Wales was strongly correlated with the previous prevalence of endemic goiter in the towns. The authors concluded that current high dietary intakes of iodine (largely through the milk supply) caused toxic nodular goiter in people made susceptible by a lack of iodine early in life.

Acute Iodine Toxicity

Large single doses of iodine-containing solutions may have extreme side effects and may result in death. A 56-year-old female who attempted suicide with Lugol's solution showed gastrointestinal irriation and ulceration, chemical pneumonitis, hyperthyroidism, hemolytic anemia, acute renal failure (due to tubular necrosis), and metabolic acidosis (124). A fatal case of iodine poisoning in a 57-year-old male showed symptoms of weak pulse, urinary retention, delirium, stupor, and collapse (125).

ACKNOWLEDGMENTS

The author acknowledges the chemists and laboratory technicians of the FDA Kansas City District Office, who completed the iodine analyses of Total Diet Study foods; Dennis Wilson, who provided statistical and computer assistance; and Denise Hughes, who typed this manuscript.

REFERENCES

1. C. M. Stowe, Iodine, iodides, and iodism, *J. Am. Vet. Med. Assoc.*, *179*: 334 (1981).
2. M. S. Croxson, T. D. Hall, O. A. Kletzky, J. E. Jaramillo, J. T. Nicoloff, Decreased serum thyrotropin induced by fasting, *J. Clin. Endocrinol. Metab.*, *45*: 560 (1977).
3. D. F. Gardner, M. M. Kaplan, C. A. Stanley, R. D. Utiger, Effect of tri-iodothyronine replacement on the metabolic and pituitary responses to starvation, *N. Engl. J. Med.*, *300*: 579 (1979).
4. K. Toyoshima, Y. Matsumoto, M. Nishida, H. Yabuuchi, Five cases of absence of iodide concentrating mechanism, *Acta Endocrinol.*, *84*: 527 (1977).
5. A. Kapoor, M. Godbole, S. Chauhan, B. K. Ghosh, B. Rao, Congenital goitrous hypothyroidism due to organification defect (a case report), *Indian Pediatr.*, *14*: 573 (1977).
6. V. Chan, C. Wang, R. T. T. Yeung, Dissociated thyroxine,

triiodothyronine and reverse triiodothyronine levels in patients with familial goitre due to iodide organification defects, *Clin. Endocrinol.*, *11*: 257 (1979).

7. J. W. F. Elte, Causes of non-toxic goitre other than mere iodine deficiency, *Neth. J. Med.*, *24*: 79 (1981).

8. E. J. Underwood, Iodine, in *Trace Elements in Human and Animal Nutrition*, 4th ed., Academic Press, New York, 1977.

9. Food and Nutrition Board (FNB), National Research Council, *Iodine Nutriture in the United States*, National Academy of Sciences, Washington, D.C. (1970).

10. National Academy of Sciences (NAS), *Recommended Dietary Allowances*, 9th revised ed., Washington, D.C. (1980).

11. M. A. L. Odne, S. C. Lee, L. P. Jeffrey, Rationale for adding trace elements to total parenteral nutrient solutions—a brief review, *Am. J. Hosp. Pharm.*, *35*: 1057 (1978).

12. Code of Federal Regulations (CFR), Title 21 Food and Drugs, 107.100 Infant Formula. Nutrient specifications, Office of the Federal Register National Archives and Records Administration (1987).

13. I. Ilyes, F. Peter, Serum reverse triiodothyronine (rT3) in children with goitre, *Acta Paediatr. Acad. Sci. Hung.*, *21*: 65 (1980).

14. J. Matovinovic, Endemic goiter and cretinism at the dawn of the third millennium, *Ann. Rev. Nutr.*, *3*: 341 (1983).

15. S. R. Aston, P. H. Brazier, Endemic goitre, the factors controlling iodine deficiency in soils, *Sci. Total Environ.*, *11*: 99 (1979).

16. Y. Ingenbleek, C. Beckers, Evidence for intestinal malabsorption of iodine in protein-calorie malnutrition, *Am. J. Clin. Nutr.*, *26*: 1323 (1973).

17. Y. Ingenbleek, C. Beckers, Thyroidal iodide clearance and radioiodide uptake in protein-calorie malnutrition, *Am. J. Clin. Nutr.*, *31*: 408 (1978).

18. A. K. Osman, A. A. Fatah, Factors other than iodine deficiency contributing to the endemicity of goitre in Darfur Province (Sudan), *J. Human Nutr.*, *35*: 302 (1981).

19. A. K. Osman, Endemic goitre, *Pennisetum* ssp. and iodine (letter), *Trans. R. Soc. Trop. Med. Hyg.*, *75*: 474 (1981).

20. A. Papas, J. R. Ingalls, L. D. Campbell, Studies on the effects of rapeseed meal on thyroid status of cattle, glucosinolate and iodine content of milk and other parameters, *J. Nutr.*, *109*: 1129 (1979).

21. O. O. Oyelola, S. O. Ayangade, O. L. Oke, The possible role of cassava on the thiocyanate level of pregnant women, *Nutr. Rep. Int.*, *28*: 585 (1983).

22. R. L. Savage, Drugs and breast milk, *J. Human Nutr.*, *31*: 459 (1977).

23. Code of Federal Regulations (CFR), Title 21 Food and Drugs, 74.303 FD&C Red No. 3; 178.1010 Sanitizers; 184.1206 Calcium iodate; 184.1265 Cuprous iodide; 184.1643 Potassium iodide; 184.1635 Potassium iodate, Office of the Federal Register National Archives and Records Administration (1987).

24. D. Hightower, Bioavailability study of iodine from erythrosine and iodophors, Final report of contract No. 223-79-2279 with the Food and Drug Administration, Texas A&M Research Foundation (undated, probably 1983).

25. S. Katamine, Y. Mamiya, K. Sekimoto, N. Hoshino, K. Totsuka, M. Suzuki, Differences in bioavailability of iodine among iodine-rich foods and food colors, *Nutr. Rep. Inter.* *35*: 289 (1987).

26. J. A. T. Pennington, B. E. Young, D. B. Wilson, R. D. Johnson, J. E. Vanderveen, Mineral content of foods and total diets: The Selected Minerals in Foods Survey, 1982–84, *J. Am. Dietet. Assoc.*, *86*: 876 (1986).

27. M. E. Dellavalle, D. M. Barbano, Iodine content of milk and other foods, *J. Food Prot.*, *47*: 678 (1984).

28. P. S. Kidd, F. L. Trowbridge, J. B. Goldsby, M. Z. Nichaman, Sources of dietary iodine, *J. Am. Dietet. Assoc.*, *65*: 420 (1974).

29. D. G. Dunsmore, Iodophors and iodine in dairy products: I. The iodine content of Australian dairy products, *Aust. J. Dairy Technol.*, *31*: 125 (1976).

30. R. W. Wenlock, D. H. Buss, R. E. Moxon, N. G. Bunton, Trace nutrients 4. Iodine in British foods, *Br. J. Nutr.*, *47*: 381 (1982).

31. P. Varo, E. Saari, A. Paaso, P. Koivistoinen, Iodine in Finnish foods, *Int. J. Vit. Nutr. Res.*, *52*: 80 (1982).

32. J. C. Bruhn, A. A. Franke, R. B. Bushnell, H. Weischeit, G. H. Hutton, G. C. Gurtle, Sources and content of iodine in California milk and dairy products, *J. Food Prot.*, *46*: 41 (1983).

33. M. Allegrini, J. A. T. Pennington, J. T. Tanner, Total Diet Study: Determination of iodine intake by neutron activation analysis, *J. Am. Dietet. Assoc.*, *83*: 18 (1983).

34. H. J. Bakker, Gas-liquid chromatographic determination of total inorganic iodine in milk, *J. Assoc. Off. Anal. Chem.*, *60*: 1307 (1977).

35. A. Costa, P. Lorenzini, O. Brambati-Testori, F. Cottino, G. De Sanso, Thyroid performance in an iodine-rich environment (Salsomaggiore Spa), *J. Nucl. Med. Allied Sci.*, *21*: 30 (1977).

36. J. C. Bruhn, A. A. Franke, An indirect method for the estimation of the iodine content in raw milk, *J. Dairy Sci.*, *61*: 1557 (1978).

37. L. M. Conrad, III, R. W. Hemken, Milk iodine as influenced by an iodophor teat dip, *J. Dairy Sci.*, *61*: 776 (1978).

38. R. W. Hemken, Factors that influence the iodine content of milk and meat: A review, *J. Anim. Sci.*, *48*: 981 (1979).

39. C. Bergerioux, J. Boisvert, Rapid neutron activation method for the determination of minerals in milk, *Int. J. Nucl. Med. Biol.*, *6*: 128 (1979).

40. R. F. Sheldrake, R. J. T. Hoare, S. C. Chen, J. McPhillips, Post-milking iodine teat skin disinfectants 3. Residues, *J. Dairy Res.*, *47*: 33 (1980).

41. D. E. Lacroix, N. P. Wong, Determination of iodide in milk using the iodide specific ion electrode and its application to market milk samples, *J. Food Prot.*, *43*: 672 (1980).

42. R. E. D. Moxon, E. J. Dixon, Semi-automatic method for the determination of total iodine in food, *Analyst*, *105*: 344 (1980).

43. S. M. Wheeler, L. R. Fell, G. H. Fleet, R. J. Ashley, The evaluation of two brands of ion-selective electrode used to measure added iodide and iodophor in milk, *Aust. J. Dairy Technol.*, *35*: 26 (1980).

44. S. Ohno, Simple and rapid determination of iodine in milk by radioactivation analysis, *Analyst.*, *105*: 246 (1980).

45. D. Hillman, A. R. Curtis, Chronic iodine toxicity in dairy cattle: Blood chemistry, leukocytes, and milk iodide, *J. Dairy Sci.*, *63*: 55 (1980).

46. P. W. F. Fischer, M. R. L'Abbe, Acid digestion determination of iodine in foods, *J. Assoc. Off. Anal. Chem.*, *64*: 71 (1981).

47. A. R. Curtis, P. Hamming, Differential pulse polarographic determination of total iodine in milk, *J. Assoc. Off. Anal. Chem.*, *65*: 20 (1982).

48. S. M. Wheeler, G. H. Fleet, R. J. Ashley, Effect of processing upon concentration and distribution of natural and iodophor-derived iodine in milk, *J. Dairy Sci.*, *66*: 187 (1983).

49. J. F. Lawrence, R. K. Chadha, H. B. S. Conacher, The use of ion-exchange filters for the determination of iodide in milk by X-ray fluorescence spectrometry, *Intern. J. Environ. Anal. Chem.*, *15*: 303 (1983).

50. G. J. Ruegsegger, P. Kuehn, L. H. Schultz, Iodine in field milk samples and effects on mastitis organisms, *J. Dairy Sci.*, *66*: 1976 (1983).

51. L. L. Larson, S. E. Wallen, F. G. Owen, S. R. Lowry, Relation of age, season, production, and health indices to iodine and beta-carotene concentrations in cow's milk, *J. Dairy Sci.*, *66*: 2557 (1983).

52. A. A. Franke, J. C. Bruhn, R. B. Osland, Factors affecting iodine concentration of milk of individual cows, *J. Dairy Sci.*, *66*: 997 (1983).

53. J. J. Fardy, G. D. McOrist, Determination of iodine in milk products and biological standard reference materials by epithermal neutron activation analysis, *J. Radioanal. Nucl. Chem.*, *87*: 239 (1984).

54. W. G. Olson, J. B. Stevens, J. Anderson, D. W. Haggard, Iodine toxicosis in six herds of dairy cattle, *J. Am. Vet. Med. Assoc.*, *184*: 179 (1984).

55. N. Etling, F. Gehin-Fouque, Iodinated compounds and thyroxine binding to albumin in human breast milk, *Pediatr. Res.*, *18*: 901 (1984).

56. J. C. Bruhn, A. A. Franke, Iodine in human milk, *J. Dairy Sci.*, *66*: 1396 (1983).

57. C. A. Gushurst, J. A. Meuller, J. A. Green, F. Sedor, Breast milk iodide: Reassessment in the 1980s, *Pediatrics*, *73*: 354 (1984).

58. P. Miles, Determination of iodide in nutritional beverage products using an ion selective electrode, *J. Assoc. Off. Anal. Chem.*, *61*: 1366 (1978).

59. W. T. Binnerts, The iodine content of milk: No reason for concern yet, *Neth. Milk Dairy J.*, *33*: 12 (1979).

60. R. W. Hemken, J. D. Fox, C. L. Hicks, Milk iodine content as influenced by feed sources and sanitizer residues, *J. Food Prot.*, *44*: 476 (1981).

61. R. J. Eberhart, P. L. LeVan, L. C. Griel, Jr., E. M. Kesler, Germicidal teat dip in a herd with low prevalence of *Streptococcus agalactiae* and *Staphylococcus aureus* mastitis, *J. Dairy Sci.*, *66*: 1390 (1983).

62. J. W. Pankey, W. N. Philpot, R. L. Boddie, Efficacy of low concentration iodophor teat dips against *Staphylococcus aureus*, *J. Dairy Sci.*, *66*: 155 (1983).

63. D. G. Dunsmore, C. Nuzum, B. Dettman, Iodophors and iodine in dairy products: 3. Teat dipping, *Aust. J. Dairy Technol.*, *32*: 45 (1977).

64. P. A. Lewis, R. W. Hemken, W. L. Crist, The effect of teat dip viscosity on milk iodine levels, *J. Dairy Sci.*, *63* (suppl. 1): 182 (1980).

65. D. G. Dunsmore, C. Nuzum, Iodophors and iodine in dairy products: 2. Udder washes and salves, *Aust. J. Dairy Technol.*, *32*: 42 (1977).

66. C. J. McCaughan, K. W. Laurie, M. C. Martin, M. W. Hooper, Iodine in milk of cows after intrauterine infusion of Lugol's solution, *Aust. Vet. J.*, *61*: 200 (1984).

67. J. A. Pittman, Jr., G. E. Dailey, III, J. R. Beschi, Changing normal values for thyroidal radioiodine uptake, *N. Engl. J. Med. 280*: 1431 (1969).

68. G. Freund, W. C. Thomas, Jr., E. D. Bird, R. N. Kinman, A. P. Black, Effect of iodinated water supplies on thyroid function, *J. Clin. Endocrinol.*, *26*: 619 (1966).

69. W. C. Thomas, Jr., A. P. Black, G. Freund, R. N. Kinman, Iodine disinfection of water, *Arch. Environ. Health*, *19*: 124 (1969).

70. M. B. Block, S. J. DeFrancesco, Hyperthyroidism possibly induced by iodine in a multivitamin, *Ariz. Med.*, *36*: 510 (1979).

71. D. C. Postellon, Iodine in mother's milk (letter), *J. Am. Med. Assoc.*, *247*: 463 (1982).

72. S. P. Pyati, R. S. Ramamurthy, M. T. Krauss, R. S. Pildes, Absorption of iodine in the neonate following topical use of povidone iodine, *J. Pediatr.*, *91*: 825 (1977).

73. A. Gruters, D. l'Allemand, P. H. Heidemann, P. Schurnbrand, Incidence of iodine contamination in neonatal transient hyperthyrotropinemia, *Eur. J. Ped.*, *140*: 299 (1983).

74. J. R. Fisher, Effect of iodine treatment on thyroid function (letter), *N. Engl. J. Med.*, *297*: 171 (1977).

75. J. P. Chabrolle, A. Rossier, Danger of iodine skin absorption in the neonate (letter), *J. Pediatr.*, *93*: 158 (1978).

76. J. P. Chabrolle, A. Rossier, Transient neonatal hypothyroidism (letter), *Pediatrics*, *62*: 857 (1978).

77. E. M. Prager, R. E. Gardner, Iatrogenic hypothyroidism from topical iodine-containing medications, *West. J. Med.*, *130*: 553 (1979).

78. H. A. Miller, J. A. Farley, D. A. Major, Topical iodine and hyperthyroidism, *Ann. Intern. Med.*, *95*: 121 (1981).

79. R. Rajatanavin, M. Safran, W. A. Stoller, J. P. Mordes, L. E. Braverman, Five patients with iodine-induced hyperthyroidism, *Am. J. Med.*, *77*: 378 (1984).

80. W. Mertz, Our most unique nutrients, *Nutr. Today*, *18*: 6 (1983).

81. F. L. Trowbridge, K. A. Hand, M. Z. Nichman, Findings relating to goiter and iodine in the Ten-State Nutrition Survey, *Am. J. Clin. Nutr.*, *28*: 712 (1975).

82. G. M. Owen, K. M. Kram, P. J. Garry, J. E. Lowe, A. H. Lubin, A study of nutritional status of preschool children in the United States, 1968—70, *Pediatrics*, *53*: 597 (1974).

83. F. L. Trowbridge, J. Matovinovic, G. D. McLaren, M. Z. Nichaman, Iodine and goiter in children, *Pediatrics*, *56*: 82 (1975).

84. S. Sostre, Changing values for the normal radioactive iodine uptake test, *J. Am. Med. Assoc.*, *239*: 1035 (1978).

85. E. T. Wong, A. L. Schultz, Changing values for the normal thyroid radioactive iodine uptake test, *J. Am. Med. Assoc.*, *238*: 1741 (1977).

86. W. C. Culp, W. T. Huskison, Changing normal values for thyroid uptake of radioactive iodine, *South. Med. J.*, *71*: 674 (1978).

87. J. A. T. Pennington, D. B. Wilson, R. F. Newell, B. F. Harland, R. D. Johnson, J. E. Vanderveen, Selected minerals in foods surveys, 1974 to 1981/82, *J. Am. Dietet. Assoc.*, *84*: 771 (1984).

88. H. Dige-Petersen, The pathogenetic significance of low iodine intake in non-endemic goiter—Absolute iodine uptake, *Nukl. Med. 16*: 174 (1977).

89. M. Abdulla, I. Anderson, N-G. Asp., K. Berthelsen, D. Birkhed, I. Dencker, C-G. Johansson, M. Jagerstad, K. Kolar, B. M. Nair, P. Nilsson-Ehle, A. Norden, S. Rassner, B. Akesson, P-A. Ockerman, Nutrient intake and health status of vegans. Chemical analyses of diets using the duplicate portion sampling technique, *Am. J. Clin. Nutr.*, *34*: 2464 (1981).

90. W. Van Dokkum, R. H. De Vos, F. A. Cloughley, K. F. A. M. Hulshof, F. Dukel, J. A. Wijsman, Food additives and food components in total diets in the Netherlands, *Br. J. Nutr.*, *48*: 223 (1982).

91. T-Y. Huang, G. H. Peterson, Pulmonary edema and iododerma induced by potassium iodide in the treatment of asthma, *Ann. Allergy*, *46*: 264 (1981).

92. K. Kohri, S. Miyoshi, A. Nagahara, M. Ohtani, Bilateral parotid enlargement ("iodide mumps") following excretory urography, *Radiology*, *122*: 654 (1977).

93. J. C. Cohen, D. M. Roxe, R. Said, G. Cummins, Iodide mumps after repeated exposure to iodinated contrast media, *Lancet*, *1*: 762 (1980).

94. J. C. Cohen, Radiopaque contrast agents as cause of iodide mumps (letter), *Postgrad. Med.*, *71*: 44 (1982).

95. G. Heydenreich, P. O. Larsen, Iododerma after high dose urography in an oliguric patient, *Br. J. Dermatol.*, *97*: 567 (1977).

96. P. Lieberman, R. L. Siegle, W. W. Taylor, Jr., Anaphylatoid reactions to iodinated contrast material, *J. Allergy Clin. Immunol.*, *62*: 174 (1978).

97. C-D. Agardh, B. Arner, S. Ekholm, E. Boijsen, Desensitisation as a means of preventing untoward reactions to ionic contrast media, *Acta Radiol. Diagnosis*, *24*: 235 (1983).

98. G. Patriarca, A. Venuti, D. Schiavino, A. Romano, Specific desensitizing treatment in allergy to iodine-containing contrast media: Observation in one patient, *Ann. Allergy*, *40*: 200 (1978).

99. A. Kint, L. Van Herpe, Iododerma, *Dermatologica*, *155*: 171 (1977).

100. J. G. Curd, H. Milgrom, D. D. Stevenson, D. A. Mathison, J. H. Vaughan, Potassium iodide sensitivity in four patients with hypocomplementenic vasculitis, *Ann. Intern. Med.*, *91*: 853 (1979).

101. M. C. Kincaid, W. R. Green, R. E. Hoover, E. R. Farmer, Iododerma of the conjunctiva and skin, *Ophthalmology*, *88*: 1216 (1981).

102. S. C. Kurtz, R. C. Aber, Potassium iodide as a cause of prolonged fever, *Arch. Intern. Med.*, *142*: 1543 (1982).

103. J. M. Hitch, B. G. Greenburg, Adolescent acne and dietary iodine, *Arch. Dermatatol.*, *84*: 898 (1961).

104. J. E. Rasmussen, Diet and acne, *Int. J. Dermatol.*, *16*: 488 (1977).

105. M. Tai, Y. Zhi-heng, L. Ti-zhang, W. Shi-ying, D. Cheng-fang, H. Xuan-yang, Z. Hui-cheng, L. Rong-ning, Y. Cheng-yun, W. Guo-qiang, C. Hui-zhen, W. Qi, High-iodide endemic goiter, *Chin. Med. J.*, *95*: 692 (1982).

106. H. Herxheimer, Effect of iodide treatment on thyroid function (letter), *N. Engl. J. Med.*, *297*: 171 (1977).

107. J. L. Penfold, C. C. Pearson, J. P. Savage, L. L. Morris, Iodide induced goitre and hypothyroidism in infancy and childhood, *Aust. Paediatr. J.*, *14*: 69 (1978).

108. S. Korsager, H. P. O. Kristensen, Iodine-induced hypothyroidism and its effect on the severity of asthma, *Acta Med. Scand.*, *205*: 115 (1979).

109. B. J. Rosenstein, L. P. Plotnick, P. A. Blasco, Iodide-induced hypothyroidism without a goitre in an infant with cystic fibrosis, *J. Pediatr.*, *93*: 261 (1978).

110. K. C. Verma, G. Dhar, Iodine-induced neonatal congenital goitre with hypothyroidism, *J. Indian Med. Assoc.*, *68*: 80 (1977).

111. G. R. Melvin, T. Aceto, Jr., J. Barlow, D. Munson, D. Wierda, Iatrogenic congenital goiter and hypothyroidism with respiratory distress in a newborn, *S. D. J. Med.*, *31*: 15 (1978).

112. N. Takasu, T. Mori, Y. Koizumi, S. Takeuchi, T. Yamada, Transient neonatal hypothyroidism due to maternal immunoglobulins that inhibit thyrotropin-binding and post-receptor processes, *J. Clin. Endocrinol. Metab.*, *59*: 142 (1984).

113. B. Thorsteinsson, C. Kirkegaard, Iodine-induced hyperthyroidism and bronchial asthma (letter), *Lancet*, *2*: 294 (1977).

114. T. M. Boehm, J. McLain, K. D. Burman, R. deShazo, L. Wartofsky, Iodine treatment of iodine-induced thyrotoxicosis, *J. Endocrinol. Invest.*, *4*: 419 (1980).

115. D. R. Gutknecht, Asthma complicated by iodine-induced thyrotoxicosis (letter), *N. Engl. J. Med.*, *296*: 1236 (1977).

116. G. Dickstein, S. Amikam, E. Riss, D. Barzilai, Thyrotoxicosis induced by amiodarone, a new efficient antiarrhythmic drug with high iodine content, *Am. J. Med. Sci.*, *288*: 14 (1984).

117. S. M. K. Paindakhel, N. Begum, H. Shah, I. Ahmad, Iodine induced thyrotoxicosis—case reports and review of literature, *J. Pakistan Med. Assoc.*, *30*: 122 (1980).

118. I. S. Salti, N. O. Kronfol, Aggravation of thyrotoxicosis by an iodinated contrast medium, *Br. J. Radiol.*, *50*: 670 (1977).

119. J. M. Jacobson, G. V. Hankins, J. M. Murray, R. L. Young, Self-limited hyperthyroidism following intravaginal iodine administration, *Am. J. Obstet. Gynecol.*, *140*: 472 (1981).

120. L. Wartofsky, Guideline for the treatment of hyperthyroidism, *Am. Fam. Physician*, *30*: 199 (1984).

121. I. C. Lewis, Commentary on iodine (letter), *J. Am. Vet. Med. Assoc.*, *180*: 1396 (1982).

122. D. P. Livadas, D. A. Koutras, A. Souvatzoglou, C. Beckers, The toxic effects of small iodine supplements in patients with autonomous thyroid nodules, *Clin. Endocrinol. (Oxf.)*, 7: 121 (1977).
123. D. J. P. Barker, D. I. W. Phillips, Current incidence of thyrotoxicosis and past prevelance of goitre in 12 British towns, *Lancet*, 2: 567 (1984).
124. R. F. Dyck, R. A. Bear, M. B. Goldstein, M. L. Halperin, Iodine/iodide toxic reaction: case report with emphasis on the nature of the metabolic acidosis, *Can. Med. Assoc.*, *120*: 704 (1979).
125. M. N. Clark, A fatal case of iodine poisoning, *Clin. Toxicol.*, *18*: 807 (1981).

9
Tin and Aluminum

JANET L. GREGER / University of Wisconsin, Madison, Wisconsin

TIN

Beneficial Effects of Dietary Tin

In 1970 Schwarz et al. (1) reported that low levels of tin (0.5 to 2 µg Sn/gm diet) promoted growth in suboptimally growing rats fed purified amino acid-based diets and housed in plastic isolator systems. These observations have not been confirmed and a number of experts doubt whether tin is essential (2–4).

Several investigators have studied the cariostatic properties of tin in diets, dentifrices, and mouthwashes (5–9). The results were not always consistent. However, rats fed diets supplemented with tin (i.e., 15 to 75 µg Sn/gm diet) were found to develop fewer caries in some studies (5–7). Tin flouride was found to have more antiplaque properties against *Streptococcus mutans* that other fluoride compounds (8,9).

Dietary Exposure to Tin

Sources of Tin in Food and Water

Tin is widely distributed in plant and animal materials (2). However, most fresh and frozen foods probably contain less than 1 µg Sn/gm food (10).

Food additives are also only a minor source of dietary tin. Stannous chloride is used as a coloring/decoloring agent, preservative or sequestrant (11,12). According to the 1977 Survey of Industry on the Use of Food Additives, the average American consumed 2.7 mg of this additive daily in 1977 (11). Organotin compounds are used as polymerization aids in plastics that come in contact with food (10). The migration of tin from these compounds into foods is thought to be generally very small (13).

Tin is present in natural waters in only trace amounts, but there are a variety of potential sources of tin in water supplies (14). Various inorganic tin compounds are used in dyeing fabrics, weighting silks, tinning vessels, and producing lacquers, nail polishes, and varnishes. Pewter is about 90% tin (10). Organic tin compounds are used in fungicides, insecticides, herbicides, and anthelmintics. Stannous fluoride is used in many toothpastes and consequently reaches municipal sewers. Little of this tin is believed to remain in the water supply, because many tin salts are insoluble in water (14).

The major source of dietary tin is canned foods (2,10,15,16). The tin content of canned foods, however, can vary greatly (Table 1). Foods packed in cans that are totally coated with lacquer generally contain less that 4 μg tin/gm food and thus contribute little tin to the diet.

Some foods (i.e., pineapple, grapefruit and oranges juices, applesauce, tomato sauce) are often packed in cans that are not coated with lacquer. Greger and Baier (15) found that the tin content of these foods when cans were first opened was between 40 to 150 μg Sn/gm food. Sherlock and Smart (10) report similar levels of tin in these types of canned foods in Great Britain.

The amount of tin in canned foods can be affected by storage conditions. Canned foods accumulate more tin when stored for several months (10,15,17−19). This process is accelerated when ambient temperatures are elevated above 40°C, as can occur in warehouses during summer (17,19). The composition of the food (i.e., high nitrate levels and low pH) and residual oxygen in the headspace of the can can also increase the rate of pitting of cans and increase the migration of tin from unlacquered cans into food (10,20).

Capar (21) estimated that 15% of households in the United States store foods in the refrigerator in cans that have been opened. This can result in two- to tenfold increases in the tin levels in foods (15, 22). Greger and Baier (15) found that the levels of tin in several foods stored in this manner for one week exceeded 250 μg Sn/gm food, a level considered to be excessive by experts worldwide (10,23).

Dietary Intakes of Tin

During the last 45 years the tin content of typical Western diets have been reported to range from 1 to 38 mg tin daily (2,10,15,19,24−26). The differences in the estimates primarily reflect differences in the amounts of canned foods, particularly foods packed in unlacquered cans, included in the diet composites. Currently canned vegetable and fruits account for less than 5% (w/w) of the food included in the

Table 1. Estimated Tin Concentrations of Canned Foods

Food	Type of can	Average concentration of tin (μg/gm food)	
		Can just opened	Opened can stored in refrigerator for 1 week
Apple juice	lacquered	0.1	0.1
Cranberry sauce	lacquered	1.4	1.8
Pineapple-grapefruit drink	lacquered	1.7	1.8
Tomatoes, stewed	lacquered	2.8	3.7
Applesauce	partially lacquered	51	137
Grapefruit sections	partially lacquered	96	462
Orange juice	partially lacquered	53	175
Pineapple, crushed	not lacquered	89	544
Tomato sauce	partially lacquered	150	669

Source: Ref. 15.

British Total Diet Study menu (10), and foods canned in tin-plated cans account for less that 5% (w/w) of the food included in the U.S. Total Diet Study menu (27).

Although most American and Europeans consume a limited amount of canned foods, there are exceptions. Low-income individuals and institutions, such as nursing homes and schools, often select canned fruits, vegetables, and juices because of economy and ease of storage, i.e., no refrigerator or freezer space is required. Some elderly individuals are accustomed to or even prefer the texture of canned fruits and vegetables. Some mothers routinely leave the juice in opened cans in the refrigerator for their children's snacks. Individuals who routinely consume canned fruits, vegetables, and juices from unlacquered cans could ingest 50 to >200 mg tin daily (10,15).

Metabolism of Inorganic Tin

Absorption

Generally both animals (28—31) and humans (19,24,32) fed moderate (>100 µg Sn/gm dry weight) or large doses of tin have been found to excrete more than 90% of the tin in feces. However, human subjects fed very low levels often (0.11 mg Sn/day) lost only 50% of their tin intake in feces (32).

Little is known about factors that affect the absorption of tin. Absorption of tin does not appear to be sensitive to changes in the anion component of dietary tin salts (28). Hiles (28) observed that rats absorbed a single dose of tin II more efficiently than tin IV (2.85 vs. 0.64%). Fritsch et al. (29) found that administration of different food components with single doses of ^{113}Sn had little effect on the efficiency of tin absorption. However, Kojima et al. (33) noted that organic acid increased the absorption of tin. Johnson and Greger (31) also observed that a threefold increase in dietary zinc levels resulted in greater fecal losses of tin when animals were fed 100 to 200 µg Sn/gm diet.

Some tin lost in the feces may be of endogenous origin. Hiles (28) noted that 12.1% of a single intravenous dose of ^{113}Sn(II) appeared in the feces while only 3.1% of a single intravenous dose of ^{113}Sn(IV) appeared in the feces. Moreover when ^{113}Sn(II) was injected into rats with the bile duct canulated, 11.5% of the dose was collected in the bile. Virtually none of the ^{113}Sn from an intravenous dose of ^{113}Sn(IV) was collected in the bile.

Little tin is excreted in the urine of rats (28,29) or humans (24,32,34). Urinary losses of tin will reflect large differences in tin intake. Eight human subjects excreted four times as much tin (122 vs. 29 µg Sn/day) when fed 50 mg rather than 0.11 mg tin daily (3).

Retention of Tin in Tissues

Although the absorption and overall apparent retention of tin by human subjects in balance studies is low (19,24,32), tin has been found in at least trace amounts in most mammalian tissues (2,35,36).

Rats fed diets supplemented with tin have been found to accumulate tin in their tibias, kidneys, and livers in proportion to their dietary intake of tin as shown in Fig. 1 (31). Johnson and Greger (31) found that the concentration of tin in the tibias of rats fed diets supplemented with tin were more than five times greater than the concentration of tin in kidneys and nearly 20 times greater than the concentration of tin in livers. Other investigators have also observed that both animals (28,30,37) and humans (25) accumulated more tin in bone than soft tissues.

It is not clear whether tin ingested by females can accumulate in fetuses. Schroeder et al. (2) observed virtually no tin in the tissues of stillborn human infants. Theuer et al. (38) found fetal tin values were elevated when the maternal rat diet contained tin salts, but without apparent correlation to dietary tin levels. Hiles (28) found little accumulation of [113]Sn in fetal tissues when female rats were dosed with tin.

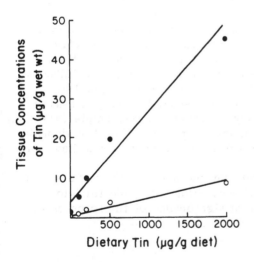

Fig. 1. Concentrations of tin (μg Sn/gm wet weight) in tibias ($\bullet$) and kidneys (o) of rats fed five different levels of tin in three different studies (31).

Disease states may also affect the retention of tin in tissues. Nunnelley et al. (39) observed the accumulation of tin in liver samples from 44 uremic patients. The average level of tin in livers from control, nondialyzed uremics, and dialyzed uremics were 2.1, 6.6, and 17.1 µg/gm dry weight, respectively. However, urinary tin excretion of nondialyzed uremic patients appeared to be in a normal range.

Toxic Effects of Organic Tin

A variety of organic tin compounds are used commercially; a number are extremely toxic (40,41). The toxicity of these compounds is dependent upon the organic constituents of the tin compounds, the manner of exposure, and the animal species studied. This topic has been reviewed thoroughly and will be discussed only briefly here (40, 41).

Trimethyl and triethyltin compounds are the most toxic organotin compounds partially because they are well absorbed from the gastrointestinal tract (40,42). Mushak et al. (41) observed significant quantities of tin accumulated in the livers, kidneys, and brains of rats dosed with a variety of organotin compounds, not just trimethyltin and triethyltin. However, only rats dosed with trimethyltin and triethyltin had elevated levels of tin in their blood.

Trimethyltin and triethyltin differ in their effects. Triethyltin causes cerebral edema, myelinopathies, and spongy degeneration of the brain (41,43). Trimethyltin produces brain damage which is primarily restricted to limbic system structures (41,44). However, the region-specific pathology of trimethyltin cannot be correlated to regional deposition of tin in the brains of animals dosed with trimethyltin (45). Furthermore these compounds produce dissimilar behavioral effects following accidental human exposure (40,41).

Other organs besides the nervous system, inclusing the liver and thymus, are affected adversely by organotin compounds (41). However, certain organotin compounds, e.g., tri-n-butyltin fluoride (46) and a di-n-butyltin complex (47), have been found to have antitumor activity.

Hallas et al. (48) have noted that microorganisms in the silt at the bottom of bays are capable of converting inorganic tin into organic forms of tin. The potential significance of this process as a source of organic tin in water is not clear.

Toxic Effects of Inorganic Tin

Acute Effects

Animals and humans are fairly resistant to single large oral doses of tin and reports of acute responses to dietary tin are rare (49—51). Symptoms of acute tin toxicity appear rapidly and include nausea, abdominal cramping, diarrhea and vomiting. Generally, individuals have developed these symptoms after consuming canned juices or acidic punches prepared in tinned vessels. The level of tin in the contaminated foods have ranged from 500 to 2000 μg Sn/ml. This suggests that other factors in the beverages may have exacerbated the local irritation of the gastrointestinal tract by tin in some cases.

Chronic Effects

Growth Depression. Several incestigators have observed that the effects of inorganic tin on growth are dependent on the dose and the form of the tin salts fed. Growth of rats was generally not affected by dietary levels of tin less than 300 μg Sn/gm diet (30,52,54). As dietary tin levels were elevated above 500 μg Sn/gm diet, growth depression becomes more severe (52–55). DeGroot et al. (53) found that ingestion of soluble tin compounds (e.g., stannous chloride, stannous sulfate, and stannous oxalate) affected growth more than the ingestion of insoluble tin compounds (e.g., stannous oxide, stannous oleate, and stannous sulfide).

Changes in Enzyme and Immune Function. The mechanism by which tin affects cellular functions are not known. Some of the effects may be indirect and due to interactions between tin and essential minerals such as zinc, copper, iron, and selenium (26); some of the effects may be direct. McLean et al. (56,57) have found that in vitro tin (II) but not tin (IV) was readily taken up by white blood cells and ovary cells and damaged the DNA in the cells. In any case, exposure to tin does cause changes in the activity of certain enzymes and in immune function. Ingestion of tin has been demonstrated to depress the activity of serum alkaline phosphatase (37,55) and serum lactic dehydrogenase (37). Both are zinc metalloproteins. A single injection of stannous chloride has been found to depress hepatic azo-reductase and aromatic hydroxylase activity (58). In contrast, in-jections of tin have been found to induce heme oxygenase activity in the kidneys of rats (59,60).

Chiba et al. (61) and Zareba and Chmielnicka (62) have demon-strated that animals injected with tin had decreased acitvity of blood δ amino levulinic dehydratase (δALAD), However, Johnson and Greger (31) demonstrated that very high levels of dietary tin (>2000 μg Sn/gm diet), but not moderate levels of dietary tin (≅200 μg Sn/gm diet), inhibited blood δALAD activity in rats. Zinc partially reversed in vitro the inhibition of δALAD by tin (63), but injected zinc did not counteract the effect of tin on δALAD activity in vivo (62). Similarly a threefold increase in dietary zinc levels did not alter the response of blood δALAD to tin exposure (31).

Only a limited amount of work has been done on the effects of tin on immune function. Immunological responses have varied with the number and site of doses of tin. Dimitrov et al. (64) found that a single injection of stannic chloride decreased the formation of plaque-forming cells in mice. Levine and Sowinski (65) observed marked proliferation of plasma cells and Russel body cells in draining lymph nodes of Lewis rats inoculated once with metallic tin. Repeated injection of tin produced splenomegaly in rats and mice (66,67). However, peripheral inoculations of tin prevented the splenomegaly that occurred in response to repeated IV doses of tin (66). Prior but not concurrent administration of tin chloride in drinking water prevented plasma cell hyperplasia in response to a single injected dose of tin (65). Pretreatment with immunosuppressive drugs did not mimic the effects of pretreatment with tin chloride (65). The mechanisms are unclear at this time.

Interactions with Zinc and Selenium. Potentially some of the effects of tin on growth, enzyme levels, and even immune function can be explained on the basis of interactions between tin and zinc or selenium. It is well established that growth depression is a common sympton of zinc deficiency (3). Several of the enzymes affected by tin are zinc metalloproteins. Moreover immune function is sensitive to changes in the nutritional status of both zinc and selenium (68).

Johnson and Greger (30,54) demonstrated repeatedly that rats fed $\geqslant 500$ µg tin/gm diet had depressed levels of zinc in bone and soft tissues. Tibia zinc levels were even sensitive to moderate doses of tin (100–200 µg Sn/gm diet). Rats repeatedly injected with tin also retained less zinc in most soft tissues, except liver (69).

At least part of this effect is due to the effect of dietary tin on apparent absorption of zinc. Johnson et al. (70) found that human subjects lost an additional 2 mg of zinc daily in the feces when fed 50 mg tin rather than 0.1 mg tin daily; this resulted in significantly poorer overall retention of zinc by these subjects. Valberg et al. (71) confirmed these results and found inorganic tin depressed the absorption by humans of ^{65}Zn both from zinc chloride and from a test meal containing turkey. However, Solomons et al. (72) could not demonstrate a tin-zinc interaction in subjects fed load doses of both tin and zinc.

Tin also adversely affected the apparent absorption of zinc in rats (30,54). The mechanism appears to depend on dose. When rats were fed high levels of tin ($\cong 2000$ µg Sn/gm diet), their gastrointestinal tracts were hypertrophied and endogenous losses of zinc in the feces were significantly increased (54). When moderate levels of tin ($\cong 200$ and 500 µg Sn/gm diet) were fed, endogenous losses of zinc in the feces were constant but the true absorption of zinc tended to be depressed.

Less is known about the interaction between tin and selenium. Hill and Matrone (73) showed that high dietary levels of tin depressed the apparent absorption of selenium from chick intestinal segments.

Greger et al. (74) demonstrated that human subjects fed 50 mg versus
0.11 mg tin daily apparently absorbed significantly less selenium.
Chiba et al. (75) found the simultaneous injection of sodium selenite
with tin prevented a decrease in δALAD activity in mice.

Anemia and Interactions with Copper and Iron. The ingestion of high
level of dietary tin can induce anemia in rats (52,53,55). There are
several potential mechanisms. One of these involves copper. Generally dietary tin did not depress tissue levels of iron in rats (30,31),
but ingestion of 200 μg Sn/gm diet usually depressed copper levels
in soft tissues (30,31). The plasma copper levels of animals fed high
levels of tin (500 and 2000 μg Sn/gm diet) were depressed to less
than 20% of the levels found in control animals (31). Kidney and liver
levels of copper were also severely depressed in these animals. It
is well established that copper deficiency can induce anemia (3).
Moreover deGroot (52) demonstrated that the addition of copper to the
diets of rats eliminated the anemia induced by feeding 150 μg tin/gm
diet.

Ingestion of high levels of dietary tin (>1000 μg Sn/gm diet),
but probably not moderate levels of tin, may induce anemia in other
ways too. Animals injected with tin exhibited alterations in the activity of at least two enzymes involved in heme metabolism (δALAD, an
enzyme involved in heme synthesis, is depressed; heme oxygenase,
an enzyme involved in heme catabolism, is increased) (59–62). The
oral administration of high doses (2000 μg Sn/gm diet), but not low
doses of tin (100 or 200 μg Sn/gm diet), depressed the activity of
blood δALAD (31).

Although all of these factors may affect the development of anemia
in laboratory animals that are fed high levels of tin, their significance
to humans fed moderate levels of tin is questionable. Johnson et al.
(70) found that the addition of 50 mg tin daily (equivalent to about
100 μg Sn/gm dry diet) to the diets of human subjects for 20 days
had no effect on the apparent absorption of copper or iron or on
plasma copper, ceruloplasmin, or ferritin levels. Similarly no changes
in copper or iron metabolism were observed in rats fed only 100 μg
Sn/gm diet (31).

Changes in Calcium and Bone Metabolism. A group of Japanese workers have reported that the ingestion of tin affected calcium metabolism
in that it: reduced the calcium content of bone (37,76,77), reduced
the calcium level of serum (37,76,78), increased kidney calcium levels
(78), and increased biliary calcium losses (79). They found that
even dietary levels of tin as low as 50 μg Sn/gm diet had significant
effects on the calcium content of the femoral epiphysis (76). Johnson
and Greger (31) also observed that low levels of dietary tin (≅100
μg Sn/gm diet) depressed the calcium content of bone but observed
no changes in plasma calcium levels. The addition of 50 mg tin daily

to the diets of humans had no effects on calcium or magnesium excretion and retention (32,70). Differences between studies may be due to difference in how tin was administered, the dietary levels of calcium and phosphorus, and the size of animals.

Ogoshi et al. (80) have observed that compressive strength of femurs of rats given tin (300 μg Sn/ml) in their drinking water was significantly decreased. Yamaguchi et al. (81) also observed that collagen synthesis was depressed in bones of rats orally dosed with tin.

Conclusion

Most clinicians will never see patients with acute toxicity symptoms due to the ingestion of organic forms of tin in pesticides or due to the ingestion of large doses of inorganic tin from improperly stored canned fruit juices. Some individuals, however, regularly ingest moderately large doses (i.e., 50–200 mg) of tin daily. Exposure to this amount of tin would chiefly be important to those individuals who consume low levels of essential elements (e.g., zinc, copper, and perhaps calcium) and who are already in marginal nutritional status in regard to these elements. For example, many elderly individuals and some children who are picky eaters routinely consume one-half to two-thirds of the Recommended Dietary Allowances (RDA) for zinc (82,83). Most women in the U.S. consume less than two-thirds of the RDA for calcium (84). For these individuals the routine consumption of foods packed in unlacquered cans may result in excessive exposure to tin.

ALUMINUM

Beneficial Effects of Aluminum

Aluminum is the third most abundant element in the Earth's crust. However, there is no conclusive evidence that aluminum is essential for growth, reproduction or health of men and animals (3).

Kleber and Putt (85) reviewed about 100 articles on aluminum and dental caries. They concluded that topical application of aluminum: 1) reduced dental caries, 2) enhanced the systemic and topical effect of fluoride, 3) decreased the acid solubility of enamel, 4) inhibited dental plaque formation and acidogenicity, and 5) prevented dental fluorosis.

Exposure to Aluminum

Most foods contain some aluminum (Table 2). Estimates of the aluminum content of any biological sample should be viewed with skepticism. During the last 20 years, estimates of the amounts of aluminum in

Table 2. Estimated Aluminum Concentrations of Selected Food

Food	Aluminum concentrations (mg/100 gm)	Food	Aluminum concentrations (mg/100 gm)
Animal Products		Nuts	
Beef, cooked	0.02	Peanut butter	0.02
Cheese, natural	1.57	Walnuts	0.02
Cheese, processed	29.7		
Fish (cod), cooked	0.04	Vegetables	
Milk	0.07	Beans, green cooked	0.34
		Cucumber	0.17
Fruits		Lettuce	0.06
Applesauce	0.01	Peas, cooked	0.19
Bananas	0.04	Potatoes, unpeeled, boiled	0.01
Orange juice	0.04	Potatoes, with skin, baked	0.24
		Spinach, cooked	2.52
Grains		Tomatoes, cooked	0.01
Bran, wheat	1.28		
Bread, white	0.30	Other	
Bread, whole wheat	0.54	Baking powder	2300
Rice, cooked	0.17	Cocoa	4.50
Spaghetti, cooked	0.04	Coffee, brewed	0.04
		Pickles with aluminum additives	3.92
Herbs and spices		Salt with aluminum additives	16.4
Bay	43.6	Tea bag, dry	128
Cinnamon	8.2	Tea, steeped	0.46
Oregano	60.0		
Pepper, black	14.3		
Thyme	75.0		

Source: Ref. 87. Adapted and reprinted from Food Technology (1985) 39: 73, 74, 76, 78–80. Copyright by Institute of Food Technologists.

plasma have decreased by more than 50-fold because of improvements in methodology (86). It is possible that some "old" values on the aluminum content of foods are artificially high, too.

Natural Sources of Aluminum in Food and Water

Grain and vegetable products, especially herbs and tea leaves, contain more aluminum from natural sources than animal products (87–89). In fact, Eden (90) reported that some tea leaves contained as much as 17,000 µg Al/gm dry leaf. The aluminum content of many of those products will vary greatly because of differences in plant varieties and soil conditions and pH (90,91).

Although most foods contain some aluminum, the amount of aluminum naturally present in the diets of Americans probably ranges from 2 to 10 mg daily (87). This reflects two points. Generally Americans consume only small quantities of herbs daily. Much of the aluminum in tea leaves does not dissolve in tea infusates (88). Thus ingestion of 8 oz. of tea with each meal would add only 1 to 4 mg aluminum to the daily diet (87).

Another "natural" source of aluminum is water. Miller et al. (92) surveyed 186 water utilities. They found that the median concentration of aluminum in finished water was 0.0017 mg Al/100 ml. Thus, an individual consuming 2 L of water daily would consume less than 0.04 mg of aluminum in water daily. Some aluminum in finished water might be there because aluminum flocculants were used to clarify the water (93). However, according to a report from the National Academy of Science modern purification practices usually result in the presence of lower concentrations of aluminum in drinking water than in raw water (93).

Food Additives as a Source of Aluminum

Committees of the National Research Council (i.e., GRAS List Survey Data Committees) regularly review the use of food additives, including aluminum-containing food additives, in the United States (94). In 1982 the most commonly used aluminum-containing food additives were bentonite, sodium aluminum phosphate-acidic, sodium aluminum silicate, sodium aluminum phosphate-basic, aluminum lakes of various food dyes and colors, and aluminum sulfate (94). In total, approximately 4.0 million pounds of aluminum were used in food additives in the United States in 1982. This means that the average U.S. citizen theoretically consumed 21.5 mg aluminum daily in food additives in 1982. Food disappearance data of this sort tends to overestimate food intake somewhat.

Americans probably vary greatly in their intakes of aluminum from food additives. The Committee on the GRAS List Survey—Phase III (11) estimated that 5% of adult Americans consumed ⩾95 mg aluminum daily in food additives. Those foods which contribute the greatest amounts of aluminum in food additives to the diets of Americans are baked goods prepared with chemical leavening agents and processed cheeses (87).

Packaging and Utensils as a Source of Aluminum

Recently, several physicians have suggested that aluminum cooking utensils were a major source of aluminum in food (95—96). Generally, investigators have found that most foods stored or cooked in aluminum pans, trays, or foil accumulated some aluminum. The amounts of aluminum that accumulated in foods during preparation depended on the pH of the foods, the length of the cooking periods, the type of utensils, and how they had been used previously (97–103). Even so, most foods accumulated less than 0.2 mg Al/100 gm of food during preparation and storage.

One group of foods that accumulated more aluminum when cooked in aluminum pans was tomato products. Greger et al. (97) found that tomato sauces cooked for three hours in aluminum pans accumulated 5.7 mg Al/100 gm serving. Similarly Lione (103) observed that tomatoes cooked for two hours accumulated 3.2 mg Al/100 gm serving.

Obviously food choices and preparation methods will greatly affect aluminum contamination of food during preparation. However, a "typical" individual would not add more than 3.5 mg aluminum daily to their diet through the use of aluminum foil, trays, and pans to prepare and heat food and beverages (87).

Pharmaceutical Sources of Aluminum

Most Americans probably consume 20 to 40 mg aluminum in food and beverage daily. A few may consume as little as 3 mg aluminum daily and a few may consume as much as 100 mg aluminum daily (87). All of these quantities of aluminum are small compared to the amounts of aluminum that can be ingested in pharmaceutical products, such as antacids, buffered analgesics, antidiarrheals and certain antiulcer drugs (104). Lione (105) estimated that 840 to 5000 mg aluminum and 126 to 728 mg aluminum were possible daily doses of aluminum in antacids and in buffered analgesics, respectively.

Metabolism of Aluminum

Absorption and Urinary Excretion

When human subjects were fed pharmacological doses of aluminum ($\cong$2000 mg Al daily), fecal losses of aluminum generally were less than aluminum intake (106–108). However when subjects were fed 5 to 125 mg aluminum daily, aluminum losses in the feces approximated dietary intake (24,107–109). This does not mean that aluminum absorption did not occur in this range; it means balance techniques were not sensitive enough to detect differences.

Both humans (24,34,109,112–114) and animals (110,111) excrete little aluminum in their urine. Generally humans excrete less than 100 µg Al/day (34,109,113,114). However, urine, not bile, appears to be the major excretory route for injected aluminum (115).

Urinary aluminum levels do reflect changes in dietary intake (109,112,113). In fact subjects excreted three times as much aluminum in urine when they were fed 125 mg rather than 5 mg aluminum daily (109).

Generally investigators attempting to asses the effect of dietary or hormonal factors on aluminum absorption have not used balance techniques (113,114,116—122) because data from balance studies are difficult to interpret for aluminum (24,106—111). There are also no convenient radioactive isotopes of aluminum. Thus several investigators have attempted to demonstrate aluminum absorption by monitoring changes in urinary losses or tissue levels of aluminum (109—114, 116, 121).

Data on the relative bioavailability of aluminum from various aluminum compounds, including aluminum hyproxide, aluminum citrate, aluminum phosphate salts, aluminum chloride, aluminum palmitate, and aluminum lactate, are not consistent (113,116,119—122). Differences in other dietary components, in the means of administering the aluminum (i.e., diet or gavage) and in the variables monitored may account for the inconsistencies in the data.

Other factors may also influence aluminum absorption. The evidence in most cases is not conclusive. Mayor et al. (117) have found that rats injected with parathyroid hormone absorbed aluminum more efficiently than control rats. Cam et al. (107) suggested that some renal patients may absorb aluminum more efficiently than healthy subjects. However Feinroth et al. (123) reported that aluminum absorption in rat everted gut sacs was not influenced by renal failure or changes in parathyroid hormone levels in the rats from which the sacs were prepared. Furthermore they believed that aluminum absorption from rat jejunum was energy-dependent and carrier-mediated.

Very young children may also absorb aluminum more efficiently than more mature animals (124). Dietary levels of magnesium and boron may also influence aluminum absorption (125). The administration of very high levels of fluoride will increase elimination of aluminum in both the feces and urine (101).

Retention of Aluminum in Tissues

Aluminum, at least in trace amounts, has been found in most tissues of men and animals (3,88). Generally more aluminum has been found to accumulate in bone than in soft tissues (101,116,120,121,126,127). For example, rats fed only 300 μg Al/gm diet as aluminum hydroxide for 18 days accumulated 7 to 15 times as much aluminum in bone and about 1½ times as much aluminum in their kidneys as control animals (116).

As already indicated, a variety of factors, most of which are not well understood, affect the accumulation of aluminum in tissues. Tissue (including kidney, liver, serum, and bone) levels of aluminum respond to changes in aluminum intake (101,108,110,113,116,118,120, 121,126—129), but the response does not appear to be directly proportional to intake levels or length of exposure (116,126,129).

Disease states and their treatments, including kidney dialysis, intravenous feeding and the use of phosphate binders, can greatly affect tissue aluminum levels. Many investigators have reported that aluminum accumulated in the tissues of renal dialysis patients (114, 130—145). For example, Alfrey et al. (130) found that dialysis patients who were the most severely affected by aluminum toxicity, i.e., those with dialysis encephalopathy syndrome, accumulated 10 times the normal levels of aluminum in muscle and brain and more than 20 times the normal level of aluminum in bone. Dialysate fluids were probably the major source of aluminum exposure for many patients during the 1960s and 1970s (131—133). More recently several groups of investigators have reported the accumulation of aluminum in tissues, particularly bone, of uremic patients who were dialyzed with aluminum-free fluids (138,140) and in tissues of children who were never dialyzed (118,141—146). Generally the source of this aluminum was aluminum-containing, phosphate binders (138—145); however, in one situation infant formula was the reported source of the aluminum (146). Deferoxamine has been used successfully to remove aluminum from the bones of some uremic patients (147,148) but not others (149).

Patients receiving intravenous solutions have also been found to accumulate aluminum in their tissues when the solutions contain aluminum, i.e., casein-hydrolysate solutions for total parenteral nutrition (150—154). Some ulcer patients also have been found to accumulate aluminum in bone (114). Generally these patients have accumulated less aluminum than dialysis patients (114). This may reflect the amount and duration of exposure, differences among the patients and the route of aluminum administration.

Aluminum levels tend to increase in tissues, including brain, with age (155,156). Some (157—160) but not all experts (155,156) found increased aluminum levels in the brains of patients with Alzheimer's disease. However, serum aluminum levels are not elevated in these patients (161). Furthermore the sites of aluminum deposits in the brains of patients with Alzheimer's disease and dialysis encephalopathy have been reported to be different (162). Brain aluminum levels are also elevated in patients with amyotrophic lateral sclerosis and Parkinsonism dementia in Guam (163,164).

Toxic Effects of Aluminum

Recognition that aluminum was part of the etiology of dialysis encephalopathy and dialysis osteodystrophy and perhaps was part of the etiology of amyotrophic lateral sclerosis and Parkinsonism dementia in Guam and of Alzheimer's disease has aroused a lot of interest on the biological effects of aluminum during the last 10 years. Although some of the symptoms of these syndromes are caused by the direct toxic effects of aluminum on cell components, many of the toxic effects of aluminum are at least partially ascribable to interactions of aluminum with essential nutrients, such as phosphorus and fluoride (88,131).

Interactions With Essential Minerals

Several investigators have observed that large oral doses of aluminum interfered with phosphorus absorption (101,127,128,165—167) and lowered tissue phosphorus levels (106,107,128,165,167—169). The bone pain and fractures observed in patients who have used large doses of aluminum-containing antacids for years has been related to a phosphorus depletion syndrome (167—169). The effect of moderate doses of aluminum on phosphorus metabolism is less clear (116,126,129,170). Although the additions of 120 mg of aluminum to the diets of young adults depressed phosphorus absorption initially, subjects appeared to adjust so that no effect could be observed after two weeks (170). Similarly rats fed ≅1000 μg Al/gm diet for 30 days absorbed phosphorus less efficiently than control rats but the effect was not observed after 60 days (126).

The interactions between aluminum and calcium are complex. Spencer et al. (166,171) have observed that subjects fed low dietary levels of calcium excreted more calcium in the feces when antacids containing aluminum and magnesium salts were administered. Valdivia et al. (128) observed sheep apparently absorbed calcium less efficiently when fed 2000 μg Al/gm diet. Greger et al. (126) observed rats fed ≅260 or ≅1000 μg Al/gm diet for 30 days, but not 60 days, apparently absorbed less calcium than control animals. Other investigators have found that oral administration of aluminum did not affect calcium absorption (106,116,126,170) or the levels of calcium in soft tissues or bone (116,126,128,129,165,170). The elevation of urinary and serum calcium levels observed in patients given pharmaceutical doses of aluminum have sometimes been attributed to changes in phosphorus metabolism induced by aluminum (166,169,171—174). The role of parathyroid hormone in these interactions is debatable (172—174).

Animal scientists have demonstrated that high dietary levels of aluminum will sometimes depress tissue levels of magnesium (128,165, 175). The effects of aluminum on serum magnesium levels and the excretion of magnesium by humans are less clear because either low levels of aluminum were fed (170) or antacids containing both aluminum and magnesium were administered (176). Moreover Greger et al. (116, 126) found that rats red ≅260 μg Al/gm diet had no changes in tissue magnesium levels but absorbed magnesium less efficiently after 30 days of aluminum exposure but not after 60 days of exposure (126).

It is well documented that dietary aluminum will depress fluoride absorption in humans and animals (177—179). Although fluoride levels are also depressed, overall retention of fluoride is depressed when aluminum is administered (170,177–179). The importance of aluminum-fluoride interaction in patients suffering from dialysis osteodystrophy has not been assessed.

The effects of aluminum on iron, zinc, and copper metabolism is debatable. Some investigators have noted that the administration of aluminum compounds to animals and human subjects altered iron (128, 165,180,181), zinc (116,129,165) and copper (116,126,128,129,165, 182) metabolism. Others have noted no changes in iron (116,126,129, 170,176,183), zinc (126,128,170), and copper (170) metabolism. The difference in results may reflect differences in the measures of mineral retention (i.e., absorption or tissue levels), the length of the studies, and the general composition of the diets. It might be anticipated that aluminum hydroxide, because of its acid neutralizing capacities, might affect the absorption of minerals, particularly iron, more than other aluminum-containing compounds, but this has not been documented (116).

Cellular and Clinical Changes

Aluminum not only reacts readily with anions and cations in the diet, it also forms complexes with a variety of biomolecules, including nucleic acids (184–186), membranes and lipids (184,187), carboxylic acids (184), peptides such as enkephalin (188), and proteins such as calmoldulin (184,189). Haug (184) recently reviewed these interactions extensively. However, at this time the importance of these interactions in the toxicity of aluminum can only be surmised because most have been studied only in isolated systems.

Even the seemingly direct effects of aluminum on enzyme activity often involve other elements. Activation of the purified guanine nucleotide binding regulatory component of adenyl cyclase by fluoride requires the presence of aluminum (191). The inhibitory action of aluminum on human plasma cholinesterase activity in isolated systems can be prevented by calcium (191). Furthermore, Ondreička et al. (101) claimed that acute and chronic oral intoxication by aluminum chloride of rats resulted in decreased incorporation of ^{32}P into phospholipid, DNA, and RNA fractions of a variety of tissues and depressed levels of ATP but elevated levels of ADP and AMP in blood. Further work is needed to confirm each of these observations.

The two main tissues in which the effects of aluminum toxicity have been documented in humans are bone and brain. As already noted, aluminum has been found to accumulate in the bones of patients: dialyzed with aluminum contaminated fluids; dosed with aluminum-containing phosphate binders and ulcer medications; and treated with aluminum-containing TPN solutions. Clinical symptoms include bone pain and an increased rate of fractures and resistance to vitamin D therapy (131,135).

Aluminum accumulated in the bones of these patients at the interface between the thickened osteoid and calcified bone (172,192). Possible mechanisms of aluminum toxicity are changes in vitamin D and parathyroid hormone metabolism, phosphorus depletion, and direct (but undefined) toxic actions of aluminum (193). The lack of consistent changes in serum parathyroid hormone levels, calcium level, and alkaline phosphosphatase activities and bone aluminum levels in response to aluminum exposure makes interpretation of the data difficult (131,135,172−174,192−195). However, the primary action of aluminum toxicity is probably not via interference with vitamin D metabolism (193,194). Moreover aluminum probably reduces osteoblast numbers in bone directly and through reduced parathyroid hormone action (195).

As already noted several neurological conditions, i.e., dialysis encephalopathy, amyotrophic lateral schlerosis, and Parkinson's dementia (ALS-PD) on Guam, and *perhaps* Alzheimer's disease, are associated with elevated levels of aluminum in brain tissues. The symptoms of dialysis encephalopathy include speech difficulties, motor abnormalities, dementia, and eventually coma and death (131). Moreover behavioral changes have been reported in animals fed very high levels of aluminum (196). Thus clinical symptoms of these syndromes and those in animal models may appear similar but histological and chemical analyses of brains have revealed differences.

Patients with dialysis encephalopathy have been found to accumulate aluminum in most tissues, not just brain (131,138−140). Patients with Alzheimer's disease do not have elevated serum aluminum levels (161). However, the two uremic infants who died after consuming aluminum-contaminated formulas exhibited elevated brain, but not bone, aluminum levels (146).

The site of aluminum deposition within cells also differs between syndromes. The aluminum content of the nuclear and heterochromatin fractions were elevated in brains from patients with Alzheimer's disease or with ALS-PD in Guam but not in those with dialysis encephalopathy (162,163). DeBoni et al. (197) found that rabbits injected subcutaneously with aluminum developed neurofibrillary degeneration with aluminum being concentrated in the brain's nuclear chromatin.

Furthermore the activity of choline acetyltransferase has been found to be reduced by 60 to 90% in the cerebral cortex and hippocampus formations of brains of patients with Alzheimer's disease (198). Hertnarski et al. (199) observed that rabbits injected with aluminum developed neurofibrillary degeneration, but the activity of choline acetyltransferase in brain samples was unaltered.

The mechanism by which aluminum induces neurological damage in animal models and patients with dialysis encephalopathy and the pathogenic mechanism causing Alzheimer's disease and ALS-PD in Guam are unknown (200). However, several mechanisms have been advanced.

Banks and Kastin (201) observed aluminum increased the permeability of the blood brain barrier to peptides such as β-endorphin. Garruto et al. (164) hypothesized that secondary hyperparathyroidism resulting from low environmental levels of calcium and magnesium resulted in excess aluminum accumulation in the brains of individuals with ALS-PD in Guam. Recently Gajdusek (200) hypothesized that anything (trauma, aluminum, or subviral pathogens) that interfered with the slow axonal transport of neurofilaments down the axons of nerve cells could lead to amyloid accumulations and degeneration of the central nervous system. Obviously much work is needed.

Conclusion

A great deal has been published in the popular press on the toxicity of aluminum. Unfortunately these articles focus attention on minor sources of dietary aluminum but ignore major pharmaceutical sources of aluminum. At this time dietary exposure to aluminum does not appear to have adverse effects on healthy individuals. However, the long-term consequences of chronic use of pharmacological doses of aluminum by sensitive individuals (i.e., those with impaired kidney function, including the elderly and low-birth-weight infants) need further evaluation.

ACKNOWLEDGMENTS

The author appreciates the support of the College of Agriculture and Life Sciences, University of Wisconsin, Madison, WI, project no. 2623.

REFERENCES

1. K. Schwarz, D. B. Milne, and E. Vinyard, Growth effects of tin compounds in rats maintained in a trace element-controlled environment, *Biochem. Biophys. Res. Comm.*, *40*: 22–29 (1970).
2. H. A. Schroeder, J. J. Balassa, and I.H. Tipton, Abnormal trace metals in man: tin, *J. Chron. Dis.*, *17*: 483–502 (1964).
3. E. J. Underwood, Tin, in *Trace Elements, in Human and Animal Nutrition*, Academic Press, New York, 1977, pp. 56–108, 196–242, 430–433, 449–451.
4. F. H. Nielsen, Possible functions and medical significance of obtuse trace metals, in *Inorganic Chemistry in Biology and Medicine*, American Chemical Society Symposium Series No. 140 (A. E. Martell, ed.), American Chemical Society, Washington, D.C., 1980, pp. 23–42.

5. J. L. McDonald, and G. K. Stookey, Influence of whole grain products, phosphates and tin upon dental caries in the rat, *J Nutr.*, *103*: 1528–1532 (1973).

6. G. K. Stookey, J. L. McDonald Jr., Further studies of the cariostatic properties of tin (II) and oat hulls in the rat, *J. Dental Res.*, *53*: 1398–1403 (1974).

7. G. K. Stookey, J. L. McDonald, S. B. Hughes, R. E. Smith, and R. D. Stange, The influence of tin (II) and oat hulls upon dental caries in the rat, *Arch. Oral Biol.*, *19*: 107–112 (1974).

8. N. Tinanoff, J. M. Brady, and A. Gross, The effect of NaF and SnF_2 mouthrinses on bacterial colonization of tooth enamel. TEM and SEM studies, *Caries Res.*, *10*: 415–426 (1976).

9. G. A. Ferretti, J. M. Tanzer, and N. Tinanoff, The effect of fluoride and stannous ions on *Striptoccoccus mutans*, *Caries Res.*, *16*: 298–307 (1982).

10. J. C. Sherlock, and G. A. Smart, Tin in foods and the diet, *Food Additives and Contaminants*, *1*: 277–282 (1984).

11. Committee on the GRAS List Survey—Phase III, The 1977 Survey of Industry on the Use of Food Additives. National Academy of Sciences, Washington, D.C., 1979, pp. 473, 1786, 1789, 1793, 1918, 1922, 1923, 1955.

12. W. A. Sistrunk, and H. L. Gascoigne, Stability of color in 'Concord' grape juice and expression of color, *J. Food Sci.*, *48*: 430–433, 440 (1983).

13. J. Kumpulainen, and P. Koivistoinen, Advances in tin compound analysis with special reference to organotin pesticide residues, *Residue Rev.*, *66*: 1–18 (1977).

14. Safe Drinking Water Committee, Drinking Water and Health. National Academic of Sciences, Washington, D.C., 1977, pp. 292–296.

15. J. L. Greger, and M. Baier, Tin and iron content of canned and bottled food, *J. Food Sci.*, *46*: 1751–1754, 1765 (1981).

16. G. W. Monier-Williams, Tin in *Trace Elements in Food*, John Wiley and Sons, Inc., New York, 1949, pp. 138–161.

17. S. Nagy, R. Rouseff, and S. V. Ting, Effects of temperature and storage on the iron and tin contents of commercially canned single-strength orange juice, *J. Agric. Food Chem.*, *28*: 1166–1169 (1980).

18. M. L. Woolfer, and W. Manu-Tawiak, Tin content of canned evaporated milk manufactured in West Africa, *Ecol. Food & Nutr.*, *6*: 133–135 (1977).

19. D. H. Calloway, and J. J. McMullen, Fecal excretion of iron and tin by men fed stored canned foods, *Am. J. Clin. Nutr.*, *18*: 1–5 (1966).

20. D. R. Davis, C. W. Cockrell, and K. J. Wiese, Can pitting in green beans: relation to vacuum, pH, nitrate, phosphate, copper and iron content, *J. Food Sci.*, *45*: 1411–1415 (1980).

21. S. G. Capar, Changes in lead concentration of foods stored in their opened cans, *J. Food Safety*, *1*: 241−245 (1978).
22. S. G. Capar, and K. W. Boyer, Multielement analysis of foods stored in their opened cans, *J. Food Safety*, *2*: 105−118 (1980).
23. S. G. Capar, and K. W. Boyer (1973) Trace Elements in Human Nutrition, World Health Organization Tech. Rep. Series No. 532, Geneva, 1973, pp. 38−39.
24. I. H. Tipton, P. L. Steward, and J. Dickson, Patterns of elemental excretion in long term balance studies, *Health Phys.*, *16*: 455−462 (1969).
25. R. A. Kehoe, J. Cholak, and R. V. Story, A spectrochemical study of the normal ranges of concentration of certain trace metals in biological materials, *J. Nutr.*, *19*: 579−592 (1940).
26. J. L. Greger, Newer understanding of tin metabolism, *Int. Med. for the Specialist*, *5*: 173−178 (1984).
27. J. A. T. Pennington, Revision of the Total Diet Study food list and diets, *J. Am. Dietet. Assoc.*, *82*: 166−173 (1983).
28. R. A. Hiles, Absorption, distribution and excretion of inorganic tin in rats, *Toxicol. & Appl. Pharmacol.*, *27*: 366−379 (1974).
29. P. Fritsch, G. deSaint Blanquat, and R. Derache, Effect of various dietary components on absorption and tissue distribution of orally administered inotganic tin in rats, *Fd. Cosmet. Toxicol.*, *15*: 147−149 (1977).
30. J. L. Greger, and M. A. Johnson, Effect of dietary tin on zinc, copper, and iron utilization by rats, *Fd. Cosmet. Toxicol.*, *19*: 163−166 (1981).
31. M. A. Johnson, and J. L. Greger, Tin, copper, iron and calcium metabolism of rats fed various dietary levels of inorganic tin and zinc, *J. Nutr.*, *115*: 615−624 (1985).
32. M. A. Johnson, and J. L. Greger, Effects of dietary tin on tin and calcium metabolism of adult males, *Am. J. Clin. Nutr.*, *35*: 655−660 (1982).
33. S. Kojima, K. Saito, and M. Kiyozumi, Studies on poisonous metals: IV Absorption of stannic chloride from rat alimentary tract and effect of various food components on its absorption, *Yakugaku Zasshi*, *98*: 495−502 (1978).
34. H. M. Perry, and E. J. Perry, Normal concentrations of some trace metals in human urine: changes produced by ethylenediaminetetra acetate, *J. Clin. Invest.*, *38*: 1452−1463 (1959).
35. S. G. Schäfer, and U. Femfert, Tin-A toxic heavy metal? A review of the literature, *Regulatory Toxicol. & Pharmacol.*, *4*: 57−69 (1984).
36. H. A. Schroeder, and J. J. Balassa, Arsenic germanium, tin and vanadium in mice: effects on growth, survival and tissue levels, *J. Nutr.*, *92*: 245−252 (1967).

37. M. Yamaguchi, R. Saito, and S. Okada, Dose-effect of inorganic tin on biochemical indices in rats, *Toxicology, 16*: 267–273 (1980).
38. R. C. Theuer, A. W. Mahoney, and H. P. Sarett, Placental transfer of fluoride and tin in rats given various fluoride and tin salts, *J. Nutr., 101*: 525–532 (1971).
39. L. L. Nunnelley, W. R. Smythe, A. C. Alfrey, and L. S. Ibels, Uremic hyperstannum; elevated tissue tin levels associated with uremia, *J. Lab Clin. Med., 91*: 72–75 (1978).
40. R. D. Kimbrough, Toxicity and health effects of selected organotin compounds: a review, *Env. Health Perspectives, 14*: 51–56 (1976).
41. P. Mushak, M. R. Krigman, and R. R. Mailman, Comparative organotin toxicity in the developing rat: somatic and morphological changes and relationship to accumulation of total tin, *Neurobehav. Toxicol. Teratol., 4*: 209–215 (1982).
42. M. R. Krigman, and A. P. Silverman, General toxicology of tin and its organic compounds, *Neurotoxic, 5*: 129–140 (1984).
43. R. E. Squibb, N. G. Carmichael, and H. A. Tilson, Behavioral and neuromorphological effects of triethyl tin bromide in adult rats, *Toxicol. & Appl. Pharmacol., 55*: 188–197 (1980).
44. R. S. Dyer, T. J. Walsh, W. F. Wonderlin, and M. Bercegeay, The trimethytin syndrom in rats, *Neurobehav. Toxicol. Teratol., 4*: 127–133 (1982).
45. L. L. Cook, K. E. Stine, and L. W. Reiter, Tin distribution in adult rat tissues after exposure to trimethyltin and triethytin, *Toxicol. & Appl. Pharmacol., 76*: 344–348 (1984).
46. N. F. Cardarelli, B. M. Quitter, A. Allen, E. Dobbins, E. P. Libby, P. Hager, and L. R. Sherman, Organotin implications in anti carcinogenesis background and thymus involvement, *Aust. J. Exp. Biol. Med. Sci., 62*: 199–208 (1984).
47. A. Saxena and J. P. Tandon, Antitumor activity of some diorganotin and tin (iv) complexes of schiff bases, *Cancer Lett., 19*: 73–76 (1983).
48. L. E. Hallas, J. C. Means, and J. J. Cooney, Methylation of tin by estuarine microorganisms, *Science, 215*: 1505–1507 (1982).
49. S. Warburton, W. Udler, R. M. Ewert, and W. S. Haynes, Outbreak of foodborne illness attributed to tin, *U.S. Public Health Rep., 77*: 798–800 (1962).
50. C. J. Benoy, P. A. Hooper, and R. Schneider, The toxicity of tin in canned fruit juices and solid foods, *Fd. Cosmet. Toxicol., 9*: 645–656 (1971).
51. W. H. Barker Jr., and V. Runte, Tomato juice-associated gastroenteritis, Washington and Oregon, 1969, *Am. J. Epidemiol. 96*: 219–226 (1977).
52. A. P. deGroot, Subacute toxicity of inorganic tin as influenced by dietary levels of iron and copper, *Fd. Cosmet. Toxicol., 11*: 955–962 (1973).

53. A. P. deGroot, V. J. Feron, and H. P. Til, Short-term toxicity studies on some salts and oxides of tin in rats, *Fd. Cosmet. Toxicol.*, *11*: 19–30 (1973).
54. M. A. Johnson, and J. L. Greger, Absorption distribution and endogenous excretion of zinc by rats fed various dietary levels of inorganic tin and zinc, *J. Nutr.*, *114*: 1843–1852 (1984).
55. H. C. Dreef-Van Der Meulen, V. J. Feron, and H. P. Til, Pancreatic atrophy and other pathological changes in rats following feeding of stannous chloride, *Path. Europ.*, *9*: 185–192 (1974).
56. J. R. N. McLean, H. C. Birnboim, R. Pontefact, and J. G. Kaplan, The effect of tin chloride on the structure and function of DNA in human white blood cells, *Chem. Biol. Interactions*, *46*: 189–200 (1983).
57. J. R. N. McLean, D. H. Blakey, G. R. Douglas, and J. G. Kaplan, The effect of stannous and stannic (tin) chloride on DNA in Chinese hamster ovary cells, *Mutation Res.*, *119*: 195–201 (1983).
58. J. V. Burba, Inhibition of hepatic azo-reductase and aromatic hydroxylase by radiopharmaceuticals containing tin, *Toxicol. Lett.*, *18*: 269–272 (1983).
59. A. Kappas, and M. D. Maines, Tin: a potent inducer of heme oxygenase in kidney, *Science*, *192*: 60–62 (1976).
60. R. K. Kutty, and M. D. Maines, Effects of induction of heme oxygenase by cobalt and tin on the in vivo degradation of myoglobin, *Biochem. Pharmacol.*, *33*: 2924–2926 (1984).
61. M. Chiba, K. Ogihara, and M. Kikuchi, Effect of tin on porphyrin biosynthesis, *Arch. Toxicol.*, *45*: 189–195 (1980).
62. G. Zareba, and Chmielnicka, Aminolevulinic acid dehydratase activity in the blood of rats exposed to tin and zinc, *Ecotoxicol. Environ. Safety*, *9*: 40–46 (1985).
63. M. Chiba, and M. Kikuchi, The in vitro effects of zinc and manganese on δ-aminolevulinic acid dehydratase activity inhibited by lead or tin, *Toxicol. & Appl. Pharmacol.*, *73*: 388–394 (1984).
64. N. V. Dimitrov, C. Meyer, F. Nakhas, C. Miller, and B. A. Averill, Effect of tin on immune responses of mice, *Clin. Immunol. Immunopathol.*, *20*: 39–48 (1981).
65. S. Levine, and R. Sowinski, Tin salts prevent the plasma cell response to metallic tin in Lewis rats, *Toxicol. Appl. Pharmacol.*, *68*: 110–115 (1983).
66. S. Levine, R. Sowinski, and S. Koulish, Plasmacellular and granulomatous splenomegaly produced in rats by tin, *Expt. Molec. Pathol.*, *39*: 364–376 (1983).
67. O. Hayashi, M. Chiba, and M. Kikuchi, The effect of stannous chloride on the humoral immune response of mice, *Toxicol. Lett.*, *21*: 279–285 (1984).

68. W. R. Beisel, Single nutrients and immunity, *Am. J. Clin. Nutr.*, *35*: 417–468 (1982).

69. J. Chmielnicka, J. A. Szymanska, and J. Sniecv, Distribution of tin in the rat and disturbances in the metabolism of zinc and copper due to repeated exposure to $SnCl_2$, *Arch. Toxicol.*, *47*: 263–268 (1981).

70. M. A. Johnson, and J. L. Greger, Effects of dietary tin on zinc, copper, iron, manganese and magnesium metabolism of adult males, *Am. J. Clin. Nutr.*, *35*: 1332–1338 (1982).

71. L. S. Valberg, P. R. Flanagan, and M. J. Chamberlain, Effects of iron, tin, and copper on zinc absorption in humans, *Am. J. Clin. Nutr.*, *40*: 536–541 (1984).

72. N. W. Solomons, J. S. Marchini, R. M. Duarte-Favaro, H. Vannuchi, and J. E. Dutra de Oliveira, Studies on the bioavailability of zinc in humans: intestinal interaction of tin and zinc, *Am. J. Clin. Nutr.*, *37*: 566–571 (1983).

73. C. H. Hill, and G. Matrone, Chemical parameters in the study on in vivo and in vitro interactions of transition elements, *Federation Proceed.*, *29*: 1474–1481 (1970).

74. J. L. Greger, S. A. Smith, M. A. Johnson, and M. J. Baier, Effects of dietary tin and aluminum on selenium utilization by adult males, *Biol. Tr. Element. Res.*, *4*: 269–278 (1982).

75. M. Chiba, N. Fujimoto, and M. Kikuchi, Protective effect of selenium on the inhibition of erythrocyte 5-aminolevulinate dehydratose activity by tin, *Toxicol. Lett.*, *24*: 235–241 (1985).

76. M. Yamaguchi, K. Sugii, and S. Okada, Inorganic tin in the diet affects the femur in rats, *Toxicol. Lett.*, *9:* 207–209 (1981).

77. M. Yamaguchi, K. Sugii, and S. Okada, Tin decreases femoral calcium independently of calcium homeostasis in rats, *Toxicol. Lett.*, *10*: 7–10 (1982).

78. T. Yamamoto, M. Yamaguchi, and H. Sato, Accumulation of calcium in kidney and decrease of calcium in serum of rats treated with tin chloride, *J. Toxicol. & Env. Health*, *1*: 749–756 (1976).

79. M. Yamaguchi, and Yamamoto, Effect of tin on calcium content in the bile of rats, *Toxicol. Appl. Pharmacol.*, *45*: 611–616 (1978).

80. K. Ogoshi, N. Kurumatani, Y. Aoki, T. Moriyama, and Y. Nanzai, Decrease in compressive strength of the femoral bone in rats administered stannous chloride for a short period, *Toxicol. Appl. Pharmacol.*, *58*: 331–332 (1981).

81. M. Yamaguchi, K. Sugii, and S. Okada, Inhibition of collagen synthesis in the femur of rats orally administered stannous chloride, *J. Pharm. Dyn.*, *5*: 388–393 (1982).

82. K. M. Hambidge, P. A. Walravens, R. M. Brown, J. Webster, S. White, M. Anthony, and M. L. Roth, Zinc nutrition of preschool children in the Denver Head Start Program, *Am. J. Clin. Nutr.*, *29*: 734–738 (1976).

83. H. H. Sandstead, L. K. Henriksen, J. G. Greger, A. Prasad, and R. A. Good, Zinc nutritive in the elderly in realtion to taste acuity, immune response and wound healing, *Am. J. Clin. Nutr.*, *36*: 1046–1059 (1982).

84. Science and Education Administration, Nationwide Food Consumption Survey 1977–78, Preliminary Rep. No. 2., U.S. Dept. of Agriculture, Washington D.C., 1980, pp. 75.

85. C. J. Kleber, and M. S. Putt, Aluminum and dental caries: a review of the literature, *Clin. Prev. Dentistry, 6*: 14–25 (1984).

86. J. Versieck, and R. Cornelis, Measuring aluminum levels, *New Engl. J. Med.*, *302*: 468 (1980).

87. J. L. Greger, Aluminum content of the American diet, *Food Tech.*, *39*: 73,74,76,78–80 (1985).

88. J. R. J. Sorenson, I. R. Campbell, L. B. Tepper, and R. D. Lingg, Aluminum in the environment and human health, *Env. Health Perspectives, 8*: 3–95 (1974).

89. D. Schlettwein-Gsell, and S. Mommsen-Straub, Spurenelemente in Lebensmitteln. XII Aluminium, *Internat. Z. Vit.-Ern.-Forschung, 43*: 251–263 (1973).

90. T. Eden, Climate and soils, in *Tea*, Longman Group Ltd., London, 1976, pp. 8–15.

91. H. Hopkins, and J. Eisen, Mineral elements in fresh vegetables from different geographic areas, *Agric. Food Chem.*, *7*: 633–638 (1959).

92. R. G. Miller, F. C. Kopfler, K. C. Kelty, J. A. Stober, and N. S. Ulmer, The occurrence of aluminum in drinking water, *J. Am. Water Works Assoc.*, *76*: 84–91 (1984).

93. Safe Drinking Water Committee, Dringing Water and Health, Vol. 4, National Academy Press, Washington, D.C., 1982, pp. 155–177.

94. Committee on Food Additive Survey Data, Poundage Update of Food Chemicals, 1982, PB 84-16214, National Academy Press, Washington, D.C., 1984.

95. S. E. Levick, Dementia from aluminum pots? *New Engl. J. Med.*, *303*: 164 (1980).

96. G. A. Trapp, and J. B. Cannon, Aluminum pots as a source of dietary aluminum, *New Engl. J. Med.*, *304*: 172 (1981).

97. J. L. Greger, W. Goetz, and D. Sullivan, Aluminum levels in foods cooked and stored in aluminum pans, trays and foil, *J. Food Prot.*, *48*: 772–777 (1985).

98. J. H. Koning, Aluminum pots as a source of dietary aluminum, *New Engl. J. Med.*, *304*: 172–173 (1981).

99. A. Lione, P. V. Allen, and J. C. Smith, Aluminum coffee percolators as a source of dietary aluminum, *Fd. Chem. Toxicol.*, *22*: 265–268 (1984).

100. P. Mattsson, Aluminum from cooking vessels, *Var Foda, 33*: 231–236 (1981).

101. R. Ondreička, J. Kortus, and E. Ginter, Aluminum, its absorption, distribution and effects on phosphorus metabolism, in *Intestinal Absorption of Metal Ions, Trace Elements, and Radionuclides* (S. C. Skoryna and D. Waldron-Edward, eds.), Permagon Press, Oxford, 1971, pp. 293–305.

102. C. F. Poe, and J. M. Leberman, The effect of acid foods on aluminum cooking utensils, *Food Tech.*, *3*: 71–74 (1949).

103. A. Lione, Letters to the editor, *Nutri. Rev.*, *42*: 31 (1984).

104. A. Lione, The prophylactic reduction of aluminum intake, *Fd. Chem. Toxicol.*, *21*: 103–109 (1983).

105. A. Lione, Aluminum intake form non-prescription drugs and sucralfate, *Gen. Pharmac.*, *16*: 223–228 (1985).

106. E. M. Clarkson, V. A. Luck, W. V. Hynson, R. R. Bailey, J. B. Eastwood, J. S. Woodhead, V. R. Clements, J. L. H. O'Riordan, and H. E. deWardener, The effects of aluminum hydroxide on calicum, phosphorus and aluminum balances, the serum parathyroid hormone concentration and the aluminum content of bone in patients with chronic renal failure, *Clin. Sci.*, *43*: 519–531 (1972).

107. J. M. Cam, V. A. Luck, J. B. Eastwood, and H. E. deWardener, The effect of aluminum hydroxide orally on calcium, phosphorus and aluminum metabolism in normal subjects, *Clin. Sci. Molec. Med.*, *51*: 407–414 (1976).

108. J. E. Gorsky, A. A. Dietz, H. Spencer, and D. Osis, Metabolic balance of aluminum studies in six men, *Clin. Chem.*, *25*: 1739–1743 (1979).

109. J. L. Greger, and M. J. Baier, Excretion and retention of low or moderate levels of aluminum by human subjects, *Fd. Chem. Toxic.*, *21*: 473–477 (1983).

110. K. Mackenzie, The biochemistry of aluminum. I Excretion and absorption of aluminum in the pig, *Biochem. J.*, *24*: 1433–1441 (1930).

111. K. Mackenzie, The biochemistry of aluminum. II Excretion and absorption of aluminum in the rat, *Biochem. J.*, *25*: 287–291 (1931).

112. F. P. Underhill, and F. I. Peterman, Studies in the metabolism of aluminum, *Am. J. Physiol.*, *90*: 40–61 (1929).

113. W. D. Kaehny, A. P. Hegg, and A. C. Alfrey, Gastrointestinal absorption of aluminum from aluminum-containing antacids, *New Engl. J. Med.*, *296*: 1389–1390 (1977).

114. R. R. Recker, A. J. Blotcky, J. A. Leffler, and E. P. Rack, Evidence for aluminum absorption from the gastrointestinal tract and bone deposition by aluminum carbonate ingestion with normal renal function, *J. Lab. Clin. Med.*, *90*: 810–815 (1977).

115. M. T. Kovalchik, W. D. Kaehny, A. P. Hegg, J. T. Jackson, and A. C. Alfrey, Aluminum kinetics during hemodialysis, *J. Lab. Clin. Med.*, *92*: 712–720 (1978).

116. J. L. Greger, E. N. Bula, and E. T. Gum, Mineral metabolism of rats fed moderate levels of various aluminum compounds for short periods of time, *J. Nutr.*, *115*: 1708–1716 (1985).
117. G. H. Mayor, J. A. Keiser, D. Makdani, and P. K. Ku, Aluminum absorption and distribution; effect of parathyroid hormone, *Science*, *197*: 1187–1189 (1977).
118. S. P. Andreoli, J. M. Bergstein, and D. J. Sherrard, Aluminum intoxication from aluminum-containing phosphate binders in children with azotemia not undergoing dialysis, *New Engl. J. Med.*, *310*: 1079–1084 (1984).
119. A. C. Katz, D. W. Frank, M. W. Sauerhoff, G. M. Zwicker, and R. I. Freudenthal, A six-month dietary toxicity study of acidic sodium aluminum phosphate in beagle dogs, *Fd. Chem. Toxic.*, *22*: 7–9 (1984).
120. P. Slanina, Y. Falkiborn, W. Frech, and A. Cedergren, Aluminum concentrations in the brain and bone of rats fed citric acid, aluminum citrate or aluminum hydroxide, *Fd. Chem. Toxic.*, *22*: 391–397 (1984).
121. P. Slanina, W. Frech, A. Bernhardson, A. Cedergren, and P. Mattsson, Influence of dietary factors on aluminum absorption and retention in the brain and bone fo rats, *Acta Pharmacol et Toxicol.*, *56*: 331–336 (1985).
122. N. L. Storer, and T. S. Nelson, The effect of various aluminum compounds on chick performance, *Poult. Sci.*, *47*: 244–247 (1968).
123. M. Feinroth, M. V. Feinroth, and G. M. Berlyne, Aluminum absorption in the rat everted gut sac, *Mineral Electrolyte Metab.*, *8*: 29–35 (1982).
124. J. Van der Meulen, P. D. Bezmer, P. Lips, and P. L. Oe, Individual differences in gastrointestinal absorption of aluminum, *New Engl. J. Med.*, *310*: 1322 (1984).
125. F. H. Nielsen, Effect of boron nutriture on the response of rats to high dietary aluminum, in *Trace Substances in Environmental Health XVIII* (D. D. Hemphill ed.), University of Missouri, Columbia, 1984, pp. 47–52.
126. J.L. Greger, E. T. Gum, and E. N. Bula, Mineral metabolism of rats fed various level of aluminum hydroxide, *Biol. Trace. Ele. Res.*, *9*: 67–77 (1986).
127. R. Ondreička, E. Ginter, and J. Kortus, Chronic toxicity of aluminum in rats and mice and its effects on phosphorus metabolism, *Brit. J. Industr. Med.*, *23*: 305–312 (1966).
128. R. Valdivia, C. B. Ammerman, P. R. Henry, J. P. Feaster, and C. J. Wilcox, Effect of dietary aluminum and phosphorus on performance, phosphorus utilization and tissue mineral composition in sheep, *J. Am. Sci.*, *55*: 402–410 (1982).

129. R. Valdivia, C. B. Ammerman, C. J. Wilcox, and P. R. Henry, Effect of dietary aluminum on animal performance and tissue mineral levels in growing steers, *J. Am. Sci.*, *47*: 1351–1356 (1978).

130. A. C. Alfrey, G. R. LeGendre, and W. D. Kaehny, The dialysis encephalopathy syndrome: possible aluminum intoxications, *New Engl. J. Med.*, *294*: 184–188 (1976).

131. S. W. King, J. Savory, and M. R. Wills, The clinical biochemistry of aluminum, *CRC Crit. Rev. Clin. Lab. Sci.*, *14*: 1–20 (1981).

132. I. S. Parkinson, T. G. Feest, M. K. Ward, R. W. P. Fawcett, D. N. S. Kerr, Fracturing dialysis osteodystrophy and dialysis encephalopathy, *Lancet*, *I*: 406–409 (1979).

133. M. T. Schreeder, M. S. Favero, J. R. Hughes, N. J. Peterson, P. H. Bennett, and J. E. Maynard, Dialysis encephalopathy and aluminum exposure: an edipemiologic analysis, *J. Chron. Dis.*, *36*: 581–593 (1983).

134. P. S. Smith, J. McClure, Localisation of aluminum by histochemical and electron probe x-ray microanalytical techniques in bone tissue of cases of renal osteodystrophy, *J. Clin. Pathol.*, *35*: 1283–1293 (1982).

135. B. Ihle, M. Buchanan, B. Stevens, A. Marshal, R. Plomley, A. d'Apice, and P. Kincaid-Smith, Aluminum associated bone disease: clinico-pathlogic correlation, *Am. J. Kid. Dis.*, *II*: 255–263 (1982).

136. N. A. Maloney, S. M. Ott, A. C. Alfrey, N. L. Miller, J. W. Coburn, and D. J. Sherrard, Histological quantitation of aluminum in iliac bone from patients with renal failure, *J. Lab. Clin. Med.*, *99*: 206–216 (1982).

137. A. H. Verbucken, F. Van de Vyver, R. E. VanGrieken, G. J. Paulus, W. J. Visser, P. D. D'Haese, and M. C. DeBroe, Ultrastructural localization of aluminum in patients with dialysis-associated osteomalacia, *Clin. Chem.*, *30*: 763–768 (1984).

138. J. D. McKinney, M. Basinger, E. Dawson, and M. M. Jones, Serum aluminum levels in dialysis dementia, *Nephron*, *32*: 53–56 (1982).

139. J. G. Heaf, L. P. Nielsen, Serum aluminum in haemodialysis patients: relation to osteodystrophy, encephalopathy and aluminum hydroxide consumption, *Mineral Electrolyte Metab.*, *10*: 345–350 (1984).

140. I. B. Salusky, J. W. Coburn, L. Paunier, D. J. Sherrard, and R. N. Fine, Role of aluminum hydroxide in raising serum aluminum levels in children undergoing continuous ambulatory peritoneal dialysis, *J. Pediatrics*, *105*: 717–720 (1984).

141. E. Nathan, and S. E. Pederson, Dialysis encephalopathy in a non-dialyzed uraemic boy treated with aluminum hydroxide orally, *Acta Paediatr Scand*, *69*: 793–796 (1980).

142. W. R. Griswold, V. Reznik, S. A. Mendoza, D. Trauner, and A. C. Alfrey, Accumulation of aluminum in a nondialyzed uremic child receiving aluminum hydroxide, *Pediatrics, 71*: 56–58 (1983).

143. M. E. Randall, Aluminum toxicity in an infant not on dialysis, *Lancet, I*: 1327–1328 (1983).

144. A. B. Sedman, N. L. Miller, B. A. Warady, G. M. Lum, and A. C. Alfrey, Aluminum loading in children with chronic renal failure, *Kidney Internat., 26*: 201–204 (1984).

145. A. B. Sedman, G. N. Wilkening, B. A. Warady, G. M. Lum, and A. C. Alfrey, Encephalopathy in childhood secondary to aluminum toxicity, *J. Pediatrics 105*: 836–838 (1984).

146. M. Freundlich, C. Abitbol, G. Zilleruelo, J. Strauss, M. C. Faugere, and H. H. Mallucke, Infant formula as a case of aluminum toxicity in neonatal uraemia, *Lancet, II*: 527–529 (1985).

147. P. Ackrill, A. J. Ralston, J. P. Day, and K. C. Hodge, Successful removal of aluminum from patient with dialysis encephalopathy, *Lancet, II*: 692–693 (1980).

148. H. H. Malluche, A. J. Smith, K. Abreo, and M. C. Faugere, The use of deferoxamine in the management of aluminum accumulation in bone in patients with renal failure, *New Engl. J. Med., 311*: 140–144 (1984).

149. D. J. Brown, K. N. Ham, J. K. Daeborn, and J. M. Xipell, Treatment of dialysis osteomalacia with desferrioxamine, *Lancet, II*: 343–345 (1982).

150. G. L. Klein, A. C. Alfrey, N. L. Miller, D. J. Sherrard, T. K. Hazlet, M. E. Ament, and J. W. Coburn, Aluminum loading during total parenteral nutrition, *Am. J. Clin. Nutr., 35*: 1425–1429 (1982).

151. S. M. Ott, N. A. Maloney, G. L. Klein, A. C. Alfrey, M. E. Ament, J. W. Coburn, and D. J. Sherrard, Aluminum is associated with low bone formation in patients receiving chronic parenteral nutrition, *Ann. Int. Med., 98*: 910–914 (1983).

152. M. C. deVernejoul, B. Messing, D. Modrowski, J. Bielakoff, A. Buisine, and L. Miravet, Multifactoral low remodeling bone disease during cyclic total parenteral nutrition, *J. Clin. Endocrinol. Metab., 60*: 109–112 (1985).

153. A. B. Sedman, G. L. Klein, R. J. Merritt, N. L. Miller, K. O. Weber, W. L. Gill, H. Anand, and A. C. Alfrey, Evidence of aluminum loading in infants receiving intravenous therapy, *New Engl. J. Med., 312*: 1337–1343 (1985).

154. D. S. Milliner, J. H. Shinaberger, P. Shuman, and J. W. Coburn, Inadvertent aluminum administration during plasma exchange due to aluminum contamination of albumin-replacement solutions, *New Engl. J. Med., 312*: 165–167 (1985).

155. J. R. McDermott, I. Smith, K. Igbal, and H. M. Wisniewski,

Brain aluminum in aging and Alzheimer disease, *Neurology, 29*: 809—814 (1979).

156. W. R. Markesbery, W. D. Ehmann, T. I. M. Hossain, M. Alauddin, and D. T. Goodin, Instrumental neutron activation analysis of brain aluminum in Alzheimer disease and aging, *Ann. Neurol., 10*: 511—516 (1981).

157. D. R. Crapper, S. S. Krishnan, and S. Quittkat, Aluminum neurofibrillary degeneration and Alzheimer's disease, *Brain, 99*: 67—80 (1976).

158. N. P. Perl, and A. R. Brody, Alzeimer's disease: x-ray spectrometric evidence of aluminum accumulation in neurofibrillary tangle-bearing neurons, *Science, 208*: 297—299 (1980).

159. R. D. Traub, T. C. Rains, R. M. Garruto, D. C. Gajdusek, and C. J. Gibbs Jr., Brain destruction alone does not elevate brain aluminum, *Neurology 31*: 986—990 (1981).

160. D. R. Crapper-McLachlan, B. Farnell, H. Galin, S. Karlik, G. Eichhorn, and U. DeBoni, Aluminum in human brain disease, in *Biological Aspects of Metals and Metal-Related Disease* (B. Sarkar, ed.), Raven Press, New York, 1983, pp. 209—218.

161. D. Shore, M. Millson, J. L. Holtz, S. W. King, T. P. Bridge, and R. J. Wyatt, Serum aluminum in primary degenerative dementia, *Biol. Psych., 15*: 971—977 (1980).

162. D. R. Crapper, S. Quittkat, S. S. Krishnan, A. J. Dalton, U. DeBoni, Intranuclear aluminum content in Alzheimer's disease, dialysis encepalopathy, and experimental aluminum encephalopathy, *Acta Neuropathol., 50*: 19—24 (1980).

163. D. P. Perl, D. C. Gajdusek, R. M. Garruto, R. T. Yanagihara, and C. J. Gibbs Jr., Intraneuronal aluminum accumulation in amyotrophic lateral sclerosis and Parkinsonism-dementia of Guam, *Science, 217*: 1053—1055 (1982).

164. R. M. Garruto, R. Fukatsu, R. Yanagihara, D. C. Gajdusek, G. Hook, and C. E. Fiori, Imaging of calicum and aluminum in neurofibibrillary tangle-bearing neurons in parkinsonism-demential of Guam, *Proc. Natl. Acad. Sci., 81*: 1875—1879 (1984).

165. I. V. Rosa, P. R. Henry, and C. B. Ammerman, Interrelationship of dietary phosphorus, aluminum and iron on performance and tissue mineral composition in lambs, *J. An. Sci., 55*: 1231—1240 (1982).

166. H. Spencer, L. Kramer, C. Norris, and D. osis, Effect of small doses of aluminum-containing antacids on calcium and phosphorus metabolism, *Am. J. Clin. Nutr., 36*: 32—40 (1982).

167. M. Lotz, E. Zisman, and F. C. Bartter, Evidence for a phosphorus-depletion syndroe in man, *New Engl. J. Med., 278*: 409—415 (1968).

168. C. E. Dent, and C. S. Winter, Osteomalacia due to phosphate depletion from excessive aluminum hydroxide ingestion, *Br. Med. J., 1*: 551—552 (1974).

169. K. L. Insogna, D. R. Bordley, J. F. Caro, and D. H. Lockwood, Osteomalacia and weakness from excessive antacid ingestion, *J. Am. Med. Assoc.*, *244*: 2544–2546 (1980).
170. J. L. Greger, and M. J. Baier, Effect of dietary aluminum on mineral metabolism of adult males, *Am. J. Clin. Nutr.*, *38*: 411.–419 (1983).
171. H. Spencer, and L. Kramer, Antacid-induced calcium loss, *Arch. Internal. Med.*, *143*: 657–659 (1983).
172. B. F. Boyce, H. Y. Elder, H. L. Elliot, I. Fogelman, G. S. Fell, B. J. Junors, G. Beastall, and I. T. Boyle, Hypercalcaemic osteomalacia due to aluminum toxicity, *Lancet*, *II*: 1009–1113 (1982).
173. J. B. Cannata, B. J. R. Junor, J. D. Briggs, G. S. Fell, and G. Beastall, Effect of acute aluminum overload on calcium and parathyroid hormone metabolism, *Lancet*, *I*: 501–503 (1983).
174. C. K. Biswas, R. S. Arze, J. M. Ramos, M. K. Ward, J. H. Dewar, D. N. S. Kerr, and D. H. Kenward, Effect of aluminum hydroxide on serum ionised calcium, immunoreactive parathyroid hormone and aluminum in chronic renal failure, *Brit. Med. J.*, *284*: 776–778 (1982).
175. D. J. Jerry, C. H. Noller, and L. J. Wheeler, Effect of solubility of aluminum compounds on blood serum calcium, phosphorus and magnesium and fecal pH, *J. An. Sci.*, *57* (*suppl*): 445 (1983).
176. P. Herzog, K. F. Schmitt, T. Grendahl, and J. vanderLinden, Evaluation of serum and urine electrolyte changes during therapy with magnesium-aluminum containing antacid: results of a prospective study, in *Antacids in the Eighties* (F. Halter, ed.), Urban & Schwarzenberg, Munich, 1982, pp. 123–135.
177. A. N. Said, P. Slagsvold, H. Bergh, and B. Laksesvela, High fluorine water to wether sheep maintained in pens, *Nord. Vet. Med.*, *29*: 172–180 (1977).
178. H. Spencer, L. Kramer, C. Norris, and E. Wiatrowski, Effect of aluminum hydroxide on fluoride metabolism, *Clin. Pharmacol. Ther.*, *28* 529–535 (1980).
179. H. Spencer, L. Kramer, C. Norris, and E. Wiatrowski, Effect of aluminum hydroxide on plasma fluoride and fluoride excretion during a high fluoride intake in man, *Toxicol. Appl. Pharmacol.*, *38*: 140–144 (1981).
180. S. Freeman, and A. C. Ivy, The influence of antacids upon iron retention by the anemic rats, *Am. J. Physiol.*, *137*: 706–709 (1942).
181. B. S. Skikne, S. R. Lynch, and J. D. Cook, Role of gastric acid in food iron absorption, *Gastroenterology*, *81*: 1068–1071 (1981).
182. Conditioned copper deficiency due to antacids, *Nutr. Rev.*, *42*: 319–321 (1984).

183. A. Blumberg, Der Einfluss aluminiumhydroxyd-haltiger Antazida auf die enterale Eisenresportion von Langzeitdialysepatienten, *Schweiz. Med. Wschr.*, *107*: 1064–1066 (1977).

184. A. Haug, Molecular aspects of aluminum toxicity, *Crit. Rev. Plant Sci.*, *1*: 345–373 (1984).

185. A. Kushelevsky, R. Yaght, Z. Alfasi, and M. Berlyne, Uptake of aluminum ion by the liver, *Biomedicine*, *25*: 59–60 (1976).

186. S. J. Karlik, G. L. Eichhorn, P. N. Lewis, and D. R. Crapper, Interaction of aluminum species with deoxyribonucleic acid, *Biochem.*, *19*: 5991–5998 (1980).

187. D. M. Smith Jr., J. A. Pitcock, and W. M. Murphy, Aluminum-containing dense deposits of the glomerular basement membrane, *Am. J. Clin. Pathol.*, *77*: 341–346 (1982).

188. H. Mazarguil, R. Haran, and J. P. Laussac, The binding of aluminum to [Leu^5]-enkephalin: an investigation using ^{1}H, ^{13}C and ^{27}Al NMR spectroscopy, *Biochem. Biophys Acta*, *717*: 465–472 (1982).

189. N. Siegel, and A. Haug, Aluminum interaction with calmodulin, *Biochem. Biophys. Acta*, *744*: 36–45 (1983).

190. P. C. Sternweis, and A. G. Gilman, Aluminum: arequirement for activation of the regulatory component of adenylate cyclase by fluoride, *Proc. Natl. Acad. Sci. USA*, *79*: 4888–4891 (1982).

191. J. K. Marquis, Aluminum inhibition of human serum cholinesterase, *Bull. Environ. Contam. Toxicol.*, *31*: 164–169 (1983).

192. G. Cournot-Witmer, J. Zingraff, J. J. Plachat, F. Escaig, R. Lefevere, P. Boumati, A. Bourdeau, M. Garabedian, P. Galle, R. Bourdon, T. Drüeke, and S. Balsan, Aluminum localization in bone from hemodialized patients: relationship to matrix mineralization, *Kidney Intl.*, *20*: 375–385 (1981).

193. T. Drüeke, Dialysis osteomalacia and aluminum intoxication, *Nephron.*, *26*: 207–210 (1980).

194. Y. L. Chan, A. C. Alfrey, S. Posen, D. Lissner, E. Hills, C. R. Dunstan, and R. A. Evans, Effect of aluminum on normal and uremic rats: tissue distribution, vitamin D metabolites and quantitative bone histology, *Calcif. Tissue Int.*, *35*: 344–351 (1983).

195. C. R. Dunstan, R. A. Evans, E. Hills, S. Y. P. Wong, and A. C. Alfrey, Effect of aluminum and parathyroid hormone on osteoblasts and bone mineralization in chronic renal failure, *Calcif. Tissue Intl.*, *36*: 133–138 (1984).

196. N. C. Bowdler, D. S. Beasley, E. C. Fritze, A. M. Gaulette, J. D. Hatton, J. Hession, D. L. Ostman, D. J. Rugg, and C. J. Schmittdiel, Behavioral effects of aluminum ingestion on animal and human subjects, *Pharmacol. Biochem. & Behavior*, *10*: 505–512 (1979).

197. V. DeBoni, A. Otvos, J. W. Scott, and D. R. Crapper,

Neurofibrillary degeneration induced systemic aluminum, *Acta neuropath. (Berl.)*, *35*: 285–294 (1976).

198. J. T. Coyle, D. L. Price, and M. R. DeLong, Alzheimer's disease: a disorder of cortical cholinergic innervation, *Science*, *219*: 1184–1190 (1983).

199. B. Hetnarski, H. M. Wisniewski, K. Iqbal, J. D. Dziedzic, and A. Lajtha, Central cholinergic activity in aluminum-induced neurofibrillary degeneration, *Ann. Neurol.*, *7*: 489–490 (1980).

200. D. C. Gajdusek, Hypothesis, interference with axonal transport of neurofilament as a common pathogenetic mechanism in certain diseases of the central nervous system, *New Engl. J. Med.*, *312*: 714–719 (1985).

201. W. A. Banks, and A. J. Kastin, Aluminum increases permeability of the blood-brain barrier to labelled DSIP and Bendorphin: possible implications for senile and dialysis dementia, *Lancet*, *II*: 1127–1229 (1983).

10
Selenium

APRIL C. MASON / Purdue University, West Lafayette, Indiana

INTRODUCTION: THE IMPORTANCE OF SELENIUM
IN HEALTH AND NUTRITION

The importance of selenium in animal nutrition has been known since
1957. Animal feeds are now supplemented with selenium at nutritional
levels, a practice which has virtually eliminated economic losses due
to selenium deficiency diseases. Our understanding of the function
of selenium in human nutrition is still incomplete. The demonstration
of selenium's role in glutathione peroxidase helped to identify selenium
as a nutritionally important trace element. However, the pivotal
discovery was that selenium supplementation could reverse symptoms
of Keshan's disease, a human cardiomyopathy limited to the Keshan
region of China. The discovery that selenium deficiency could lead
to a human disease raised concerns that persons living in low-selenium
areas of the world could be susceptible to selenium deficits. These
deficits might not be manifested as severe deficiency diseases, but
potentially as other nutritionally compromising states.

   Current selenium research concentrates on the identification of
the role of selenium in nutrition. There are two major areas of
research: function and utilization. Functionally, how selenium acts
in the body, with what compounds selenium is associated, and how it
is incorporated are key biochemical questions. More clinically oriented
are the questions of how selenium is absorbed and utilized from food
sources. The two areas greatly overlap. The intent of this chapter

is to describe the biochemistry and bioavailability of selenium in order to understand the relationship of selenium in foods to health and nutrition.

CHEMISTRY OF SELENIUM

Inorganic Chemistry of Selenium

Selenium is in the VIA group of the Periodic Table. It is located between sulfur and tellurium and demonstrates properties common to both while exhibiting unique properties of its own. Selenium has metallic properties similar to tellurium and nonmetallic properties similar to sulfur. Like the other VIA group elements, selenium is allotropic that is, it can exist in different crystalline states (1).

Selenium is one of only a few nonmetals that exhibit variable valence states within the redox range of biological systems (2). The oxidation states of selenium are +6, +4, 0, and -2. In the +6 oxidation state, selenium can be in the form of selenic acid (H_2SeO_3, a strong acid) or a selenate salt (SeO_4^{-2}). In the +4 state selenium exists as selenium dioxide (SeO_2), selenious acid (H_2SeO_3), or a selenite salt (SeO_3^{-2}). Selenite salts are normally less soluble than selenate salts, especially at alkaline pH (3). Selenite is easily reduced at low pH to elemental selenium, whereas selenate is not readily reduced to selenite. The -2 oxidation state of selenium is a selenide form. Hydrogen selenide (H_2Se) is a strong acid.

All four oxidation states of selenium are present in soils. Depending on the pH and aeration of the soil, different selenium forms will predominate. Selenides and elemental selenium are present in acid soils with little aeration. These forms are insoluble and would not be available for plant uptake. Both selenite and selenate salts are found in well-aerated soil. These forms are soluble and readily absorbed by plants.

Comparison of Selenium and Sulfur Chemistry

As members of the same group in the Periodic Table, selenium and sulfur share some chemical properties. The size of the atoms of selenium and sulfur are similar, as are the two elements' bond energies, ionization potentials, and electron affinities. Selenium and sulfur are set apart, however, by comparing their quadrivalent ions. Quadrivalent selenium is in the reduced state, whereas quadrivalent sulfur is oxidized. The strengths of the respective acids for selenium and sulfur are also different. The pKa for H_2Se is 3.73, a stronger acid than H_2S with a pKa of 6.93 (2).

Selenium is a better leaving group than sulfur. With acid hydrolysis (6 N HCl at 110°C), all selenocysteine is destroyed after six

hours of digestion, whereas only a fraction of cysteine is destroyed
(4). It has also been reported that seleno-DL-cystine is six times
as soluble as DL-cystine, although no difference was seen in the
solubilities of selenocysteine and cysteine (4). Another difference in
the amino acid analogues is that, at physiological pH, the side chain
of cysteine is predominately in the sulfhydryl form. In contrast the
side chain of selenocysteine is found as the selenide ion (4).

Measurement of Selenium

The characterization of selenium compounds in living systems has been
hampered by its normally low concentrations and its high chemical
reactivity. Fortunately, the sensitivity of methods used to measure
selenium has increased markedly in recent years. There are numerous
methods used for selenium quantitation, including neutron activation
analysis; fluorescence spectroscopy, atomic absorption spectrometry
with flame, flameless graphite or hydride; inductive coupled plasma;
gas chromatography; X-ray flourescence spectrometry; and differential
pulse cathode stripping voltametry (5). The methods most commonly
used are neutron activation, fluorometry, and atomic absorption
spectroscopy. The technique of gas chromatography is gaining popularity due to improved column packing materials and automated sampling
instruments. The main drawback to fluorometry, atomic absorption
spectroscopy, or gas chromatography remains the lengthy sample
preparation procedures. Table 1 compares the sensitivity of techniques
used to quantitate selenium.

Methods for Increasing Selenium Compound
Stability

Much of the selenium in living systems has been identified in proteins
as selenocysteine. As discussed earlier, this selenium amino acid is
less stable to acid hydrolysis than its sulfur analogue. Proteins must
be subjected to harsh treatment in amino acid analysis studies, therefore methods have been devised to protect the labile selenium group
by blocking it with a stable group. These methods capitalize on the
similarity of the S—H and Se—H bond reactivity.
  Under mild reaction conditions, selenocysteine residues can be
Se-methylated using the method normally used to S-methylate cysteine
(6). The method reacts protein with methyl-p-nitrobenzenesulfonate.
This method and standard reduction and alkylation procedures are
necessary for isolation and characterization studies of selenium-containing proteins (7).
  Selenium contained in the protein glycine reductase isolated from
Clostridium was identified as selenocysteine (8). Selenocysteine
residues were chemically protected by reduction with potassium

Table 1. Most Commonly Used Methods for Determining Selenium in the ppm to ppb Range (5)

Method	Approximate sensitivity (ng)
Neutron activation analysis (NAA)	10
Fluorescence spectroscopy (FS)	2−5
Atomic absorption spectrometry (AAS)	
Flame	500[a]
Flameless (graphite)	0.5[a]
Hydride: flame	2
quartz-cell	0.02[a]
Inductive coupled plasma (ICP)	50[a]
Gas chromatography (GC)	1
X-ray fluorescence spectrometry (XRFS)	2500
XRFS after preconcentration	10
Differential pulse cathode stripping voltametry (DPCSV)	5

[a]Approximate sensitivity per mililiter of sample solution.

borohydride to form the Se-carboxyamido-methyl and Se-aminoethyl derivatives by procedures similar to those used to form the S-alkyl derivatives of cysteine.

In animal systems, glutathione peroxidase also contains selenium as selenocysteine. Identification of selenocysteine in rat liver glutathione peroxidase was made after reduction of the protein with glutathione and dithiothriotol in guanidinium hydrochloride and derivitizing selenocysteine to the carboxymethyl-selenocysteine and aminoethylselenocysteine derivitives (9,10). The selenocysteine residues in ovine erythrocyte glutathione peroxidase were identified after reduction with sodium borohydride, then reaction with 1-fluoro-2,4-dinitrobenzene in 4 M guanidine (11).

As more is learned about the chemistry of selenium, isolation and characterization of selenium-containing compounds will become more routine. Experiments in the future will capitalize on increased sensitivity in selenium measurement and the reactivity of selenium compounds.

SELENIUM IN THE FOOD CHAIN

Availability of Soil Selenium to Plants

Interest in the distribution of selenium in soils was sparked by the discovery that plants containing selenium in high levels were responsible for the toxicity seen in some grazing animals. After selenium was found to be a necessary nutrient for animals and then for man, identification of selenium-rich and -poor soils became important in predicting selenium-deficiency problems.

Soil levels of selenium vary greatly depending on the origin of soil material and the climate conditions in the area (5). Selenium is widely distributed in the Earth's crust, with many areas containing high selenium levels with high sulfur levels. Igneous and sedimentary rock have the greatest concentration of selenium (3). Sedimentary rocks are important as soil-forming materials and therefore contribute greatly to the selenium content of soils. Seleniferous soils in the United States have been linked to sedimentary rocks of the Cretaceous period (3). Normal selenium levels in soils are in the range of 0.1–2.0 μg/gm dry weight. Areas with seleniferous soils contain between 30–324 μg/gm selenium (12).

The concentration of selenium in soils does not, however, adequately estimate the selenium content of plants grown on that soil. Other physical and chemical factors of soil determine the form of selenium in the soil, which in turn affects the uptake of selenium by plants. Such factors include aeration of soil, pH, moisture levels, and level of ferric oxide (13).

The forms of selenium commonly present in soils are selenide, elemental selenium, selenate, selenite, and organic selenium compounds. Under anaerobic soil conditions, selenium will be reduced to selenide and become unavailable for plant uptake. Aerobic conditions favor the oxidation of elemental selenium to the selenite or selenate salts. In alkaline, semiarid soils, selenate is the predominant soil form (14). Selenate is readily available to plants but can be leached from the soil by rainfall. Acidic soils contain selenite. Selenite is available to plants unless bound as a ferric selenite complex in high iron soils (12). Organic selenium compounds are also present in soils as the result of the decomposition of selenium-containing plant tissue (5). These forms are easily taken up by plants.

Seleniferous soils are found in well-drained, subhumid areas where selenate and organic selenium compounds predominate in the soil. These forms of selenium are responsible for the toxic levels seen in some plants (3). Other seleniferous soils may contain the same amount of selenium per gram of soil, but in a form unavailable for plant uptake. Plants grown on such soils will not contain toxic levels of selenium.

Occurrence of Selenium in World Soils

When selenium was first implicated as a toxic agent, a survey of selenium contents of range plants was undertaken to predict areas where selenium toxicity could be a problem. The states of South Dakota, Wyoming, Montana, North Dakota, Nebraska, Kansas, Colorado, Utah, Arizona, and New Mexico contain areas where toxic levels of selenium can accumulate in plants (15). The selenium content of certain plants is a better predictor of soil selenium levels than is that of other plants. Rosenfeld and Beath distinguished these plants by applying the terminology "primary and secondary indicator plants" to plants that accumulate high and intermediate levels of selenium (16). Primary indicator plants accumulate thousands of parts per million selenium and require selenium for growth. Twenty-four species and varieties of the genus *Astragalus* are primary indicator plants which have great diversity of form and extensive geographic distribution. Secondary selenium indicators absorb moderately large levels of selenium (100—500 ppm) when grown on high-selenium soil, but do not require selenium for growth as primary indicators do.

After the identification of selenium deficiency diseases in livestock, it became important to identify not only areas with high selenium soil content but also those areas producing vegetation with low selenium levels. Forage samples from across the United States were analyzed. Areas of very low selenium levels were designated as those areas in which 80% of the forage samples analyzed contained 0.05 ppm selenium or less. Low-level areas were identified as those from which 80% of samples contained 0.10 ppm. Areas of adequate selenium levels were identified as those from which samples contained between 0.1 and 1.0 ppm selenium (17).

In the United States, soils in parts of the Atlantic and Pacific seaboard states and in the Midwest are deficient in available selenium. Globally, selenium-deficient soil is found in New Zealand, Scotland, Finland, Sweden, Denmark, Australia, Germany, Turkey, Greece, China, and Alberta, Canada. Vegetation containing toxic levels of 50 ppm selenium or more has been found in three provinces of Canada, in Mexico, England, Wales, Ireland, Israel, Colombia, Argentina, Venezuela, Nigeria, Kenya, Japan, India, South Africa, and areas of Australia and China (3,5,14). The United States, China, Canada and Australia are unique, in that within each country, areas of both high and low selenium soil levels are found.

Selenium Compounds in Plants

Work on the identification of selenium-containing compounds in plants was initially undertaken in primary indicator plant species. The modified amino acids selenomethylselenocysteine and glutamylselenomethylselenocysteine were isolated from *Astragalus bisulcatus* (18). Sulfur analogues of these peptides were also identified in *A. bisulcatus*, but not in nonaccumulator species of *Astragalus*. Selenomethylselenocysteine was the major acid-soluble organic selenium compound identified in six species of *Astragalus*, *Stanleya pennata*, and *Oonopsis condensata* (19, 20). Selenocystathionine and selenomethylselenomethionine were also identified in many of these plants (21).

As the nutritional importance of selenium became clearer, research in the area of plants switched from indicator species to those plants important to animal and human nutrition. Radioactive selenium (^{75}Se) has been used in studies to follow the absorption, transport, and deposition of selenium in plants. When ^{75}Se was supplied to growing corn as selenite, 80—90% of the label was recovered in the amino acid fraction of the xylem exudate. When ^{75}Se-labeled selenate was supplied to corn, 90% was recovered in the xylem sap as free selenate (22). However, this study did not determine whether the transported form of selenium influenced the form deposited in the seed. When wheat was supplied with ^{75}Se-selenite in hydroponic culture, incorporation of label into the protein fraction of wheat was found (23). One-half of the selenium in naturally seleniferous wheat (31 ppm) has been identified as selenomethionine (24).

The level of selenium accumulated in crop plants can be very high. In our laboratory soybeans grown to maturity hydroponically with 0.1 ppm sodium selenate in the nutrient solution contained 128 ppm selenium. This high level of selenium did not cause toxicity to the plants. In a separate experiment, wheat was grown with 0.1 ppm sodium selenite in the nutrient solution. The grain contained 80 ppm selenium. Interestingly, wheat grown with sodium selenate at the same level accumulated only 10 ppm selenium. This further emphasizes the importance of selenium form to uptake by plants.

Just as the level of selenium in soil does not adequately predict the amount of the mineral in plants, the amount in plants does not predict the absorption and utilization by humans. Plant research in the future should include studies to isolate and purify selenium containing compounds from crop plants as well as the feeding of these products to animals or humans. In this way a complete picture of selenium utilization from food can be determined.

REQUIREMENTS AND ESSENTIALITY

The ninth edition of the Recommended Dietary Allowances (RDAs) published in 1980 (25) included selenium as an essential mineral for humans. A range of selenium intake of 50—200 µg/day was suggested as adequate and safe for adults. No sex differentiation was suggested and the recommendations for other age groups were extrapolated from the adult range depending on expected food consumption. The inclusion of selenium in the RDAs was based primarily on the well-defined essential nature of the mineral in animal species. The adequate and safe range was extrapolated from human balance studies and mammalian animal studies.

Since the 1980 publication of the RDAs, two selenium-deficiency related diseases have been identified in humans: Keshan's disease, a cardiomyopathy prevalent in children, and Kaschin-Beck disease, an osteoarthropathy-causing ossification of affected joints (26–28). Both of these diseases have been observed only in a limited region of China where soil levels of selenium are extremely low and the food consumed by residents is exclusively grown locally. Both diseases are responsive to selenium supplementation. In the case of Keshan's disease, selenium as sodium selenite supplementation decreased morbidity rates from 10% in 1973 to 0% in 1977 for children 1–9 years old. No other selenium deficiency disorders have been identified in free-living individuals outside of China. The occurrence of the disease is linked directly to the soil and food levels of the mineral.

Cases of selenium deficiency have been reported in patients receiving total parenteral nutrition for long periods of time. Case studies include autopsy results showing selenium plasma levels only 7% of normal in a patient who died of cardiomyopathy after eight years of total parenteral formula diet (29); muscle pain and tenderness in a young child on total parenteral nutrition (TPN) for 18 months with very low serum selenium concentration (30); and 6% of normally erythrocyte glutathione peroxidase activity in a patient on TPN for 18 months (31). In the last two cases, signs of deficiency were reversed with selenium supplementation. A short-term prospective study has shown that plasma selenium levels significantly drop after only two weeks on a selenium-deficient TPN formula (32).

With information on selenium-deficiency diseases and cases of deficiency-like symptoms in nutritionally limited patients, the next revision of the RDA will have more human data to rely on when evaluating the established adequate and safe range. Whether the range will be changed or an actual recommended dietary allowance will be established is up to the Committee on Dietary Allowances. The recent human data has supported the committee's decision in 1980 to include selenium as an essential mineral for humans.

SELENIUM IN FOODS

Selenium Content of Food

When discussing the selenium content of foods, one must consider the geographic origin of the product. Generally, some trends of the selenium content of foods can be shown. Using the seven-day diet record method, total selenium intake in a group of Finnish women has

been estimated (33). The main sources of selenium in that diet
were fish and eggs, accounting for 29.5% of the total diet selenium,
cereals (28.1%), milk (18.7%), and meat (13%). The source of
selenium for each of these foods determined the level in the food and
therefore its contribution to the total diet. Selenium levels in Finnish
soil are among the lowest in the world. It is not surprising that meat
and milk provided relatively small percentages of the total selenium
content of the diet. Finland imports much of the total cereal consumed
(34). The contribution of cereal to the selenium content of the overall diet is not, therefore, a reflection of native soil levels of the
mineral. Chicken feeds are supplemented with selenium. The level
of selenium in eggs can be a major contribution to selenium in the
diet, as was shown here, but does not reflect Finnish soil levels of
selenium.

In contrast to the selenium levels in Finnish foods, a study of
foods in Maryland showed that, in the United States, meat and fish
contain the highest selenium levels ($0.1-0.7$ µg/gm) (35). Fruits
and vegetables were generally low in selenium (<0.01 µg/gm) except
for garlic (0.25 µg/gm) and mushrooms (0.13 µg/gm). The contribution of cereals to the total selenium in the diet fluctuated depending
on the cereal consumed. The level of selenium in corn was low (0.25
µg/gm), but the level in barley was high (0.66 µg/gm). These
differences reflect not only a difference in the variety of plant, but
also in the geographic location of plant growth. The selenium content
of corn and wheat products from Ontario, Canada, and the midwestern
United States have very low levels of selenium ($0.07-0.08$ µg/gm)
(36). This is in direct contrast to the level of selenium in these same
foods grown in western Canada, where levels average 0.56 µg/gm.

In the United States and Canada, grains are transported all over
the country for processing, thus providing a mixing of potentially
high- and low-selenium grains to all locations. This does not, however, occur for all types of foods. Most notably, dairy and meat
products are usually consumed close to the site of production. These
food sources may reflect more local soil levels of the mineral.

In general, taking into account regional variability in the United
States food supply, fish contains the greatest amount of selenium,
representing the most concentrated food source of the mineral.
Selenium levels are high in oysters, red snapper, crab, and tuna
(35,37,38). Egg yolks and meat represent the foods containing the
next highest level of selenium. Organ meats are particularly high
sources of the mineral (pork kidney—1.9 µg/gm) (35). Cereal grains
represent the next highest selenium source.

As cereal grains are variable in their selenium content, so are
dairy products. Two samples of nonfat dry milk analyzed from the
Beltsville, Maryland, area were very different in their selenium
content—0.243 and 0.098 µg/gm (35). A more recent study, using

nonfat dry milk samples from Finland, showed an even wider range of selenium levels 0.033—0.420 µg/gm (39). It is therefore difficult to determine whether milk in the form of nonfat dry milk could contribute substantially to the total diet selenium. The levels of selenium in other dairy products in Maryland (cheese, skim and whole milk) were also variable.

Finally, fruits and vegetables, with few exceptions, provide very little selenium to the diet. In conclusion, selenium is most concentrate in high-protein foods, but the content is greatly influenced by growth conditions.

Effect of Processing on Selenium Content

Measuring the amount of selenium in a food product before it has been processed, packaged and cooked for serving does not reflect the quantity of selenium actually consumed by humans. In this section two major areas of mineral loss from food will be addressed: processing and cooking.

Loss of selenium from foods occurs first in the refining procedure. Using wheat as an important example, high-selenium Canadian whole grain wheat was found to contain 1.09 µg/gm. After refining into flour it contained 0.78 µg/gm (40). Similar selenium losses attributable to processing were seen in wheat and corn samples from Beltsville, Maryland, ranging from 10—30% of total selenium (41). The milling of oats and rice did not cause appreciable selenium losses. The authors pointed out that the 10—30% loss of selenium in wheat and corn was not as high as 80% losses reported for manganese, magnesium and cobalt, or the loss of 65% of the copper, iron, and zinc due to milling. Also, these samples were nutritionally adequate even after the 30% selenium loss due to milling.

The selenium level in cow's milk is variable depending on the animal's diet. In Japan, cow's milk has been reported to contain 23 ngSe/ml, with 12% of that selenium contained in glutathione peroxidase (42). The enzyme activity is destroyed in the pasteurization process and the form of selenium altered before consumption. The effect of this on bioavailability is not known.

The variability of selenium content in different samples of nonfat dry milk has already been mentioned. Differences in the dairy animal' diet is most likely a cause of this variability. However, one cannot discount the possibility that drying conditions used in processing might lead to volitilization losses of selenium (35).

Canning is a process used extensively in the food industry. It is unknown whether the high heat and pressure conditions of canning affect the selenium content of foods. Again referring to the Maryland study, it was found that canned carrots contained lower levels of selenium than fresh carrots. Conversely, canned tomatoes and potatoes contained more selenium than their fresh counterparts. It is unlikely that the canned and fresh vegetables analyzed in this study were grown under similar selenium soil conditions, therefore no consistent relationship between fresh and canned vegetables can be inferred from this study (35). One food item has shown more consistent behavior in response to the canning process. Tuna packed in water has been shown to have a higher level of selenium than oil-packed varities (38).

Very little work has been done to identify the effect of cooking and general preparation on the content of selenium in foods. One study that has addressed this issue showed that most ordinary cooking techniques do not cause major losses of selenium (43). For example, broiling meats, baking seafoods, frying eggs, and boiling cereals did not lead to selenium loss. Some exceptions were identified, however. Boiling asparagus or mushrooms led to a 29—44% loss of selenium, and the dry heating of cereal grains led to a loss of 7—24%. Whether these losses could be significant to the total selenium level in the diet is doubtful.

In conclusion, the selenium content of food is dependent on many factors. Continued highly controlled studies are necessary to determine the effect of processing and cooking procedures on that content.

SELENIUM METABOLISM

Function

The importance of selenium in animal nutrition has been appreciated since 1957. Schwarz and Foltz identified selenium as a part of Factor 3, which is important in preventing liver necrosis in rats (44). The functional or biochemical role of selenium in preventing the disease was not classified for another 16 years. In the early 1970s it was shown that selenium was involved in a glutathione-dependent reduction of hydroperoxides (45). In 1973, selenium was shown to be an integral part of the enzyme glutathione peroxidase (glutathione; H_2O_2 oxidoreductase, E.C.1.11.1.9;) (46,47). The reaction catalyzed is:

$$2GSH + H_2O_2 \longrightarrow GSSG + 2H_2O$$

This initial work was done in an in vitro erythrocyte system, where the presence of selenium in cell cultures was found to help prevent oxidative damage of the cells. The protective effect of selenium was expressed only when glucose was also present to maintain reduced glutathione (45). Selenium fits into cellular metabolism as a component of glutathione peroxidase to reduce lipid peroxides.

Reduced glutathione is regenerated by glutathione reductase and NADPH. These reactions are shown in Fig. 1 (48).

Glutathione peroxidase has been isolated and purified from animal (49) and human erythrocytes (50). The enzyme varies somewhat in molecular weight from system to system, within the range of 85–95 kilodaltons. A molecule of glutathione peroxidase has four subunits, each contains one g-atom of selenium per enzyme subunit. Amino acid analysis studies have identified the form of selenium in glutathione peroxidase as selenocysteine covalented incorporated into the primary structure of the enzyme (9).

The critical biochemical function of glutathione peroxidase, and therefore selenium as a part of the enzyme, is that of the reduction of hydroperoxides, destroying these reactive intermediates. A similar function is seen for the enzyme catalase. The two enzymes do not compete for H_2O_2, but are thought to be compartmentalized in many cells. Glutathione peroxidase is contained in the cytosol and mito-chondrial matrix, and catalase in the peroxisomes of the cell (51).

Glutathione peroxidase activity is measured in vitro by a detecting absorbance loss of NADPH as it becomes oxidized (52).

$$2GSH + H_2O_2 \xrightarrow{\text{GSH-Px}} H_2O + GSSG$$

$$GSSG + 2NADPH \xrightarrow{\text{GSSG-reductase}} 2GSH + 2NADP$$

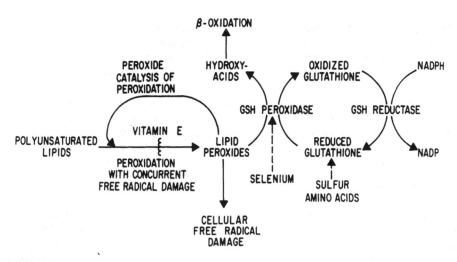

Figure 1. The function of selenium in cellular metabolism (48).

For the measurement of glutathione peroxidase activity in blood or tissue samples, the choice of substrate for the reaction is important. Glutathione peroxidase activity has been detected in selenium-deficient rat liver (53). This led to the isolation of a selenium-independent glutathione peroxidase enzyme in liver. This enzyme has been identified as glutathione-S-transferase and is active with cumene hydroperoxide as a substrate in vitro (54). Selenium-dependent glutathione peroxidase is best measured with H_2O_2 as the substrate.

The future of research in the metabolism of selenium will be in the isolation and identification of other selenium-containing proteins in mammalian tissues. Already selenium accumulation in rat testes has been shown to be high and localized in the midpiece of sperm (55). A 17,000 dalton selenoprotein has been isolated from sperm (56), and three others have been isolated from rat testis cytosol (57). The level of selenium in the male gonads of rats is maintained with priority over the selenium content of other tissues (58) and is thought to play an integral part in development and functioning of normal sperm. This function of selenium is independent of glutathione peroxidase, as the activity of glutathione peroxidase is low in testes and poorly correlated with selenium concentration.

Another selenoprotein has been identified in plasma (59). This protein may function as a selenium transport protein in the plasma. Further characterization of this protein may lead to a better understanding of the transport and deposition of selenium in tissues and the incorporation of the mineral into amino acids.

Assessment of Selenium Status

The establishment of a safe and adequate range of selenium intake instead of an RDA by the Food and Nutrition Board in 1980 reflected an uncertainty of the human nutritional requirement for selenium. Since 1980, studies measuring the selenium status of various population groups have advanced the understanding of human needs for the mineral. Selenium status in humans has been measured by balance studies, blood selenium levels, glutathione peroxidase activity in blood components and absorption and retention of stable isotopes. A number of these studies are described below.

Balance studies have shown that free-living North American adults have an average selenium intake of 90 µg/day for men and 74 µg/day for women (60). These intake levels were adequate for positive balance. When average intakes were expressed as a function of body weight, there was no significant difference in the intake of men and women. These investigators observed that selenium balance became more positive with increasing intake. This conservation of selenium might have serious implications for persons consuming high-selenium diets or selenium supplements.

The level of dietary selenium intake measured in the above American study was somewhat lower than that determined in a Canadian study of free-living young women (61). The Canadian women averaged 131 μg Se/day in their self-selected diets. The main sources of selenium in the diets of these women were meat, bread, and cereal. Both the American and Canadian study showed selenium intakes well within the range of the adequate and safe recommendation.

Positive selenium balance in North American males was not maintained when subjects were fed a depletion formula diet containing 33–36 μg Se/day (62). Fecal and urinary selenium losses were measured at 54 μg Se/day. These results contrast with results of balance studies using New Zealand women as subjects. Lower levels of selenium (24 μg Se/day) were adequate to maintain positive balance (63). As has been mentioned, New Zealand is an area with very low selenium soil content. Long-time residents may have developed an adaptive mechanism for conserving selenium.

Within the United States the differences seen in selenium intake are influenced more by food choice than by geographical region. This is demonstrated by two studies. The first study measured the selenium intake of a rural (71% Amish) population in Ohio dependent on food grown on low-selenium soil. In this study, rural intakes were not significantly different from intakes of an urban Columbus, Ohio, population (64). Both populations had selenium intakes of 82–83 μg Se/day. The second study was performed in South Dakota, an area with very high soil selenium levels (65). The level of selenium intake was positively correlated with the amount of protein consumed. Four different diets were tested: high animal protein, ovo-lacto vegetarian, vegetarian, and seafood-lacto vegetarian. The high-animal-protein diet provided the highest level of selenium: 331 μg Se/day. The other three diets provided roughly the same amount of selenium: 159 μg Se/day.

In humans, blood selenium levels have been used to evaluate selenium status. Selenium is routinely measured in plasma, serum, erythrocytes, whole blood, and, most recently, in platelets. Studies measuring selenium levels in blood components complement intake and balance studies by identifying the amount of selenium that is actually absorbed from a dietary source. The amount of correlation between intake and blood selenium levels varies. The study mentioned above conducted with Ohio rural and urban residents showed no significant difference in the intake levels of the two groups (64). However, rural subjects had significantly lower whole blood, plasma, and erythrocyte selenium levels than urban residents. The correlation between selenium intake and blood selenium concentration was poor in this study. Selenium levels in blood have also been shown to reach a plateau in response to high selenium intakes. A supplementation study in New Zealand women demonstrated a leveling off of whole blood erythrocyte, and plasma selenium levels after 6−8 weeks of 200 μg Se/day intakes (66). Therefore the correlation of selenium intake and blood selenium may be dependent on the levels of the mineral consumed.

Measurement of glutathione peroxidase takes the evaluation of selenium status one step further by evaluating the amount of selenium incorporated into the active selenium-containing enzyme. In rat platelets, it was shown that glutathione peroxidase activity increased in a stepwise manner when selenium-depleted rats were fed graded levels of sodium selenite (67). Because platelet selenium levels are high and the turnover of these cells is rapid, they reflect recent dietary selenium levels. The liver has the largest labile selenium pool in the body which undergoes rapid changes in response to dietary levels. This study suggests that platelet selenium and glutathione peroxidase levels correlate with the liver levels.

Increasing the reliability and safety of evaluating selenium status in humans has allowed research into the requirements of individuals under special metabolic conditions. One example of this is the effect of pregnancy on selenium status. Whole blood and plasma selenium levels were shown to decrease during pregnancy (68). The decline began around the beginning of the second trimester of pregnancy. However, in this study, erythrocyte and plasma glutathione peroxidase activities were shown to increase. A separate study on pregnant women showed high levels of selenium absorption but lower glutathione peroxidase levels in plasma and higher levels in platelets (69). These studies show a lack of correlation between selenium intake and glutathione peroxidase activity and may suggest a specialized selenium metabolism during pregnancy.

The final method of evaluating selenium status is that of stable isotope tracer methodology. Because stable isotopes present no harmful side effects and can be selectively differentiated from other isotopes, this methodology is well suited to human research. A recent study has described the absorption of selenium measured by balance and by stable isotope methodology in pregnant and nonpregnant women (69). The absorption of selenium from food was not significantly different when measured by balance or by stable isotope methodology. The apparent absorption was 80% for all three subject groups from an intrinsically labeled egg albumin source. Analysis of urinary excretion showed pregnant women had decreased selenium urinary output, thus conserving selenium, suggesting a possible increased metabolic need.

A metabolic study conducted with young North American males using the stable selenium isotope $^{74}SeO_3^{-2}$ has revealed some interesting aspects of selenium metabolism in humans (70,71). Subjects consumed a controlled diet containing 107 μg selenium for 10 days, then a low-selenium diet containing 11.4 μg selenium for an additional 35 days. With the change in diet, fecal selenium excretion decreased rapidly. Urinary selenium excretion also declined, but more gradually, indicating that the immediate contribution of ingested selenium to the urinary pool is small. All subjects went into negative balance on the low-selenium diet. Retention data and urine isotope enrichment curves were used to estimate the body pool size for selenium at 10.36 mg. However, this estimate may be dependent on the form of stable isotope used (selenite vs. selenomethionine) and may be limited by the short-term nature of this study as some body pools turn over very slowly.

Excretion

The major route of selenium excretion is in the urine. Body homeo-stasis for selenium is controlled by the kidney. The selenium metabolite has been identified as trimethylselenonium. This metabolite was first identified in rat urine (72).

The appearance of trimethylselenonium in rat urine has been shown to be dependent on the level of selenium given to the animal. Higher levels of the ion are found in urine of animals consuming higher levels of selenium. The amount of trimethylselenonium found in the urine is also dependent on the form of selenium consumed. Under conditions of either low or high selenium, giving a dose of selenomethioine resulted in a lower level of trimethylselenonium in the urine than when similar levels of selenocystine or selenite were given (73).

The importance of trimethylselenonium as the excretion product of selenium in humans is not known. Currently, analytical techniques are being developed to better characterize selenium-containing com-pounds in urine. Ion exchange chromatography and two-dimensional paper chromatography are two techniques showing promise in this area (73,74).

SELENIUM ABSORPTION

Absorption Mechanisms for Specific Forms of Selenium

There is evidence that the form of selenium present in the gastro-intestinal tract of laboratory animals determines the mode of selenium absorption. Using the everted intestinal-sac method in golden hamsters ^{75}Se-labeled selenomethionine was transported against a concentration gradient, whereas labeled selenite and selenocystine were not (75). When methionine was introduced into the medium in a concentration higher than that of selenomethionine, transport of selenomethionine was inhibited. The transport of selenite and selenocystine was not inhibited by their corresponding sulfur analogues. The conclusion from these experiments was that selenomethionine is absorbed by an active transport mechanism utilizing the same transport sites as methionine. Selenite and selenocystine, on the other hand, are absorbed by a passive absorption mechanism.

These conclusions were substantiated by subsequent work in the same laboratory. Selenomethionine absorption was shown to be dependent on the Na^+ ion pump, as indicated by inhibition with ouabain, a potent Na^+ pump inhibitor (76). In a separate study using ligated segments of rat intestine, selenomethionine and selenite were shown to be absorbed at similar rates (77). No mechanism of absorption was tested in this study, therefore it is not known whether this study and the one mentioned above are in conflict.

More recently, the absorption of selenate and selnite have been compared using in vivo perfusion of rat intestine (78,79). These studies showed that absorption of selenate was greater than selenite in the ileum. Sulfate, thiosulfate, and ouabain significantly inhibited selenate absorption. Increasing the concentration of selenate in the perfusate increased absorption to a saturation point. The authors concluded that selenate absorption in the ileum is a carrier-mediated mechanism. They also concluded that the slower selenite absorption was by diffusion.

Another important aspect of selenium absorption is the initial amount absorbed from a sample and the retention of that selenium in the body over time. Absorption and retention of radiolabeled selenium by rats has been studied. The absorption of selenium from intragastric intubations of selenite and selenomethionine solutions was calculated by three methods: cumulative fecal excretion, total fecal excretion, and whole-body radioactivity monitoring (80). Regardless of the method used, the absorption of selenite was lower than that of selenomethionine, but over 90% of each was absorbed. The urinary excretion of selenite was significantly higher than that of selenomethionine, making initial retention of selenite 13—15% less than selenomethionine. Whole-body turnover of ^{75}Se after the first week was independent of the chemical form, perhaps indicating a common metabolic pool of selenium.

These results can be compared to those obtained when selenite, selenate, and selenomethionine were fed to rats in a soy isolate protein-based diet (81). Initial absorption was similar for selenite and selenomethionine as calculated by whole-body radiolabel determination. Selenate was absorbed more efficiently. After 14 days, however there was no difference in the retention of the radionuclide from the three different forms.

In the previous two studies, the absorption of ^{75}Se has been high—90—96% of administered dose. To determine if the level of dietary selenium can affect that absorption, selenium as sodium selenite was added at 0.5 and 4.0 ppm to a torula yeast-based diet (82). As the dose increased, so did the amount of selenium absorbed even at the 4.0 ppm toxic level.

Selenium absorption in humans has been studied using both radioactive selenium and stable isotopes. One study using ^{75}Se-selenite and another using ^{75}Se-selenomethionine were done utilizing female volunteers (83,84). Selenite or selenomethionine was given as an oral dose with 10 μg and 2 μg selenium carrier, respectively. Intestinal absorption of selenite was 70, 64, and 44% for three experimental subjects. In contrast, intestinal absorption of selenomethionine, in a separate group of four subjects, was between 95.5 and 97.3%. Greater absorption of selenomethionine was followed by increased retention and less urinary and fecal loss than was found with selenite. This difference in retention of selenite and selenomethionine in humans is different from results described above for rats. Whether the utilization of selenomethionine is more efficient than the utilization of selenite is unknown.

Absorption of the stable isotope [74]Se as selenite in adult males was shown to be 76.0% (85). There were four subjects in this study and the variability of absorption was high, ranging from 55 to 99%. Similar variability of selenite absorption was seen in the studies with females described above. An explanation for the wide variability in absorption was not given by either author. The selenium status of the individuals may contribute to the variability in absorption.

In summary, in animals there are differences in the mode of absorption for different forms of selenium, but these differences do not seem to play a major role in determining the amount of selenium retained from a solution or from food. Overall, the absorption and retention of selenium from a test dose in both experimental animals and humans is high.

For some minerals, absorption is a predictor of bioavailability or metabolic use of a mineral. This is not the case for selenium. Determining bioavilability of selenium must be done utilizing other techniques in conjunction with absorption.

Bioavailability of Selenium

As described in the previous section, absorption of selenium from a dietary source is critical to selenium availability, but absorption represents only the first step of selenium utilization. Unlike the minerals zinc and iron, absorption of selenium is not equivalent to utilization. Just as dietary intake does not predict absorption, absorption does not predict bioavailability.

What than is bioavailability? Young has defined it as the fraction of element or compound in a food that is used in a given biological process (86). A more specific definition is provided by Spivey-Fox: the quantitative measure of the utilization of a nutrient under specific conditions to support the organism's normal structure and physiological processes (87). Spivey-Fox goes on to describe an important aspect of bioavailability research—the selection of indices for evaluating bioavailability. This selection is made based on knowledge of nutrient requirements, metabolism, and function. As knowledge of requirements, metabolism, and function of selenium has developed, so have the methods of determining bioavailability. Three different methods of determining bioavailability will be described here.

Selenium-responsive Diseases

Selenium deficiency causes diseases of economic concern in farm animals. Two such diseases in chicks are exudative diathesis and pancreatic fibrosis. Exudative diathesis is a severe subcutaneous edema caused by increased permeability of capillaries. Chicks are anemic and show reduced growth. In pancreatic fibrosis, acinar cells undergo vacuolation and hyaline body formation, leading to the loss of pancreatic lipase and protease synthesis and therefore impaired lipid absorption (3). Both diseases have been used as indices to measure bioavailability of different selenium forms and the bioavailability of selenium from various food sources. The relative effectiveness of sodium selenite, selenomethionine, and selenocystine for promoting weight gain and preventing exudative diathesis has been tested in chicks (88). Results showed that sodium selenite and selenocystine were equivalent in preventing exudative diathesis and selenomethionine was less effective. The chicks fed selenomethionine in a corn and torula yeast-based diet had higher levels of selenium in the breast muscle and pancreas and lower levels of selenium in the heart, kidney, and liver than the sodium selenite- and selenocystine-fed animals. Two results are of interest here: 1) There is a tissue-specific distribution of different chemical forms of selenium, and 2) the level of selenium in the heart, kidney, and liver is important in the prevention of exudative diathesis.

Using sodium selenite as a standard, Cantor determined the bioavailability of selenium from different animal feeds. The ability of selenium in the feed to prevent exudative diathesis was compared to the same level of selenium as sodium selenite (89). In this study, feeds of both plant and animal origin were tested. The bioavailability of selenium to prevent exudative diathesis was very different for plant and animal selenium sources. Animal sources of selenium were only 20—30% as effective as plant sources in preventing exudative diathesis (Table 2). These differences may reflect the prevalence of different selenium forms in plant and animal tissue.

In addition to bioavailability studies measuring exudative diathesis, Cantor has shown that selenomethionine is more effective than other selenium forms in the prevention of pancreatic fibrosis (90). Sodium selenite, selenocystine, selenomethionine, wheat, and tuna were tested for their ability to prevent pancreatic fibrosis. Wheat and seleno-

Table 2. Biological Availability of Selenium from Plant and Animal Sources (89)

Diet Source	Biological activity
Wheat	70.7 ± 14.1
Brewer's yeast	88.6 ± 5.5
Brewer's grain	79.8 ± 14.9
Corn	86.3 ± 14.6
Soybean meal	59.8 ± 13.9
Cottonseed meal	86.4 ± 18.6
Dehydrated alfalfa meal	210.0 ± 14.7
Tuna meal	22.4 ± 7.0
Poultry by-product meal	18.4 ± 4.6
Menhaden meal	15.6 ± 4.4
Fish solubles	8.5 ± 1.5
Herring meal	24.9 ± 6.7
Meat and bone meal	15.1 ± 4.2

methionine were the most effective sources of selenium to prevent histological indications of pancreatic fibrosis. Selenomethionine was four times as effective as selenite or selenocystine in preventing the disease.

Animal models of selenium responsive diseases provide a specific and practical method for determining the bioavailability of selenium from animal feeds.

Absorption, Retention, and Tissue Distribution

Selenium absorption, retention, and tissue distribution can be used to determine selenium bioavailability in experimental animals fed a radioactive selenium compound. Whole-body retention of ^{75}Se was greater in rats fed ^{75}Se-selenomethionine than in rats fed intrinsically ^{75}Se-labeled rabbit kidney (91). Tissue distribution of the ^{75}Se was similar for the two groups, as was retention of the whole body label after one week. The tissues assayed for radioactivity were heart, lungs, spleen, liver, kidneys, adrenals, ovaries, and portions of the rat thigh muscle. This uniform distribution of selenium is in contrast to that seen in chick studies described earlier and may reflect a species difference.

Fish products generally contain high levels of selenium. Whether this selenium is bioavailable is a question not fully answered. Fish muscle has been intrinsically and extrinsically labeled with both selenite and selenomethionine incorporated in diets and fed to rats (92). The absorption of selenium from intrinsically labeled was found to be lower than that for extrinsically labeled fish, as determined by measuring whole-body radioactivity. For both fish-labeling methods, selenite was less efficiently absorbed than selenomethionine by rats. Again, the tissue distribution and long-term retention of the selenium label were not different for the four groups.

Balance studies have also been used to measure selenium absorption and retention. Comparison of the absorption and retention of selenite, selenomethionine, and selenium from seleniferous wheat has been made in rats (93). There was no significant difference in selenium retention from the different selenium sources when the level in the diet was less than 0.1 ppm selenium. However, skeletal muscle content of selenium was higher in rats supplemented with selenomethionine than selenite in this study. Selenium levels in other tissues (skin, spleen, kidney, liver, and blood) were not affected by the selenium source. The higher level of selenium retained in skeletal muscle tissue from selenomethionine suggests that there is a difference in utilization of the different chemical forms of selenium. However, measurement of retention cannot explain functionally why there is a difference.

Measuring absorption and retention are important in bioavailability studies, but they cannot give a complete picture of the physiologic function of selenium. Absorption measurements may be used in conjunction with other analyses to give a better total picture of availability and utilization. Currently, the method most used in bioavailability studies of selenium is that of glutathione peroxidase activity.

Glutathione Peroxidase Activity Regeneration

Glutathione peroxidase contains selenium and is inactive if selenium is not present. A quantitative relationship between dietary selenium and the amount of glutathione peroxidase activity was shown in 1974 (94), only one year after the discovery that glutathione peroxidase was a selenium-containing enzyme. Weanling male rats were fed torula yeast-based diets with 0, 0.05, 0.1, 0.5, 1.0, or 5.0 ppm selenium added as sodium selenite. Glutathione peroxidase levels in the liver fell to zero within 24 days in animals fed the 0 ppm diet. Diets containing 0.1 ppm or more caused increases in liver glutathione peroxidase over the control level. Similar, but more gradual results were found in rat erythrocytes. The authors concluded that tissue glutathione peroxidase activity could be used as an indicator of animal selenium status.

There are limitations that must be kept in mind when evaluating data utilizing glutathione peroxidase activity. Other factors have been shown to affect glutathione peroxidase activity besides dietary selenium. Some of these factors include age, sex, oxidant stresses (for example, vitamin E deficiency), starvation, and iron and vitamin B_{12} deifciency (95). These factors may not play a large role in controlled animal atudies, but certainly must be considered if glutathione peroxidase activity is to be used in human studies.

Given its limitations, glutathione peroxidase has been and continue to be used to measure the bioavailability of selenium. In chicks, both dietary selenite and selenomethionine increased glutathione peroxidase levels (96). Glutathione peroxidase activity was logarithmically related to the dietary selenium level. In rats, glutathione peroxidase levels were increased at a similar rate by selenite and selenomethionine (97). This similarity in selenite and selenomethionine was unexpected by the authors in light of the differences seen in the prevention of pancreatic fibrosis. They postulated that the level of selenium and other dietary components may have led to the difference seen. Selenate was also shown to be effective at raising glutathione peroxidase levels in turkey Selenate, selenite, Se-DL-cystine, Se-DL-methionine, and Se-DL-ethionine were added at 0.2 ppm to individual turkey diets and the level of glutathione peroxidase activity measured (98). The effectiveness of these compounds to increase enzyme activity was selenate > Se-DL-cystine > Se-DL-methionine > Se-DL-ethionine. The effect of selenite was significantly lower than selenate but not different from the other three forms.

Free-living healthy individuals do not rely on pure selenium compounds to fulfill dietary requirements, but instead consume foods containing selenium. An important issue in selenium bioavailability research is the determination of selenium availability from foods. The availability of selenium to chicks from both animal and plant food sources was determined by measuring glutathione peroxidase activity (99). When compared to the absorption of selenite as a control (100%) the bioavailability of selenium from capelin fish was 48.0%, mackerel fish meal 34.1%, soybean meal 17.5%, corn gluten meal 25.7%, and selenomethionine 78.3%. Unfortunately, the level of protein and selenium varied substantially between diets and makes interpretation of these results difficult. In a separate study, Manhaden fish meal was found to be 45% as effective as selenite at raising glutathione peroxidase levels in turkey plasma when equal amounts of selenium were in the diets (98).

In another study, selenium from tuna was 57% as effective as selenite at raising rat liver glutathione peroxidase activity (100). Selenium from kidney and wheat were more effective than tuna at raising glutathione peroxidase activity. However, the selenium levels measured in the blood and liver of these animals were not found to be different between tuna-, kidney-, or wheat-fed animals. In a similar study, the bioavailability of selenium from wheat and tuna were compared to selenite in the rat model (101). Feeding rats selenium from wheat resulted in a higher liver, kidney, and whole blood selenium level than did feeding rats selenium from tuna. Glutathione peroxidase levels were also higher when selenium from wheat was eaten. This study again shows a lower bioavailability for fish selenium than for plant selenium. Investigations with other foods are necessary to identify good sources of selenium for the diet.

Glutathione peroxidase is the best index of selenium availability currently in use, however, this index is only as good as its applicability to humans for research. Measurements of liver and kidney glutathione peroxidase levels is not possible in human studies. However, measurement of enzyme activity in blood fractions provides an alternative. Platelets contain high selenium levels, demonstrate glutathione peroxidase activity, and their turnover rate is 8—14 days in humans. Levander has found that in rats, platelet glutathione peroxidase is sensitive to changes in dietary selenium up to a saturable endpoint and that platelet enzyme levels reflect hepatic levels (67). Because of these observations, glutathione peroxidase activity in platelets shows promise as a method for determining short-term availability of dietary selenium in humans. In a human study conducted in Finland, selenium bioavailability was determined by platelet glutathione peroxidase activity (102). Groups of Finnish men were fed 200 µg of selenium from Se-rich wheat, Se-rich yeast, or sodium selenate. These men had naturally low plasma selenium concentrations due to low selenium levels in the Finnish food supply. Platelet glutathione peroxidase activity increased rapidly in subjects supplemented with wheat or sodium selenate. A slower increase in enzyme activity was seen in the yeast-supplemented group.

Another human bioavailability study was conducted in a low-selenium area of China (103). After supplementation with sodium selenite or selenomethionine, glutathione peroxidase levels increased significantly in the plasma followed by an increase in red blood cells. No difference in bioavailability between sodium selenite and selenomethionine was observed in this study. The authors point out that this may have been due to the relatively large (150 µg Se) supplement of selenium given.

Our ability to determine the bioavailability of selenium has progressed dramatically in recent years. Future study must include investigation of bioavailability of selenium from different foods. As our knowledge of selenium metabolism increases, other methods of measuring selenium bioavailability may be developed. Only a portion of selenium in the body is associated with glutathione peroxidase, and the amount is species-specific (104). Other selenium proteins are being isolated and purified. These may provide additional or better indices of selenium bioavailability.

CONCLUSION

Selenium clearly has an important role in human nutrition, although its function is not fully understood. The identification of selenium-containing compounds in tissues and the determination of its bioavailability from food are areas of research that will improve our understanding. Bioavailability of selenium to humans is of particular importance as studies using animal models can only approximate the bioavailability of selenium from food. Bioavailability is also important as supplementation of restricted nutrition formula may be warranted in the future, especially in cases where patients may receive formula for long periods of time. Determination of the best form and the amount of selenium to be used in supplementation must be made.

The difference between nutritional and potentially toxic levels of selenium has been estimated at only 20-fold (105). This is not a wide margin of safety and may leave the public confused about nutritional selenium requirements. As with any vitamin or mineral, dietary selenium sources should be supplied in the RDA or safe and adequate range of amounts for physiologic function without toxic overdoses.

REFERENCES

1. W. C. Cooper, K. G. Bennet, and F. C. Croxton, The history, occurrence and properties of selenium, in *Selenium* (R. A. Zingaro and W. C. Cooper, eds.), Van Nostrand Reinhold Co., New York, 1974.
2. C. C. Reddy, E. J. Massaro, Biochemistry of Selenium: A Brief Overview, *Fundamentals of Applied Toxicology, 3:* 431 (1983).
3. National Research Council, *Selenium in Nutrition*, revised edition, National Academy Press, Washington, D. C., 1983.
4. R. E. Huber and R. S. Criddle, Comparison of the chemical properties of selenocysteine and selenocystine with their sulfur analogs, *Archives of Biochemistry and Biophysics, 122:* 164 (1967).
5. G. Gissel-Nielsen, U. C. Gupta, M. Lamand, and T. Westermarck, Selenium in soils and plants and its importance in livestock and human nutrition, *Advances in Agronomy, 37:* 397 (1984).
6. R. L. Henrikson, Selective S-methylation of crysteine in proteins and peptides, *Biochemical and Biophysical Research Communication, 41:* 967 (1970).

7. M. Hermodson, G. Schmer, and K. Kurachi, Isolation, crystalli-
zation, and primary amino acid sequence of human platelet factor
4, *Journal of Biological Chemistry*, 252: 6276 (1977).
8. J. E. Cone, R. Martin del Rio, J. N. Davis, and T. C. Stadtman,
Chemical characterization of the selenoprotein component of
clostridial glycine reductase: Identification of selenocysteine as
the organoselenium moiety, *Proc. Natl. Acad. Sci. U. S.*, 73:
2659 (1976).
9. J. W. Forstrom, J. J. Zakowski, and A. L. Tappel, Identification
of the catalytic site of rat liver glutathione peroxidase as
selenocysteine, *Biochemistry*, 17: 2639 (1978).
10. J. J. Zakowski, J. W. Forstrom, R. A. Condell, and A. L.
Tappel, Attachment of selenocysteine in the catalytic site of
glutathione peroxidase, *Biochemical and Biophysical Research
Communications*, 84: 248 (1978).
11. R. J. Kraus, S. J. Foster, and H. E. Ganther, Identification
of selenocysteine in glutathione peroxidase by mass spectroscopy, *Biochemistry*, 22: 5853 (1983).
12. C. A. Girling, Selenium in agriculture and the environment,
Agriculture, Ecosystems and Environment, 11: 37 (1984).
13. B. Bisbjeig and G. Gissel-Nielsen, The Uptake of applied selenium
by agricultural plants: The influence of soil type and plant species,
Plant and Soil, 31: 287 (1969).
14. H. W. Lakin, Selenium accumulation in soils and its absorption
by plants and animals, *Geological Society of Am. Bulletin*, 83:
181 (1972).
15. W. H. Allaway, An overview of distribution patterns of trace
elements in soils and plants, *Ann. New York Acad. Sci.*, 199:
17 (1972).
16. I. Rosenfeld and O. A. Beath, *Selenium: Geobotany, Biochemistry,
Toxicity, and Nutrition*, Academic Press, New York, 1964.
17. J. Kubota, W. H. Allaway, D. L. Carter, E. E. Cary, and
V. A. Lazar, Selenium in crops in the United States in relation
to selenium-responsive diseases of animals, *J. Agric. Food Chem.*,
15: 448 (1967).
18. S. N. Nigam, Jan-I. Lu, and W. B. McConnell, Distribution of
selenomethylselenocysteine and some other amino acids in species
of *Astragalus*, with special reference to their distribution during
the growth of *A. bisulcatus*, *Phytochemistry*, 8: 1161 (1969).
19. A Shrift and T. K. Virupaksha, Selenoamino acids in selenium
accumulating plants, *Biochimica et Biophysica Acta*, 100: 65 (1965).
20. J. L. Martin, A. Shrift, and M. L Gerlach, Use of ^{75}Se-selenite
for the study of selenium metabolism in *Astragalus*, *Phytochemistry*,
10: 945 (1971).
21. A. Shrift, Selenium compounds in nature and medicine: Metabolism
of selenium by plants and microorganisms, in *Organic Selenium*

Compounds: Their Chemistry and Biology (D. L. Klayman and
W. H. Gunther, eds.), Wiley Interscience, New York, 1973.

22. G. Gissel-Nielsen, Uptake and translocation of selenium-75 in
Zea Mays, Int'l Atomic Energy Agency, 427 (1979).

23. P. J. Peterson and G. W. Butler, The uptake and assimilation
of selenium by higher plants, Aust. J. Biol. Sci., 15: 126
(1962).

24. O. E. Olson, E. J. Novacek, E. I. Whitehead, and I. S. Palmer,
Investigations on selenium in wheat, Phytochemistry, 9: 1181
(1970).

25. Recommended Dietary Allowances, ninth revised edition, National
Academy of Sciences, Washington, D. C., 1980.

26. Keshan Disease Research Group, Observations on effect of sodium
selenite in prevention of Keshan disease, Chinese Medical Journal,
92: 471 (1979).

27. Keshan Disease Research Group, Epidemiologic studies on the
etiologic relationship of selenium and Keshan disease, Chinese
Medical Journal, 92: 477 (1979).

28. M. Dongxa, Pathology and selenium deficiency in Kaschin-Beck
disease, in Selenium in Biology and Medicine (G. F. Combs, Jr,
O. A. Levander, J. E. Spallholz, J. E. Oldfield, eds.) AVI
Publishing, New York, 1987.

29. R. A. Quercia, S. Korn, D. O'Neill, J. E. Dougherty, M. Ludwig,
R. Schweizer, and R. Sigman, Selenium deficiency and fatal
cardiomyopathy in a patient receiving long-term home parenteral
nutrition, Clin. Pharm., 3: 531 (1984).

30. C. L. Kien and H. E. Ganther, Manifestations of chronic selenium
deficiency in a child receiving total parenteral nutrition, Am. J.
Clin. Nutr., 37: 319 (1983).

31. S. S. Baker, R. H. Lerman, S. H. Krey, K. S. Crocker, E.
F. Hirsh, and H. Cohen, Selenium deficiency with total parenteral
nutrition: reversal of biochemical and functional abnormalities by
selenium supplementation: a case report, Am. J. Clin. Nutr.,
38: 769 (1983).

32. C. D. McGee, M. J. Ostro, R. Kurian, and K. N. Jeejeebhoy,
Vitamin E and selenium status of patients reveiving short-term
total parenteral nutrition, Am. J. Clin. Nutr., 42: 432 (1985).

33. M. Mutanen, Dietary intake and sources of selenium in young
Finnish women, Human Nutrition: Applied Nutrition, 38A: 265
(1984).

34. M. Mutanen and P. Koivistoinen, The role of imported grain on
the selenium intake of Finnish population in 1941—1981, International J. Vit. Nutr. Res., 53: 102 (1983).

35. V. C. Morris and O. A. Levander, Selenium content of foods,
J. Nutr., 100: 1383 (1970).

36. D. Arthur, Selenium content of Canadian foods, Can. Inst. Food
Sci. Technol. J., 5: 165 (1972).

37. C. S. Wilson, H. B. Chin, P. A. Lerke, and S. A. Gellert, Selenium in commonly consumed U. S. foods as determined by fluorimetry and x-ray excitation spectrometry: A preliminary report, in *Selenium in Biology and Medicine* (J. E. Spallholz, J. L. Martin, and H. E. Ganther, eds.), AVI Publishing, New York, 1981.

38. H. W. Lane, B. J. Taylor, E. Stool, D. Servance, and D. C. Warlen, Selenium content of selected foods, *J. Am. Dietetic Association, 82*: 24 (1983).

39. P. Varo, M. Nuurtamo, P. Koivistoinen, Selenium content of nonfat dry milk in various countries, *J. Dairy Science, 67*: 2071 (1984).

40. K. Lorenz, Selenium in U. S. and Canadian wheats and flours, in *Selenium in Biology and Medicine* (J. E. Spallholz, J. L. Martin, and H. E. Ganther, eds.), AVI Publishing, New York, 1981.

41. R. J. Ferretti and O. A. Levander, Effect of milling and processing on the selenium content of grains and cereal products, *J. Agric. Food Chem., 22*: 1049 (1974).

42. Y. Hojo, Selenium concentration and glutathione peroxidase activity in cow's milk, *Biological Trace Element Research, 4*: 233 (1982).

43. D. J. Higgs, V. C. Morris, and O. A. Levander, Effect of cooking on selenium content of foods, *J. Agric. Food Chem., 20*: 678 (1971).

44. K. Schwarz and C. M. Foltz, Selenium as an integral part of factor 3 against dietary necrotic liver degeneration, *J. Am. Chem. Soc., 79*: 3292 (1957).

45. J. T. Rotruck, A. L. Pope, H. E. Ganther, and W. G. Hoekstra, Prevention of oxidative damage to rat erythrocytes by dietary selenium, *J. Nutr., 102*: 689 (1972).

46. J. T. Rotruck, A. L. Pope, H. E. Ganther, A. B. Swanson, D. G. Hafeman, and W. G. Hoekstra, Selenium: Biochemical role as a component of glutathione peroxidase, *Science, 179*: 588 (1973).

47. L. Flohe, W. A. Gunzler, and H. H. Schock, Glutathione peroxidase: A selenoenzyme, *FEBS Letters, 32*: 132 (1973).

48. C. K. Chow and A. L. Tappel, Response of glutathione peroxidase to dietary selenium in rats, *J. Nutr., 104*: 444 (1974).

49. S. H. Oh, H. E. Ganther, and W. H. Hoestra, Selenium as a component of glutathione peroxidase isolated from ovine erythrocytes, *Biochemistry, 13*: 1825 (1974).

50. Y. H. Awasthi, E. Beutler, and S. K. Srivastava, Purification and properties of human erythrocyte glutathione peroxidase, *J. Biol. Chem., 250*: 5144 (1975).

51. L. Flohe, W. A. Gunzler, R. Ladenstein, Glutathione peroxidase, in *Glutathione: Metabolism and function* (I. M. Arias and W. B. Jakoby, eds.), Raven Press, New York, 1976.
52. D. E. Paglia and W. N. Valentine, Studies on the quantitative and qualitative characterization of erythrocyte glutathione peroxidase, *J. Lab. Clin. Med., 70*: 158 (1967).
53. R. A. Lawrence and R. F. Burk, Glutathione peroxidase activity in selenium-deficient rat liver, *Biochem. Biophys. Res. Comm., 71*: 952 (1976).
54. C. C. Reddy, C. P. Tu, J. R. Burgess, C. Y. Ho, R. W. Scholz, and E. J. Massaro, Evidence for the occurrence of selenium-independent glutathione peroxidase activity in rat liver microsomes, *Biochem. Biophys. Res. Comm., 101*: 970 (1981).
55. D. H. Brown and R. F. Burk, Selenium retention in tissues and sperm of rats fed a torula yeast diet, *J. Nutr., 102*: 102 (1972).
56. H. I. Calvin, Selective incorporation of selenium-75 into a polypeptide of the rat sperm tail, *J. Exp. Zool., 204*: 445 (1978).
57. K. P. McConnell, R. M. Burton, T. Kute, and P. J. Higgins, Selenoproteins from rat sperm testis cytosol, *Biochimica Biophysica Acta, 588*: 113 (1979).
58. D. Behne, T. Hofer, R. Berswordt-Wallrabe, and W. Elger, Selenium in the testis of the rat: Studies on its regulation and its importance for the organism, *J. Nutr., 112*: 1682 (1982).
59. R. F. Burk and P. E. Gregory, Some characteristics of ^{75}Se-P, a selenoprotein found in rat liver and plasma, and comparison of it with selenoglutathione peroxidase, *Archives Biochem. Biophys., 213*: 73 (1982).
60. O. A. Levander and V. C. Morris, Dietary selenium levels needed to maintain balance in North American adults consuming self-selected diets, *Am. J. Clin. Nutr., 39*: 809 (1984).
61. R. S. Gibson and C. A. Scythes, Trace element intakes of women, *Br. J. Nutr., 48*: 241 (1982).
62. O. A. Levander, B. Sutherland, V. C. Morris, and J. C. King, Selenium balance in young men during selenium depletion and repletion, *Am. J. Clin. Nutr., 34*: 2662 (1981).
63. C. D. Thomson and M. F. Robinson, Selenium in human health and disease with emphasis on those aspects of peculiar to New Zealand, *Am. J. Clin. Nutr., 33*: 303 (1980).
64. J. T. Snook, D. L. Palmquist, A. L. Moxon, A. H. Cantor, and V. M. Vivian, Selenium status of a rural (predominantly Amish) community living in a low selenium area, *Am. J. Clin. Nutr., 38*: 620 (1983).
65. I. S. Palmer, O. E. Olson, L. M. Ketterling, and C. E. Shank, Selenium intake and urinary excretion in persons living near a high selenium area, *J. Am. Dietetic Assoc., 82*: 511 (1983).

66. C. D. Thomson, L. K. Ong, and M. F. Robinson, Effects of supplementation with high-selenium wheat bread on selenium, glutathione peroxidase and related enzymes in blood components of New Zealand residents, *Am. J. Clin. Nutr.*, *41*: 1015 (1985).

67. O. A. Levander, D. P. Lehoach, V. C. Morris, and P. B. Moser, Platelet glutathione peroxidase activity as an index of selenium status in rats, *J. Nutr.*, *113*: 55 (1983).

68. J. D. Butler, P. D. Whanger, and M. J. Tripp, Blood selenium and glutathione peroxidase activity in pregnant women: comparative assays in primates and other animals, *Am. J. Clin. Nutr.*, *36*: 15 (1982).

69. C. A. Swanson, D. C. Reamer, C. Veillon, J. C. King, and O. A. Levander, Quantitative and qualitative aspects of selenium utilization in pregnant and nonpregnant women: an application of stable isotope methodology, *Am. J. Clin. Nutr.*, *38*: 169 (1983).

70. M. Janghorbani, L. J. Kasper, and V. R. Young, Dynamics of Selenite metabolism in young men: studies with the stable isotope tracer method, *Am. J. Clin. Nutr.*, *40*: 208 (1984).

71. L. J. Kasper, V. R. Young, and M. Janghorbani, Short-term dietary selenium restriction in young adults: quantitative studies with the stable isotope $^{74}SeO_3^{-2}$, *Br. J. Nutr.*, *52*: 443 (1984).

72. I. S. Palmer, R. P. Gunsalus, A. W. Haverson, and O. E. Olson, Trimethylselenonium ion as a general excretory product from selenium metabolism in the rat, *Biochimica et Biophysica Acta*, *208*: 260 (1970).

73. A. T. Nahapetian, V. R. Young, and M. Janghorbani, Measurement of trimethylselenonium ion in human urine, *Analytical Biochemistry*, *140*: 56 (1984).

74. R. F. Burk, Selenium in man, in *Trace elements in Human Health and Disease II. Essential and Toxic Elements* (A. S. Prasad and D. Oberlas, eds.), Academic Press, New York, 1976.

75. K. P. McConnell and G. J. Cho, Transmucosal movement of selenium, *Am. J. Physiol.*, *208*: 1191 (1965).

76. K. P. McConnell and G. J. Cho, Active transport of L-selenomethionine in the intestine, *Am. J. Physiol.*, *213*: 150 (1967).

77. P. D. Whanger, N. D. Pederson, J. Hatfield, and P. H. Weswig, Absorption of selenite and selenomethionine from ligated digestive tract segments in rats, *Proc. Soc. Exp. Biol. Med.*, *153*: 295 (1976).

78. S. Wolffram, F. Arduser, and E. Scharrer, In vivo intestinal absorption of selenate and selenite by rats, *J. Nutr.*, *115*: 454 (1985).

79. F. Arduser, S. Wolffram, and E. Scharrer, Active absorption of selenate by rat ileum, *J. Nutr.*, *115*: 1203 (1985).

80. C. D. Thomson and R. D. H. Stewart, Metabolic studies of ^{75}Se selenomethionine and ^{75}Se selenite in the rat, *Br. J. Nutr.*, *30*: 139 (1973).

81. A. C. Mason and C. M. Weaver, The metabolism in rats of selenium from intrinsically and extrinsically labeled isolated soy protein, *J. Nutr.*, *116*: 1883 (1986).

82. D. G. Brown, R. F. Burk, R. J. Seely, and K. W. Kiker, Effect of dietary selenium on the gastrointestinal absorption of $^{75}SeO_3^=$ in the rat, *Int'l. J. Vit. Nutr. Res.*, *42*: 588 (1972).

83. D. C. Thomson and R. D. H. Stewart, The metabolism of ^{75}Se-selenite in young women, *Br. J. Nutr.*, *32*: 47 (1974).

84. N. M. Griffiths, R. D. H. Stewart, and M. F. Robinson, The metabolism of ^{75}Se-selenomethionine in four women, *Br. J. Nutr.*, *35*: 373 (1976).

85. M. Janghorbani, M. J. Christensen, A Nahapetian, and V. R. Young, Selenium metabolism in healthy adults: quantitative aspects using the stable isotope $^{75}SeO_3^{-2}$, *Am. J. Clin. Nutr.*, *35*: 647 (1982).

86. V. R. Young, A. Nahapetian, and M. Janghorbani, Selenium bioavailability with reference to human nutrition, *Am. J. Clin. Nutr.*, *35*: 1076 (1982).

87. M. R. Spivey-Fox, R. M. Jacobs, A. O. L. Jones, B. E. Fry, M. Rakowaska, R. P. Hamilton, B. F. Harland, C. L. Stone, and S. H. Tao, Animal models for assessing bioavailability of essential and toxic elements, *Cereal Chem.*, *58*: 6 (1981).

88. M. Osman and J. D. Latshaw, Biological potency of selenium from sodium selenite, selenomethionine, and selenocystine in the chick, *Poultry Science*, *55*: 987 (1976).

89. A. H. Cantor, M. L. Scott and T. Noguchi, Biological availability of selenium in feedstuffs and selenium compounds for prevention of exudative diathesis in chicks, *J. Nutr.*, *105*: 96 (1975).

90. A. H. Cantor, M. L. Langevin, T. Noguchi, and M. L. Scott, Efficacy of selenium in selenium compounds and feedstuffs for prevention of pancreatic fibrosis in chicks, *J. Nutr.*, *105*: 106 (1975).

91. C. D. Thomson, R. D. H. Stewart, and M. F. Robinson, Metabolic studies in rats of ^{75}Se-selenomethionine and of ^{75}Se incorporated in vivo into rabbit kidney, *Br. J. Nutr.*, *33*: 45 (1975).

92. M. Richold, M. F. Robinson, and R. D. H. Stewart, Metabolic studies in rats of ^{75}Se incorporated in vivo into fish muscle, *Br. J. Nutr.*, *38*: 19 (1977).

93. E. E. Cary, W. H. Alloway, and M. Miller, Utilization of different forms of dietary selenium, *J. of Animal Science*, *36*: 285 (1973).

94. D. G. Hafeman, R. A. Sunde, and W. G. Hoekstra, Effect of dietary selenium on erythrocyte and liver glutathione peroxidase in rats, *J. Nutr.*, *104*: 580 (1974).

95. W. G. Hoekstra, Glutathione peroxidase activity of animal tissues as an index of selenium status, in *Trace Substances in Environ-*

mental Health IX (D. D. Hemphill, ed.), University of Missouri Press, Columbia, 1975.

96. S. T. Omaye and A. L. Tappel, Effect of dietary selenium on glutathione peroxidase in the chick, *J. Nutr.*, *104*: 747 (1974).

97. S. Pierce and A. L. Tappel, Effects of selenite and seleno-methionine on glutathione peroxidase in the rat, *J. Nutr.*, *107*: 475 (1977).

98. A. H. Cantor, and J. Z. Tarino, Comparative effects of inorganic and organic dietary sources of selenium on selenium levels and selenium-dependent glutathione peroxidase activity in blood of young turkeys, *J. Nutr.*, *112*: 2187 (1982).

99. B. O. Gabrielson and J. Opstvedt, Availability of selenium in fish meal in comparison with soybean meal, corn gluten meal and selenomethionine relative to selenium in sodium selenite for restoring glutathione peroxidase activity in selenium-depleted chicks, *J. Nutr.*, *110*: 1096 (1980).

100. J. S. Douglass, V. C. Morris, J. H. Soares, Jr., and O. A. Levander, Nutritional availability to rats of selenium in tuna, beef kidney, and wheat, *J. Nutr.*, *111*: 2180 (1981).

101. A. R. Alexander, P. D. Whanger, and L. T. Miller, Bioavail-ability to rats of selenium in various tuna and wheat products, *J. Nutr.*, *113*: 196 (1983).

102. O. A. Levander, G. Alfthan, H. Arvilomoni, C. G. Gref, J. K. Hultunen, M. Kataja, P. Koivistoinen, and J. Pekkarainen, Bioavailability of selenium to Finnish men as assesed by platelet glutathione peroxidase activity and other blood parameters, *Am. J. Clin. Nutr.*, *37*: 887 (1983).

103. X. Luo, H. Wei, C. Yang, J. Xing, X. Lui, C. Qiao, Y. Feng, J. Liu, Y. Lui, Q. Ulu, X. Lui, J. Guo, B. J. Stoecker, J. E. Spallholz, and S. P. Yang, Bioavailability of selenium to residents in a low selenium area of China, *Am. J. Clin. Nutr.*, *42*: 439 (1985).

104. D. Behne and W. Wolters, Distribution of selenium and glutathione peroxidase in the rat, *J. Nutr.*, *113*: 456 (1983).

105. T. H. Jukes, Nuggets on the surface: Selenium, an "essential poison", *Journal of Applied Biochemistry*, *5*: 233 (1983).

11
The Ultratrace Elements

FORREST H. NIELSEN / Unites States Department of Agriculture,
Grand Forks Human Nutrition Research Center, Grand Forks, North Dakota
North Dakota

INTRODUCTION

Since 1970, scientists have suggested that at least 11 elements could
be added to the list of trace elements that are essential nutrients.
Estimated animal dietary requirements for those elements usually are
less than 1 μg/gm, and often are less than 50 ng/gm dry diet. They
have been designated ultratrace elements. The 11 elements are
arsenic, boron, bromine, cadmium, fluorine, lead, lithium, nickel,
silicon, tin, and vanadium. A review (1) of the experimental evidence
supporting the suggestion of nutritional essentiality for the 11 ultratrace elements indicated that only arsenic, nickel, and silicon met the
definition of essentiality. At the time of that review, it was thought
that there was evidence suggesting that boron, lithium, and vanadium
are also essential nutrients. Only limited evidence supported the
nutritional essentiality of bromine, cadmium, fluorine, lead, and tin.
Since the review, further support for essentiality has come forth only
for boron.

This chapter will emphasize the four recently established essential
ultratrace elements (As, B, Ni, Si) plus molybdenum, an element
found essential about 30 years ago (see Ref. 125). Vanadium also
will recieve some attention because it may have some pharmacological
importance. Bromide and lithium will not be discussed because the
evidence for them being of practical nutritional importance is very
limited. Fluorine, tin, cadmium, and lead are discussed elsewhere
in this volume.

ARSENIC

History

Since ancient times, arsenicals have been characterized by actions both benevolent and malevolent. Because arsenic has such a long history, only a few highlights can be presented here. Very early in the history of arsenic it was found that some arsenic compounds were convenient scentless and tasteless instruments for homicidal purposes. Thus, for about 1100 years, up to the last century, arsenicals reigned as the king of poisons. Even today arsenic is often thought of as being synonymous with poison. Nonetheless, the bad reputation of arsenic did not prevent it from becoming an important pharmaceutical agent. According to Schroeder and Balassa (2), medicinal virtues of arsenicals were acclaimed by Hippocrates (460—357 B.C.), Aristotle (384—322 B.C.(, Theophrastus (370—288 B.C.), and Pliny the Elder (A.D. 23—70). In 1905, Ehrlich first synthesized organic arsenicals and demonstrated their chemotherapeutic action against trypanosomes (see Ref. 2). Subsequently, the pharmacology of arsenicals was extended to more than 8000 compounds by 1937. Arsenicals were considered at various times to be specific remedies in the treatment of anorexia, other nutritional disturbances, neuralgia, rheumatism, asthma, chorea, malaria, tuberculosis, diabetes, skin diseases, and numerous hematologic abnormalities. The use of arsenic for these disorders has fallen into disrepute or has been replaced by more effective alternatives. The first evidence for arsenic essentiality appeared in 1975—1976 (vide infra). Thus, it is only recently that arsenic has been studied from the biochemical, nutritional, and physiological, and not the toxicological or pharmacological, points of view.

Chemistry of Arsenic

Both the trivalent and pentavalent states of arsenic apparently exist in biological material. Which oxidation state is most important in the biochemistry, nutrition, and physiology of living organisms is not known. However, the organic arsenic compounds that are probably of most biochemical importance are those that contain methyl groups. The methylation of inorganic oxyarsenic anions occurs in organisms ranging from microbial to mammalian.

Challenger (3) reviewed the early history of studies on biomethylation of arsenic. The methylated compound first identified was the poisonous gas trimethylarsine, which is synthesized by various molds. Challenger (3) proposed the following metabolic pathway for the production of trimethylarsine from arsenic:

$$AsO_3^{3-} \xrightarrow{[CH_3^+]} CH_3AsO_3^{2-} \xrightarrow[b)[CH_3^+]]{a)2e^-} (CH_3)_2AsO^{2-}$$

arsenite　　　　　　　　methylarsonate　　　　　dimethylarsinate

$$\xrightarrow[b)[CH_3^+]]{a)2e^-} (CH_3)_3AsO \xrightarrow{2e^-} (CH_3)_3As$$

　　　　　　　　　　　　　　　　　　　　　　trimethylarsine

Findings of Cullen et al. (4) indicated that S-adenosylmethionine or some related sulphonium compound is involved in the methylation of arsenic. This supports Challenger's proposed use of the carbocation (CH_3^+) rather than the carbanion (CH_3^-) or the radical $(CH_3^{\cdot})$ in the methyl transfer. However, other findings of Cullen et al. (5) indicated that the proposed pathway is an oversimplification of the many processes involved. Using *Candida humicola* as the test organism, they found that methylarsonate apparently does not occur as a free intermediate in the arsenate-to-trimethylarsine pathway.

Since 1977, methylated arsenicals have been identified in a number of other organisms. In aquatic organisms, arsenic was found to be present as both lipid-soluble and water-soluble compounds. Some of the lipid-soluble organic arsenicals probably are formed through a biomethylation pathway similar to that proposed by Challenger (3). Benson et al. (6) suggested that, in *Dunaliella tertiolecta*, trimethyl-arsine condenses with phosphoenolpyruvate to produce trimethylarson-iumlactate. Then this trimethylarsoniumlactate is used to make o-phosphatidyltrimethylarsoniumlactate, an arsenophospholipid of algae, by a mechanism analogous to that used for the production of phos-phatidylserine or other plant lipids. This arsenolipid differs from an arsenolecithin in that it possesses a carboxyl group which contributes to the physical and chemical properties of the hydrophilic moiety of the amphiphatic lipid. However, Wrench and Addison (7) reported that *D. tertiolecta* synthesizes arsenolipids distinctly different from those identified by Benson and co-workers. According to Wrench and Addison (7), this organism synthesizes three acid-labile arseno-lipids, one of which behaved as an anionic phosphatide similar to a compound synthesized from phosphatidylinositol and arsenite. Thus, they proposed that one of the major arsenolipids in *D. tertiolecta* was a complex between arsenite and phosphatidylinositol. They also suggested that, because arsenic has rapid access to the lipid pool by a single biochemical process of reduction and because a wide range of organisms, from bacteria to mammals, carry out this reductive step, arsenite-lipid complexes may be widely distributed in nature.

Some of the water-soluble organic arsenicals in marine organisms may be precursors, or degradation products, of the lipid-soluble organic arsenicals. According to Benson et al. (6), degradation of

O-phosphatidyltrimethylarsoniumlactate yields trimethylarsoniumlactate, a neutral zwitterionic molecule with a structure closely related to arsenocholine and arsenobetaine. Edmonds and Francesconi (8) suggested that arsenobetaine found in some higher organisms, such as western rock lobster, dusty shark, and school whiting, may evolve through the cycling in the marine ecosystem of arseno-sugars synthesized by the brown kelp (*Ecklonia radiata*). These sugars are 2-hydroxyl-3-sulfopropyl-5-deoxy-5-(dimethylarsenoso) furanoside and 2,3-dihydroxy-propyl-5-deoxy-5-(dimethylarsenoso) furanoside. Bacterial action on the arsenic-containing sugars produce dimethyloxarsylethanol, a possible precursor of arsenobetaine (9).

The biochemistry of arsenic at trophic levels higher than phytoplankton in the marine food chain is an unsettled topic. Some findings indicate that higher organisms themselves transform inorganic arsenic to organic forms, whereas other findings suggest that most organic arsenic that appears in these organisms comes from preformed sources. Both mechanisms probably occur, with the predominant one being dependent upon the organism and its environment. Although the ability to methylate arsenic is uncertain in marine animals, there is no doubt that mammals have this ability. The methylation apparently occurs through the complexion of arsenous acid with adjacent thiol groups in enzyme(s) (10,11). This arsenous acid is reduced, methylated in steps by CH_3^+ and finally de-complexed after addition of the second methyl group. Oxidation finally produces dimethylarsinic acid, a compound found in urine (10). The finding that periodate-oxidized adenosine, an inhibitor of certain methyltransferases, depressed the production of dimethylarsinic acid in mice and rabbits suggests that S-adenosylmethionine is the methyl-donor in the methylation of inorganic arsenic in vivo (12).

Arsenic compounds other than those of the methylated form that are of interest in biochemistry are those possibly formed when arsenate replaces phosphate in biological molecules. This may occur because arsenates and phosphates are both tetrahedral in structure and often isomorphous in crystals. Generally, the phosphoryl compounds are more stable than the arsenyl compounds. The less stable nature of the arsenyl compounds apparently is the reason that direct evidence for existence of sugar or similar arsenate esters has not been reported. In other words, the synthesis and characterization of such compounds has not been done. However, indirect evidence has been reported for the existence of glucose-6-arsenate and ADP-arsenate (13,14).

Arsenic Content of Foods

Because of the reputation of arsenic as a poison or carcinogen, the arsenic content of foods has received some attention. However, except for foods of marine origin, the arsenic content of most foods is very low. Analysis of a great number of Finnish foods seldom found an item containing more than 50 ng As/gm fresh weight (15–18). According to a U. S. total diet study (19), almost all dietary arsenic came from the meat, fish, and poultry classes of foods. A small quantity of dietary arsenic also came from grain and cereal products. Arsenic was essentially nondetectable by the methodology used with all other classes of foods. Table 1 shows the arsenic content of some selected marine organisms. The values given are not absolute. Falconer et al. (21) found that fish and shellfish caught in different areas vary greatly in their arsenic content. Table 1 also indicates that much of the arsenic in these organisms is in an organic form. This organic form apparently passes through humans without being utilized, accumulated, transformed, or further metabolized (vide infra). Thus, the amount of arsenic in seafood that can be metabolized or transformed probably is not much higher than that found in most other foods.

The arsenic content of foods of plant origin is quite low because root uptake of arsenic generally is quite low. The uptake is low because arsenic has to compete with phosphorus not only at the root level but for absorption on clay surfaces. Factors that influence the small amount of arsenic found in foods of plant origin include the following, according to Woolson (22); 1) Plants grown on sands and

Table 1. Inorganic and Organo-Arsenic in Some Marine Organisms

| Species | As, μg/gm dry weight[a] | | As, μg/gm wet weight[b] |
	Inorganic	Organic	
Cod	1.3	23.0	0.50
Haddock	0.9	12.0	
Herring	1.0	5.2	0.51
Mackerel	1.1	8.9	
Prawn	1.8	36.5	
Perch			0.56
Salmon			1.20
Rainbow trout			0.27
Flounder			0.95
Redfish			2.50
Whitefish			0.02
Pike			0.03
Burbot			0.02

[a] Data from Ref. 20.
[b] Data from Ref. 16.

sandy loams have higher contents at equivalent arsenic levels than those grown on heavier textured soils (silts, clays). 2) Fruit has lower arsenic levels than leaves, stems, or roots. 3) Roots contain the highest levels of arsenic. 4) The skin of root crops has higher levels of arsenic than the inner flesh. 5) Various crops are different in their uptake of arsenic. 6) The application of agricultural arsenicals can influence arsenic levels in plants.

It is apparent from the preceding that daily arsenic ingestion varies widely. A diet high in fish and seafood would be much higher in arsenic than a diet based on leafy vegetables, fruits, and some meats. Several reports (18,19,21,23) indicate that arsenic intake probably is in the range of 8–125 µg/day for most people.

Arsenic Metabolism

The dynamics of arsenic distribution in the body as well as the kinetics of its elimination in urine and feces vary substantially depending on the chemical form and mode of administration. In normal metabolism, oral intake of arsenic is of primary importance, and thus is emphasized here.

Inorganic arsenate and arsenite are well absorbed by higher animals, including humans. The form of organic arsenic determines the extent of its absorption. Table 2 indicates that compounds like arsenobetaine or arsenocholine are very well absorbed, sodium-p-N-glycolylarsanilate are poorly absorbed, and dimethylarsinic acid are intermediately absorbed.

Only a limited number of studies have examined the mechanisms involved in the gastrointestinal absorption of arsenic. Arsenic in the form of arsenate apparently is absorbed in a manner similar to phosphorus as phosphate (32). Some forms of organic arsenic (carbarsone, tryparsamide, and dimethylarsinic acid) were found to be absorbed at rates directly proportional to their intestinal concentration over a 100-fold range in rat intestine (33). This finding indicates that some organic arsenicals are absorbed mainly by simple diffusion through lipoid regions of the intestinal boundary.

As indicated in Table 2, the excretion of ingested arsenic is quite rapid, with both urine and feces serving as major routes of elimination. Some of the fecal arsenic probably represents that absorbed and excreted via the bile (34–36), but most of the fecal arsenic probably is of exogenous origin. Absorbed organic arsenic, such as that found in seafood, apparently undergoes limited chemical change and mixes very little with the inorganic arsenic pool in higher organisms (30,37, 38). Early studies on this subject suggested no chemical change or mixing, but recent reports indicate that arsenocholine can be converted into arsenobetaine by mice and rats (25), dimethylarsinic acid can be further methylated to a trimethylarsenic compound in hamsters (26),

Table 2. Some Reported Excretory Routes of Organic Arsenicals Administered Orally

Organic Arsenical	Species	Days postadministration	% Arsenic found		Reference
			Urine	Feces	
Arsenobetaine	Mouse	3	98.1	1.7	24
Arsenocholine	Rat	3	68.4	5.8	25
	Mouse	3	64.4	12.2	25
Dimethylarsinic acid	Hamster	1	45.0	34.7	26
Trimethylarsenic in prawns	Human	3	86—89	—	27
Sodium-p-N-glycolylarsanilate	Rat or Human	3	4—5	>90	28
Plaice organoarsenic (arsenobetaine)	Human	5	69—85	—	29
Witch flounder organoarsenic	Human	8	76	0.33	30
Flounder organoarsenic	Human	8—9	77	—	31

and trimethylarsenic can be converted to inorganic arsenic, methylarsonic acid, and dimethylarsinic acid in humans (27).

Arsenic ingested as an inorganic form appears in urine in both the inorganic and methylated forms. However, the proportion of the forms of arsenic in urine is species-dependent. Tam et al. (39) found that the arsenic in human urine was 51% dimethylarsinic acid, 21% monomethylarsenic compound, and 27% inorganic arsenic after an oral dose of inorganic arsenic. Dimethylarsinic acid was much more prevalent in urine of mice (about 80% total arsenic) given inorganic arsenic orally (4) and in urine of dogs (about 90% total arsenic) given inorganic arsenic intravenously (41). Rats given an oral dose of arsenate excreted only 4% of the dose as dimethylarsinic acid and 0.3% as methylarsonic acid after 48 hours (40). About 13% of the dose was excreted as inorganic arsenic. The marmoset monkey apparently does not methylate inorganic arsenic at all (42).

The proportions of the forms of arsenic in urine may also be dependent on intake. Foa et al. (38) found that with environmental exposure to arsenic, urinary arsenic consisted of 10% each of inorganic arsenic, monomethylarsonic acid, and dimethylarsinic acid. The remaining 70% consisting of other forms of organic arsenic. Upon exposure to larger amounts of inorganic arsenic, the proportion of dimethylarsinic acid increased.

The site of arsenic methylation has not been clearly established. Tam et al. (43) found that human or dog, urine, plasma, and red blood cells did not methylate arsenic in vitro. Shirachi et al. (44) found that, in vitro, rat liver methylated sodium arsenate. Furthermore, they suggested that two different enzymes were involved in the methylation because monomethylarsonic acid was formed in all subcellular fractions of the liver, whereas dimethylarsinic acid was formed mainly in the supernatant fraction. On the other hand, Lerman et al. (45) obtained findings suggesting that the liver is not the primary site for the metabolism for arsenate. They thought the kidney may be involved in the methylation of arsenate.

When the retention of arsenic by higher animals is discussed, it is difficult to make generalizations because they do not all metabolize arsenic in the same manner. For example, rats, unlike other mammals, concentrate arsenic in their red blood cells. This accumulated arsenic apparently is mainly associated with the hemoglobin, or hematin (99%), fraction (46). In the marmoset monkey, which does not methylate arsenic, retention of arsenic is relatively long, with the liver a major site of retention (42). About 50% of the arsenic, as an inorganic form, is strongly bound to the rough microsomal membranes. A similarly strong binding of arsenic to microsomes has not been observed in other animal species. Species differences were also demonstrated by Marafante et al. (47), who compared the metabolism of arsenic in rats and rabbits after an intraperitoneal injection of [74]As-labeled arsenite. The highest arsenic concentrations at 12 and 48 hours after injection were found in liver, kidney, and lung of rabbits. The arsenic in these tissues was rapidly cleared. In corresponding tissues of rats, the rate of arsenic decline was significantly lower, because of the higher binding of arsenic to tissue constituents. Poor binding of arsenic to plasma proteins was seen in rabbits, while in rats it was totally bound to this fraction. The amount of dimethyl-arsinic acid in the tissues was lower in the rat than in the rabbit, reflecting the total amount of diffusible arsenic, which was also much lower in tissues of rats than in rabbits. In contrast to the above, rabbits retain arsenic longer than mice (48).

Vahter and Marafante (48) suggested that the differences in metabolism of arsenic between animal species, e.g., retention time, may be partly due to differences in the rate of methylation. They also indicated that differences in the binding sites of arsenic in the tissues may be partly responsible. Rowland and Davies (49) found that the biotransformation of arsenic in the rat is similar to other animals. Thus, they concluded that differences in biotransformation of arsenic compounds between animal species are not the main reason for interspecies differences in arsenic excretion and tissue distribution.

A study of arsenic retention over a 103-day period in persons given oral [74]As as arsenic acid gave data that were best represented by a three-component exponential function, with the values of the coefficients being 65.9%, with a half-life of 2.09 days; 30.4%, with a half-life of 9.5 days; and 3.7%, with a half-life of 38.4 days (50). There is no question that homeostatic mechanisms are present in humans to prevent unnecessary accumulation, or retention, of arsenic ingested at physiological levels. (For more extensive review of the metabolism of arsenic, see Ref. 51.)

Signs of Abnormal Arsenic Nutrition

In 1975—1976, the first findings indicating that arsenic is essential came from two laboratories. As a result of those and subsequent investigations, signs of arsenic deprivation were described for 4 animal species. The signs of deficiency for minipigs and goats were recently reviewed by Anke et al. (52,53) and those for chicks and rats by Uthus et al. (54). In the goat, minipig, and rat, the most consistent signs of arsenic deprivation were depressed growth and abnormal reproduction characterized by impaired fertility and elevated perinatal mortality. Other notable signs of deprivation in goats were depressed serum triglycerides and death during lactation. Histologic examination revealed myocardial damage in the lactating goats that died. The organelle of the myocardium most profoundly affected by arsenic deprivation was the mitochondria at the membrane level. In advance stages the membrane actually ruptured, allowing mitochondrial materials to lie free in the cytoplasm. There have been other reported signs of arsenic deprivation. However, definitively stating deficiency signs seems risky because studies with chicks indicated that the extent, severity, and direction of the signs of arsenic deprivation were affected by the arginine, methionine, and zinc status of the animal. Thus, until more is known about the essential function of arsenic, listing the signs of arsenic deficiency must be done with care.

Because there are mechanisms for the homeostatic regulation of arsenic (vide supra), the toxicity of arsenic through oral intake is relatively low. It is actually much less toxic than selenium, a trace element with a well-established nutritional value. Toxic quantities of arsenic generally are measured in milligrams and the ratio of the toxic to nutritional dose for rats apparently is near 1250. The toxicity of a given arsenical depends upon its rate of excretion from the body and, thus, the degree to which it accumulates in tissues. Because species metabolize arsenic differently (vide supra), it is not surprising that there apparently are interspecies differences in resistance to arsenic poisoning. Vallee et al. (55) estimated that the acute fatal dose of arsenic trioxide for humans is 70–180 mg (about 0.76–1.95 mg As/kg body weight of a 70-kg human). This dose, expressed on the body weight basis, is much less than that for rats.

It is impossible to list all signs of arsenic toxicity here because they vary extensively depending upon the species examined and the form, dose, and length of exposure to the arsenical. A recent review of the subject was done by Squibb and Fowler (56). Briefly, the signs of subacute and chronic high arsenic exposure in humans include the development of dermatoses of various types (hyperpigmentation, hyperkeratosis, desquamation, and loss of hair, etc.), hemapoietic depression, anhydremia caused by loss of fluids from blood into tissue and the gastrointestinal tract, liver damage characterized by jaundice, portal cirrhosis, and ascites, sensory distrubances, peripheral neuritis, anorexia, and loss of weight.

Numerous epidemiological studies have suggested an association between chronic arsenic overexposure and certain diseases, such as cardiovascular disease or cancer. However, the role of arsenic in these disorders, especially carcinogenesis, remains controversial, because laboratory studies have not succeeded in producing tumors in animals. A recent examination of this controversy (57) resulted in the conclusion that there apparently is some influence of arsenic on carcinogenesis. However, this influence is not that of a direct carcinogenic action, but that of indirectly influencing other metabolic systems (e.g., immune system) or nutrients (e.g., selenium) that may have a more direct role in the carcinogenic process.

Arsenic Function

The evidence showing that arsenic is essential does not clearly define its metabolic function. Thus the mode of action of arsenic is open to conjecture. Recent findings (58) suggest that arsenic might have a role that affects taurine or sulfate production from methionine. Such a role seems reasonable because arsenic affects, or is affected by, methyl, and thus methionine, metabolism (vide supre). Perhaps the effect is through a function in some enzyme system because arsenic has been shown to activate or inhibit a number of enzymes in vitro.

Another possible role for arsenic is one related or similar to lipid phosphorus in biological systems. Marafante et al. (25) suggested that the long retention of ^{73}As following (^{73}As) arsenocholine administration was probably due to the incorporation of the arsenocholine into phospholipids in the same way that choline is incorporated. This may explain the high levels of arsenic found in some tissues rich in lipids. In higher animals, arsenocholine can replace choline in some of its functions (59). Arsenocholine is antiperotic and growth promoting in the choline-deficient fowl.

Arsenic Requirements

An arsenic requirement of less than 50 ng/gm of diet and probably near 25 ng/gm was suggested for chicks and rats fed an experimental diet containing 20% protein, 9% fat, 60% carbohydrate, and 11% fiber, minerals, and vitamins (60). An arsenic requirement of less than 50 ng/gm of diet was also suggested for goats (53). Thus, the arsenic requirement apparently was somewhere between 6.25 and 12.5 µg/1000 kcal. Extrapolating from animal data, one might conclude that a possible arsenic requirement for humans eating 2000 kcal/day would be about 12−25 µg daily.

Pharmacologic Actions of Arsenic

In the history section (vide supra), it was mentioned that, in 1937, more than 8000 arsenic compounds were considered to be pharmacologic. Fowler's solution, a 1% solution of As_2O_3 as potassium arsenite, which was first formulated in 1786, was still considered the best medicinal in the Pharmacopoeia in 1912. Thus, to try to elaborate briefly on the pharmacological actions of arsenic would be futile. A limited review of the numerous pharmacological actions of various arsenicals was given by Klevay (32). It should be suffice to say here that in biochemical, clinical, and nutritional studies, the investigator should be aware that arsenic can act pharmacologically in numerous ways, including through binding to sulhydryl groups, replacing the phosphate group, or by interacting with other nutrients such as selenium, iodine, and molybdenum (61,62). These pharmacological actions should not be construed as manifestations of an essential physiological function of arsenic.

Nutritional Implications of Arsenic

It is inappropriate to suggest specific disorders in which subnormal arsenic nutrition is a contributing factor until more is known about its physiological function. At present, it is probably most important to be aware of the likelihood that arsenic is essential for humans. Beliefs that any form or amount of arsenic is unnecessary, toxic, or carcinogenic might lead to efforts for a zero-based exposure to arsenic or for elimination of as much arsenic as possible from dietary sources for humans. This could be catastrophic because the amount of arsenic in some diets (8 µg/day; vide supra) may be too low already based upon the dietary requirements suggested by data from animals (12–25 µg/day). Because of this there is a slight sense of urgency for the clarification of the need of arsenic for optimal health and performance so that safe and adequate intakes of the element can be established.

BORON

History

In 1857, Wittstein and Apoiger (see Ref. 63) detected the presence of boron in *Maesa picta* seeds. According to a review by Pfeiffer et al. (64), boron toxicity was first described in the late 1800s to early 1900s. However, it was not until after 1910 that boron was recognized as an element of physiological importance. In that year Agulhon (65) reported findings indicating that boron was essential for higher plants. Conclusive evidence and acceptance of the essentiality of boron probably dates from the work of Warington reported in 1923 (66). Between 1939 and 1944, several attempts to induce a boron deficiency in rats were unsuccessful, although the diets used apparently contained only 155–163 ng B/gm (67–69). In 1945, there was a report (70) that supplemental dietary boron enhanced survival and maintenance of body fat and elevated liver glycogen in potassium-deficient rats. Those findings were not confirmed in a subsequent study (71) in which rats were fed a different diet with an unknown boron content and different levels of boron supplementation. After those reports, boron was generally accepted as being essential for plants, but not for animals. In 1981–1985, however, evidence was accumulated that indicated that boron is an essential nutrient (vide infra). Thus, studies on the biochemical, nutritional, and physiological roles of boron in higher animals are quite limited.

Chemistry of Boron

Boron exhibits bonding and structural characteristics intermediate to metals and nonmetals. Like carbon, boron has a tendency to form double bonds and macromolecules. Like elements such as aluminum and germanium, boron complexes with organic compounds containing hydroxyl groups. The complexion occurs best when the hydroxyl groups are adjacent and *cis*. Compounds with more than two hydroxyl groups react more strongly, and the intensity of the reaction increases with an increase in the number of the adjacent hydroxyl groups (glycerol < erythritol < adonitol < arabitol < mannitol) (67). Complexes may also be formed with diols that do not have *cis* hydroxyl groups if the angle between the adjacent hydroxyl group relative to the carbon-carbon axis is suitable. In other words, certain ring configurations make it possible for boron to complex with hydroxyls of *trans* 1,2-diol groups.

Because of its ability to complex with hydroxyl groups, boron complexes with many substances of biological interest. These substances include numerous sugars and polysaccharides, adenosine 5-phosphate, pyridoxine, riboflavin, dehydroascorbic acid, and pyridine nucleotides (67,68). It is thought that the compounds formed among sugars, sugar alcohols, and borate are one of several types (Fig. 1). The structure of the only biologically systhesized compounds containing boron identified to date contains boron bound to four oxy groups as shown by III. These compounds are aplasmomycin, a novel ionophoric macrolide antibiotic which was isolated from strain SS-20 of *Streptomyces griseus*, and boromycin, an antibiotic produced by *Streptomyces antibioticus* (69,70).

Boron Content of Foods

A recent extensive study of the mineral content of over 200 Finnish foods included boron (15–18). The average boron content ($\mu g/gm$ dry weight) in different food groups was: cereals, 0.92; meat, 0.16; fish, 0.36; dairy products, 1.1; vegetable foods, 13; and other, 2.6. Foods that contained the highest levels of boron ($\mu g/gm$ wet weight) included soy meal, 28; prune, 27; raisin, 25; almond, 23; rose hips, 19; peanut, 18; hazelnut, 16; date, 9.2; and honey, 7.2. Wines contained up to 8.5 μg B/gm. The high level of boron in nuts, fruit kernels, and honey was also reported elsewhere (71–73). Other reports of high levels of boron in fruits and vegetables are those of Ploquin (63), Schlettwein-Gsell and Mommsen-Straub (74), and Szabo (75). Examples of values reported by Szabo are (in $\mu g/gm$ dry weight): apple, 468; pear, 709; tomato, 1258; and red pepper, 440. Like other elements, the boron content of foods from plants can be

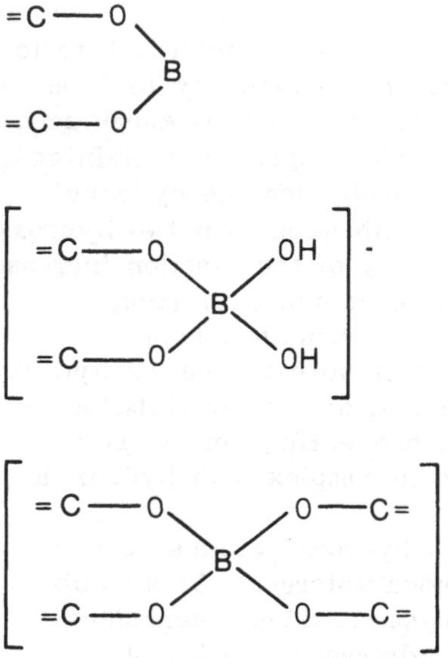

Fig. 1 Proposed types of boron structures in biological material.
The only structure identified to date in biologically synthesized
compounds is shown by III.

influenced by several factors, such as boron content of the medium
on which the plant is grown, stage of maturity of the plant when the
edible portion is harvested, and the portion of the plant to be consumed (63). Nonetheless, it is obvious that foods of plant origin are
rich sources of boron. Meat or fish are apparently poor sources of
boron.

It is apparent from the preceding that the daily intake of boron
by humans can vary widely depending upon the proportions of various
food groups in the diet. Ploquin (63) estimated that, excluding
beverages (wine, cider, beer), the daily intake of boron would be
near 7 mg; with beverages included, the daily intake would be much
higher. These values are higher than others recently reported. The
reported average daily intake of Finnish people is 1.7 mg of boron
(76). In their study of the minerals in total diets made from various
sources to supply 4200 kcal/day, Zook and Lehmann (77) measured
an overall average boron content of 3.1 mg; with individual composites
varying relatively little from 2.1 to 4.3 mg/day. These levels conform
with the mean of 2.8 ± 1.5 mg/day reported for English total diets
(78).

Boron Metabolism

Boron in food, sodium borate, or boric acid is rapidly absorbed and excreted largely in the urine (79–82). Because of variations in the boron dose and length of the study, reported recoveries of ingested boron from the urine range from 30 to 92%. Kent and McCance (79) did human balance studies in which a 352-mg dose of boron as boric acid was ingested on day one. At the end of 1 week, over 90% of the boron was recovered from the urine. Over 40 years later, Jansen and Schou (82) fed to six male volunteers each a single dose of boric acid (750–1473 mg) as either a water solution or a 3% waterless, water-emulsifying ointment. After 96 hours, mean urinary recovery was 93.9% and 92.4%, respectively. Pfeiffer et al. (64) recovered 40% of the boron ingested by dogs in a 24-hour urine collection. Owen (80) found that, in two dairy cows, 57% of the boron from a control ration (271 mg B/day) was excreted in the urine, but 71% of the boron from boronated rations (2058 and 2602 mg B/day) was recovered from urine. Akagi et al. (83) reported that after 48 hours over 70% of an oral dose of boron as sodium borate was excreted in the urine by rabbits and guinea pigs; after 120 hours over 80% of the dose was excreted in the urine. Jansen et al. (84) infused into eight male volunteers each a single dose of boric acid (562–611 mg) within 20 minutes. The plasma concentrations curves, followed for three days, best fitted a three-compartment open model.

Signs of Abnormal Boron Nutrition

In 1981, Hunt and Nielsen (85) reported that boron deprivation depressed growth and elevated plasma alkaline phosphatase activity in chicks fed inadequate cholecalciferol. Some subsequent experiments indicated that cholecalciferol deficiency enhanced the need for boron and that boron might interact in some manner other than through a direct effect on cholecalciferol metabolism, with the metabolism of calcium, phosphorus, or magnesium (86). The relationship seemed strongest between boron and magnesium, because boron tended to normalize the abnormalities associated with magnesium deficiency in chicks. Boron did not consistently alleviate signs of calcium and phosphorus deficiency. Nonetheless, because the B:Mg ratio was quite low in both plasma and diet, boron apparently indirectly affected magnesium metabolism.

Recent experiments with rats suggest that the indirect influence of boron was the result of altered parathormone activity (87). Dietary boron markedly affected the response of rats to treatments that supposedly cause changes in parathormone activity. For example, magnesium deficiency, which causes an apparent hyperparathyroid state in rats, depressed growth and elevated the spleen weight/body weight, liver weight/body weight, and kidney weight/body weight ratios. The changes were more marked in boron-deprived than boron-supplemented rats. Furthermore, the differences due to dietary boron were the greatest when dietary methionine was marginal or possibly deficient.

Studies on the signs of boron deprivation are still in the formative stages. Therefore, definitively stating deficiency signs is difficult. The most consistent sign of deficiency is depressed growth. Almost all other reported signs of deficiency vary in extent, severity, and direction when diets were varied in content of calcium, phosphorus, magnesium, cholecalciferol, aluminum, and methionine. Until more is known about the essential function of boron, it seems inappropriate to list these other signs, because they probably occur only under certain conditions.

Boron has a low order of toxicity when administered orally. Three excellent reviews of the toxicity of boron (63,88,89) indicate that toxicity signs generally occur only after the dietary boron concentration exceeds 100 μg/gm. Weir and Fisher (90) found that, after two months, rats fed 1170 μg B/gm diet exhibited coarse hair coats, scaly tails, a hunched position, swelling and desquamation of the pads of the paws, abnormally long toenails, bloody discharge of the eyes, and depressed hemoglobin and hematocrit. After 38 weeks, dogs fed 1170 μg B/gm diet exhibited reversible testicular degeneration and cessation of spermatogenesis. Green et al. (91) found that when boron exceeded 150 mg/liter in drinking water, rats exhibited depressed growth, continued prepubescent fur, lack of incisor pigmentation, aspermia, and impaired ovarian development. When the boron content of drinking water was 300 mg/L, rats also exhibited depressed bone fat and calcium (92). In humans, the signs of acute toxicity are well known and include nausea, vomiting, diarrhea, dermatitis, and lethargy (88,93). In addition, high boron ingestion includes riboflavinuria (93). The association between riboflavin and boron is consistent with Landauer's (94) observation that newly hatched chicks treated with boric acid at 96 hours of incubation exhibited "curled toe paralysis," an abnormality associated with riboflavin deficiency. Landauer (94,95) also found that boron-induced teratogenic abnormalities, including several types of skeletal abnormalities, were reduced by the administration of riboflavin. Other polyhydroxy compounds (D-ribose, pyridoxine·HCl, D-sorbitol hydrate) also reduced or abolished the teratogenic effects of boric acid on chick embryos (95).

Boron Function

Although the data need clarification by further experimentation many

findings support the hypothesis that boron has an essential function
that influences parathormone action, and therefore indirectly influences
the metabolism of calcium, phosphorus, magnesium, and cholecalciferol.
In addition to findings already described (vide supra), others have
reported evidence to support this hypothesis. Elsair and co-workers
(96,97) found that high dietary boron partially alleviated the fluoride-
induced secondary hyperparathyroidism signs of hypercalcemia,
hypophosphatemia, and depressed renal absorption of phosphorus in
rabbits. Seffner and co-workers (98,99) found that dietary borate,
and magnesium silicate to a lesser extent, reduced the fluoride-
induced thickening of the cortices of the long bones in pigs, and
that dietary boron altered the histology of the parathyroid. Baer
et al. (100) reported that boron corrected radiographic and histologic
changes caused by fluoride toxicity in bone. It would not be surprising to find that boron has a regulatory role involving a hormone such
as parathormone because boron is suspected of having a regulatory
role in the metabolism of such plant hormones as auxin, gibberellic
acid, and cytokinin (101,102). perhaps through control of the production of a second messenger, such as cyclic AMP, at the cell
membrane level (101,103).

Other reports also suggest that boron has a function at the cell
membrane level. Palytoxin, an extremely poisonous animal toxin from
coral, raises the permeability of excitable and nonexcitable membranes
of animals. The binding of palytoxin to membranes is potentiated by
borate (104). Aplasmomycin, a novel ionophoric macrolide antibiotic,
is a boron-containing compound (70). Several recent reviews have
presented evidence consistent with the view that boron is directly
associated with membranes and is involved in maintaining their
functional efficiency (105-107). In other words, many symptoms of
boron deficiency in plants are secondary effects caused by changes
in membrane permeability. Tanada (108) found that a major part of
the boron of mung bean seedlings was localized in the membranes.

The possibility that boron has a role in some enzymatic reaction
cannot be overlooked because boron has been shown to affect the
activity of numerous enzymes in vitro and in plants. Wolny (109)
briefly reviewed the in vitro enzyme work and found that borate
competitively inhibits two classes of enzymes. One class is the
pyridine or flavin nucleotide-requiring oxidoreductases, such as
yeast alcohol dehydrogenase, aldehyde dehydrogenase, xanthine
dehydrogenase, and cytochrome b_5 reductase. Borate apparently
competes with the enzyme for NAD or flavin because of its great
affinity for cis-hydroxy groups. The other class of borate-inhibited
enzymes are those in which borate and boronic acid derivatives bind
to the active enzyme site. These enzymes include chymotrypsin,
subtilism, and glyceraldehyde-3-phosphate dehydrogenase. Plant

enzymes affected by boron nutriture have been reviewed by Duggar (107) and Lewis (102).

Boron Requirements

The minimum amounts of boron required by animals to maintain health, based on the addition of graded increments to a deficient diet in the conventional manner, have not been determined. However, based on the finding that rats and chicks sometimes have altered mineral metabolism when fed diets containing 0.3 to 0.4 μg B/gm (85—87), requirements probably exceed this level of boron.

Pharmacologic Actions of Boron

The ability of boron to complex with a large number of biologically important substances containing cis-hydroxy groups (vide supra) indicates that this element could be a pharmacologically active element. Boron could complex with substrates, end products, or enzymes themselves, thus inhibiting or stimulating various metabolic pathways.

In the past, boric acid has been used as an eyewash and as an antiseptic in pharmaceutical preparations, although it has been shown to have limited bacteriostatic activity (88). Recently, some organic boron compounds have been synthesized that apparently have beneficial pharmacological effects. Amine cyanoboranes and amine carboxy-boranes (boron analogs of α-amino acids) were shown to inhibit inflammation and induced arthritis in rodents (110). Boron analogues of betaine were shown to have antitumor and hypocholesteremic effects in mice (111). Tablets containing magnesium carbonate and sodium borate have been touted as a remedy for arthritis (112). These limited studies suggest that the pharmacology of boron may be a fruitful area of research.

Nutritional Implications of Boron

It is too soon to suggest specific disorders in which subnormal boron nutrition is a contributing factor. More needs to be known about its physiological function. However, it seems safe to state that boron is a dynamic ultratrace element that affects major mineral metabolism in higher animals. Thus, boron may have a role in some disorders of unknown etiology that exhibit disturbed major mineral metabolism (e. g., osteoporosis). Further research on the nutritional, biochemical, and clinical aspects of boron is needed to evaluate the nature and importance of the physiologic and pharmacologic actions of boron.

MOLYBDENUM

History

Although the presence of molybdenum in biological material was discovered in 1900, studies concerned with its biological properties did not receive much attention until after 1930 (113), when molybdenum was discovered to be required for the growth of the nitrogen-fixing organism *Azotobacter*. Shortly thereafter (114), a long-recognized severe diarrhea of cattle (teart scours) in England was found to be caused by molybdenum toxicosis, and attention turned to the metabolic significance of molybdenum in animals. Evidence for the essentiality of molybdenum first appeared in 1953 when xanthine oxidase was identified as a molybdenum enzyme. Subsequently, attempts to produce molybdenum deficiency signs in rats and chickens were successful only when the diet contained massive levels of tungsten, an antagonist of molybdenum metabolism. These studies (vide infra) showed that the dietary requirement to maintain normal growth of animals was less than 1 μg molybdenum/gm diet, a level substantially lower than requirements for other trace elements recognized as essential at the time. Thus, molybdenum was not considered to be of much practical importance in animal and human nutrition. Consequently, over the past 30 years, relatively little effort was devoted to studying the metabolic and pathologic consequences of molybdenum deficiency in monogastric animals or humans. Recently, however, interest in molybdenum has been stimulated by reports indicating that it may affect cancer incidence, and that deficiency was produced in a human receiving total parenteral nutrition.

Chemistry of Molybdenum

Molybdenum is a transition element that readily changes its oxidation state and can thus act as an electron transfer agent in oxidation-reduction reactions. Because molybdenum can exist in oxidation states from -2 to +6 and coordination ranging from 4 to 8, it has a very complex chemistry. Animal enzymes known to contain molybdenum are shown in Table 3. Spence (116) and Rajagopalan (115) have reviewed the chemistry of those enzymes. In the oxidized form of those enzymes, molybdenum is probably present as the +6 state. Although the enzymes during electron transfer are probably first reduced to the +5 state, the oxidation state of the completely reduced enzyme is uncertain. There is evidence that one or more of the enzymes, in the presence of excess substrate, can have molybdenum present in either the +4 or +3 state. The molybdenum apparently is present at the active site of the enzymes as molybdopterin, a small nonprotein cofactor containing a pterine nucleus (115,117,118). This

Table 3. Animal Molybdenum Enzymes[a]

Enzyme	Substrate	Electron donor or acceptor
Aldehyde Oxidase (E.C.1.2.3.1)	Aldehydes	O_2
Sulfite oxidase (E.C.1.8.3.1)	SO_3^{2-}	O_2
Xanthine dehydrogenase	Purines	NAD
Xanthine dehydrogenase	Purines	Ferredoxin
Xanthine oxidase (E.C.1.2.3.2)	Purines	O_2

[a]Other enzymes found in plants and microorganisms include pyridoxal oxidase, nicotinic acid hydroxylase, purine hydroxylase, CO dehydrogenase, nitrate reductase, formate dehydrogenase, and nitrogenase (115).

cofactor, comprising over 50% of the nonenzymic form of molybdenum in the liver, is bound to the mitochondrial outer membrane, and can be transferred to an apoenzyme of xanthine oxidase or sulfite oxidase, forming an active enzyme molecule.

As indicated in two recent reviews (119,120), another important biological form of molybdenum is molybdate. In herbage, molybdenum is present as sodium, calcium, and ammonium molybdate (119). Plants also contain some molybdenum sulfide and molybdenum oxide. There is evidence (120) suggesting that molybdenum in blood and urine exists mainly as the molybdate ion (MoO_4^{2-}). Thus, molybdate in food and water apparently is not radically changed by absorption and transport in the blood.

Molybdenum Content of Foods

The analysis of foods for molybdenum content has received limited attention apparently because molybdenum has been considered to be of relatively little practical importance in nutrition. Three major studies that have been done (15–18,76,121,122) showed that the richest food sources of molybdenum include milk and milk products, dried legumes, organ meats (liver and kidney), cereals, and baked goods (Table 4); the poorest sources of molybdenum include vegetables other than legumes, fruits, sugars, oil, fats, and fish. In the more recent studies (15–18,122) foods were quite consistent in molybdenum content considering the possible variations in the source of the foods and processing methods.

As with other elements, daily intake of molybdenum varies depending upon the makeup of the diet. Most estimates for the average daily intake of molybdenum range between 80 and 350 μg (120). For

Table 4. Selected Foods High in Molybdenum Content

Food	Molybdenum, μg/gm fresh wt.[a]
Liver, beef	1.60
Liver, pork	2.00
Kidney, beef	0.60
Kidney, pork	0.70
Butter lima beans (canned)	1.70
Lima beans (dry)	8.70
Small white beans (dried)	4.5
String beans (fresh)	0.60
Green split peas (dry)	1.50
Sweet peas (canned)	0.20
Bakery (sweets, all types)	0.27
Bread (all types)	0.27
Breakfast cereal (Quaker Oats)	1.80
Rice	0.29
Cheese	0.11
Milk (fresh)	0.05
Pork and beans	0.44
Macaroni	0.38
Whole milk powder	0.40

[a]The liver and kidney values are from Nuurtamo et al. (16), the milk powder value from Varo et al. (15), and all other values are from Tsongas et al. (122).

example, using USDA estimates of food consumption and composition as a guide, Tsongas et al. (122) estimated an average daily intake of 180 μg/day. A study of total diets from different regions of the United Kingdom led to a reported average daily intake of 128 μg Mo (78).

Molybdenum Metabolism

Molybdenum (except as MoS_2) in foods and in the form of soluble complexes is readily absorbed. In humans, between 25 and 80% of ingested molybdenum is absorbed. In his review, Winston (120) stated that studies with rats indicated that molybdenum absorption takes place in the stomach and small intestine, the rate of absorption being higher in the proximal than in the distal parts of the small intestine. No absorption of molybdenum takes place in the large intestine. Whether an active or a passive mechanism is most important in the absorption of molybdenum is uncertain. One study (123) produced evidence indicating that at low concentrations of molybdenum, its absorption was carrier-mediated and active. Another study (124) showed that in vivo absorption rates were essentially the same over a 10-fold range of molybdenum concentrations. This finding and the finding that the rate of absorption in both stomach and the small intestine was high suggest that the molybdenum was absorbed via diffusion only. Winston (120) raised the possibility that molybdate is moved both by diffusion and by active transport, but that at high concentrations the relative intensity of the latter is small. After absorption, molybdenum is rapidly turned over and eliminated as molybdate via the kidney, thus indicating this, rather then regular absorption, is the major homeostatic mechanism for this element (120).

Signs of Abnormal Molybdenum Nutrition

The signs of molybdenum deficiency have been reviewed by Mills and Bremner (125). In rats and chickens, molybdenum deficiency, aggravated by excessive dietary tungsten, results in depression of the molybdenum enzymes (xanthine oxidase, aldehyde oxidase, and sulfite oxidase), disturbances of uric acid metabolism, and increased susceptibility to sulfite toxicity. Deficiency uncomplicated by tungsten has been produced in goats and minipigs fed diets containing less then 60 ng molybdenum/gm, Deficiency signs were depressed feed consumption and growth, impaired reproduction characterized by elevated mortality in both mothers and offspring, and elevated copper concentrations in liver and brain.

Under field conditions, a molybdenum-responsive syndrome was found in hatching chicks. This syndrome was characterized by a high incidence of late embryonic mortality, mandibular distortion, anophthalmia, and defects in leg bone development and feathering. Skeletal lesions, subsequently detected in older birds, included separation of the proximal epiphysis of the femur, osteolytic changes in the femoral shaft, and lesions in the overlying skin that were ultimately attributed to intense irritation in these areas. The incidence of this syndrome was particularly high in commercial flocks reared on diets containing high concentrations of copper (a molybdenum antagonist) as a growth stimulant. These apparently dissimilar pathologic changes could possibly be explained by a defect in sulfur metabolism. Recognition of the role of molybdenum as a component of sulfite oxidase and evidence that sulfite oxidase deficiency markedly deranges cysteine metabolism suggest that the metabolic consequences of molybdenum deprivation should be reappraised.

The need for this reappraisal is further supported by two recent human studies. A lethal inborn error in metabolism that markedly deranged cysteine metabolism in two patients was determined to result in a sulfite oxidase deficiency (115). Another patient receiving prolonged total parenteral nutrition (TPN) therapy acquired a syndrome described as acquired molybdenum deficiency (126). This syndrome, exacerbated by methionine administration, was characterized by hypermethioninemia, hypouricemia, hyperoxypurinemia, hypouricosuria, and very low urinary sulfate excretion. In addition, the patient suffered mental disturbances that progressed to coma. The symptoms were indicative of a defect in sulfur amino acid metabolism at the level of sulfite transformation to sulfate (sulfite oxidase deficiency), and a defect in uric acid production at the level of xanthine and hypoxanthine transformation to uric acid (xanthine oxidase deficiency). Supplementation of the patient with ammonium molybdate improved the clinical condition, reversed the sulfur-handling defect, and normalized uric acid production.

Large oral doses are necessary to overcome the homeostatic control of molybdenum. Thus, molybdenum is a relatively nontoxic element. In nonruminants an intake of 100–5000 mg/kg of food or water is required to produce clinical toxicity symptoms (119,120). Ruminants are much more susceptible to elevated dietary molybdenum. The mechanisms of molybdenum toxicity are uncertain. However, most toxicity signs are similar or identical to those of copper deficiency (i.e., growth depression, anemia). Signs obviously due to a direct action of molybdenum are not known. However, both occupational and high level dietary exposure to molybdenum have been linked through epidemiologic methods to elevated uric acid levels in blood and increased incidence of gout.

Molybdenum Function

Molybdenum functions as an enzyme cofactor (see Table 3). Animal molybdoenzymes catalyze the transfer of an oxygen atom from water to a variety of compounds. Aldehyde oxidase oxidizes and detoxifies various pyrimidines, purines, pteridines, and related compounds. Xanthine oxidase/dehydrogenase catalyzes the transformation of hypoxanthine to xanthine, and xanthine to uric acid. Sulfite oxidase catalyzes the transformation of sulfite to sulfate.

Molybdenum Requirements

Minimum dietary requirements for molybdenum to maintain optimal health and performance of animals and humans are presently unknown. Deficiency studies not utilizing molybdenum antagonists have not been helpful in determining the dietary requirements of molybdenum because, although molybdoenzyme levels are affected, health and performance has not been noticeably altered in these studies. Thus, human requirements can be estimated only on the basis of balance studies.

The National Academy of Sciences (127) has estimated that an adequate and safe intake of molybdenum is 0.15 to 0.5 mg/day. This amount is easily furnished by most diets consumed in the United States (vide supra).

Pharmacologic Actions of Molybdenum

Molybdenum as a pharmacologic substance has not received much attention. However, the epidemiologic finding that molybdenum levels in foods and water were lower in areas with high incidence than in areas with normal incidence of esophageal cancer (128) apparently has stimulated research in the possible anticarcinogenic properties of molybdenum. One research group has found that, in rats, 2 or 20 ppm molybdenum in the drinking water significantly inhibited N-nitroso-sarcosine ethyl ester-induced esophageal and forestomach carcinogenesis (128), and 10 ppm molybdenum in the drinking water inhibited N-methyl-N-nitroso-urea-induced mammary gland carcinogenesis (129).

Nutritional Implications of Molybdenum

Except for the molybdenum-responsive patient with TPN "acquired molybdenum deficiency" (126), there is no indication that molybdenum deficiency is of clinical importance. However, the existence of this patient, and of patients with genetically induced sulfite oxidase deficiency (115), suggests that further studies examining for possible molybdenum-responsive syndromes in humans are warrented. The absence of a general clinical deficiency syndrome, despite the depression of molybdoenzyme activities associated with low molybdenum intakes, only suggests that the right questions are not being asked.

It is likely that situations can occur in which molybdoenzyme activities can be changed by dietary manipulations. Several reviews (115,117—121,125) have discussed the evidence that numerous nutrients and substances affect molybdenum metabolism, and therefore molybdenum requirement and toxicity. These substances include copper, sulfate, manganese, zinc, iron, lead, tungstate, ascorbic acid, methionine, cystine, and protein. The basis for many of these interactions are yet unexplained. The most important interaction is between molybdenum and copper, and it can be modified by dietary sulfur. The formation of copper tetrathiomolybdates in the gastrointestinal tract apparently is the basis for this interaction. In this compound, copper and molybdenum are unavailable for biological action. Another important interaction is between molybdenum and sulfate. Inorganic sulfate inhibits the intestinal absorption of molybdate in the chicken, rat, and sheep. Other findings indicate that elevated dietary sulfate enhances the excretion of molybdenum. Thus, the possibility that high dietary sulfate or copper might induce a molybdenum responsive syndrome in humans cannot be ignored.

NICKEL

History

The first study of the biological action of nickel was reported in 1826 when the oral toxicity signs exhibited by rabbits and dogs were described. Between 1853 and 1912 numerous studies on the pharmacologic and toxicologic actions of various nickel compounds were described. The findings from these studies were summarized by Nriagu (130). The first reports on the presence of nickel in plant and animal tissues appeared in 1925 (131,132). Although Bertrand and Nakamura (133) first suggested in 1936 that nickel might be an essential element, conclusive evidence for essentiality did not appear until after 1970. Thus, most of the studies on the biochemical, nutritional, and physiological roles of nickel were done after 1970.

Chemistry of Nickel

Both the divalent and trivalent oxidation states of nickel are important in biochemistry. Like other ions of the first transition series, Ni^{2+} has the ability to complex, chelate, or bind with many substances of biological interest (134). In the hydrogenases of several microorganisms (135), a substantial amount of the nickel has been found in the Ni^{3+} state. It has been suggested (135,136) that redox-sensitive nickel may be a primary site of catalytic action in the hydrogenases.

The binding of divalent nickel by various ligands probably is important in the extracellular transport of nickel, intracellular binding of nickel, and excretion of nickel in urine and bile. In human serum, amino acids were found to be components of the low-molecular-weight Ni^{2+}-binding fraction, and L-histidine was found to be the main Ni^{2+}-binding amino acid (137). At physiological pH, nickel coordinates with histidine via the imidazole nitrogen. In rabbit serum, cysteine, histidine, and aspartic acid may be involved in the binding of Ni^{2+} (138). The binding of Ni^{2+} by cysteine probably occurs at N (amino) and S (sulfhydryl) sites. Computer approaches have predicted that the predominant interaction with naturally occurring low-molecular-weight ligands would occur with histidine and cysteine (139).

Albumin is the principal Ni^{2+}-binding protein in human, bovine, rabbit, and rat serum. Nickel apparently is bound to albumin by a square planar ring formed by the terminal amino group, the first two peptide nitrogen atoms at the N-terminus, the imidazole nitrogen of the histidine residue, which is located at the third position from the

N-terminus, and the side-chain carboxyl oxygen of the NH_2-terminal aspartic acid residue (140,141). Canine and porcine albumins, which contain tyrosine instead of histidine at the third position, have less affinity for Ni^{2+} than albumins from other' species (138,142). However, initial binding of nickel was found to take place at the N-terminal tripeptide in canine albumin (142). At higher nickel-to-albumin molar ratios, circular dichroism spectra indicated the presence of sulfur-containing ligands but not any involvement of histidine.

Another substance in human and rabbit serum that contains nickel is a macroglobulin that has been named nickeloplasmin (138). Nickel in nickeloplasmin is not readily exchangeable with $^{63}Ni^{2+}$ in vivo or in vitro. Nickeloplasmin has been suggested to be a ternary complex of serum α_1-macroglobulin and a nickel constituent of serum such as a 9.55 α_1-glycoprotein which has been found to strongly bind Ni^{2+} (138).

Another substance in which Ni^{2+} is tightly bound is urease, a nickel metalloenzyme found in several plants and microorganisms (134). Jack bean urease is stable and fully active in the presence of 0.5 mM EDTA at neutral pH (143). The nickel ion can be removed only upon exhaustive dialysis in the presence of chelating agents or EDTA at low pH, and then it is not possible to restore nickel to reconstitute enzymatic activity. The findings of Dixon et al. (143) were consistent with an octahedral coordination of Ni^{2+} with an unreactive cysteine. Hasnain and Piggott (144) suggested that the nickel ion in urease may be bound in part to a histidine residue in a distorted octahedral complex. Thus, further work is needed to definitely characterize the ligand and coordination geometry of the two nickel ions in urease. Binding of the substrate urea to at least one nickel ion in the enzyme molecule is an integral part of the mechanism in the hydrolysis reaction catalyzed by urease. Blakeley and Zerner (145) have postulated that both nickel ions in urease are involved in the attack on urea, resulting in a tetrahedral intermediate. This hypothesis would require nickel to be present in urease as a binuclear cluster, and analogous to the di-iron center at the active site of the violet acid phosphatase (uteroferrin) from pig allantoic fluid.

Ni^{2+} ions bound to organic acids are probably involved in the uptake and translocation of nickel in plants. Nickel translocates in plants as a stable anionic organic complex. Early suggestions were that these complexes were formed by amino acids. However, recent phytochemical studies showed that in nickel-accumulating plants, nickel was contained as anionic citrate or malate complexes (146).

In microorganisms, there have been three types of enzymes identified that contain tightly bound stoichiometric nickel. These enzymes are the multienzyme complex that has carbon monoxide dehydrogenase activity, methyl coenzyme M methylreductase, and hydrogenase.

Both acetogenic and methanogenic bacteria contain an oxygen-sensitive carbon monoxide dehydrogenase that converts carbon monoxide to carbon dioxide (147). The synthesis of carbon monoxide dehydrogenase apparently requires nickel (147,148). Moreover, the enzyme contains stoichiometric nickel (147,149). The anaerobic dimeric CO dehydrogenase from *Clostridium thermoaceticum* contains 2 nickel, 1 to 3 zinc, 11 iron, and 14 acid-labile sulfur, but, unlike aerobic CO dehydrogenase, no molybdenum (150). Nickel dissociates from CO dehydrogenase as a small molecular weight factor (150), the structure of which remains to be elucidated. However, the nickel atom in CO dehydrogenase reacts with CO and forms a Ni(III)-carbon species (151).

Much of the nickel taken up by methanogenic bacteria is incorporated into a low-molecular-weight compound called factor F430. In a short review (152), factor F430 was described as being nickel porphinoid biosynthetically derived from eight molecules of 5-amino-levulinic acid, with uro porphyrinogen III and sirohydrochlorin being intermediates. Factor F430 is present in all methanogenic bacteria in both protein-bound and free forms. The bound form is a component of methyl-CoM reductase to which it is tightly but not covalently bound.

Nickel-dependent or nickel-containing hydrogenases have been described for organisms such as methanogenic or Knallgas bacteria (153), sulfate-reducing bacteria (135), photosynthetic bacteria (154), aerobic hydrogen bacteria (155), and blue-green alga (156). Hydrogenases, which catalyze the simplest oxidation-reduction process of $H_2 \rightleftharpoons 2H^+ + 2e^-$, are iron-sulfur proteins. The Ni(III) state has been identified in several hydrogenases (135,136,154,155,157). It has been suggested (135,136,157) that redox-sensitive nickel probably is the binding site for the substrate, H_2, even though the hydrogenases contain iron-sulfur centers. In the *Nocardia opaca 1b* hydrogenase, nickel apparently has two functions (155). Two atoms of nickel are loosely bound and are required to hold together the two dimers of the hydrogenase. Two more nickel atoms are firmly bound to the smaller subunit dimer and may function in hydrogen activation. Albracht et al. (157) proposed that in the hydrogenase of *Chromatium vinosum*, the nickel ion has five ligands provided by the protein in a square-pyramidal coordination. Lindahl (158) suggested that the nickel in F_{420}-reducing hydrogenase has equatorial sulfurs with either one or two loosely held axial ligands in roughly a tetragonally distorted octahedron or square pyramid.

Whether or not a Ni(III)/Ni(II) redox couple is important in higher animal metabolism is unknown. However, Niebor et al. (159) have postulated that human serum albumin should stabilize Ni(III) because it is attached to this protein at a nitrogen chelation center similar to that of the deprotonated peptide nitrogens in tri- and tetrapeptides where Ni(II) can be reduced to Ni(III) by air.

Nickel Content of Foods

The nickel content of foods has attracted increased attention since nickel was described as nutritionally essential, and oral nickel was found to produce a positive skin reaction in some nickel-sensitive individuals. Schlettwein-Gsell and Mommsen-Straub (160) adequately summarized the reports up to 1971 on the nickel contents of foods. Some of the more extensive and reliable recent listings of the nickel content in foods have been prepared by Ellen et al. (161), Thomas et al. (162), Brun (163), and Anke et al. (164). Casey (165) reported the nickel content of infant milk foods and supplements. The nickel content of foods from plants can be influenced by several factors, including stage of maturity of the plant when the edible portion is harvested, and the nickel concentration of, and availability from, the soil or medium in which the plant is grown (130,164,166). The nickel content of foods or animal origin may be influenced by the level of nickel in the animal's diet and by exposure to nickel-containing materials during processing (130,164). Nonetheless, foods that generally contain high concentrations (>0.3 μg/gm) of nickel include nuts, leguminous seeds, shellfish, cacao products, and hydrogenated solid shortenings (see Table 5). Grains, cured meats, and vegetables are generally intermediate (0.1–0.3 μg/gm) in nickel content. Foods of animal origin, such as fish, milk, and eggs, are generally low (<0.1 μg/gm) in nickel. The canning of some fruits and vegetables apparently increases their nickel content. For example, Thomas et al. (162) found fresh tomatoes, plums, and rubarb contained 0.09, 0.16, and 0.15 μg Ni/gm respectively, whereas the canned products contained 0.49, 0.36, and 0.49 μg Ni/gm respectively. Canned pineapple contained 0.85 μg Ni/gm.

Total dietary nickel intakes of humans vary greatly with the amounts and proportions of foods of animal (Ni-low) and plant (Ni-high) origin consumed, and with the amounts of refined and processed foods in the diet. However, several reports indicate that nickel intake probably is in the range of 150–700 μg/day (78,168–170).

Nickel Metabolism

Most ingested nickel remains unabsorbed by the gastrointestinal tract and is excreted in the feces (134,138). Limited studies suggest that typically less than 10% of ingested nickel is absorbed. However, a higher percentage may be absorbed in an iron-deficient (171), gravid (172,173), or lactating state (174). Kirchgessner et al. (173) found that pigs absorbed over 19% of nickel ingested from day 21 of gravidity until delivery. Nickel that is absorbed from the intestine is excreted primarily via the urine. Some absorbed nickel may be excreted via the bile, as nickel has been found in the bile of rodents injected with ^{63}Ni^{2+} (175). The nickel content of sweat is high (170,176),

Table 5. Foods High in Nickel Content

	Nickel		
Food	µg/gm, dry wt[a]	µg/gm, fresh wt[b]	µg/gm, fresh wt[b,c,d]
Nuts			
Almond	1.6	1.3	
Black walnut	4.8		
Brazil nut	5.8		
Butter nut	4.3		
Cashew	5.0	5.1	
English walnut	1.1	3.6	
Filbert	1.8	1.6	
Hickory nut	9.8		
Pecan	1.6		
Pistachio	1.1	0.8	
Peanuts		1.6	
Coconut	2.1		
Cacao products			
Cacao powder			9.8
Bittersweet chocolate			2.6
Milk chocolate			1.2
Leguminous seeds			
Red kidney beans			0.45
Broad beans, frozen			0.55
Peas, frozen			0.35
Beans, frozen			0.35
Other			
Spinach			0.39
Asparagus, frozen			0.42
Shortening, solid			0.592–2.772

[a]From Furr et al. (72).
[b]From Ellen et al. (161).
[c]From Thomas et al. (162).
[d]From Nash et al. (167).

pointing to active nickel secretion by the sweat glands. Thus, under conditions of excessive sweating, dermal losses of nickel could be high.

Becker et al. (177) reported that the transport of nickel across the mucosal epithelium apparently is an energy-driven process rather than a simple diffusion and suggested that nickel ions use the iron transport system located in the proximal part of the small intestine. Once in the blood, the transport of nickel is accomplished by serum albumin and by ultrafiltrable serum amino acids ligands. The binding of nickel by these substances was described (vide supra).

The kinetics of $^{63}Ni^{2+}$ metabolism in rodents apparently fits a two-component model (175). A summary of the tissue retention and clearance of $^{63}Ni^{2+}$ administered by all routes of entry has been presented by Kasprzak and Sunderman (178). This summary shows that kidney retains significant levels of nickel shortly after $^{63}Ni^{2+}$ is given. The retention probably reflects the role of the kidney in nickel excretion. The level of $^{63}Ni^{2+}$ in kidney falls quickly over time. After six days, the lung apparently has the highest affinity for nickel (179). Studies with $^{63}Ni^{2+}$ show that nickel readily passes through the placenta. Jacobsen et al. (180) found that, two days after administration, embryonic tissue retained greater amounts of parenteral administered nickel than did maternal tissue. Also, amniotic fluid retains relatively high amounts of orally administered nickel (173). The level of nickel in the fetus does not fall quickly after parenteral administration to the dam, thus suggesting retention or inhibited clearance by the fetus (180,181).

Since it was established that the kidney has a major role in nickel metabolism, identification of the form of nickel in kidney and urine has been an active research area. Sunderman et al. (182) found that renal cytosol and microsomes contained several macromolecular nickel constituents in addition to low-molecular-weight components. Abdulwajid and Sarkar (183) found that nickel in kidney was primarily bound to a glycoprotein with a molecular weight of about 15,000—16,000, containing 10% carbohydrate (high mannose with glactose/glucose and glucosamine) and a protein moiety with a molecular weight of about 12,000. The protein contained high glycine and low amounts of cysteine and tyrosine, thus indicating the protein was not metallothionein. Urine also contained this nickel-binding glycoprotein. Sayato et al. (184) found that the ^{63}Ni metabolite in urine after oral and intravenous administration of $^{63}NiCl_2$ to rats behaved like a complex of ^{63}Ni with creatine phosphate.

Signs of Abnormal Nickel Nutrition

The first description of possible signs of nickel deprivation appeared in 1970. However, those findings, and others that followed shortly thereafter, were obtained under conditions that produced suboptimal growth in the experimental animals (185). Also, some reported signs of nickel deprivation appeared inconsistent. During the years 1975–1983, diets and environments that allowed for apparently optimal growth and survival of experimental animals were prefected for the study of nickel nutrition. As a result, what was thought to be clear signs of nickel deprivation were described for six animal species - chick, cow, goat, minipig, rat, and sheep. Unfortunately, the described signs probably will have to be redefined because recent studies indicate that many of the reported signs of nickel deprivation may have been misinterpreted and might be manifestations of pharmacologic actions of nickel (186). That is, a high dietary level of nickel was alleviating an abnormality caused by something other than a nutritional deficiency of nickel, or was causing a change in a parameter that was not necessarily subnormal. In many studies in which the nickel-deprived animals were compared to controls fed 5–20 µg Ni/gm diet, the iron content of the diet was inadequate. This is of concern because nickel can partially alleviate many manifestations of iron deficiency by pharmacological mechanisms (186).

The suggestion that some of the reported signs of nickel deprivation are misinterpreted manifestations of a pharmacological action of nickel does not necessarily detract from the conclusion that nickel is an essential nutrient. The following signs apparently are representative of nickel deficiency. Iron utilization is imparied. As a consequence, the trace element profile of both the femur and liver changes. In the femur, the concentrations of calcium and manganese are depressed, and the concentrations of copper and iron are elevated. In the liver, the concentrations of copper and zinc are elevated. If the nickel deficiency is severe, growth and hematopoiesis are depressed, especially in marginally iron-adequate animals. Other possible deficiency signs for specific species have been described in recent reviews (134,135).

Because there are mechanisms for the homeostatic regulation of nickel, life-threatening toxicity of nickel through oral intake is low, ranking with such elements as zinc, chromium, and manganese. Nickel salts exert their toxic action mainly by gastrointestinal irritation and not by inherent toxicity. Large oral doses of nickel salts are necessary to overcome the homeostatic control of nickel. Generally, 250 µg or more of nickel/gm of diet is required to produce signs of nickel toxicity in rats, mice, chicks, rabbits, and monkeys (187). The ratio of the minimum toxic dose and the minimum dietary requirement for chicks and rats is apparently near 5000. If animal data can be extrapolated to humans this translates into a daily dose of 250 mg of soluble nickel to produce toxic symptoms in humans.

Recent findings, however, suggest that oral nickel in not particularly high doses can adversely affect health under certain conditions.

Relatively low levels of dietary nickel can exacerbate the manifestations of copper deficiency and severe iron deficiency (vide infra).
Christensen and Möller (188) presented evidence that suggested the
ingestion of small amounts of nickel may be of greater importance
than external contacts in maintaining hand eczema caused by a
sensitivity to nickel. Cronin et al. (189) observed that an oral dose
of 0.6 mg of nickel as $NiSO_4$ produced a positive reaction in some
nickel-sensitive individuals. That dose is only 12 times as high as
the human daily requirement postulated from animal studies.

The metabolism and toxicity of the carcinogen nickel carbonyl
differs markedly from the Ni^{2+}. Nickel carbonyl is highly volatile
and is absorbed readily by the lungs. The inhalation route is the
most important in respect to nickel carbonyl toxicity. Also, the
metabolism and toxicity of relatively insoluble nickel compounds such
as Ni_3S_2 (usually administered intramuscularly or intrarenally) differ
from Ni^{2+} in their metabolism. A review on the toxicity of nickel
carbonyl and relatively insoluble nickel compounds was given by
Sunderman (138).

Nickel Function

To date, there is no firmly established biological function of nickel
in man and animals. However, the discovery that nickel functions
either as a cofactor or structural component in four types of enzymes
from plants and microorganisms (vide supra) suggests that a similar
role exists for nickel in higher organisms. Further support for this
role is that nickel can activate many enzymes in vitro. Some of the
more interesting examples follow. Calcineurin, a Ca^{2+} and calmodulin-
dependent phosphoprotein phosphatase, was found to be markedly
activated by Ni^{2+} ions (190,191). Although they found calcineurin
was a zinc and iron metalloenzyme (192), King and Huang found that
an additional metal ion was vital to the structural and catalytic
properties of calcineurin. Pallen and Wang (191) found that it was
possible to manipulate the activity, and possibly the conformation,
of calcineurin through provision or withdrawal of particular cations
to or from the enzyme. They detected four distinct states of
calcineurin activity, depending upon whether Ni^{2+}, Mn^{2+}, or Ca^{2+}
was present. Moreover, King and Huang (190) found that the
activation was not reversed by high concentrations of chelators, thus
indicating nickel tightly binds the enzyme. These findings suggest
that nickel might have a role in regulating the substrate specificity
of phosphoprotein phosphatases.

Fishelson and co-workers (193,194) found that Ni^{2+} could replace,
and was more efficient than, Mg^{2+} in the formation of the two C3
convertases of the complement system—C3b, Bb of the alternative
pathway, and C4b, 2a of the classical pathway. Furthermore, up

to nine times more factor B was specifically bound to C3b-bearing
sheep erythrocytes (EC3b) in the presence of nickel than in the
presence of magnesium under identical conditions. To form one
effective hemolytic site per EC3b cell with nickel, three times less
Factor B, 12 times less Factor D, and 66 times less metal ions were
required than when using magnesium. These findings suggest that
nickel has a role in the complement pathway.

Pullarkat et al. (195) presented evidence suggesting that ATP
and Ni^{2+} are part of a reaction that converts phosphatidic acid to
pyrophosphatidic acid, which is a precursor of phosphatidyl serine
biosynthesis by rat brain microsomes. In isotope studies, the
specific activity of the $ATP-Ni^{2+}$-dependent phosphatidyl serine was
increased more than twofold during active myelination.

Another possible function for nickel is part of a bioligand cofactor
facilitating the intestinal absorption of the Fe^{3+} ion. Nickel was
found to interact with iron to affect hematopoiesis in rats fed dietary
iron as ferric sulfate only, but not as a mixture of ferric and
ferrous sulfates (196). Furthermore, when only ferric sulfate was
supplemented to the diet, liver content of iron was depressed in
nickel-deprived rats (197). On the other hand, when a ferric-
ferrous mixture was supplemented to the diet, nickel deprivation
elevated the liver content of iron (197). The idea that nickel might
act in an enzyme mechanism that converts Fe^{3+} to Fe^{2+} for absorption is attractive because of the recent findings of redox-sensitive
nickel in enzymes of microorganisms, and perhaps human serum
albumin (vide supra). However, the possibility that nickel promotes
the absorption of Fe^{3+} per se by enhancing its complexion to a
molecule that can be absorbed cannot be overlooked. Such a molecule
could be similar to the nickel porphinoid molecule found in microorganisms.

Nickel Requirements

The minimum amounts of nickel required by animals to maintain health,
based on the addition of graded increments to a known deficient diet
in the conventional manner, are not yet known. Nielsen et al. (198)
reported that 50 ng of nickel/gm satisfied the dietary nickel requirements of chicks, and Kirchgessner and co-workers (173) reported
that rats had a similar requirement for growth. Subsequently,
Kirchgessner and co-workers (199,200) obtained findings which they
thought suggested an even higher nickel requirement for rats.
Although growth was not affected, rats fed 60 or 150 ng Ni/gm diet
exhibited lower activities of the enzymes, pancreatic α-amylase,
hepatic glucose-6-phosphate dehydrogenase, and hepatic lactate
dehydrogenase than rats fed 20 μg Ni/gm diet, thus suggesting a
requirement higher than 150 ng/gm diet (199). Kirchgessner et al.

(200) also found that the iron concentration in liver, serum, femur, spleen, and muscle increased when dietary nickel was increased from 60 to 10,000 or 15,000 ng/gm. Thus they suggested that the nickel requirement of growing rats is much higher than the 50–100 ng/gm, which prevents the appearance of clinical signs of deficiency. The preceding suggestions must be accepted with caution considering the pharmacological action of nickel on iron and amylase activity (vide infra). The nickel requirement of pigs apparently is between 100 and 120 ng/gm diet. Anke and co-workers (164) indicated that minipigs must have a nickel requirement above 100 ng/gm diet, because when that species was fed a diet containing that level of nickel, nickel deprivation signs appeared. On the other hand, Spears et al. (201) obtained results suggesting that 120–160 ng Ni/gm diet was adequate for growth of neonatal pigs. They found, however, that additional nickel may improve the iron and zinc status of the young pig. Anke and co-workers (164) found that goats fed 100 ng Ni/gm of diet exhibited deprivation signs. Spears and Hatfield (202) found that a dietary nickel supplement of 5 μg/gm markedly increased ruminal urease activity in lambs fed a corn-based diet containing 320 ng Ni/gm and 9.5% crude protein. Similar findings were obtained with cattle (203). Reidel et al. (204) found a nickel supplement of 5 μg/gm diet did not increase growth, nickel content in organs, or the activity of several enzymes in cattle fed a diet containing 500 ng Ni/gm. Considering the preceding, the suggestion of Anke et al. (164) that ruminants have a dietary nickel requirement in the range of 300 to 350 ng/gm, and that monogastric animals have a dietary nickel requirement of less than 200 ng/gm seems prudent. Calculated from data from chicks, Nielsen (1) suggested that the dietary nickel requirement of humans would be near 35 μg daily.

Pharmacologic Actions of Nickel

Numerous studies indicate that nickel should be able to act pharmacologically in animals. Like other ions of the first transition series, Ni^{2+} has the ability to complex, chelate, or bind with many substances of biological interest (134). Through this ability, nickel could substitute beneficially for other ions if they were lacking for their needed role in such things as enzyme action or membrane structure. Further evidence that this is not an unlikely possibility is that in animals, plants, and microorganisms, nickel interacts directly or indirectly with at least 13 essential elements: Ca, Cr, Co, Cu, I, Fe, Mg, Mn, Mo, P, K, Na, and Zn, (134,205). The interactions of most pharmacologic significance probably are those with iron, copper, and zinc.

Nielsen et al. (186) indicated that dietary nickel in quantities not considered particularly high could have pharmacologic effects in iron-deficient rats. That is, the most evident actions of nickel on iron metabolism occurred when nickel intake far exceeded amounts suggested to meet the requirements of the rat. For example, dietary nickel at 100—1000 times the suggested requirement apparently stimulated hematopoiesis in marginally iron-deprived rats to a greater extent than dietary nickel at 2—20 times the suggested requirement. Other studies have also shown similar findings (200,201). Nielsen et al. (186) also stated that the upper dietary level at which nickel still pharmacologically benefits iron-deprived rats most likely depended on the severity of the iron deprivation. In one study, 20 to 50 μg Ni/gm diet alleviated many signs of mild iron deficiency. In another study, 20 μg Ni/gm diet apparently exacerbated some signs of severe iron deficiency.

Chung et al. (206,207) found that nickel supplementation (27 or 54 μg/gm diet) alleviated many of the signs of zinc deprivation in swine and chicks fed practical type diets. Spears et al. (208) found that 50 μg Ni/gm diet did not affect the growth of rats fed severely or marginally zinc-deficient purified diets. However, nickel partially alleviated some signs of zinc deficiency, including depressed leukocyte counts and elevated hematocrits, hemoglobin, and erythrocyte counts.

Spears et al. (209) reported that 20 μg Ni/gm diet improved growth, hematocrit, and hemoglobin, but tended to reduce tissue copper concentrations in rats fed a copper-deficient diet. However, like iron, the point at which dietary nickel is pharmacologically beneficial to copper-deprived rats apparently depends upon the severity of copper-deprivation and the level of nickel supplementation. Nielsen and co-workers (210,211) found that in copper-deficient rats with significant but mild anemia, the copper deficiency signs of elevated heart weight and plasma cholesterol and depressed hemoglobin were exacerbated by nickel supplementation. The effect was greater when dietary nickel was 50 μg/gm rather than 5 μg/gm.

The preceding discussion and previous reports (205), indicating that nickel in not particularly high quantities can interact with numerous essential nutrients and thus can alter the cellular bio-chemistry in which the nutrients are involved, suggests that nickel has many pharmacological effects. Perhaps many of the early reported signs of nickel deprivation were manifestations of pharmacologic actions of nickel, because many of the nickel-deprived animals were compared to controls fed relatively high levels of supplemental nickel, i.e., 3--20 μg/gm diet (212). Obviously, the nature of most of the pharma-cologic effects of nickel needs to be defined.

Nutritional Implications of Nickel

The preceding discussion indicates that the safe and adequate intake of nickel probably is in the range of 35—500 μg/day. Thus, abnormal nickel nutrition, especially occurrence of simple acute deficiency, seems unlikely. However, a situation other than simple deficiency might make nickel of nutritional significance. For example, alterations in nickel metabolism and/or biochemistry as a secondary consequence to malnutrition, disease, injury, or stress might change the dietary need of nickel. The findings that nickel affects the absorption and metabolism of iron and that iron status affects nickel metabolism and nutrition may be helpful in defining situations in which nickel would have nutritional significance. At present, however, it is safe only to state that emerging evidence indicates that nickel is a dynamic ultratrace element in living organisms. Also, knowledge of the biochemistry of nickel is very limited. Thus, further research on the nutritional, biochemical, and clinical aspects of nickel is needed to help evaluate the nature and importance of the physiological, pharmacological, and toxicological actions of nickel.

SILICON

History

Silicon was first found in the ash of animals in 1848 (213). In 1901, Schultz (214) reported that high concentrations of silicon were present in tendons, aponeurosis, and eye tissues. As early as 1911, it was suggested that silicon might have an anti-atheroma action (215). However, until 1972, silicon was generally considered to be nonessential except in some lower classes of organisms (diatoms, radiolarians, and sponges), where silica serves a structural role. In that year the first substantial evidence was published that indicated silicon is an essential element for chickens and rats. Most of the limited studies on the biochemical, nutritional, and physiological roles of silicon have been published since 1974.

Chemistry of Silicon

The chemistry of silicon, which was recently reviewed from the pharmacological viewpoint (216), is similar to the chemistry of carbon, its sister element. Silicon forms silicon-silicon, silicon-hydrogen, silicon-oxygen, silicon-nitrogen, and silicon-carbon bonds. Thus, organosilicon compounds are analogues of organocarbon compounds. However, the substitution of silicon for carbon, or vice versa, in organocompounds results in molecules with quite different properties because silicon is larger and less electronegative than carbon.

In animals, silicon is found both in the free and bound forms. Silicic acid probably is the free form. The bound form has never been rigorously identified. However, it has been suggested that silicon is present in biologic material as a silanolate—an ether (or esterlike) derivative of silicic acid, and that R_1—O—Si—O—R_2 or

R_1—O—Si—O—Si—O—R_2 bridges play a role in the structural organization of some mucopolysaccharides (217). The chemical nature of silicon found in high levels in certain fibers (vide infra) is not known. Here, too, the silicon could possibly exist as organically bound derivatives of silicic acid (silanolates). However, it is just as likely to be present as orthosilicic acid, polymeric silicic acid, colloidal silica (opal), or dense silica concentrations.

Silicon Content of Foods

The systematic examinations of a variety of foods as consumed by humans has been done recently in two laboratories (15–18,76,218). These studies show that silicon is ubiquitous and is supplied by many foods. Rich sources of silicon are unrefined grains of high fiber content, cereal products, and root vegetables (see Table 6). Foods of animal origin, except skin (e.g., chicken), are relatively low in silicon.

Table 6 indicates that total dietary silicon intake of humans varies greatly with the amounts and proportions of foods of animal (Si-low) and plant (Si-high) origin consumed, and with the amounts of refined and processed foods in the diet. However, Bowen and Peggs (218) estimated that the average British diet supplied 31 mg of silicon/day. A human balance study (219) indicated that the oral intake of silicon could be about 21 to 46 mg/day.

Silicon Metabolism

Little is known about the metabolism of silicon. Increasing the intake of silicon increases urinary silicon output up to fairly well-defined limits in humans, rats, and guinea pigs. However, the upper limits of urinary silicon excretion apparently are not set by the excretory ability of the kidney because urinary excretion can be elevated above these upper limits by peritoneal injections of silicon (220). Thus, the limits apparently are set by the rate and extent of silicon absorption from the gastrointestinal tract.

The form of dietary silicon determines whether it is well absorbed or not. A French group found that humans absorbed only about 1% of a large single dose of an alumino silicate compound (221), but absorbed over 70% of a single dose of methylsilanetriol salicylate, a drug used in the treatment of circulatory ischemias and osteoporosis (222). Benke and Osborn (223) also found that, depending upon the form of silicon ingested, daily urinary excretion can vary from a few to close to 50% of daily silicon intake. Silicon absorption has been also found to be affected in rats by age, sex, and the activity of various endocrine glands (224).

Table 6. Silicon Content of Selected Foods

Food	μg Si/gm, dry weight[a]	μg Si/gm, fresh weight[b]
Barley	2610	1600–2100
Barley, malt	—	2100
Beer	—	30–60
Maize	44	30
Oats	4310	3400–5800
Rolled oats	—	130
Rice, whole grain	—	360
Rice, polished	57	70
Rye	70	70–120
Sorghum, red	762	—
Sorghum, white	431	—
Wheat	65	50–80
Wheat flour, white	103	20–40
Cabbage	115	<2.0
Carrot	68	5.0
Turnip	—	120
Green bean	—	100
Dried pea	—	30
Peanut	50	—
Raisin	141	140
Milk	25	<2.0
Potato	59	2.0
Beef	—	5–10
Pork	—	5–10
Fresh fish	—	<5–5

[a]Atomic adsorption spectrometric analysis by Bowen and Peggs (218).
[b]Data from Varo et al. (15,17) and Nuurtamo et al. (16).

Regardless of the number of factors that can affect silicon absorption, humans on a normal diet excrete relatively high amounts of silicon in urine. Reported daily urinary outputs of silicon have ranged between 9 to 16 mg (219,225,226). This output supports the estimate that humans absorb 9 to 14 mg of silicon daily (227), or about 30 to 50% of the estimated daily intake (vide supra). Thus, although the mechanisms involved in intestinal absorption and in blood transport of silicon are unknown, it is apparent that food silicon is excreted mainly through the kidney.

Signs of Abnormal Silicon Nutrition

The signs of silicon deficiency have been reviewed by Carlisle (227) and Nielsen (1,60). Most of the signs of silicon deficiency in chickens and rats indicate aberrant metabolism of connective tissue and bone. Animals in early studies were fed crystalline amino acid diets that did not give optimal growth in controls. The signs of silicon deficiency were not exactly the same in those animals as in animals fed a semi-synthetic diet that produced near optimal growth in chicks. With this latter diet, in contrast to amino acid diet, silicon deprivation did not affect chick growth or outward appearance but still did affect connective tissue and bone (228,229). Abnormalities included skull structure abnormalities associated with depressed collagen content in bone, and long bone abnormalities characterized by small, poorly formed joints and defective endochondrial bone growth. Silicon-deficient chicken tibias exhibit depressed contents of articular cartilage, water, hexosamine, and collagen.

Silicon is essentially nontoxic when taken orally. Evidence for its nontoxicity is that magnesium trisilicate, an over-the-counter antacid, has been used by humans for more than 40 years. Also, silicates are added to foods as anticaking and antifoaming agents. However, ruminants consuming plants with a high silicon content may develop silicious renal calculi. Renal calculi in humans may also contain silicates (227). An extensive review of the toxicity of silicon through oral and respiratory intake was recently done by Carlisle (227).

Silicon Function

Both the distribution of silicon in animals, and the effect of silicon deficiency on the form and composition of connective tissue support the view that silicon functions as a biologic crosslinking agent that contributes to the architecture and resilience of connective tissue (1,227). The connective tissue components in which silicon apparently plays a fundamental role in the crosslinking mechanism are collagen, elastin, and mucopolysaccharide. Another finding indicating silicon is important in collagen metabolism is that silicon is required for maximal bone prolylhydroxylase activity (227). This suggests silicon is required for bone and cartilage collagen biosynthesis.

Silicon apparently is involved in bone calcification (227), however, the mechansim of involvement remains unclear. Some findings suggest a catalytic function for silicon. On the other hand, the suggestion has also been made that the marked influence of silicon on collagen and mucopolysaccharide formation and structure results in its indirect influence on bone calcification. In support of this latter view is the finding, in silicon-deficient animals, that the formation of organic matrix, whether cartilage or bone, is apparently more impaired than the mineralization process.

Silicon Requirements

Although the essentiality of silicon was suggested almost 15 years ago, the form of silicon needed and the minimum requirement for it have not been ascertained for any animal; therefore nothing can be said about possible human requirements. The estimated requirement of chickens for silicon, as sodium silicate, is in the range of 100 to 200 μg/gm diet, or about 26 to 52 mg/1000 kcal of an experimental diet (1). However, other silicon compounds might be 5 to 10 times as effective, per atom of silicon, in preventing nutritional deficiency (230). Thus, the absolute requirement for chickens (and humans) probably is much lower than 26 to 52 mg/1000 kcal.

Pharmacologic Actions of Silicon

The use of silicon-containing compounds as pharmacological agents dates back to antiquity (231). Pasteur predicted that silicon would be of importance for the therapy in many diseases. At the turn of the century German and Frecnh Scientists reported that silicon was an effective therapeutic agent for such diseases as arteriosclerosis, hypertension, malignant neoplasms, and dermatitis.

During the past 40 years, thousands of organosilicon chemicals and polymers have been produced. A silicon atom can be bonded directly to essentially any organic chemical presently available. However, research designed to determine the biological activity and uses of organosilicon has been rekindled in only the past 15 to 20 years. Thus, much careful research must be done to understand the sites, mechanisms of action, and beneficial properties of organosilicon compounds in higher organisms. The spectulations that silicon lack is involved in the causation of several human disorders, including atherosclerosis, osteoarthritis, hypercholesteremia, and hypertension, appear to be the result of studies that are pharmacological, not physiological, in design.

The studies suggesting that silicon may be an important factor in the prevention of atherosclerosis have either used relatively high levels of dietary silicon, injected silicon, or fed silicon as an organosilicon compound. For example, in studies done by Loeper and co-workers (232,233), silicon was injected intravenously as sodium silicate or lysine silicate, or fed as sodium monomethyltrisilanol salicylate when silicon was described as having antiatheromatous action in rabbits fed a high cholesterol atherogenic diet. Gendre and Cambar used sodium monomethyltrisilanol O-hydroxybenzoate to prevent the development of atherosclerosis in rabbits fed a high-cholesterol diet (234). The idea that the rabbits were responding pharmacologically to the organosilicon compounds is supported by the finding that silicon fed as sodium silicate or sodium monomethyl-trisilanol-O-hydroxybenzoate had opposite effects when it altered the calcium and magnesium contents of blood and various organs (235).

Studies on the role of silicon in preventing hypercholesteremia or hypertension also have used high dietary silicon. Becker and co-workers (231) found that 2.5% sodium silicate in water had only a mild hypotensive effect in the spontaneous hypertensive rat. Becker et al. (236) fed similarly high levels of inorganic silicon or organosilicon in their studies that showed silicon decreased serum cholesterol and triglycerides. Ki et al. (237) fed 400 ppm silicon as an acid-treated form of soluble silicate to turkey poults fed high levels of cholesterol in their studies showing that silicon depressed the levels of cholesterol in serum, aorta, cardiac, and hepatic tissues.

Nutritional Implications of Silicon

Ample evidence indicates that silicon can be accepted as an essential nutrient, but more work is needed to clarify the consequences of silicon deficiency in humans. Although the role of silicon in athero-sclerosis and hypertension may have been examined experimentally more from the pharmacological point of view (vide supra), there are some reports that inadequate silicon nutrition might have a role in those diseases and in the process of aging. In human atherosclero-sis, the concentration of silicon in the aortic wall decreases, and this decrease precedes the appearance of lipid deposits (232). Acute hypertensive rats were found to have two to three times more silicon in blood vessel walls than controls (238). Schwarz (239) hypothesized that the high level of silicon in fiber and hard water could be the active agent in these antiatherosclerotic factors. Burton et al. (240) used epidemiological methods to show that high SiO_2 in water supplies protects against cancer. Carlisle (227) reviewed the evidence showing that the silicon content of aorta, other arterial vessels, and skin decline with age. These findings demonstrate the need for further study of the importance of silicon nutrition, especial-ly in aging humans.

VANADIUM

History

In 1876, Priestley (241) reported on the toxicity of sodium vanadate. Following his studies, interest in the biological actions of vanadium lay dormant until the turn of the 19th century, when various French physicians used vanadium as a panacea for a number of human disorders (242). Shortly thereafter, a classical paper on the pharmacological and toxicological actions of vanadium was presented by Jackson (243). This was about the time that extremely high vanadium concentrations were found in the blood of ascidian worms (244,245).

The hypotheses that vanadium may play a physiological role in higher animals has had a long and inconclusive history. In the 1940s Bertrand (246) stated that "we are completely ignorant of the physiological role of vanadium in animals, where its presence is constant." In 1949 Rygh (247) reported that vanadium might be needed by animals because it markedly stimulated the mineralization of bones and teeth and prevented caries formation in rats and guinea pigs. In 1963 Schroeder et al. (248) stated that, although vanadium behaves like an essential trace metal, final proof of essentiality for mammals was still lacking. In 1974 Hopkins and Mohr (249) stated "we are secure in the concept that vanadium is an essential nutrient." However, a review in 1984 (1) presented a convincing argument that the evidence for the nutritional essentiality of vanadium is still inconclusive. Nearly all evidence presented as proof for the essentiality of vanadium is most likely nothing more than evidence that vanadium is a very active pharmacological substance in vivo and in vitro (212). It seems that the statement of Schroeder et al. (248) that "no other trace metal has so long had so many supposed biological activities without having been proved to be essential" still holds true today. Nonetheless, there is a large interest in vanadium. This was aroused by the finding in vitro that vanadate is a potent inhibitor of (Na^+, K^+)-ATPase and other phosphoryl-transfer enzymes. Thus, biochemical, nutritional, and pharmacological roles of vanadium will be discussed here.

Chemistry of Vanadium

The chemistry of vanadium is complex because the element can exist in oxidation states from -1 to +5 and can form polymers. The rich and varied solution chemistry was recently reviewed (250,251). Based on these reviews, the tetravalent and pentavalent valence states are the most important forms of vanadium in biological systems. The tetravalent state appears most simply in aqueous solution as a blue divalent cation, VO^{2+}, which is also known as the vanadyl ion or oxovanadium. The vanadyl ion is stable in acid solution, but near neutral pH it becomes hydroxylated and sensitive to oxidation by oxygen from air. However, vanadyl easily forms complexes with other substances, such as transferrin or hemoglobin, which stabilizes it against oxidation. The structures of vanadyl chelates are square pyramidal with the oxo ligand at the apex, or trigonal bipyramidal with the oxo ligand in the equatorial plane. The pentavalent state of vanadium is known as vanadate. Above pH 13 vanadate is VO_4^{3-}, and is analogous to inorganic phosphate, PO_4^{3-}. Below pH 13 vanadium acquires one proton, and below pH 8 it acquires another. Thus, at neutral pH vanadate is $H_2VO_4^-$ or more simply VO_3^-. In solution, vanadate forms oligomers. In the millimolar concentrations, vanadate forms dimers, trimers, and tetramers. This oligomerization, plus the slowness of equilibration between oligomers, makes it difficult to determine the exact nature of the species being studied. Below pH 3.5, vanadate becomes the monovalent cation, VO_2^+. Vanadate also forms complexes with other biological substances, in particular, substances with cis diols. This behavior is similar to that of other oxyanions such as molybdate and tungstate. This behavior is not shared by phosphate; the ion vanadate bears a closer resemblance in other properties than it does with metal-containing oxoanions. Vanadate is easily reduced by ascorbate, glutathione, or NADH.

Kustin and Macara (250) stated that three types of behavior can be predicted from a consideration of the aqueous chemistry of vanadium. First, as vanadate, the element will compete with phosphate at the active sites of phosphate transport proteins, phosphohydrolases, and phosphotransferases. Second, as vanadyl, the element will compete with other transition metal ions for binding sites on metalloproteins and for small ligands such as ATP. Third, vanadium will participate in redox reactions within the cell, particularly with relatively small molecules that can reduce vanadate nonenzymatically, such as glutathione. To date, only one naturally occurring organovanadium compound has been identified. A pale blue compound from the fly agaric mushroom, *Amanita muscaria*, called amavadine, was found to have a composition of $C_{12}H_2ON_2VO_{11} \cdot H_2O$ and molecular weight of 415 (252). Contrary to initial beliefs, vanadium apparently exists in the vanadocyte of ascidians as a small complex, not as a vanadium-protein complex (253). Hemovanadin, once thought to be the functional form of vanadium in vivo, apparently was an artifactual finding.

Vanadium Content of Foods

Recent studies based on fairly reliable analytical methodology show that the vanadium content of most foods is low. Myron et al. (254) used atomic absorption spectrometry to show that beverages, fats and oils, and fresh fruits and vegetables contained the least vanadium, ranging from <1 to 5 ng/gm. Whole grains, seafood, meats, and dairy products were generally within a range of 5 to 30 ng/gm. Prepared foods ranged from 11 to 93 ng/gm while dill seed and black pepper contained 431 and 987 ng/gm, respectively. Byrne and Kosta (255) obtained similar results using neutron activation analyses. They found that most fats, oils, fruits, and vegetables contained <1 ng/gm. Cereals, liver, and fish tended to have intermediate levels of about 5 to 40 ng/gm. Only a few items, such as spinach, parsley, mushrooms, and oysters, contained relatively high amounts of vanadium. Shellfish apparently are a rich source of vanadium because several types were found to contain over 100 ng vanadium/gm on a fresh basis (256), or over 400 ng vanadium/gm on a dry basis (255, 257).

The vanadium content of foods from plants can be increased by increasing available vanadium in the growing medium. Vanadium concentrations of 28 to 55 ng/gm were found in the seeds of wheat, barley, oats, and peas from plants grown in nutrient solution, and these were substantially increased (75 to 175 ng/gm) when vanadate was added to the nutrient solution (258).

From the preceding, it is obvious that the daily intake of vanadium is relatively low in comparison to other essential trace elements. Byrne and Kosta (255) stated that their analyses indicated that the daily dietary intake of vanadium is in the order of a few tens of μg and may vary widely. Myron et al. (168) found that nine institutional diets supplied 12.4 to 30.1 μg of vanadium daily, and intake averaged 20 μg. Unreliable analytical procedures probably were behind the apparent erroneous report that the daily vanadium intake for humans is 1 to 4 mg (248).

Vanadium Metabolism

Limited information exists about vanadium metabolism at physiologic levels in animals. However, the very low concentration of vanadium normally in urine (<0.2 to 0.8 $\mu g/L$) (255,259) compared with the estimated dietary intake and fecal level of vanadium indicates that $\leq 1\%$ of vanadium ingested is absorbed; but more evidence is needed for confirmation, as shown by the following. In one study, only 0.1–1.0% of the vanadium in 100 mg of soluble diammonium oxytartarovanadate was found to be absorbed from the human gut (260). Hansard et al. (261) estimated that only 0.13 to 0.75% (mean of 0.34%) of ingested vanadium as ammonium metavanadate was absorbed from the sheep gut. On the other hand, two studies with rats indicated much greater vanadium absorption from the gut. Wiegmann et al. (262) fasted rats overnight, then gavaged them with 5 μmol of Na_3VO_4 in 1.0 ml 0.9% NaCl containing 1 μC ^{48}V. After four days, 86.6 ± 2.4% of the administered ^{48}V was recovered in the feces and urine.

Only 69.1 ± 1.8% of the dose was recovered in the feces; recovery
from feces increased to 85.7 ± 1.5% if 1.0 ml of a suspension of
Al(OH)$_3$ was administered simultaneously with the vanadium. In
either case, the findings indicated an absorption of greater than 10%.
Bogden et al. (263) found that rats retained 39.7 ± 18.5% and
excreted in the feces 59.1 ± 18.8% of vanadium ingested as sodium
metavanadate supplemented at 5 or 25 ppm in a casein—sucrose—
dextrin-based diet. Dietary composition and vanadium form probably
affect the percentage of ingested vanadium absorbed from the gut.
Regardless, the rat studies suggest caution in assuming that ingested
vanadium will always be poorly absorbed from the gastrointestinal
tract.

From the preceding, it is apparent that most ingested vanadium
is unabsorbed and is excreted via the feces. Urine is the major
excretory route for absorbed vanadium, and bone apparently is a
major sink for retained vanadium. This is demonstrated by studies
in which vanadium is injected into the animal body. For example,
at 96 hr, 30 to 46% of an intravenously injected dose of ^{48}V had been
excreted in the urine and 9 to 10% in the feces (264,265). In urine,
^{48}V was mainly associated with high-molecular-weight components 72
hours after ^{48}V administration (265). Other studies showed that as
the length of time subsequent to injection increased, bone and, in
some cases, spleen and liver became major sites of vanadium retention.
For example, Roschin et al. (266) found that 30 minutes after intraperitoneally injecting rats with ^{48}V, the kidney retained 7.2%, and
the bone retained 2.1% of the dose; at 48 hours, the kidney retained
1.6%, and the bone retained 3.45% of the dose. In mice one day after
an intraperitoneal injection of ^{48}VOCl$_3$, the highest amount of ^{48}V
was found in the kidney; 4 to 12 days after injection, bone, spleen,
and liver contained markedly higher levels of ^{48}V than kidney (267).

Evidence to date suggests that the binding of the vanadyl ion
to iron-containing proteins is important in vanadium metabolism.
Regardless of the oxidation state administered to animals, vanadium
apparently is converted into vanadyl-transferrin and vanadyl-ferritin
complexes in plasma and body fluids (264,268). After intravenous
injection, vanadium in blood was mainly present in the plasma and
disappeared from the bloodstream with different rates corresponding
to the clearance of three plasma components (265). Initially, only a
small amount of vanadium in the plasma was associated with proteins,
but after 96 hours it was present only as vanadium-transferrin
bicomplex. Sabbioni and Marafante (269) found that one day after
intravenous administration of ^{48}VO^{2+}, 29% of ^{48}V incorporated in rat
liver cytosol existed as a vanadium—low-molecular-weight complex
(<5000 MW). By day nine, however, they found the low-molecular-
weight complex had disappeared and vanadium was present only as
V-ferritin (15%) and V-transferrin (85%) in rat liver cytosol. Nine

days after the administration of vanadium, partial purification of heart myoglobin, liver mitochondrial and microsomal cytochromes b, b5, and c, ferriporphyrin, and red blood cell hemoglobin showed no significant incorporation of ^{48}V into these proteins (270). These findings suggest that nonheme iron metalloproteins are more involved than iron hemoproteins in the metabolism of vanadium in vivo. It remains to be determined whether ferritin is a storage vehicle for vanadium as well as for iron in the liver and whether vanadyl-transferrin can transfer vanadium to cells through the transferrin receptor.

Signs of Abnormal Vanadium Nutrition

There has been no demonstration to date that vanadium deficiency reproducibly and consistently impairs a biological function in any animal. Between 1971 and 1974, a number of findings reported by four different research groups led many to conclude that vanadium is an essential nutrient. However, close examination of the findings from the vanadium deprivation studies reveals that they were confusing and inconsistent. For example, in rat studies, findings were as follows:

In 1971, Strasia (271) reported that rats fed less than 100 ppb vanadium exhibited slower growth, higher plasma and bone iron, and higher hematocrits than controls fed 500 ppb vanadium. Williams (272) however, could not duplicate those findings in the same laboratory under similar conditions. Schwarz and Milne (273) reported that a vanadium supplement of 0.25–0.50 µg/gm of diet gave a positive growth response in suboptimally growing rats fed a purified diet with an unknown vanadium content. On the other hand, Hopkins and Mohr (249) reported that the only effect of vanadium deprivation on rats was impaired reproductive performance (decreased fertility and increased perinatal mortality) that became apparent only in the fourth generation. Nielsen reported on an attempt to establish a reproducible set of signs of vanadium deprivation (274). In several experiments with rats fed diets of different composition, vanadium deprivation adversely affected perinatal survival, growth, physical appearance, hematocrit, plasma cholesterol, and lipids and phospholipids in liver. Unfortunately, no sign of deficiency was found consistently throughout all experiments. Findings from chick studies were similarly inconsistent.

Recent studies suggest that the reported differences between vanadium-deprived and vanadium-supplemented animals in early deficiency studies were the consequence of high vanadium supplements (0.5–3.0 ppm) that resulted in pharmacological-type changes (212). The dose of available vanadium given to supplemented animals probably ranged from 10 to 100 times that normally found in a diet under natural conditions. Vanadium is a relatively toxic element and, therefore, it would not be surprising that a few ppm would exert a pharmacologic effect in vivo, especially if the nutritional status of the organism was suboptimal.

Many of the early studies were done with animals fed imbalanced diets, thus resulting in suboptimal performance, such as depressed growth. Moreover, the diets used in the early studies had widely varying contents of protein, sulfur amino acids, ascorbic acid, iron, copper, and perhaps other nutrients that affect vanadium metabolism (212,251). Under these conditions, if vanadium were acting pharmacologically, the response to dietary vanadium might be expected to be quite variable. In other words, the nature of the difference between vanadium-deprived and vanadium-supplemented animals would depend upon the imbalanced condition that is being affected by vanadium. On the other hand, if vanadium were acting as an essential nutrient, the response to dietary vanadium, although it might change in intensity, should be relatively consistent. As stated (vide supra), the early findings were inconsistent.

Vanadium can be a relatively toxic element to some animal species. Some years ago Franke and Moxon (275) found the relative toxicity of five different elements, fed at the level of 25 µg/gm diet to rats, to be in the ascending order As, Mo, Te, V, and Se. Dietary vanadium concentrations of 25 µg/gm were toxic to rats and at the vanadium intake of 50 µg/gm diet the animals exhibited diarrhea and mortality. The toxicity of ingested vanadium (as vanadate) is similar in chicks. In chicks, calcium vanadate fed at the level of 30 µg V/gm of practical ration depressed growth and at the level of 200 µg/gm caused high mortality (276). However, the concentration of dietary vanadium that is toxic to these species can be affected by dietary composition. In his reviews, Nielsen (212,277) mentioned a number of substances that can ameliorate vanadium toxicity, including ascorbic acid, EDTA, chromium, protein, ferrous iron, chloride, and perhaps aluminum hydroxide. Nielsen (277) also reviewed the evidence showing that age and animal species influenced toxicity. Mature animals are more tolerant of high vanadium than young animals. Animal species more tolerant to vanadium than the chick and rat apparently include Coturnix (quail), sheep, and perhaps humans. The latter species, especially quail and sheep, apparently can tolerate without much distress levels of vanadium that cause mortality in rats and chicks.

From their in-depth study of vanadium toxicity, Proescher et al. (278) concluded that vanadium was a neurotoxic and a hemorrhagic endotheliotoxic poison with a nephrotoxic, hepatotoxic, and probably a leucocytactic and hemtoxic component. Thus, it is not surprising that there can be a variety of signs of vanadium toxicity and that they can vary among species and with the dose. Some of the more consistent signs include depressed growth, elevated organ vanadium, diarrhea, depressed food intake, and death (251).

Vanadium Function

Because vanadium is such an active in vitro or pharmacological element due to its rich varied chemistry (vide supra), there have been numerous suggestions as to possible biochemical and physiological functions for the element. Recently, a number of reviews (250, 25,277,279,280) have discussed the evidence behind these suggestions. Thus, those reviews document the original publications that reported the findings described in the following discussion of possible vanadium functions.

    1. *Regulation of (Na,K)-ATPase and the sodium pump.* (Na, K)-ATPase plays a role in the maintenance of membrane potential and sodium gradient across the plasma membrane. Shortly after vanadate was identified as the impurity in a commercial preparation of horse muscle ATP that interfered with the in vitro studies of (Na,K)-ATPase, it was hypothesized that in vivo vanadium might function as a physiological regulator of sodium pump activity. Vanadate is the form that inhibits (Na,K)-ATPase activity, and reduction of vanadate to vanadyl reverses that inhibition. The finding that the predominant form of vanadium in tissue is vanadyl has caused some to dismiss the possibility that vanadium regulates sodium pump activity. Others still allow for the possibility that vanadium plays a regulatory role, especially in organs, such as the kidney, which often contains significant amounts of vanadium. Such a role would be supported by the conclusive finding of an in vivo mechanism whereby vanadium in tissue is converted from vanadyl to vanadate.

    2. *Regulation of an ATPase.* There are a number of ATPases that are inhibited by physiologic concentrations of vanadium in vitro. These include plasma membrane Ca^{2+}-ATPase, sarcoplasmic reticulum Ca^{2+}-ATPase, fungal H^+-ATPase, gastric mucosa and colon epithelium H^+, K^+-ATPase, myosin ATPase, and dynein-1-ATPase. However, like (Na,K)-ATPase, a regulatory role with these ATPases would seem more likely if an in vivo mechanism existed whereby vanadium in tissue is converted from vanadyl to vanadate.

    3. *Regulation of phosphoryl transfer enzymes.* Vanadium is present in tissues at concentrations that might inhibit phosphoryl transfer enzymes in vivo, possibly by forming a transition state analogue (trigonal bipyramidal species) of phosphate in the phosphoryl-enzyme intermediate; such inhibition could reflect a regulation function for vanadium in addition to, or other than, inhibition of (Na,K)-ATPase or other ATPases. Phosphohydrolases and phosphotransferases inhibited by vanadate in vitro include glucose-

6-phosphatase, alkaline phosphatase, acid phosphatase, 2,3-biphos-
phoglycerate-dependent-phosphoglycerate mutase, and phospho-
glucomutase. The inhibitory action of vanadium on phosphoryl transfer enzymes is quite selective. It is this selectivity that prompts
the belief that vanadium has a regulatory role in vivo.

   4. *Regulation of adenylate cyclase and/or protein kinase.*
Vanadium stimulates the synthesis of cyclic AMP in a variety of cell
membranes through the activation of adenylate cyclase. Vanadium
apparently activates the adenylate cyclase complex by promoting its
association with an otherwise inactive guanine nucleotide regulatory
protein. Thus, the idea that vandium as vanadate is bound to
enzyme·GDP to form a stable, ternary enzyme·GDP·vandate complex
so that the G protein behaves as though GTP were bound and
adenylate cyclase is activated seems reasonable. The elevation of
cyclic AMP probably explains why vanadium stimulates cyclic AMP-
dependent protein kinase activity in rat liver. However, a physiological role for vanadium in regulating adenylate cyclase activity is
questionable because the effective concentration needed to affect the
enzyme is relatively high—much higher than that required for ATPase
inhibition. Also, like the hypotheses of a regulatory role for vanadium
in phosphoryl transfer, adenylate cyclase regulation would need a
mechanism whereby vanadyl is converted to vanadate in vivo.

   5. *Vanadyl as an enzyme cofactor.* Other than a possible
regulatory role, vanadium as the vanadyl cation might have a catalytic
function or might be a cofactor for some enzyme. When vanadyl
replaces other metals in metalloproteins, the metals replaced include
Zn^{2+}, Cu^{2+}, and Fe^{3+}, so it is possible that vanadyl has a role
similar to these cations.

   6. *Hormone expression.* There is a possibility that vanadium
might be involved in the expression of some endocrine function. In
rats, vanadium metabolism is disturbed in endocrine deficiency induced
by hypophysectomy or thyroidectomy-parathyroidectomy. Hypophys-
ectomized rats retain elevated amounts of injected ^{48}V in most of
their organs. Normal or near normal retention of ^{48}V in most organs
occurred upon growth hormone or thyroxine replacement. Another
study suggesting that vanadium can affect endocrine function involved
the cell-free activation of progesterone receptor from the avian oviduct.
Vanadate inhibited the rate of transformation of receptor-hormone
complex to the activated form.

   7. *Role in glucose metabolism.* In vitro studies have shown
that vanadium might affect glucose metabolism by altering the action
of insulin, by mimicking the action of insulin, or by altering the
activity of the multifunctional enzyme glucose-6-phosphatase. Insulin-
mimetic actions of vanadium contribute to an increased utilization of
glucose. Vanadium stimulates glucose oxidation and transport in
adipocytes and glycogen synthesis in liver and diaphragm, and

inhibits hepatic gluconeogenesis and intestinal glucose transport.
Vanadium seems to act like insulin by altering membrane function
for ion transport processes. However, whether vanadyl or vanadate
is most important in this action is controversial. Insulin stimulatory
actions of vanadium include the enhancement of the stimulatory effect
of insulin on DNA synthesis in cultured cells. Although a potent
inhibitor of glucose-6-phosphatase, vanadate up to a concentration
of 65 μM has no effect on either rat hepatic glucokinase or hexokinase.
This selective inhibition of vanadate on cellular glucose phosphoryla-
tion/dephosphorylation mechanism might be physiologically important
in modifying blood glucose levels or maintaining glucose homeostasis.
Because of such a range of in vitro actions on glucose metabolism, it
is not surprising that pharmacologic levels of dietary vanadium
improved oral glucose tolerance in guinea pigs. Vanadium apparently
improved glucose utilization in the guinea pig by enhancing glucose
removal from plasma.

8. *Role in lipid metabolism.* Based on the effects of vanadium
on glucose metabolism, hormone expression, and enzyme action, it
is not surprising to find that vanadium affects lipid metabolism. An
inhibition of cholesterol synthesis by vanadium was observed in vivo
in human and animal organs. Vanadium was used at pharmacological
levels. This inhibition was accompanied by decreased plasma phoso-
lipid and cholesterol levels and by reduced aortic cholesterol
concentrations. In older individuals and in patients with hyper-
cholesterolemia or ischemic heart disease, no such effect from
vanadium was apparent, while in older rats the inhibition could be
demonstrated in vitro but not in vivo. The site of the inhibition
by vanadium apparently was the microsomal enzyme system referred
to as squalene synthetase. In contrast to the lowering of plasma
phospholipids and cholesterol findings, another study showed that
vanadium fed as ammonium vanadate to young chicks at 100 ppm V
increased liver and plasma total lipid and cholesterol levels and
plasma cholesterol turnover rate. Also, rabbits intoxicated chronically
with 0.15 mg V/kg developed atherosclerosis.

Thus depending upon the dose, pharmacological, toxicological,
or in vitro studies show that vanadium depresses, enhances, or has
no effect on the biosynthesis or metabolism of various lipids.
Deprivation studies also gave a similar array of findings. These
confusing findings point strongly to a role for vanadium that
influences lipid metabolism. However, the findings also highlight
the need for further research to clarify that role.

9. *Role in bone and tooth metabolism.* Radiovanadium injected
subcutaneously into mice, or intraperitoneally into rats, is concentrated in the areas of rapid mineralization of bones and tooth dentine.
The radiovanadium is incorporated into the tooth structure of rats
and retained in the molars up to 90 days after injection. The

addition of vanadium to specially purified diets has also been reported to promote mineralization of the bones and teeth and to reduce the number of carious teeth in rats and guinea pigs. Aqueous solutions of V_2O_5 (0.1 mg/kg) given intraperitoneally daily to rabbits with holes drilled in their mandibles accelerated reparative regeneration of bone by stimulating ossification. The function of vanadium in developing bone and teeth, if any, is unknown.

Vanadium Requirements

Failure to define the conditions that induce reproducible deficiency signs in animals or that establish an essential function for vanadium prevents any suggestion of a vanadium requirement. However, any human requirement for vanadium would be very small. Diets containing 4−25 mg vanadium/gm apparently adversely affected chicks and rats only under certain conditions (274). However, since food contains very little vanadium, the metabolism of which apparently is affected profoundly by other dietary components, there is a possibility that vanadium intake is not always optimal. That possibility demonstrates the need to clarify the biological importance of vanadium and the conditions in which foods rich in vanadium may be beneficial.

Pharmacological Actions of Vanadium

The preceding sections indicate that vanadium is an active pharmacological substance. Almost all the findings described in the section on vanadium function demonstrate pharmacological actions of vanadium. In addition to those, there are numerous other physiological changes induced by vanadium in in vitro tissue preparations that have been described. These changes, which probably reflect a pharmacologic action of vanadium, include the following (see Refs. 250,251,277,279, 280). Vanadate inhibits renin secretion from rat kidney slices and excites Limulus photoreceptors. In a number of cardiac muscle preparations—isolated ventricular and atrial muscles of rat, dog, cat, and rabbit—vanadate induces an increase in contractile force which is called positive inotropic effect; in left atrial muscle from cat and guinea pig it produces a negative inotropic effect. In other words, the response of cardiac muscle is dependent upon the species and the area of the heart. An explanation for this variable response could be that it is dependent upon whether the stimulatory (insulinlike) or inhibitory action of vanadate predominates in the various heart cell types. Intravenous injection of vanadate causes a hypertensive vascular response in dogs and increased excretion of water, sodium, potassium, calcium, phosphate, bicarbonate, and chloride by the kidney in rats. Most likely, all the in vitro, and many in vivo, pharmacological effects of vanadate can be explained by changes in cellular sodium, potassium, and calcium.

With such a wide range of known pharmacological actions, it is surprising that no pharmaceutical has been developed that is beneficial for treating some human disorder. Perhaps it is just a matter of time. A recent study showed that high dietary vanadate (0.6—0.8 mg/ml drinking water) prevented an increase in glucose, despite low insulin, in blood, and prevented the decline in cardiac performance of rats made diabetic with streptozotocin (281). Another study indicated that high dietary vanadium as vanadyl (25 µg/gm) inhibited 1-methyl-1-nitrosourea-induced mammary carcinogenesis in rats (282).

Nutritional Implications of Vanadium

Whether vanadium has some practical nutritional importance beyond its toxicological and pharmacological aspects remains to be determined. Nonetheless, two human disorders have been suggested to be influenced by vanadium. Evidence has been reported suggesting that vanadium is a factor involved in the causation of manic-depressive illness. Mean plasma and hair vanadium concentrations were higher in manic-depressive patients than in normal controls (283). Significant negative correlations were found between plasma vanadium concentration and the ratio of Na-K-Mg ATPase to Mg-ATPase in two manic-depressive subjects, but not in normal subjects (283). Drugs that have a therapeutic value in manic-depressive illness are very effective in the reduction of vanadate to vanadyl (283). Epidemiological studies have suggested that low vanadium intakes may be associated with human cardiovascular disease (284).

CONCLUDING REMARKS

Knowledge about nutritional, biochemical, and clinical aspects of the ultratrace elements that have been identified as essential should be extended and refined. Further research might establish the essentiality of some other ultratrace elements (i.e., vanadium) thought to be essential. The research might also reveal that some ultratrace elements are more important in human health than is now generally acknowledged. Those who are concerned with ascertaining that humans receive wholesome, nutritionally adequate diets should be cognizant of this possibility, and therefore should not approach the ultratrace elements as an esoteric consideration.

REFERENCES

1. F. H. Nielsen, Ultratrace elements in nutrition, *Ann. Rev. Nutr.*, *4*: 21 (1984).
2. H. A. Schroeder and J. J. Balassa, Abnormal trace metals in man: Arsenic, *J. Chron. Dis.*, *19*: 85 (1966).
3. F. Challenger, Biosynthesis of organometallic and organometalloidal compounds, *ACS Symp. Ser.*, *82*: 1 (1978).
4. W. R. Cullen, C. L. Froese, A. Lui, B. C. McBride, D. J. Patmore, and M. Reimer, The aerobic methylation of arsenic by microorganisms in the presence of L-methionine methyl-d_3, *J. Organomet. Chem.*, *139*: 61 (1977).
5. W. R. Cullen, B. C. McBride, and A. W. Pickett, The transformation of arsenicals by *Candida humicola*, *Can. J. Microbiol.*, *25*: 1201 (1979).
6. A. A. Benson, R. V. Cooney, and R. E. Summons, Arsenic metabolism—a way of life in the sea, in *3. Spurenelement-Symposium Arsen* (M.Anke, H. -J. Schneider, and C. Bruckner, eds.), Friedrich-Schiller-Univ., Jena, 1980, p. 5.
7. J. J. Wrench and R. F. Addison, Reduction, methylation, and incorporation of arsenic into lipids by the marine phytoplankton *Dunaliella tertiolecta*, *Can. J. Fish Aquat. Sci.*, *38*: 518 (1981).
8. J. S. Edmonds and K. A. Francesconi, Arseno-sugars from brown kelp (*Ecklonia radiata*) as intermediates in cycling of arsenic in a marine ecosystem, *Nature*, *289*: 602 (1981).
9. J. S. Edmonds, K. A. Francesconi, and J. A. Hansen, Dimethyloxarsylethanol from anaerobic decomposition of brown kelp (*Ecklonia radiata*): A likely precursor of arsenobetaine in marine fauna, *Experientia*, *38*: 643 (1982).
10. R. S. Braman, Environmental reaction and analysis methods, in *Biological and Environmental Effects of Arsenic* (B. A. Fowler, ed.), Elsevier, Amsterdam, 1983, p. 141.
11. F. C. Knowles and A. A. Benson, The biochemistry of arsenic, *Trends Biochem. Sci. (Pers. Ed.)*, *8*: 178 (1983).
12. E. Marafante and M. Vahter, The effect of methyltransferase inhibition on the metabolism of [^{74}As] arsenite in mice and rabbits, *Chem.-Biol. Interactions*, *50*: 49 (1984).
13. R. Lagunas, Sugar-arsenate esters: Thermodynamics and biochemical behavior, *Arch. Biochem. Biophys.*, *205*: 67 (1980).
14. M. J. Gresser, ADP-arsenate. Formation by submitrochondrial particles under phosphorylating conditions, *J. Biol. Chem.*, *256*: 5981 (1981).
15. P. Varo, M. Nuurtamo, E. Saari, and P. Koivistoinen, Mineral element composition of Finnish Foods. III. Annual variations in the mineral element composition of cereal grains; IV. Flours and bakery products; VIII. Dairy products, eggs and margarine;

IX. Beverages, confectionaries, sugar and condiments; X. Industrial convenience foods, quantity service foods and baby foods, *Acta Agric. Scand., Suppl., 22*: 27; 37; 115; 127; 141 (1980).

16. M. Nuurtamo, P. Varo, E. Saari, and P. Koivistoinen, Mineral element composition of Finnish foods. V. Meat and meat products; VI. Fish and fish products, *Acta Agric. Scand., Suppl., 22*: 57; 77 (1980).

17. P. Varo, O. Lähelmä, M. Nuurtamo, E. Saari, and P. Koivistoinen Mineral element composition of Finnish foods. VII. Potato, vegetables, fruits, berries, nut and mushrooms, *Acta Agric. Scand., Suppl., 22*: 89 (1980).

18. P. Varo, M. Nuurtamo, E. Saari, L. Räsänen, and P. Koivistoinen Mineral element composition of Finnish foods. XI. Comparison of analytical and computed compositions in some simulated total diets, *Acta Agric. Scand., Suppl., 22*: 161 (1980).

19. D. S. Podrebarac, Pesticide, metal, and other chemical residues in adult total diet samples. (XIV). October 1977—September 1978, *J. Assoc. Off. Anal. Chem., 67*: 176 (1984).

20. G. Lunde, Occurrence and transformation of arsenic in the marine environment, *Environ. Health Perspect., 19*: 47 (1977).

21. C. R. Falconer, R. J. Shepherd, J. M. Pirie, and G. Topping, Arsenic levels in fish and shellfish from the North Sea, *J. Exp. Mar. Biol. Ecol., 71*: 193 (1983).

22. E. A. Woolson, Emissions, cycling and effects of arsenic in soil ecosystems, in *Biological and Environmental Effects of Arsenic* (B. A. Fowler, ed.), Elsevier, Amsterdam, 1983, p. 51.

23. J. P. Buchet, R. Lauwerys, A. Vandevoorde, and J. M. Pycke, Oral daily intake of cadmium, lead, manganese, copper, chromium, mercury, calcium, zinc and arsenic in Belgium: A duplicate metal study, *Fd. Chem. Toxic., 21*: 19 (1983).

24. M. Vahter, E. Marafante, and L. Dencker, Metabolism of arseno-betaine in mice, rats and rabbits, *Sci. Total Environ., 30*: 197 (1983).

25. E. Marafante, M. Vahter, and L. Dencker, Metabolism of arseno-choline in mice, rats and rabbits, *Sci. Total Environ., 34*: 223 (1984).

26. H. Yamauchi and Y. Yamamura, Metabolism and excretion of orally administered dimethylarsinic acid in the hamster, *Toxicol. Appl. Pharmacol., 74*: 134 (1984).

27. H. Yamauchi and Y. Yamamura, Metabolism and excretion of orally ingested trimethylarsenic in man, *Bull. Environ. Contam. Toxicol., 32*: 682 (1984).

28. E. W. McChesney, J. O. Hoppe, J. P. McAuliff, and W. F. Banks, Jr., Toxicity and physiological disposition of sodium

p-N-glycolylarsanilate. I. Observations in the mouse, cat, rat and man, *Toxicol. Appl. Pharmacol.*, *4*: 14 (1962).

29. J. B. Luten, G. Riekwel-Booy, and A. Rauchbaar, Occurrence of arsenic in plaice (*Pleuronectes platessa*), nature of organo-arsenic compound present and its excretion by man, *Environ. Health Perspect.*, *45*: 165 (1982).

30. G. K. H. Tam, S. M. Charbonneau, F. Bryce, and E. Sandi, Excretion of a single oral dose of fish-arsenic in man, *Bull. Environ. Contam. Toxicol.*, *28*: 669 (1982).

31. H. C. Freemen, J. F. Uthe, R. B. Fleming, P. H. Odense, R. G. Ackman, G. Landry, and C. Musial, Clearance of arsenic ingested by man from arsenic contaminated fish, *Bull. Environ. Contam. Toxicol.*, *22*: 224 (1979).

32. L. M. Klevay, Pharmacology and toxicology of heavy metals: Arsenic, *Pharmacol. Ther. A*, *1*: 189 (1976).

33. S. W. Hwang and L. S. Schanker, Absorption of organic arsenical compounds from the rat small intestine, *Xenobiotica*, *3*: 351 (1973).

34. C. D. Klassen, Biliary excretion of arsenic in rats, rabbits and dogs, *Toxicol. Appl. Pharmacol.*, *29*: 447 (1974).

35. M. Cikrt, V. Bencko, M. Tichy, and B. Benes, Biliary excretion of [74]As and its distribution in the golden hamster after administration of [74]As(III) and [74]As(V), *J. Hyg. Epidemiol. Microbiol. Immunol.*, *24*: 384 (1980).

36. J. Alexander and J. Aaseth, Excretion of arsenic in rat bile: A role of complexing ligands containing sulfur and selenium, *Nutr. Res., Suppl.*, *1*: 515 (1985).

37. E. A. Crecelius, Changes in the chemical speciation of arsenic following ingestion by man, *Environ. Health Perspect.*, *19*: 147 (1977).

38. V. Foa, A. Colombi, and M. Maroni, The speciation of the chemical forms of arsenic in the biological monitoring of exposure to inorganic arsenic, *Sci. Total Environ.*, *34*: 241 (1984).

39. G. K. H. Tam, S. M. Charbonneau, F. Bryce, C. Pomroy, and E. Sandi, Metabolism of inorganic arsenic ([74]As) in humans following oral ingestion, *Toxicol. Appl. Pharmacol.*, *50*: 319 (1979).

40. M. Vahter, Biotransformation of trivalent and pentavalent inorganic arsenic in mice and rats, *Environ. Res.*, *25*: 286 (1981).

41. G. K. H. Tam, S. M. Charbonneau, F. Bryce, and G. Lacroix, Separation of arsenic metabolites in dog plasma and urine following intravenous injection of [74]As, *Anal. Biochem.*, *86*: 505 (1978).

42. M. Vahter, E. Marafante, A. Lindgren, and L. Dencker, Tissue distribution and subcellular binding of arsenic in marmoset

monkeys after injection of [74]As-arsenite, *Arch. Toxicol.*, *51*: 65 (1982).

43. G. K. H. Tam, S. M. Charbonneau, G. Lacroix, and F. Bryce, In vitro methylation of [74]As in urine, plasma, and red blood cells of human and dog, *Bull. Environ. Contam. Toxicol.*, *22*: 69 (1979).

44. D. Y. Shirachi, J. U. Lako, and L. J. Rose, Methylation of sodium arsenate by rat liver in vitro, *Proc. West. Pharmacol. Soc.*, *24*: 159 (1981).

45. S. A. Lerman, T. W. Clarkson, and R. J. Gerson, Arsenic uptake and metabolism by liver cells is dependent on arsenic oxidation state, *Chem.-Biol. Interactions*, *45*: 401 (1983).

46. N. Fuentes, F. Zambrano, and M. Rosenmann, Arsenic contamination: Metabolic effects and localization in rats, *Comp. Biochem. Physiol. C*, *70C*: 269 (1981).

47. E. Marafante, F. Bertolero, J. Edel, R. Pietra, and E. Sabbioni, Intracellular interaction and biotransformation of arsenite in rats and rabbits, *Sci. Total Environ.*, *24*: 27 (1982).

48. M. Vahter and E. Marafante, Intracellular interaction and metabolic fate of arsenite and arsenate in mice and rabbits, *Chem.-Biol. Interactions*, *47*: 29 (1983).

49. I. R. Rowland and M. J. Davies, Reduction and methylation of sodium arsenate in the rat, *J. Appl. Toxicol.*, *2*: 294 (1982).

50. C. Pomroy, S. M. Charbonneau, R. S. McCullough, and G. K. H. Tam, Human retention studies with [74]As, *Toxicol. Appl. Pharmacol.*, *53*: 550 (1980).

51. M. Vahter, Metabolism of arsenic, in *Biological and Environment Effects of Arsenic* (B. A. Fowler, ed.), Elsevier, Amsterdam, 1983, p. 171.

52. M. Anke, B. Groppel, M. Grün, A. Hennig, and D. Meissner, The influence of arsenic deficiency on growth, reproductiveness, life expectancy and health of goats, in *3. Spurenelement-Symposium Arsen* (M. Anke, H. -J. Schneider, and Chr. Bruckner, eds.), Friedrich-Schiller-Universität, Jena, GDR, 1980, p. 25.

53. M. Anke, A. Schmidt, B. Groppel, and H. Kronemann, Further evidence for the essentiality of arsenic, in *4. Spurenelement-Symposium* (M. Anke, W. Baumann, H. Bräunlich, and Chr. Bruckner, Eds.), Friedrich-Schiller-Universität, Jena, GDR, 1983, p. 97.

54. E. O. Uthus, W. E. Cornatzer, and F. H. Nielsen, Consequences of arsenic deprivation in laboratory animals, in *Arsenic Symposium, Production and Use, Biomedical and Environmental Perspectives* (W. H. Lederer, ed.), Van Nostrand Reinhold, New York, 1983, p. 173.

55. B. L. Vallee, D. D. Ulmer, and W. E. C. Wacker, Arsenic Toxicology and biochemistry, A. M. A. Arch. Indust. Health, 21: 132 (1960).

56. K. S. Squibb and B. A. Fowler, The toxicity of arsenic and its compounds, in Biological and Environmental Effects of Arsenic (B. A. Fowler, Ed.), Elsevier, Amsterdam, 1983, p. 233.

57. F. H. Nielsen, E. O. Uthus, and W. E. Cornatzer, Arsenic possibly influences carcinogenesis by affecting arginine and zinc metabolism, Biol. Trace Element Res., 5: 389 (1983).

58. E. O. Uthus and F. H. Nielsen, Interactions among dietary arsenic, arginine and methionine in the chick, Fed. Proc., 43: 680 (1984).

59. H. J. Almquist and C. R. Grau, Interrelation of methionine, choline, betaine and arsenocholine in the chick, J. Nutr., 27: 263 (1944).

60. F. H. Nielsen, Possible future implications of nickel, arsenic, silicon, vanadium, and other ultratrace elements in human nutrition, in Clinical, Biochemical, and Nutritional Aspects of Trace Elements, Curr. Top. Nutr. Dis. (A. S. Prasad, ed.), Vol. 6, Liss, New York, 1982, p. 379.

61. F. H. Nielsen and E. O. Uthus, Arsenic, in Biochemistry of the Essential Ultratrace Elements (E. Frieden, ed.), Plenum, New York, 1984, p. 319.

62. R. Hille, R. C. Stewart, J. A. Fee, and V. Massey, The interaction of arsenite with xanthine oxidase, J. Biol. Chem., 258: 4849 (1983).

63. J. Ploquin, Le bore dans l'alimentation, Bull. Soc. Sci. Hyg. Aliment., 55: 70 (1967).

64. C. C. Pfeiffer, L. S. Hallman, and I Gersh, Boric acid ointment. A study of possible intoxication in the treatment of burns, J. A. M. A., 128: 266 (1945).

65. H. Agulhon, The presence and use of boron in plants, Ann. Inst. Pasteur., 24: 321 (1910).

66. K. Warrington, The effect of boric acid and borax on broad bean and certain other plants, Ann. Bot., 40: 27 (1923).

67. C. A. Zittle, Reaction of borate with substances of biological interest, in Advances in Enzymology (F. F. Ford, ed.), Vol. XII, Interscience, New York, 1951, p. 493.

68. S. L. Johnson and K. W. Smith, The interaction of borate and sulfite with pyridine nucleotides, Biochemistry, 15: 553 (1976).

69. J. D. Dunitz, D. M. Hawley, D. Miklos, D. N. J. White, Yu. Berlin, R. Marusic, and V. Prelog, Structure of boromycin, Helv. Chim. Acta, 54: 1709 (1971).

70. T. S. S. Chen, C. -J. Chang, and H. G. Floss, Biosynthesis of the boron-containing antiobiotic aplasmomysin. Nuclear magnetic resonance analysis of aplasmomycin and desboroaplasmomycin, *J. Antibiotics*, *33*: 1316 (1980).

71. A. K. Furr, T. F. Parkinson, C. A. Bache, W. H. Gutenmann, I. S. Pakkala, G. S. Stoewsand, and D. J. Lisk, Multi-element analysis of fruit kernels, *Nutr. Repts. Internat.*, *20*: 841 (1979).

72. A. K. Furr, L. H. MacDaniels, L. E. St. John, Jr., W. H. Gutenmann, I. S. Pakkala, and D. J. Lisk, Elemental composition of tree nuts, *Bull. Environm. Contam. Toxicol.*, *21*: 392 (1979).

73. R. A. Morse and D. J. Lisk, Elemental analysis of honeys from several nations, *Am. Bee J.*, *120*: 522 (1980).

74. D. Schlettwein-Gsell and S. Mommsen-Straub, Übersicht Spurenelemente in Lebensmitteln. IX. Bor, *Internat. Z. Vit.-Ern.-Forschung*, *43*: 93 (1973).

75. A. S. Szabo, Bestimmung des Borgehaltes in pflanzlichen Lebensmitteln, *Lebensmittelindustrie*, *26*: 549 (1979).

76. P. Varo and P. Koivistoinen, Mineral element composition of Finnish foods. XII. General discussion and nutritional evaluation, *Acta Agric. Scand.*, *Suppl. 22*, p. 165 (1980).

77. E. G. Zook and J. Lehmann, Total diet study: Content of ten minerals—aluminum, calcium, phosphorus, sodium, potassium, boron, copper, iron, manganese, and magnesium, *J. Assoc. Off. Agric. Chem.*, *48*: 850 (1965).

78. E. I. Hamilton and M. J. Minski, Abundance of the chemical elements in man's diet and possible relations with environmental factors, *Sci. Total Environ.*, *1*: 375 (1972/73).

79. N. L. Kent and R. A. McCance, The absorption and excretion of "minor" elements by man. I. Silver, gold, lithium, boron, and vanadium, *Biochem. J.*, *35*: 837 (1941).

80. E. C. Owen, The excretion of borate by the dairy cow, *J. Dairy Res.*, *13*: 243 (1944).

81. I. H. Tipton, P. L. Stewart, and P. G. Martin, Trace elements in diets and excreta, *Health Phys.*, *12*: 1683 (1966).

82. J. A. Jansen and J. S. Schou, Gastro-intestinal absorption and *in vitro* release of boric acid from water-emulsifying ointments, *Food Chem. Toxic.*, *22*: 49 (1984).

83. M. Akagi, T. Misawa, and H. Kaneshima, Metabolism of the borate. I. Excretion of the boron into the urine and distribution in some organs, *Yakugaku Zasshi*, *82*: 934 (1962).

84. J. A. Jansen, J. Anderson, and J. S. Schou, Boric acid single dose pharmacokinetics after intravenous administration to man, *Arch. Toxicol.*, *55*: 64 (1984).

85. C. D. Hunt and F. H. Nielsen, Interaction between boron and cholecalciferol in the chick, in *Trace Element Metabolism in Man*

and Animals (J. McC. Howell, J. M. Gawthorne, and C. L. White, eds.), Vol. 4, Australian Academy of Science, Canberra, 1981, p. 597.

86. C. D. Hunt, T. R. Shuler, and F. H. Nielsen, Effect of boron on growth and mineral metabolism, in *4. Spurenelement-Symposium* (M. Anke, W. Baumann, H. E. Braunlich, and Chr. Brückner, eds.), Friedrich-Schiller-Univ., Jena, 1983, p. 149.

87. F. H. Nielsen, Interactions among dietary aluminum, boron, magnesium, and methionine in the rat, *Fed. Proc.*, *44*: 752 (1985).

88. Life Sciences Research Office, *Evaluation of the Health Aspects of Borax and Boric Acid as Food Packaging Ingredients*, Fed. Amer. Soc. Exp. Biol., Bethesda, Maryland, 1980, 30 p.

89. Anon., Boron, in *Mineral Tolerance of Domestic Animals*, National Academy of Sciences, Washington, D. C., 1980, p. 71.

90. R. J. Weir, Jr. and R. S. Fisher, Toxicologic studies on borax and boric acid, *Toxicol. Appl. Pharmacol.*, *23*: 351 (1972).

91. G. H. Green, M. D. Lott, and H. J. Weeth, Effects of boron-water on rats, *Proc. West. Soc. Am. Soc. Anim. Sci.*, *24*: 254 (1973).

92. B. S. Seal and H. J. Weeth, Effect of boron in drinking water on the male laboratory rat, *Bull. Environ. Contam. Toxicol.*, *25*: 782 (1980).

93. J. Pinto, Y. P. Huang, R. J. McConnell, and R. S. Rivlin, Increased urinary riboflavin excretion resulting from boric acid ingestion, *J. Lab. Clin. Med.*, *92*: 126 (1978).

94. W. Landauer, Malformations of chicken embryos produced by boric acid and the probable role of riboflavin in their origin, *J. Exp. Zool.*, *120*: 469 (1952).

95. W. Landauer, Complex formation and chemical specificity of boric acid in production of chicken embryo malformations, *Proc. Sco. Exp. Biol. Med.*, *82*: 33 (1953).

96. J. Elsair, R. Merad, R. Denine, M. Reggabi, S. Benali, B. Alamir, M. Hanrour, M. Azzouz, K. Khalfat, M. Tabet Aoul, and J. Nauer, Action of boron upon fluorosis: An experimental study, *Fluoride*, *15*: 75 (1982).

97. J. Elsair, R. Merad, R. Denine, M. Reggabi, S. Benali, M. Azzouz, K. Khelfat, and M. Tabet Aoul, Boron as an antidote in acute fluoride intoxication in rabbits. Its action on the fluoride and calcium-phosphorus metabolism, *Fluoride*, *13*: 30 (1980).

98. W. Seffner, W. Teubener, and D. Geinitz, Zur Bor-Wirkung auf Skelett und Nedenschilddrüse, in *Mengen-und Spurenelemente* (M. Anke, C. Brückner, H. Gürtler, and M. Grün, eds.), Karl-Marx-Univ., Leipzig, 1983, p. 200.

99. W. Seffner and W. Teubener, Antidotes in experimental fluorosis in pigs morphological studies, *Fluoride*, *16*: 33 (1983).

100. H. P. Baer, R. Bech, J. Franke, A. Grunewald, W. Kochmann, F. Melson, H. Runge, and W. Wiedner, Ergebnisse tierexperimenteller Untersuchungen an Kaninchen mit Natriumfluorid unter Einwirkung von Gegenmitteln, *Ztschr. Ges. Hyg. Gresz, 23*: 14 (1977).

101. J. F. Jackson and K. S. R. Chapman, The role of boron in plants, in *Trace Elements in Soil-Plant-Animal Systems* (D. J. D. Nicholas and A. R. Egan, eds.), Academic Press, New York, 1975, p. 213.

102. D. H. Lewis, Boron, lignification and the origin of vascular plants—a unified hypothesis, *New Phytol., 84*: 209 (1980).

103. A. S. Pollard, A. J. Parr, and B. C. Loughman, Boron in relation to membrane function in higher plants, *J. Exp. Bot., 28*: 831 (1977).

104. E. Habermann, Action and binding of palytoxin, as studied with brain membranes, *Naunyn-Schmiedeberg's Arch. Pharmacol., 323*: 269 (1983).

105. A. J. Parr and B. C. Loughman, Boron and membrane function in plants, *Annu. Proc. Phytochem. Soc. Eur., 21*: 87 (1983).

106. D. J. Pilbeam and E. A. Kirkby, The physiological role of boron in plants, *J. Plant Nutrition, 6*: 563 (1983).

107. W. M. Duggar, Boron in plant metabolism, *Encyl. Plant Physiol., New Ser., 15B*: 626 (1983).

108. T. Tanada, Localization of boron in membranes, *J. Plant Nutr., 6*: 743 (1983).

109. M. Wolny, Effect of borate on the catalytic activities of muscle glyceraldehyde-3-phosphate dehydrogenase, *Eur. J. Biochem., 80*: 551 (1977).

110. I. H. Hall, C. O. Starnes, A. T. McPhail, P. Wisian-Neilson, M. K. Das, F. Harchelroad, Jr., and B. F. Spielvogel, Anti-inflammatory activity of amine cyanoboranes, amine carboxy-boranes, and related compounds, *J. Pharmaceut. Sci., 69*: 1025 (1980).

111. B. F. Spielvogel, L. Wojnowich, M. K. Das, A. T. McPhail, and K. D. Hargrave, Boron analogues of amino acids. Synthesis and biological activity of boron analogues of betaine, *J. Amer. Chem. Soc., 98*: 5702 (1976).

112. R. E. Newnham, Mineral imbalance and boron deficiency, in *Trace Element Metabolism in Man and Animals* (J. McC. Howell, J. M. Gawthorne, and C. L. White, eds.), Vol. 4, Australian Academy of Science, Canberra, 1981, p. 400.

113. H. Bortels, Molybdan als Katalysator bei der biologischen Stockstoffbindung, *Arch. Mikrobiol., 1*: 333 (1930).

114. W. S. Ferguson, A. H. Lewis, and S. J. Watson, Action of molybdenum in nutrition of milking cattle, *Nature* (London), *141*: 553 (1938).

115. K. V. Rajagopalan, Molybdenum, in *Biochemistry of the Essential Ultratrace Elements* (E. Frieden, ed.), Plenum, New York, 1984, p. 149.

116. J. T. Spence, Reactions of molybdenum coordination compounds: Models for biological systems, in *Metal Ions in Biological Systems: Reactivity of Coordination Compounds* (H. Sigel, ed.), Vol. 5, Dekker, New York, 1976, p. 279.

117. J. L. Johnson, H. P. Jones, and K. V. Rajagopalan, *In vitro* reconstitution of demolybdosulfite oxidase by a molybdenum cofactor from rat liver and other sources, *J. Biol. Chem.*, *252*: 4994 (1977).

118. J. L. Johnson, B. E. Hainline, and K. V. Rajagopalan, Characterization of the molybdenum cofactor of sulfite oxidase, xanthine oxidase, and nitrate reductase, *J. Biol. Chem.*, *255*: 1783 (1980).

119. Anon., Molybdenum, in *Mineral Tolerance of Domestic Animals*, National Academy of Sciences, Washington, D. C., 1980, p. 328.

120. P. W. Winston, Molybdenum, in *Disorders of Mineral Metabolism*, Vol. 1, Academic Press, New York, 1981, p. 295.

121. H. A. Schroeder, J. J. Balassa, and I. H. Tipton, Essential trace metals in man: Molybdenum, *J. Chronic Dis.*, *23*: 481 (1970).

122. T. A. Tsongas, R. R. Meglan, P. A. Walravens, and W. R. Chappell, Molybdenum in the diet: An estimate of average daily intake in the United States, *Am. J. Clin. Nutr.*, *33*: 1103 (1980).

123. C. J. Cardin and J. Mason, Molybdate and tungstate transfer by rat ileum. Competitive inhibition by sulfate, *Biochim. Biophys. Acta*, *455*: 937 (1976).

124. L. J. Kosarek and P. W. Winston, Absorption of molybdenum-99 (Mo-99) as molybdate with various doses in the rat, *Fed. Proc.*, *36*: 1106 (1977).

125. C. F. Mills and I. Bremner, Nutritional aspects of molybdenum in animals, in *Molybdenum and Molybdenum-Containing Enzymes* (M. P. Coughlan, ed.), Pergamon, Oxford, England, 1980, p. 519.

126. N. N. Abumrad, A. J. Schneider, D. Steel, and L. S. Rogers, Amino acid intolerance during prolonged total parenteral nutrition reversed by molybdate therapy, *Am. J. Clin. Nutr.*, *34*: 2551 (1981).

127. National Academy of Sciences, *Recommended Dietary Allowances*, 9th Ed., Food and Nutrition Board, National Academy of Sciences, Washington, D. C., 1980.

128. X. -M. Luo, H. -J. Wei, and S. P. Yang, Inhibitory effects of molybdenum on esophageal and forestomach carcinogenesis in rats, *J. Natl. Cancer Inst.*, *71*: 65 (1983).

129. C. D. Seaborn, S. P. Yang, and H. J. Wei, Inhibitory effect of molybdenum on N-methyl-N-nitrosourea-induced mammary gland carcinogenesis in female rats, Fed. Proc., 43: 793 (1984).

130. J. O. Nriagu (ed.), Nickel in the Environment, Wiley, New York, 1980.

131. R. Berg, Das Vorkommen seltener Elemente in den Nahrungsmitteln und menschlichen Ausscheidungen, Biochem. Z., 165: 461 (1925).

132. G. Bertrand and M. Macheboeuf, Sur la presence der nickel et du cobalt chez les animaux, C. R. Acad. Sci. (Paris), 180: 1380 (1925).

133. G. Bertrand and H. Nakamura, Recherches sur l'importance physiologique du nickel et due cobalt, Bull. Soc. Sci. Hyg. Aliment., 24: 338 (1936).

134. F. H. Nielsen, Nickel, in Biochemistry of the Essential Ultra-trace Elements (E. Frieden, ed.), Plenum, New York, 1984, p. 293.

135. J. J. G. Moura, M. Teixeira, I. Moura, and A. V. Xavier, Nickel—A redox catalytic site in hydrogenase, J. Mol. Cat., 23: 303 (1984).

136. N. Kojima, J. A. Fox, R. P. Hausinger, L. Daniels, W. H. Orme-Johnson, and C. Walsh, Paramagnetic centers in the nickel-containing, deazaflavin-reducing hydrogenase from Methanobacterium thermoautotrophicum, Proc. Natl. Acad. Sci. USA, 80: 378 (1983).

137. M. Lucassen and B. Sarkar, Nickel(II)-binding constituents of human blood serum, J. Toxicol. Environ. Health, 5: 897 (1979).

138. F. W. Sunderman, Jr., A review of the metabolism and toxicology of nickel, Ann. Clin. Lab. Sci., 1: 377 (1977).

139. D. C. Jones, P. M. May, and D. R. Williams, Computer simulation models of low-molecular-weight nickel(II) complexes and therapeuticals in vivo, in Nickel Toxicology (S. Brown and F. W. Sunderman, Jr., eds.), Academic Press, New York, 1980, p. 73.

140. B. Sarkar, Biological specificity of the transport process of copper and nickel, Chem. Script., 21: 1011 (1983).

141. J. -P. Laussac and B. Sarkar, Characterization of the copper (II)- and nickel(II)-transport site of human serum albumin. Studies of copper(II) and nickel(II) binding to peptide 1-24 of human serum albumin by ^{13}C and ^{1}H NMR spectroscopy, Biochemistry, 23: 2832 (1984).

142. P. Mohanakrishnan and C. F. Chignell, Copper and nickel binding to canine serum albumin. A circular dichroism study, Comp. Biochem. Physiol., 79C: 321 (1984).

143. N. E. Dixon, R. L. Blakeley, and B. Zerner, Jack bean urease (EC 3.5.1.5). III. The involvement of active-site nickel in inhibition by β-mercaptoethanol, phosphoramidate, and fluoride, *Can. J. Biochem.*, *58*: 481 (1980).

144. S. S. Hasnain and B. Piggott, An EXAFS study of jack bean urease, a nickel metalloenzyme, *Biochem. Biophys. Res. Commun.*, *112*: 279 (1983).

145. R. L. Blakeley and B. Zerner, Jack bean urease: The first nickel enzyme, *J. Mol. Cat.*, *23*: 263 (1984).

146. W. J. Kersten, R. R. Brooks, R. D. Reeves, and T. Jaffre, Nature of nickel complexes in *Psychotria douarrei* and other nickel-accumulating plants, *Phytochemistry*, *19*: 1963 (1980).

147. K. E. Hammel, K. L. Cornwell, G. B. Diekert, and R. K. Thauer, Evidence for a nickel-containing carbon monoxide dehydrogenase in *Methanobrevibacter arboriphilicus*, *J. Bacteriol.*, *157*: 975 (1984).

148. G. Diekert and R. K. Thauer, The effect of nickel on carbon monoxide dehydrogenase formation in *Clostridium thermoaceticum*, and *Clostridium formicoaceticum*, *FEMS Microbiol. Lett.*, *7*: 187 (1980).

149. H. L. Drake, Occurrence of nickel in carbon monoxide dehydrogenase from *Clostridium pasteurianum* and *Clostridium thermoaceticum*, *J. Bacteriol.*, *149*: 561 (1982).

150. S. W. Ragsdale, J. E. Clark, L. G. Ljungdahl, L. L. Lundie, and H. L. Drake, Properties of purified carbon monoxide dehydrogenase from *Clostridium thermoaceticum*, a nickel, iron-sulfur protein, *J. Biol. Chem.*, *258*: 2364 (1983).

151. S. W. Ragsdale, L. G. Ljungdahl, and D. V. DerVartanian, ^{13}C and ^{61}Ni isotope substitutions confirm the presence of a nickel(III)-carbon species in acetogenic CO dehydrogenases, *Biochem. Biophys. Res. Comm.*, *115*: 658 (1983).

152. D. Ankel-Fuchs, R. Jaenchen, N. A. Gebhardt, and R. K. Thauer, Functional relationship between protein-bound and free factor F430 in *Methanobacterium*, *Arch. Microbiol.*, *139*: 332 (1984).

153. R. K. Thauer, G. Diekert, and P. Schönheit, Biological role of nickel, *Trends Biochem. Sci. (Pers. Ed.)*, *5*: 304 (1980).

154. A. Colbeau and P. M. Vignais, The membrane-bound hydrogenase of *Rhodopseudomonas capsulata* is inducible and contains nickel, *Biochim. Biophys. Acta*, *748*: 128 (1983).

155. K. Schneider, R. Cammack, and H. G. Schegel, Content and localization of FMN, Fe-S clusters and nickel in the NAD-linked hydrogenase of *Nocardia opaca* 1b, *Eur. J. Biochem.*, *142*: 75 (1984).

156. H. Almon and P. Böger, Nickel-dependent uptake-hydrogenase activity in the blue-green alga *Anabaena variabilis*, *Z. Naturforsch.*, *39C*: 90 (1984).

157. S. P. J. Albracht, J. W. Van Der Zwaan, and R. D. Fontijn, EPR spectrum at 4, 9 and 35 GHz of hydrogenase from *Chromatium vinosum*. Direct evidence for spin-spin interaction between Ni(III) and the iron-sulphur cluster, *Biochim. Biophys. Acta, 766*: 245 (1984).

158. P. A. Lindahl, N. Kojima, R. P. Hausinger, J. A. Fox, B. K. Teo, C. T. Walsh, and W. H. Orme-Johnson, Nickel and iron EXAFS of F_{420}-reducing hydrogenase from *Methanobacterium thermoautotrophicum, J. Am. Chem. Soc., 106*: 3062 (1984).

159. E. Nieboer, P. I. Stetsko, and P. Y. Hin, Characterization of the Ni(III)/Ni(II) redox couple for the nickel(II) complex of human serum albumin, *Abstracts, Third International Conference on Nickel Metabolism and Toxicology*, Paris, 1984, p. 17.

160. D. Schlettwein-Gsell and S. Mommsen-Straub, Spurenelemente in Lebensmitteln. V. Nickel, *Int. J. Vitam. Nutr., 41*: 429 (1971).

161. G. Ellen, G. Bosch-Tibbesma, and F. F. Douma, Nickel content of various Dutch foodstuffs, *Z. Lebensm. Unters.-Forsch., 166*: 145 (1978).

162. B. Thomas, J. A. Roughan, and E. D. Watters, Cobalt, chromium and nickel content of some vegetable foodstuffs, *J. Sci. Fd. Agric., 25*: 771 (1974).

163. R. Brun, Nickel dan les aliments et eczema de contact, *Dermatologicia, 159*: 365 (1979).

164. M. Anke, M. Grün, B. Groppel, and H. Kronemann, Nutritional requirements of nickel, in *Biological Aspects of Metals and Metal-Related Diseases* (B. Sarker, ed.), Raven Press, New York, 1983, p. 89.

165. C. E. Casey, The content of some trace elements in infant milk foods and supplements available in New Zealand, *New Zealand Med. J., 85*: 275 (1977).

166. J. W. Spears, Nickel as a "newer trace element" in the nutrition of domestic animals, *J. Anim. Sci., 59*: 823 (1984).

167. A. M. Nash, T. L. Mounts, and W. F. Kwolek, Determination of ultratrace metals in hydrogenated vegetable oils and fats, *JAOCS, 60*: 811 (1983).

168. D. R. Myron, T. J. Zimmerman, T. R. Shuler, L. M. Klevay, D. E. Lee, and F. H. Nielsen, Intake of nickel and vanadium by humans. A survey of selected diets, *Am. J. Clin. Nutr., 31*: 527 (1978).

169. G. K. Murthy, U. S. Rhea, and J. T. Peeler, Levels of copper, nickel, rubidium and strontium in institutional total diets, *Environ. Sci. Technol., 7*: 1042 (1973).

170. E. Horak and F. W. Sunderman, Jr., Fecal nickel excretion by healthy adults, *Clin. Chem., 19*: 429 (1973).

171. F. H. Nielsen, Studies on the interaction between nickel and iron during intestinal absorption, in *4. Spurenelement-Symposium*

(M. Anke, W. Bauman, H. Braünlich, and C. Brückner, eds.), Friedrich-Schiller-Univ., Jena, 1983, p. 11.

172. M. Kirchgessner, R. Spörl, and D. A. Roth-Maier, Exkretion im Kot und scheinbare Absorption von Kupfer, Zink, Nickel und Mangan bei nichtgraviden und graviden Sauen mit unterschiedlicher Spurenelementversorgung, Z. Tierphysiol., Tierernaehr. Futtermittelkd., 44: 98 (1980).

173. M. Kirchgessner, D. A. Roth-Maier, and A. Schnegg, Progress of nickel metabolism and nutrition research, in Trace Element Metabolism in Man and Animals (TEMA-4) (J. McC. Howell, J. M. Gawthorne, and C. L. White, eds.), Australian Academy of Science, Canberra, 1981, p. 621.

174. M. Kirchgessner, D. A. Roth-Maier, and R. Spörl, Spurenelementbilanzan (Cu, Zn, Ni, und Mn) laktierender Sauen, Z. Tierphysiol., Tierernaehr. Futtermittelkd., 50: 230 (1983).

175. C. Onkelinx, J. Becker, and F. W. Sunderman, Jr., in Trace Element Metabolism in Animals (W. G. Hoekstra, J. W. Suttie, H. E. Ganther, and W. Mertz, eds.), Vol. 2, University Park Press, Baltimore, 1974, p. 560.

176. J. R. Cohn and E. A. Emmett, The excretion of trace metals in human sweat, Ann. Clin. Lab. Sci., 8: 270 (1978).

177. G. Becker, U. Dorstelmann, U. Frommberger, and W. Forth, On the absorption of cobalt(II)- and nickel(II)-ions by isolated intestinal segments in vitro of rats, in 3. Spurenelement-Symposium, Nickel (M. Anke, H. -J. Schneider, and C. Brückner, eds.), Friedrich-Schiller Univ., Jena, 1980, p. 79.

178. K. S. Kasprzak and F. W. Sunderman, Jr., Radioactive [63]Ni in biological research, Pure & Appl. Chem., 51: 1375 (1979).

179. M. -C. Herlant-Peers, H. F. Hildebrand, and G. Biseite, [63]Ni(II)-incorporation into ling and liver cytosol of Balb/C mouse. An in vitro and in vivo study, zbl. Bakt. Hyg., I. Abt. Orig. B, 176: 368 (1982).

180. N. Jacobsen, I. Alfheim, and J. Jonsen, Nickel and strontium distribution in some mouse tissues passage through placenta and mammary glands, Res. Comm. Chem. Pathol. Pharmacol., 20: 571 (1978).

181. C. -C. Lu, N. Matsumoto, and S. Iijima, Teratogenic effects of nickel chloride on embryonic mice and its transfer to embryonic mice, Teratology, 19: 137 (1979).

182. F. W. Sunderman, Jr., B. L. K. Mangold, S. H. Y. Wong, S. K. Shen, M. C. Reid, and I. Jansson, High-performance size-exclusion chromatography of [63]Ni-constituents in renal cytosol and microsomes from [63]$NiCl_2$-treated rats, Res. Comm. Chem. Pathol. Pharmacol., 39: 477 (1983).

183. A. W. Abdulwajid and B. Sarkar, Nickel-sequestering renal glycoprotein, *Proc. Natl. Acad. Sci. USA, 80*: 4509 (1983).

184. N. Y. Sayato, K. Nakamuro, S. Matsui, and A. Tanimura, Studies on behavior and chemical form of nickel in rat administered with $^{63}NiCl_2$, *J. Pharmacobio-Dyn., 4*: S-73 (1981).

185. F. H. Nielsen, Evidence of the essentiality of arsenic, nickel, and vanadium and their possible nutritional significance, in *Advances in Nutritional Research* (H. H. Draper, ed.), Vol. 3, Plenum, New York, 1980, p. 157.

186. F. H. Nielsen, T. R. Shular, T. G. Mcleod, and T. J. Zimmerman, Nickel influences iron metabolism through physiologic, pharmacologic and toxicologic mechanisms in the rat, *J. Nutr., 114*: 1280 (1984).

187. F. H. Nielsen, Nickel toxicity, in *Advances in Modern Toxicology: Toxicology of Trace Elements* (R. A. Goyer and M. A. Mehlman, eds.), Vol. 2, Wiley, New York, 1977, p. 129.

188. O. B. Christensen and H. Möller, External and internal exposure to the antigen in the hand eczema of nickel allergy, *Contact Derm., 1*: 136 (1975).

189. E. Cronin, A. D. DiMichiel, and S. S. Brown, Oral challenge in nickel-sensitive women with hand eczema, in *Nickel Toxicology* (S. S. Brown and F. W. Sunderman, Jr., eds.), Academic Press, New York, 1980, p. 149.

190. M. M. King and C. Y. Huang, Activation of calcineurin by nickel ions, *Biophys. Res. Comm., 114*: 955 (1983).

191. C. J. Pallen and J. H. Wang, Regulation of calcineurin by metal ions. Mechanism of activation by Ni^{2+} and an enhanced response to Ca^{2+}/calmodulin, *J. Biol. Chem., 259*: 6134 (1984).

192. M. M. King and C. Y. Huang, The calmodulin-dependent activation and deactivation of the phosphoprotein phosphatase, calcineurin, and the effect of nucleotides, pyrophosphate, and divalent metal ions. Identification of calcineurin as a Zn and Fe metalloenzyme, *J. Biol. Chem., 259*: 8847 (1984).

193. Z. Fishelson and H. J. Müller-Eberhard, C3 convertase of human complement: Enhanced formation and stability of the enzyme generated with nickel instead of magnesium, *J. Immunol., 129*: 2603 (1982).

194. Z. Fishelson, M. K. Pangburn, and H. J. Müller-Eberhard, C3 convertase of the alternative complement pathway, *J. Biol. Chem., 258*: 7411 (1983).

195. R. K. Pullarkat, M. Staschnig-Agler, and H. Reha, Biosynthesis of phosphatidylserine in rat brain microsomes, *Biochim. Biophys. Acta, 663*: 117 (1981).

196. F. H. Nielsen, Effect of form of iron on the interaction between nickel and iron in rats: Growth and blood parameters, *J. Nutr., 110*: 965 (1980).

197. F. H. Nielsen and T. R. Shuler, Effect of form of iron on nickel deprivation in the rat. Liver content of copper, iron, manganese, and zinc, *Biol. Trace Element Res.*, 3: 245 (1981).

198. F. H. Nielsen, D. R. Myron, S. H. Givand, and D. A. Ollerich, Nickel deficiency and nickel-rhodium interaction in chicks, *J. Nutr.*, 105: 1607 (1975).

199. M. Kirchgessner and A. Schnegg, Alpha-amylase und dehydrogenasenaktivität bei suboptimaler Ni-Versorgung, *Ann. Nutr. Metab.*, 25: 307 (1981).

200. M. Kirchgessner, R. Maier, and A. M. Reichlmayr-Lais, Konzentrationen von Fe, Cu, Zn, Mn, Co und Mg in verschiedenen Organen und Geweben nach unterschiedlicher Ni-Versorgung, *Z. Tierphysiol., Tierernaehr. Futtermittelkd.*, 52: 217 (1984).

201. J. W. Spears, E. E. Jones, L. J. Samsell, and W. D. Armstrong, Effect of dietary nickel on growth, urease activity, blood parameters and tissue mineral concentrations in the neonatal pig, *J. Nutr.*, 114: 845 (1984).

202. J. W. Spears and E. E. Hatfield, Nickel for ruminants. I. Influence of dietary nickel on ruminal urease activity, *J. Anim. Sci.*, 47: 1345 (1978).

203. J. W. Spears, E. E. Hatfield, and R. M. Forbes, Nickel for ruminants. II. Influence of dietary nickel on performance and metabolic parameters, *J. Anim. Sci.*, 48: 649 (1979).

204. E. Reidel, M. Anke, S. Schwarz, A. Regius, M. Szilagyi, H. -J. Löhnert, G. Flachowsky, G. Zenker, and S. Glös, The influence of the nickel offer on growth and different biochemical parameters in fattening cattle, in *3. Spurenelement-Symposium, Nickel* (M. Anke, H. -J. Schneider, and C. Brückner, eds.), Friedrich-Schiller-Univ., Jena, 1980, p. 55.

205. F. H. Nielsen, Interactions of nickel with essential minerals, in *Nickel in the Environment* (J. O. Nriagu, ed.), Wiley, New York, 1980, p. 611.

206. A. S. Chung, W. G. Hoekstra, and R. H. Grummer, Supplemental cobalt or nickel for zinc deficient G. F. pigs, *J. Anim. Sci.*, 42: 1352 (1976).

207. A. S. Chung, M. L. Sunde, R. H. Grummer, and W. G. Hoekstra, The sparing effect of cobalt or nickel on zinc in chicks, *Fed. Proc.*, 37: 668 (1978).

208. J. W. Spears, E. E. Hatfield, and R. M. Forbes, Interrelationship between nickel and zinc in the rat, *J. Nutr.*, 108: 307 (1978).

209. J. W. Spears, E. E. Hatfield, and R. M. Forbes, Nickel-copper interrelationship in the rat, *Proc. Soc. Exp. Biol. Med.*, 156: 140 (1977).

210. F. H. Nielsen and T. J. Zimmerman, Interactions among nickel, copper, and iron in rats. Growth, blood parameters, and organ wt/body wt ratios, *Biol. Trace Element Res.*, 3: 83 (1981).

211. F. H. Nielsen, T. J. Zimmerman, and T. R. Shuler, Interactions among nickel, copper, and iron in rats. Liver and plasma content of lipids and trace elements, *Biol. Trace Element Res.*, 4: 125 (1982).

212. F. H. Nielsen, The importance of diet composition in ultratrace element research, *J. Nutr.*, 115: 1239 (1985).

213. E. F. von Gorup-Besanez, Weitere Untersuchungen über die Verbreitung der Kieselerde im Thierreich, *Ann. Chem. Pharm.*, 66: 321 (1848).

214. H. Schultz, Über den Kieselsaüregehalt menschlicher und thierischer Gewebe, *Pflügers Arch. Ges. Physiol.*, 84: 67 (1901).

215. M. A. Gouget, Athérome Expérimental Et Silicate De Soude, *La Presse Medicale*, 97: 1005 (1911).

216. U. Wannagat, Sila-pharmaca, in *Nobel Symposium*, 40: 447 (1978).

217. K. Schwarz, A bound form of silicon in glycosaminoglycans and polyuronides, *Proc. Natl. Acad. Sci. USA*, 70: 1608 (1973).

218. H. J. M. Bowen and A. Peggs, Determination of the silicon content of food, *J. Sci. Food Agric.*, 35: 1225 (1984).

219. J. L. Kelsay, K. M. Behall, and E. Prather, Effect of fiber from fruits and vegetables on metabolic responses of human subjects. IV. Calcium, magnesium, iron and silicon balances, *Am. J. Clin. Nutr.*, 32: 1876 (1979).

220. F. Sauer, D. H. Laughland, and W. M. Davidson, Silica metabolism in guinea pigs, *Can. J. Biochem. Physiol.*, 37: 183 (1959).

221. Y. Mauras, J. C. Renier, A. Tricard, and P. Allain, Mise en évidence de l'absorption gastro-intestinale du silicium à partir d'un alumino-silicate, *Thérapie*, 38: 175 (1983).

222. P. Allain, A. Cailleux, Y. Mauras, and J. C. Renier, Etude de l'absorption digestive du silicium apres administration unique chez l'homme sous forme de salicylate de méthyl silane triol, *Thérapie*, 38: 171 (1983).

223. G. M. Benke and T. W. Osborn, Urinary silicon excretion by rats following oral administration of silicon compounds, *Food Cosmet. Toxicol.*, 17: 123 (1979).

224. Y. Charnot and G. Péres, Contribution á l'etude de la regulation endocrinienne du metabolisme silicique, *Anal. Endocrinol.*, 32: 397 (1971).

225. L. J. Goldwater, The urinary excretion of silica in non-silicotic humans, *J. Ind. Hyg. Toxicol.*, 18: 163 (1936).

226. J. W. Dobbie, M. J. B. Smith, and A. R. A. S. Abdullah, Silicon and the kidney, *Proc. 8th Int. Congr. Nephrol.*, 1981, p. 1030.

227. E. M. Carlisle, Silicon, in *Biochemistry of the Essential Ultratrace Elements* (E. Frieden, ed.), Plenum, New York, 1984, p. 257.

228. E. M. Carlisle, A silicon requirement for normal skull formation in chicks, *J. Nutr.*, *110*: 352 (1980).

229. E. M. Carlisle, Biochemical and morphological changes associated with long bone abnormalities in silicon deficiency, *J. Nutr.*, *110*: 1046 (1980).

230. K. Schwarz, Recent dietary trace element research exemplified by tin, fluorine, and silicon, *Fed. Proc.*, *33*: 1748 (1974).

231. C. -H. Becker, D. Matthias, H. WooBmann, A. Schwartz, and E. Engler, Investigations on a possible medical importance of silicon, in *4. Spurenelement-Symposium* (M. Anke, W. Baumann, H. Bräunlich, and C. Brückner, eds.), Friedrich-Schiller-Univ., Jena, 1983, p. 142.

232. J. Loeper, J. Loeper, and M. Fragny, The physiological role of the silicon and its antiatheromatous action, *Nobel Symp.*, *40*: 281, (1978).

233. J. Loeper, J. Emerit, J. Goy, L. Rozensztajn, and M. Fragny, Étude des acides gras et de la peroxydation lipidique dans l'athérome experimental du lapin. Role joué par le silicium, *Pathol. Biol.*, *32*: 693 (1984).

234. P. Gendre and R. Cambar, Role du silicium et son efficacite dans le traitement de l'athérosclérose, *Bull. Soc. Pharm. Bordeaux*, *111*: 3 (1972).

235. Y. Charnot, M. C. Asseko, and G. Pérés, Comparison des effets de l'ingestion de Silicium sur la teneur en Ca et en Mg de différents tissus de Rattes normales ou traitées par des oestroprogestatifs, *Ann. Endocrinol.*, *38*: 377 (1977).

236. C. -H. Becker, E. Engler, D. Hoebbel, and R. Rühlmann, Zur Reaktion des Lipidblutspiegels auf orale Siliziumapplikationen bei der Ratte, in *Mengen-und Spurenelemente* (M. Anke, C. Brückner, H. Gürtler, and M. Grün, eds.), Karl-Marx-Univ., Leipzig, 1983, p. 194.

237. P. Ki, J. A. Negulesco, R. L. Hamlin, M. A. Coleman, and R. F. Wilson, Effects of silicon on the ^{3}H-cholesterol level of serum, aorta, cardiac and hepatic tissues of young turkeys, *IRCS Med. Sci.*, *12*: 959 (1984).

238. C. H. Becker and A. G. S. Jánossy, Silicon in the blood vessel wall: A biological entity?, *Micron.*, *10*: 267 (1979).

239. K. Schwarz, Silicon, fibre, and atherosclerosis, *Lancet, 1977*: 454 (1977).

240. A. C. Burton, J. F. Cornhill, and P. B. Canham, Protection from cancer by 'silica' in the water supply of the U. S. cities, *J. Environ. Pathol. Toxicol.*, *4*: 31 (1980).

241. J. Priestley, On the physiological action of vanadium, *Phil. Tr. Roy. Soc.*, *166*: 495 (1876).

242. C. E. Lewis, The biological actions of vanadium. I. Effects upon serum cholesterol levels in man, *A. M. A. Arch. Ind. Health*, *19*: 419 (1959).

243. D. E. Jackson, The pharmacological action of vanadium, *J. Pharmacol.*, *3*: 477 (1912).
244. M. Henze, Untersuchung über das Blut der Ascidien, *Hoppe-Seyler's Z. Physiol. Chem.*, *79*: 215 (1912).
245. M. Henze, Untersuchungen über das Blut der Ascidien. I. Mitt. Die Vanadiumverbindung der Blutkörperchen, *Hoppe-Seyler's Z. Physiol. Chem.*, *72*: 494 (1911).
246. D. Bertrand, Survey of comtemporary knowledge of biogeochemistry. 2. The biogeochemistry of vanadium, *Bull. Amer. Mus. Nat. Hist.*, *94*: 403 (1950).
247. O. Rygh, Recherches sur les oligo-éléments. II. De l'importance du thallium et du vanadium, du silicium et du fluor, *Bull. Soc. Chem. Biol.*, *31*: 1403 (1949).
248. H. A. Schroeder, J. J. Balassa, and I. H. Tipton, Abnormal trace metals in man—vanadium, *J. Chron. Dis.*, *16*: 1047 (1963).
249. L. L. Hopkins, Jr. and H. E. Mohr, Vanadium as an essential nutrient, *Fed. Proc.*, *33*: 1773 (1974).
250. K. Kustin and I. G. Macara, The new biochemistry of vanadium, *Comments Inorg. Che.*, *2*: 1 (1982).
251. B. R. Nechay, L. B. Nanninga, P. S. E. Nechay, R. L. Post, J. J. Grantham, I. G. Macara, L. F. Kubena, T. D. Phillips, and F. H. Nielsen, Role of vanadium in biology, *Fed. Proc.*, *45*: 123 (1986).
252. H. Kneifel and E. Bayer, Determination of the structure of the vanadium compound, Amavadine, from the fly agaric, *Angew Chem. Internat. Edit.*, *12*: 508 (1973).
253. K. Gilbert, K. Kustin, and G. C. McLeod, Gel filtration analysis of vanadium in *Ascidia nigra* blood cell lysate, *J. Cell. Physiol.*, *93*: 309 (1977).
254. D. R. Myron, S. H. Givand, and F. H. Nielsen, Vanadium content of selected foods as determined by flameless atomic absorption spectroscopy, *J. Agric. Food Chem.*, *25*: 297 (1977).
255. A. R. Byrne and L. Kosta, Vanadium in foods and in human body fluids and tissues, *Sci. Total Environ.*, *10*: 17 (1978).
256. K. Ikebe and R. Tanaka, Determination of vanadium and nickel in marine samples by flameless and flame atomic absorption spectrophotometry, *Bull. Environ. Contam. Toxicol.*, *21*: 526 (1979).
257. A. J. Blotcky, C. Falcone, V. A. Medina, and E. P. Rack, Determination of trace-level vanadium in marine biological samples by chemical neutron activation analysis, *Anal. Chem.*, *51*: 178 (1979).
258. R. M. Welch and E. E. Cary, Concentration of chromium, nickel, and vanadium in plant materials, *Agric. Food Chem.*, *23*: 479 (1975).

259. J. Thürauf, G. Syga, and K. H. Scholler, Biological monitoring bei beruflich vanadium-exponierten Personen, in *Biological Monitoring*, A. W. Gentner, Stuttgart, 1978, p. 165.

260. G. L. Curran, D. L. Azarnoff, and R. E. Bolinger, Effect of cholesterol synthesis inhibition in normocholesteremic young men, *J. Clin. Invest.*, *38*: 1251 (1959).

261. S. L. Hansard II, C. B. Ammerman, and P. R. Henry, Vanadium metabolism in sheep. II. Effect of dietary vanadium on performance, vanadium excretion and bone deposition in sheep, *J. Anim. Sci.*, *55*: 350 (1982).

262. T. B. Wiegmann, H. D. Day, and R. V. Patak, Intestinal absorption and secretion of radioactive vanadium ($^{48}VO_3^-$) in rats and effect of Al(OH)$_3$, *J. Toxicol. Environ. Health*, *10*: 233 (1982).

263. J. D. Bogden, H. Higashino, M. A. Lavenhar, J. W. Bauman, Jr., F. W. Kemp, and A. Aviv, Balance and tissue distribution of vanadium after short-term ingestion of vanadate, *J. Nutr.*, *112*: 2279 (1982).

264. L. L. Hopkins, Jr. and H. E. Mohr, The biological essentiality of vanadium, in *Newer Trace Elements in Nutrition* (W. Mertz and W. E. Cornatzer, eds.), Dekker, New York, 1971, p. 195.

265. E. Sabbioni and E. Marafante, Metabolic patterns of vanadium in the rat, *Bioinorg. Chem.*, *9*: 389 (1978).

266. A. V. Roshcin, E. K. Ordzhonikidze, and I. V. Shalganova, Vanadium—toxicity, metabolism, carrier state, *J. Hyg. Epidemiol. Microbiol. Immunol.*, *24*: 377 (1980).

267. R. D. R. Parker, R. P. Sharma, and S. G. Oberg, Distribution and accumulation of vanadium in mice tissues, *Arch. Environ. Contam. Toxicol.*, *9*: 393 (1980).

268. E. Sabbioni, E. Marafante, L. Amantini, and L. Ubertalli, Similarity in metabolic patterns of different chemical species of vanadium in the rat, *Bioinorg. Chem.*, *8*: 503 (1978).

269. E. Sabbioni and E. Marafante, Progress in research on newer trace elements: The metabolism of vanadium as investigated by nuclear and radiochemical techniques, in *Trace Element Metabolism in Man and Animals (TEMA-4)* (J. McC. Howell, J. M. Gawthorne, and C. L. White, eds.), Australian Academy of Science, Canberra, 1981, p. 629.

270. E. Sabbioni and E. Marafante, Relations between iron and vanadium metabolism: In vivo incorporation of vanadium into iron proteins of the rat, *J. Toxicol. Environ. Health*, *8*: 419 (1981).

271. C. A. Strasia, *Vanadium: Essentiality and Toxicity in the Laboratory Rat*, Ph. D. Thesis, Purdue University, 1971, 49 pp.

272. D. L. Williams, *Biological Value of Vanadium for Rats, Chickens, and Sheep*, Ph. D. Thesis, Purdue University, 1973, 98 pp.

273. K. Schwarz and D. B. Milne, Growth effects of vanadium in the rat, *Science, 174:* 426 (1971).

274. F. H. Nielsen, Possible functions and medical significance of the abstruse trace metals, in *Inorganic Chemistry in Biology and Medicine* (A. E. Martell, ed.), ACS Symposium Series 140, American Chemical Society, Washington, D. C., 1980, p. 23.

275. K. W. Franke and A. L. Moxon, The toxicity of orally ingested arsenic, selenium, tellurium, vanadium and molybdenum, *J. Pharmacol. Exp. Ther., 61:* 89 (1937).

276. G. L. Romoser, W. A. Dudley, L. T. Machlin, and L. Loveless, Toxicity of vanadium and chromium for the growing chick, *Poult. Sci., 40:* 1171 (1961).

277. F. H. Nielsen, Vanadium, in *Trace Elements in Human and Animal Nutrition* (W. Mertz, ed.), Academic Press, New York, 1987, in press.

278. F. Proescher, H. A. Seil, and A. W. Stillians, A contribution to the action of vanadium with particular reference to syphillis, *Amer. J. Syph., 1:* 347 (1917).

279. B. R. Nechay, Mechanisms of action of vanadium, *Ann. Rev. Pharmacol. Toxicol., 24:* 501 (1984).

280. B. J. Stoecker and L. L. Hopkins, Vanadium, in *Biochemistry of the Essential Ultratrace Elements* (E. Frieden, ed.), Plenum, New York, 1984, p. 239.

281. C. E. Heyliger, A. G. Tahiliani, and J. H. McNeill, Effect of vanadate on elevated blood glucose and depressed cardiac performance of diabetic rats, *Science, 227:* 1474 (1985).

282. H. J. Thompson, N. D. Chasteen, and L. D. Meeker, Dietary vanadyl (IV) sulfate inhibits chemically-induced mammary carcinogenesis, *Carcinogenesis, 5:* 849 (1984).

283. G. J. Naylor, Vanadium and manic depressive psychosis, *Nutrition and Health, 3:* 79 (1984).

284. R. Masironi, Trace elements and cardiovascular diseases, *Bull. Wld. Hlth. Org., 40:* 305 (1969).

12
Isotopic Tracer Methodology: Potential in Mineral Nutrition

CONNIE WEAVER / Purdue University, West Lafayette, Indiana

INTRODUCTION

Isotopic tracers are useful when a specific mineral from a specific source needs to be distinguished from other sources of the mineral in a complex environment. Tracer methodology allows one to follow the fate of a mineral into the food supply or the absorption and metabolism of the mineral upon consumption or introduction into body pools of animals or humans.

Radioisotopes have been used to elucidate processes by which metals are absorbed, distributed, stored, and secreted by the body. However, our understanding of human nutrition is deficient because of ethical constraints to study those processes with radioisotopes in humans. In particular, pregnant women, infants, and children have not typically been studied with radioisotopes due to possible harm to growing tissues. Stable isotopes (nonradioactive tracers) have begun to be used in the 1980s by mineral nutritionists to fill in this gap of knowledge.

RADIOISOTOPES AND STABLE ISOTOPES

The available radioisotopes and stable isotopes are summarized in Table 1. Radioisotopes are characterized by their mode of decay. Some are gamma emitters, some are beta emitters, and others emit X-rays or alpha particles, while still others decay by electron capture. The usefulness of radioisotopes depends on the half-life and energy

Table 1. Useful Common Mineral Isotopes

Atomic number	Symbol and mass number	Half-life	β (Mev)	E (Mev)	Natural abundance (%)
			Radioisotopes		Stable isotop
			Maximum radiation energies		Natural abundance (%)
11	^{22}Na	2.6y	0.54(β^+)	0.325	—
12	^{25}Mg	—	—	—	10.13
	^{26}Mg	—	—	—	11.17
	^{28}Mg	21.2h	0.42	0.032	—
				0.40	—
				0.95	—
				1.35	—
14	^{29}Si	—	—	—	4.70
	^{30}Si	—	—	—	3.09
15	^{32}P	14.30d	1.71	—	—
	^{33}P	24.4d	0.25	—	—
16	^{33}S	—	—	—	0.76
	^{35}S	87d	0.168	—	—
17	^{36}Cl	3×10^5y	0.71	—	—
19	^{42}K	12.5h	2.00	1.53	—
20	^{42}Ca	—	—	—	0.646
	^{43}Ca	—	—	—	0.135
	^{44}Ca	—	—	—	2.083
	^{45}Ca	164d	0.255	—	—
	^{46}Ca	—	—	—	0.0033
	^{47}Ca	4.53d	1.98	1.29	—
	^{48}Ca	—	—	—	0.18
23	^{49}V	16d	4.0(β^+)	—	—
	^{50}V	—	—	—	0.24
24	^{50}Cr	—	—	—	4.31
	^{51}Cr	27.8d	E.C.	0.325	—
	^{53}Cr	—	—	—	9.55
	^{54}Cr	—	—	—	2.38
25	^{54}Mn	303d	E.C.	0.834	—
26	^{54}Fe	—	—	—	5.82
	^{55}Fe	2.6y	E.C.	—	—
	^{57}Fe	—	—	—	2.19
	^{58}Fe	—	—	—	0.33
	^{59}Fe	45.1d	0.27	1.098	—
			0.46	1.289	—

Table 1. Continued

Atomic number	Symbol and mass number	Radioisotopes			Stable isotopes
			Maximum radiation energies		Natural abundance (%)
		Half-life	β (Mev)	E (Mev)	
7	^{57}Co	270d	E.C.	(Fe X-rays) 0.014 0.122	—
	^{60}Co	5.24y	0.312	1.177	—
8	^{59}Ni	8×10^4y	E.C.	—	—
	^{60}Ni	—	—	—	26.095
	^{61}Ni	—	—	—	1.134
	^{62}Ni	—	—	—	3.593
	^{63}Ni	125y	0.067	—	—
	^{64}Ni	—	—	—	0.904
9	^{64}Cu	12.8h	0.57 0.65	— (X-rays)	—
	^{65}Cu	—	—	—	30.91
	^{67}Cu	61.88h	0.395	0.093 0.184	—
0	^{64}Zn	—	—	—	48.89
	^{65}Zn	245d	E.C.	1.119	—
	^{66}Zn	—	—	—	27.89
	^{67}Zn	—	—	—	4.11
	^{68}Zn	—	—	—	18.57
	^{70}Zn	—	—	—	0.62
4	^{72}Se	8.4d	E.C.	0.046	—
	^{74}Se	—	—	—	0.87
	^{75}Se	120.4d	0.865	0.401	—
	^{76}Se	—	—	—	9.02
	^{77}Se	—	—	—	7.58
	^{78}Se	—	—	—	23.52
	^{80}Se	—	—	—	49.82
	^{82}Se	—	—	—	9.19
8	^{84}Sr	—	—	—	0.56
	^{86}Sr	—	—	—	9.86
	^{87}Sr	—	—	—	7.02
	^{89}Sr	50.5d	1.46	—	—
	^{90}Sr	27.7y	0.545	—	—

Table 1. Continued

Atomic number	Symbol and mass number	Radioisotopes			Stable isotope
			Maximum radiation energies		Natural abundance (%)
		Half-life	β (Mev)	E (Mev)	
42	^{92}Mo	—	—	—	15.84
	^{94}Mo	—	—	—	9.04
	^{95}Mo	—	—	—	15.72
	^{96}Mo	—	—	—	16.53
	^{97}Mo	—	—	—	9.46
	^{98}Mo	—	—	—	23.78
	^{99}Mo	66h	0.41	0.140	—
			1.18	0.745	—
				0.780	—
				0.850	—
	100 Mo	—	—	—	9.63
48	^{106}Cd	—	—	—	1.22
	^{108}Cd	—	—	—	0.88
	^{109}Cd	470d	E.C.	0.087	—
	^{110}Cd	—	—	—	12.39
	^{111}Cd	—	—	—	12.75
	^{112}Cd	—	—	—	24.07
	^{113}Cd	—	—	—	12.26
	^{114}Cd	—	—	—	28.86
	^{116}Cd	—	—	—	7.58
50	^{112}Sn	—	—	—	0.96
					4.72
	^{113}Sn	119d	E.C.	0.260	—
				0.303	
	^{114}Sn	—	—	—	0.66
	^{115}Sn	—	—	—	0.35
	^{116}Sn	—	—	—	14.30
	^{117}Sn	—	—	—	7.61
	^{118}Sn	—	—	—	24.03
	^{119}Sn	—	—	—	8.58
	^{122}Sn	—	—	—	4.72
	^{124}Sn	—	—	—	5.94
56	^{138}Ba	—	—	—	71.66
	^{140}Ba	12.8d	0.48	0.030	—
			1.02	0.537	

Table 1. Continued

Atomic number	Symbol and mass number	Radioisotopes			Stable isotopes
			Maximum radiation energies		Natural abundance (%)
		Half-life	β (Mev)	E (Mev)	
80	^{200}Hg	—	—	—	23.13
	^{201}Hg	—	—	—	13.22
	^{202}Hg	—	—	—	29.80
	^{203}Hg	46.6d	0.210	0.279	—
			0.49		
	^{204}Hg	—	—	—	6.85
82	^{204}Pb	—	—	—	1.48
	^{206}Pb	—	—	—	23.6
	^{207}Pb	—	—	—	22.6

of emmission of the isotope. Stable isotopes are not radioactive, i.e., they do not decay. Stable isotopes differ in atomic weight. The stable isotopes of any element exist in nature in fixed ratios. To be useful as a tracer, an isotope must be naturally abundant at a relatively small percentage. Enriched sources of these isotopes may be purchased from Oak Ridge National Laboratory. Manganese, aluminum, cobalt, gold, arsenic, and beryllium have no useful stable isotopes because they exist in nature as a single atomic weight. Copper, silver, and antimony are difficult to study using stable isotopes because they have only two naturally occurring isotopes, both of which are abundant.

The relative advantages and disadvantages of radioisotopes and stable isotopes are given in Table 2. Radioisotopes are much less expensive than stable isotopes. The cost of stable isotopes depends on the natural abundance. For example, of the calcium isotopes listed in Table 3, ^{43}Ca is the least abundant natural isotope. This isotope of calcium is the most expensive to purchase of those listed. However, less of the isotope is required in a human study to perturb the small amount of endogenous calcium in order to measure the tracer in biological secretions with accuracy.

Radioisotope-counting techniques are relatively simple, and instruments for assessing radioactivity in small samples are available to most scientists. Stable isotopes are much more difficult to analyze than radioisotopes. The instrumentation is expensive and the analysis is laborious. The tissue being analyzed must have significant isotopic

Table 2. Relative Advantages and Disadvantages of Radioisotopes vs. Stable isotopes

Radioisotopes	Stable isotopes
Less expensive	Expensive
Analysis relatively easy and sensitive	Analysis laborious and costly
Decay	Do not decay
Potential health risk	No health risk
Handling restricted to approved laboratories	Can be used in normal food processing/handling facilities

enrichment relative to measurement precision to result in meaningful data. This may be difficult and costly to achieve or result in non-physiological levels of the mineral. For these reasons, radioisotopes are preferred when it is possible to use them. Mineral nutritionists and physicists make excellent partners for studying mineral metabolism using stable isotopes.

The decay of radioisotopes renders them easy to analyze but confers other limitations on their use. Experiments with radioisotopes with short half-lives must be planned around the production schedule of the isotopes and/or preparation of labeled compounds or tissues.

Table 3. Natural Abundance and Optimal Dosages of Minor Isotopes of Calcium

Isotope	Natural abundance	Price ($/mg)	Optimal Dose[a] (mg/person)	Price per person per study($)
^{40}Ca	96.92	—	—	—
^{42}Ca	0.64	31	10.0	310
^{43}Ca	0.13	422	2.5	1055
^{44}Ca	2.08	7.35	35.0	257.25
^{48}Ca	0.18	282	3.0	846

[a]The optimal dose varies with the study design.

Stable isotopes allow felxibility in timing of experiments and samples may be stored indefinitely.

Although radioisotopes have been used in human studies, they are usually not allowed in those very populations targeted to be potentially deficient in trace elements. Stable isotopes can be used in pregnant women with no risk to the fetus. They pose no danger to infants and children. Several stable tracers can safely be used simultaneously in the same subject who can be studied repeatedly. The innocuous nature of stable isotopes also allows them to be used in any food handling or processing facility. Determining bioavailability of minerals from foods is most meaningful if foods are prepared as they are normally consumed.

ANALYTICAL TECHNIQUES

Radioisotopes

Radioisotopes can be detected by scintillation counters. Detection of gamma emitters requires little sample preparation. Sample geometry should be uniform; liquids should be adjusted to the same volume, and tissues should be ground and adjusted to the same height in the counting vials before assaying for radioactivity. Small animal and human counters can be used to determine whole-body retention following administration of a gamma emitter. The lower energy radiation of beta emitters does not permit accurate detection without appropriate sample preparation to minimize sample self-absorption. Tissues must be ashed and dissolved in a suitable solvent or suspended in a gel mixture.

Some precautions in reporting and interpreting radioisotope data should be heeded. When reporting levels of radioactivity, it is important to give the efficiency of the counting system used or to use internal standards to convert counts per minute to absolute levels of radioactivity. Quenching of beta emitters is a severe problem in many instances and warrants correction with internal standards added to each counting vial. When partitioning samples as in chromatography or determining distribution among components, it is also important to monitor the percent recovery. Furthermore, investigators must be cognizant of exchange problems. The radioisotope must be exchanging with the mineral pool being studied rather than forming a new fraction in order to be useful as a tracer.

Stable Isotopes

Stable isotopes can be analyzed by neutron activation analysis (NAA) and by several types of mass spectrometry (MS). The particular method of choice depends on the sensitivity, precision, and simplicity desired and availability of analytical expertise and instrumentation. The common methods of analysis and typical precisions obtainable are listed in Table 4.

NAA is a technique for measuring those stable isotopes which interact with thermal neutrons to produce radioactive nuclides or X-radiation. Extensive sample preparation is typically required and precision is less than desirable. Furthermore, a nuclear reactor is required. Use of this facility dictates batch processing; convenience of batch processing depends on the number of samples to be analyzed.

Thermal ionization mass spectrometry (TIMS) is a direct ionization method that produces intense ions from hot surfaces at appropriate isotopic masses. The most precise and expensive instrument for TIMS is magnetic sector MS. TIMS can also be done on a less expensive quadrupole MS. Precision depends on the degree of sample preparation Calcium can be analyzed at 2% precision with very little sample preparation (1), whereas 0.1% precision has been achieved for zinc on samples that have undergone extensive sample preparation (2).

With both gas chromatography/mass spectrometry (GC/MS) and direct-probe mass spectrometry, stable isotopes are reacted to form volatile chelates. The sample is introduced by a solid probe or by a gas chromatograph which purifies the sample as it enters. GC/MS is preferred over direct-probe MS because the sample is cleaner, however direct-probe MS can be used when the sample will not pass through the GC. Both GC/MS and direct-probe MS are easier and faster than TIMS if a volatile chelate can be made with the mineral of interest. The precision is less than with TIMS but if enrichment is adequate, meaningful results are possible, at least with copper and zinc (3) and selenium (4). Both GC/MS and direct-probe MS suffer from memory effects. The ligand of the volatile chelate picks up minerals from the stainless steel ion source and residual contamination.

Table 4. Stable Isotope Methods

Method	% precision
Neutron activation analysis (NAA)	2-10
Mass spectrometry	
Thermal ionization (TIMS)	0.1-2
Gas chromatography (GC/MS)	1-5
Direct probe	1-5
Fast atom bombardment (FAB)	0.1-1
Inductively coupled plasma (ICP)	0.5-2
Resonance ionization (RIMS)	
Mossbauer spectroscopy	

Fast atom bombardment (FAB) uses high energy particles (fast atoms) to vaporize a sample and high resolution mass spectrometry (HRMS) to isolate and detect ions of interest. Advantages of this method include minimal sample preparation, high precision, rapid sample analysis, high sensitivity for all isotopes, and a standard high resolution mass spectrometer is used.

Inductively coupled plasma mass spectrometry (ICPMS) also requires very little sample preparation. This method employs a discharge tube with argon, which inoizes all elements in the sample at high temperatures of 2000 to 3000°C. Molecules are split into atoms and ions are sampled with a quadrupole MS. This method is particularly applicable for doing multiple samples.

Resonance ionization mass spectrometry (RIMS) uses a laser to excite an electron to a specific atomic energy level via one or more photons. The photons have just the right wavelength to effect ionization so the process is very selective. RIMS, like secondary ion mass supttering (SIMS) and laser desorption (LAMA), can analyze ultrasmall samples. Lasers or ions can be focused on a sopt within a cell. In fact, single atoms have been detected with RIMS (5).

Mossbauer spectrometry studies gamma ray emission from a nucleus. It is sensitive to changes in oxidation, spin state, and configuration of ligands around the atom. Only a limited number of elements can be measured by Mossbauer spectroscopy. The elements of nutritional importance which can be measured by this technique include cobalt, iron, iodine, and zinc.

Once the instrumentation has been selected, two methods can be used to determine the isotopic enrichment of a sample. The first method involves determining the ratio of the enriched isotope to natural isotopes and determination of the total mineral content of the sample by an independent method. Usually the total mineral content of the sample is determined by atomic absorption spectroscopy, however, this method is less accurate than methods for determining stable isotope ratios. The second method is an isotope dilution technique in which two isotopic ratios are determined (4). The amount of analyte originally present in a sample is determined from the ratios of an added spike and the enriched isotope to a third, naturally occurring isotope. The spike serves as an ideal internal standard. A major advantage of this methodology is that quantitative recovery of the analyte is unnecessary. Assuming the spike and analyte are chemically identical, incomplete recoveries affect them both in the same way. It is an absolute method since it is not necessary to calibrate the instrument response against known standards as ratio measurements and all isotopes are affected the same way. Accuracy of the method hinges on knowing the exact amount of the added spike. Typically, the second technique is used with MS techniques and not NAA, since NAA has different sensitivities for different isotopes depending on the cross section. Depending on the mineral being studied, it is often not practical to use the isotope dilution techniques for quantifying the stable isotope tracer.

TRACER APPLICATIONS

Tracers can be used as a probe to study the fate of a specific mineral through processing schemes or within biological organisms even in the presence of large amounts of the same mineral from other sources. Absorption and transport of minerals can be monitored by disappearance of the label, e.g., by whole-body counting techniques or by appearance of the label into body secretions or pools. A number can be studied simultaneously with tracer methodology (6), which enables mineral interactions to be evaluated.

Quantifying Minerals

Isotope dilution MS is the state of the art in terms of sensitivity, precision, and accuracy. The high sensitivity of some of the MS techniques for quantifying stable isotopes enables measurement of ultratrace amounts of minerals in ultrasmall samples. These techniques have been used in the geological, nuclear, and analytical sciences for decades to measure isotopes, but only recently have they been applied to biological tissues. RIMS has been used to quantify iron in water and human serum using the isotope dilution method (7). Table 5 shows the ^{56}Fe sensitivity for various human samples using HR FAB MS. Approximately 10^4 ions are required to get a 1% precision. The sensitivity is adequate to quantify iron in ferritin without further sample preparation but not red cells, transferrin, or plasma. With a simple extraction to prepare heme chloride, it is potentially possible to see how fast iron can be transported to different areas.

Table 5. ^{56}Fe Signal for Various Samples

Sample	Ions/sec
Hemin chloride	1,100,000
Ferritin	660,000
Red cells	33,000
Transferin	18,000
Plasma	5,000

Source: D. L. Smith, unpublished data.

Absorption

Traditionally, metabolic balance studies have been used to determine net absorption in humans as:

Net absorption = intake — fecal excretion

This approach requires complete fecal collection and analysis of the diet and feces. The balance technique underestimates true absorption because endogenous secretion of minerals back into the alimentary canal is not accounted for, which results in substantial error for some minerals, such as zinc. Precision for this method is poor. The balance approach combined with an intravenous tracer can be used to determine endogenous fecal loss for true absorption:

True absorption = net absorption — endogeneous secretion

Alternatively, an oral tracer can be used to distinguish between dietary minerals and those in endogeneous secretions. Some tracer can be reexcreted into the gastrointestional tract after it is absorbed, but usually it has been demonstrated that the effect is small. In the case of copper, 8.4% of intraveneously administered radiocopper was recovered in the feces (8), so that approximately 8.4% of that portion which was absorbed of an oral dose of copper might reappear in the feces. Similarly, approximately 10% of an absorbed dose of zinc was reexcreted into the intestine and appeared in the feces in a 12-day period (9). Janghorbani and Young (10) estimated an error of 0.8– 1.9% associated with ignoring reexcretion of an absorbed dose into the fecal pool with selenium. The error of early reexcretion of an absorbed isotope can be corrected with a nonabsorable fecal marker (11).

Numerous studies on the absorption of minerals by humans and animals using both radioisotopes and stable isotopes have been reported. When gamma-emitting radioisotopes are used as tracers, whole-body retention measurements are the most convenient means of determining absorption of the tracer if a small animal or human whole-body counter is available. Alternatively, incorporation of ^{59}Fe, a gamma emitter, into red blood cells has been used extensively to estimate iron absorption (12). Since 99% of body calcium is in the skeletal system, measurements of a calcium tracer in the bone following oral administration is an index of calcium absorption in animals. Plasma appearance of ^{69m}Zn has been used to estimate absorption of this short-lived gamma emitter (13). For beta emitters (other than ^{55}Fe, which can be monitored by red blood cell incorporation) or stable isotopes, complete fecal collections to determine recovery of the administered isotopes is the usual approach. Fecal monitoring of stable isotopes is less accurate than for radioisotopes because of the presence of stable isotopes as a normal component of the diet. The error of measurement increases as absorption of the oral tracer decreases.

For example, Eagles et al. (14) estimated that the error incurred when measuring absorption of [58]Fe is approximately ±0.5% when absorption of the tracer approaches 90%, whereas an error of ±4.5% was predicted when absorption is 10%. The capabilities and limitations of using stable isotope enrichment and fecal monitoring for absorption measurements has been reviewed by Janghorbani and Young (15). Plasma appearance of beta emitters and stable isotopes has also been used to estimate absorption (16).

Of the trace elements, iron has been studied the most extensively. Using radioiron and whole-body retention measurements (17) or incorporation of radioactivity into red blood cells (18), it was shown that iron absorption depends on iron status (19) and valency of the iron.

A summary of studies employing isotopes of zinc is given by Solomons (20). In the rat model, whole-body retention of radiozinc appears to be dependent upon zinc status (21), as is true for iron. Absorption of zinc approached 100% at low dietary intakes and plateaued as zinc intake increased (22,23). However, using an isotope dilution technique which accounts for the increase in specific activity of tracer in the intestine as endogenous zinc secretion decreases in zinc-deplete rats relative to zinc-adequate rats, Evans et al. (21) concluded that zinc homeostasis in rats is maintained by zinc secretion from the intestine rather than by regulation of zinc absorption. Weigand and Kirchgessner (22) attributed homeostatic controls of zinc to both regulation of absorption and endegenous secretion. Metabolism of radiozinc in man was investigated by Spencer et al. (24). A significant portion of intraveneously injected [68]$ZnCl_2$ entered the intestinal lumen via pancreatic juice.

Both fecal monitoring and whole-body retention of radioselenium indicated that selenomethionine was almost completely absorbed by adult women (25) in contrast to selenite (26). However, the retention values obtained by the two different methods were different by several percent. The authors suggested that whole-body retention data were influenced by a change in the distribution of the ^{75}Se in the body shortly after absorption. This example illustrates the pitfalls of unequal geometry associated with determining counting rates of gamma emitters.

Using stable isotopes and complete fecal collection to determine recovery of the administered isotope, average daily absorption of iron and zinc were estimated to approximate endogenous losses (27). Similar methodology revealed that elderly men absorbed less zinc but not less iron or copper than young men. Oral contraceptive use did not significantly affect absorption of iron, copper, and zinc (28), and pregnant women absorbed more copper but not more zinc than nonpregnant controls (29,30). Absorption of stable isotopes of selenium approximated 80% regardless of pregnancy status, but pregnant women conserved selenium by decreasing urinary selenium (29

An even more accurate approach than use of a single tracer for determination of true absorption employs the simultaneous use of two tracers, one fed orally and the other intraveneously. The double isotope tracer technique assumes that both isotopes enter the exchangeable pool at the same time and are subject to the same physiological processes. The validity of this assumption for calcium is illustrated by the appearance of ^{44}Ca administered orally and ^{42}Ca administered intraveneously in the urine of an adult male (Fig. 1) (31). The assumption that both tracers entered the circulating system simultaneously was not good for the first few hours, but appeared valid after 20 hours. The ratio of the oral to intraveneous tracers in plasma or urine can be used to calculate fractional absorption after correcting for differences in initial tracer doses as follows:

$$\text{Fractional true absorption} = \frac{\int_0^t T_{OR}\, dt}{\int_0^t T_{IV}\, dt} \cdot \frac{DOSE_{IV}}{DOSE_{OR}}$$

$$\cong \frac{T_{OR}}{T_{IV}} \cdot \frac{DOSE_{IV}}{DOSE_{OR}}$$

where T_{OR} is the oral tracer concentration and T_{IV} is the intraveneous tracer concentration. The ratio of the two tracer doses at some point beyond 20 hours following administration of the isotopes approximates the ratio of the integrated area under the curves for the two isotopes.

A modification of the double isotope tracer technique has been developed (32) for use in animal models when only one isotope is available. One group of animals is given the isotope orally and another group of animals received the isotope intramuscularly. The ratio of the intercepts from extrapolation of semilogarthmic plots of whole-body retention curves are used to calculate fractional absorption (Fig. 2). This approach is more realistic for use with animals, which can be more easily controlled in terms of mineral status and uniform environment.

In addition to determining the amount of mineral absorbed, tracers can be used in conjunction with ligated segments of intestine, the isolated, intestinal perfusion system (33) or in vitro methods, such as everted intestinal sacs or intestinal strips, to determine the site or mechanism of absorption or factors affecting ion transfer (34–36). For example, the mechanism (37) and site (38) of zinc absorption was studied using isolated, vascularly perfused rat intestine. Schachter and Rosen (39) used everted gut-sacs prepared from segments of the proximal small intestine of young rabbits, rats, and guinea pigs to demonstrate that calcium is transferred from mucosal to serosal surfaces against a concentration gradient and that this transfer is influenced by vitamin D.

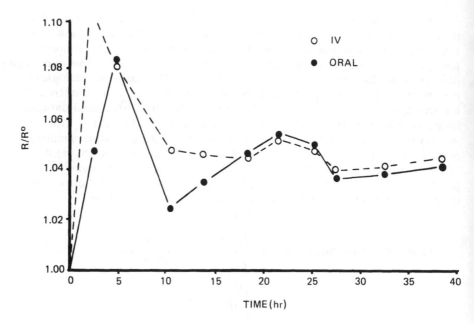

Figure 1. Enrichment of tracer isotopes in urine after a 77-kg male received 13 mg of ^{44}Ca orally and 4 mg of ^{42}Ca intravenously. Reproduced with permission from Ref. 31.

Kinetics

Intraveneously administered tracer data can be used to measure size and turnover rates of exchangeable body pools. The turnover rate of calcium in humans is illustrated in Fig. 1 (30). The turnover rate of minerals can be calculated by monitoring stable isotope enrichment or specific activity of radioisotopes in either plasma or urine. The requirement for high precision for measuring the isotopes is illustrated in Fig. 3. The solid bars indicate the natural abundance of three calcium isotopes. A 5% enrichment of ^{44}Ca in urine or plasma (open bar) is sufficient perturbation for analysis by FAB HR MS or TIMS. It is desirable to perturb the natural abundance level enough to measure the tracer over the entire study without perturbing homeostasis. The more sensitive the measuring technique, the smaller the required dose. This results in a cheaper and more physiologically normal study. For oral doses, if absorption is low or the natural abundance of the isotope is high as with copper, the normal RDA for the mineral must be exceeded or that isotope must be fed over several days.

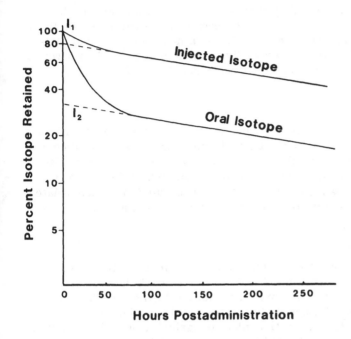

Figure 2. Percent absorption calculated by dividing I_2 (the y intercept of the extrapolated retention curve for oral isotope) by I_1 (the y intercept of the extrapolated retention curve for injected isotope) and multipling by 100. Method of Heth and Hoekstra (32).

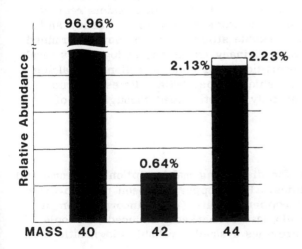

Figure 3. Isotope perturbation of 5% following administration of enriched ^{44}Ca.

In addition to determining turnover rates of minerals in whole organisms, tracers can be used to determine turnover rates in tissues or individual cellular components. For example, the turnover rate of ^{75}Se was found to be higher in liver than in the heart and whole blood, and even lower in skeletal muscle and bone (25).

Body Pool Sizes

Body pool sizes can be estimated using tracers under conditions of a controlled diet so that actual intake of the mineral being studied is known. Veillon (40) described determination of selenium pool size for a single pool compartment, circulating plasma, using data from a metabolic study of Swanson et al. (41). The equation which relates pool size to concentration of isotope in urine or plasma is:

$$(I)_t = (I)_o e^{\frac{-Ut}{P}}$$

where $(I)_t$ is the amount of isotope appearing in the plasma or urine at time t, $(I)_o$ is the amount of isotope ingested, U is the total amount of the mineral being ingested per day, and P is the quantity of mineral in the pool. Figure 4 illustrates the ln of the 24-hour urinary selenium isotope, [76]Se, excretion plotted against time. The slope of the line is $-U/P$. Since the subjects were participating in a metabolic study, U was known and P could easily calculated. The broken line indicates one rapidly changing circulating plasma selenium pool and a second, larger selenium pool after 11 days with a slower turnover rate.

Models exist which describe up to four exchanging compartments comprising the total body pool (42). More complex models require computer simulation, analysis, and modeling (43,44). Heaney and Skillman (45) reported a remarkable study in which they determined body pool size, turnover rate, urinary excretion, endogeneous fecal losses, absorption, bone resorption, and bone accretion for calcium in pregnant women using a stable isotope, [48]Ca. Rates of bone accretion and resorption can only be determined through use of isotopic tracers.

Diagnosing Diseases

Isotopic tracers are useful for diagnosing malabsorption syndrome and other absorption abnormalities. Radiocopper has been used to detect Wilson's disease, a chronic copper toxicity (46). Incorporation of radiocopper into serum ceruloplasm is markedly reduced in people with Wilson's disease and provides a simple, reliable diagnositic tool.

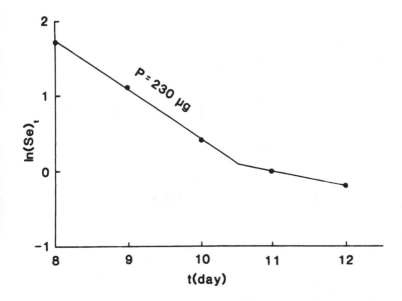

Figure 4. Plot of ln $(Se)_t$ vs. t for the nonpregnant subjects of Swanson et al. (41). Similar plots for their early-pregnant and late-pregnant subjects gave increasing values of P. Reproduced with permission from C. Veillon, GC/MS Measurement of stable isotopes of selenium for use in metabolic tracer studies, in *Stable Isotopes in Nutrition*, ACS Symposium Series 258, (J. R. Turnlund and P. E. Johnson, eds.) American Chemical Society, Washington, D. C. 1984 (40). Copyright 1984 American Chemical Society.

Bioavailability

Isotopic tracers are particularly useful for determining how well a mineral is absorbed and utilized from a specific food because they allow the endegenous mineral to be distinguished from minerals from other dietary sources or body secretions.

A typical bioavailability study (Fig. 5) involves labeling the food of interest intrinsically, i.e., biological incorporation of an isotope, or extrinsically, i.e., mixing the food with the isotope of interest. The labeled food is then fed to humans or an animal model and absorption is determined by one of the methods previously discussed.

Production of Intrinsically Labeled Foods

Intrinsic labeling of plant foods can be accomplished by administering the isotope of interest to the nutrient solution of hydroponically

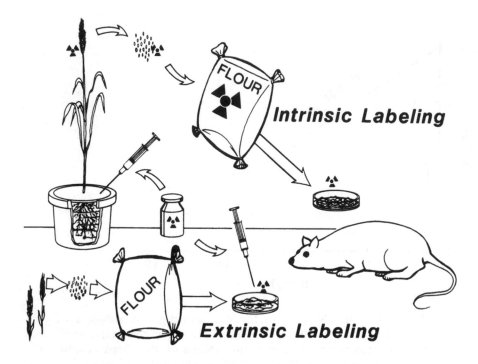

Figure 5. Scheme of a bioavailability study using intrinsic and extrinsic labeling techniques.

grown plants. A detailed discussion of intrinsic labeling methodology for plant foods is given by Weaver (47). This method simulates field-grown conditions because the minerals are taken up by the roots. Application of isotopes directly to the soil would result in inefficient incorporation of the isotope into edible plant parts because of the presence of mineral-binding ligands in the soil and absorption of isotopes by the soil matrix. Hydroponics allows more efficient transfer of isotopes to the roots. Efficiency of transfer of isotopes from the plant root to the shoot varies with the mineral.

Isotopic tracers may also be incorporated into plants by stem injection or by foliar application. These methods preclude the need for hydroponic culture. For some minerals, the efficiency of incorporation of isotopes into edible plant parts is improved because the root-shoot barrier is bypassed. The leaf cuticle can also be a major barrier to penetration of minerals; consequently, calcium, magnesium, and other minerals remain at the site of application when applied foliarly. Several of the trace elements including iron, zinc, and copper are considered to have intermediate mobility when foliarly applied (48).

A consideration in selecting the method of administration of the isotopic tracer is whether the isotope will be incorporated into edible parts in the natural form. If it is not, short-cuts in methodology undermine understanding bioavailability of minerals in foods as they are normally consumed. Validity of the stem injection and foliar application techniques in producing wheat labeled with ^{65}Zn was studied by Starks and Johnson (49). Results showed that both techniques produced seeds with ^{65}Zn distribution patterns similar to total zinc.

Animal tissue may be endogenously labeled by incorporating the isotopic tracer in the diet or drinking water of the animal, administering the isotopic tracer by gavage, or injecting the isotope intraveneously, intraperitoneally, or intramuscularly. As for plants, the ultimate goal is to produce tissue in which the isotopic tracer has been deposited in the same manner as the endogenous mineral. Bypassing the gut may result in deposition of unnatural forms of the tracer or nonphysiological levels, but does offer improved efficiency of label incorporation for some minerals. Efficiency of label incorporation is of utmost importance when using stable isotopes because of the high level of enrichment needed relative to the total mineral content.

Extrinsic vs. Intrinsic Labeling

Use of isotopic tracers added extrinsically to a food or meal prior to feeding is vastly simpler and less expensive than producing intrinsically labeled foods. The assumption that an extrinsically applied tracer can exchange with the mineral endogenous to the food has been tested thoroughly only for iron (50). Absorption of an extrinsic iron radioisotopic salt is similar to absorption of intrinsically labeled nonheme iron in the food or meal (19,51–53) and absorption of labeled hemoglobin added to a food or meal is similar to an intrinsic heme iron tracer (54). Extrinsic labeling may overestimate absorption by 10 to 20% when incomplete exchange occurs. Iron in foods that are dense (53) or of large particulate size (55,56) does not exchange as easily with an extrinsic label. Moreover, pH may determine efficiency of exchange since iron is less soluble at alkaline pH (57).

Once isotopic tracer exchange with the endogenous mineral is established, several implications ensue. Extrinsic labeling techniques can be used to estimate bioavailability of a mineral. Bioavailability of a mineral from a food consumed singly or from a complex meal can be assessed, but bioavailability of a mineral from a single food consumed as part of a complex meal cannot be determined. For example, when two biologically labeled foods, of differing bioavailability when consumed singly, were fed together in one meal, the two radioiron isotopes were absorbed equally and at a level intermediate between the two foods fed individually (53). This finding led to the concept of a common nonheme iron pool in the diet.

Studies which have compared absorption of extrinsic and intrinsic labeling techniques for assessing bioavailability of iron and other trace minerals from plant foods has been reviewed recently (47). Extrinsic labeling approximated absorption of an intrinsic zinc label for a number of foods (58—60).

Exchange of stable isotopic tracers with the endogenous mineral may be more difficult than for radioisotopes, since a much larger fraction of the total mineral pool is the tracer. Very few reports on the exchangeability of stable isotopic tracers exist. An exchangeability of 90% or greater between $MgCl_2$ and ^{26}Mg in intrinsically labeled vegetables was reported (61). Exchangability of intrinsic and extrinsic stable isotopes of zinc in chicken was less good, however (62). Absorption of intrinsic and extrinsic stable isotopic tracers of selenium was similar for eggs (63) but not for poultry (64).

Bioavailability of several minerals from one food can be determined or bioavailability of one mineral from several foods fed together can be determined if several isotopes of the mineral exist and they do not exchange. This concept is especially true for stable isotopes where there is no risk of cumulative biohazard, unlike with radio-isotopes.

Determination of Chemical Form In Foods, Tissues, and Body Secretions

The nature of the chemical form and associations of minerals in a complex environment can be studied with intrinsically labeled tissues. Knowledge of the chemical form of a mineral in foods allows prediction of bioavailability and characterization of functional sites. The chemical forms of iron and selenium in wheat were identified as monoferric phytate and selenomethionine, respectively, through the use of intrinsically labeled wheat (65,66).

Tracers can be used to isolate and characteriza mineral complexes in body fluids as well as in foods. For example, a new plasma copper transport protein, tentatively named transcuprein, was recently identified after labeling rats with ^{67}Cu (67).

Binding ligands can be labeled in addition to minerals in order to study mineral-ligand interactions. This dual labeling approach was used to partially characterize an organic chromium complex in alfalfa (68).

CONCLUSION

Radioisotopic tracers have been a powerful probe for studying mineral metabolism for years. The recent application of stable isotopic tracers to the study of nutrition problems has opened the door to further our understanding of mineral metabolism in humans. Many questions need to be addressed with stable isotopes. Analytical techniques need to be refined and methodologies for more minerals needs to be developed. Exchangability of isotopic tracers with endogenous mineral pools needs a great deal of attention. Where isotopic exchange is limited, efficient procedures for isotopic labeling of animal and plant food models need to be sought. With the exception of iron, bioavailability of minerals from different foods and diets is not well understood. However, recent strides have been made in the development of stable isotope methodology applicable to the study of metabolism of minerals other than iron. Beyond whole-body mineral metabolism, some new evolving methodologies promise to be useful in biotechnology and disease as we learn to study ultratrace metal transport among cells.

REFERENCES

1. A. L. Yergey, N. E. Vieira, D. G. Covell, and J. W. Hansen, Calcium metabolism studied with stable isotopic tracers, in *Stable Isotopes in Nutrition*, ACS Symposium Series 258 (J. R. Turnlund and P. E. Johnson, eds.), American Chemical Society, Washington, D. C., 1984.
2. J. R. Turnlund, Assessment of the bioavailability of dietary zinc in humans using the stable isotopes ^{70}Zn and ^{67}Zn, in *Nutritional Bioavailability of Zinc*, ACS Symposium Series 210 (Inglett, ed.), American Chemical Society, Washington, D. C., 1983.
3. P. E. Johnson, Stable isotopes of iron, zinc, and copper used to study mineral absorption in humans, ACS Symposium Series 256 (J. R. Turnlund and P. E. Johnson, eds.), American Chemical Society, Washington, D. C., 1984.
4. D. C. Reamer and C. Veillon, A double isotope dilution method for using stable selenium isotopes in metabolic tracer studies: Analysis by gas chromatography/mass spectrometry (GC/MS), *J. Nutr.*, *113*: 786 (1983).
5. G. S. Hurst, M. H. Nayfek, and J. P. Young, A demonstration of one-atom detection, *Appl. Phys. Letters*, *30*: 229 (1977).
6. P. E. Johnson, A mass spectrometric method for use of stable isotopes as tracers in studies of iron, zinc, and copper absorption in human subjects, *J. Nutr.*, *112*: 1414 (1982).

7. J. D. Fassett, L. J. Powell, and L. J. Moore, Determination of iron in serum and water by resonance ionization dilution mass spectrometry, *Anal. Chem.*, *56*: 2228 (1984).

8. G. E. Cartwright and M. M. Wintrobe, Copper metabolism in normal subjects, *Am. J. Clin. Nutr.*, *14*: 224 (1964).

9. H. Spencer, V. Vankinscott, K. Lewin, and J. Samachson, Zinc—65 metabolism during low and high calcium intake in man, *J. Nutr.*, *86*: 169 (1965).

10. M. Janghorbani, L. J. Kasper, and V. R. Young, Dynamics of selenium metabolism in young men: Studies with the stable isotope tracer method, *Am. J. Clin. Nutr.*, *40*: 208 (1984).

11. P. R. Flanagan, J. Cluett, M. J. Chamberlain, and L. S. Valberg, Dual-isotope method for determination of human zinc absorption: The use of a testmeal of turkey meat, *J. Nutr.*, *115*: 111 (1985).

12. L. Larsen and N. Milman, Normal iron absorption determined by means of whole body counting and red cell incorporation of ^{59}Fe, *Acta Med. Scand.*, *198*: 271 (1975).

13. M. Molokhia, G. Stumiolo, R. Shields, and L. A. Turnberg, A simple method for measuring zinc absorption in man using a short lived isotope (^{69m}Zn), *Am. J. Clin. Nutr.*, *33*: 881 (1980).

14. J. Eagles, S. J. Fairweather-Tait, and R. Self, Stable isotope ratio mass spectrometry for iron bioavailability studies, *Anal. Chem.*, *57*: 469 (1985).

15. M. Janghorbani and V. R. Young, Use of stable isotopes to determine bioavailability of minerals in human diets using the method of fecal monitoring, *Am. J. Clin. Nutr.*, *33*: 2021 (1980).

16. M. Janghorbani and V. R. Young, Stable isotopes in studies of dietary mineral bioavailability in humans, with special reference to zinc, in *Clinical, Biochemical, and Nutritional Aspects of Trace Elements* (A. S. Prasad, ed.), Alan R. Liss, Inc., New York, 1982.

17. J. D. Cook, H. E. Palmer, K. G. Pailthorp, and C. A. Finch, The measurement of iron absorption by whole-body counting, *Phys. Med. Biol.*, *15*: 467 (1970).

18. J. D. Eakins and D. A. Brown, Absorption of fortification iron in bread, *Am. J. Clin. Nutr.*, *26*: 861 (1973).

19. T. H. Bothwell, R. W. Charlton, J. D. Cook, and C. A. Finch, *Iron Metabolism in Man*, Blackwell Scientific Publications, Oxford, 1979.

20. N. W. Solomons, Biological availability of zinc in humans, *Am. J. Clin. Nutr.*, *35*: 1048 (1982).

21. G. W. Evans, C. I. Grace, and C. Hahn, Homeostatic regulation of zinc absorption in the rat, *Proc. Soc. Exp. Biol. Med.*, *143*: 723 (1973).

22. E. Weigand and M. Kirchgessner, Homeostatic adjustment in zinc digestion to widely varying dietary zinc intake, *Nutr. Metab.*, *22*: 101 (1978).

23. E. Weigand and M. Kirchgessner, Total true efficiency of zinc utilization: Determination and homeostatic dependence upon the zinc supply status in young rats, *J. Nutr.*, *110*: 469 (1980).

24. H. Spencer, B. Rosoff, and A. Feldstein, Metabolism of zinc-65 in man, *Rad. Res.*, *24*: 432 (1965).

25. N. M. Grittiths, R. D. H. Stewart, and M. F. Robinson, The metabolism of [75Se] selenomethionine in four women, *Br. J. Nutr.*, *35*: 373 (1976).

26. C. D. Thomson and R. D. H. Stewart, The metabolism of [75Se] selenite in young women, *Br. J. Nutr.*, *32*: 47 (1974).

27. J. R. Turnlund, M. C. Michel, W. R. Keys, J. C. King, and S. Margen, Use of enriched stable isotopes to determine zinc and iron absorption in elderly men, *Am. J. Clin. Nutr.*, *35*: 1033 (1982).

28. J. C. King, W. L. Raynolds, and S. Margen, Absorption of stable isotopes of iron, copper, and zinc during oral contraceptive use, *Am. J. Clin. Nutr.*, *31*: 1198 (1978).

29. C. A. Swanson, J. R. Turnland, and J. C. King, Effect of dietary zinc sources and pregnancy on zinc utilization in adult women fed controlled diets, *J. Nutr.*, *113*: 2557 (1983).

30. J. R. Turnlund, C. A. Swanson, and J. C. King, Copper absorption and retention in pregnant women fed diets based on animal and plant proteins, *J. Nutr.*, *113*: 2346 (1983).

31. D. L. Smith, C. Atkin, and C. Westenfelder, Stable isotopes of calcium as tracers: Methodology, *Clin. Chim. Acta*, *146*: 97 (1985).

32. D. A. Heth and W. G. Hoekstra, Zinc-65 absorption and turnover in rats. A procedure to determine zinc-65 absorption and the antagonistic effect of calcium in a practical diet, *J. Nutr.*, *85*: 367 (1965).

33. E. L. Krawitt, Duodenal calcium transport in hyperthyroidism, *Proc. Soc. Expt. Biol. Med.*, *125*: 417 (1967).

34. T. H. Wilson and G. Wiseman, The use of everted small intestine for the study of transference of substances from the mucosal to the serosal surface, *J. Physiol.*, *London*, *123*: 116 (1954).

35. D. L. Martin and H. F. Deluca, Calcium transport and the role of vitamin D, *Arch. Biochem. Biophys.*, *134*: 139 (1969).

36. D. Schachter, E. B. Dowdle, and H. Schenker, Accumulation of Ca45 by slices of small intestine, *Am. J. Physiol.*, *198*: 275 (1960).

37. K. T. Smith, R. J. Cousins, B. L. Silbon, and M. L. Failla, Zinc absorption and metabolism by isolated, vascularly perfused rat intestine, *J. Nutr.*, *108*: 1849 (1978).

38. D. L. Antonson, A. J. Barak, and J. A. Vanderhoof, Determination of the site of zinc absorption in rat small intestine, *J. Nutr.*, *109*: 142 (1979).

39. D. Schachter and S. M. Rosen, Active transport of Ca^{45} by the small intestine and its dependence on vitamin D, *Am. J. Physiol.*, *196*: 357 (1959).

40. C. Veillon, GC/MS Measurement of stable isotopes of selenium for use in metabolic tracer studies, in *Stable Isotopes in Nutrition*, ACS Symposium Series 258, (J. R. Turnlund and P. E. Johnson, eds.), American Chemical Society, Washington, D. C., 1984.

41. C. A. Swanson, D. C. Reamer, C. Veillon, J. C. King, and O. A. Levander, Quantitative and qualitative aspects of selenium utilization in pregnant and nonpregnant women: An application of stable isotope methodology, *Am. J. Clin. Nutr.*, *38*: 169 (1983).

42. R. Neer, M. Berman, L. Fisher, and L. E. Rosenberg, Nulti-compartmental analysis of calcium kinetics in normal adult males, *J. Clin. Invest.*, *46*: 1364 (1967).

43. M. Berman, E. Shahn, and M. F. Weiss, The routine fitting of kinetic data to models: a mathematical formalism for digital computers, *Biophys. J.*, *2*: 275 (1962).

44. M. Berman, M. F. Weiss, and E. Shahn, Some formal approaches to the analysis of kinetic data in terms of linear compartmental systems, *Biophys. J.*, *2*: 289 (1962).

45. R. P. Heaney and T. G. Skillman, Calcium metabolism in normal human pregnancy, *J. Clin. Endocr.*, *33*: 661 (1971).

46. I. Sternlieb and I. H. Scheinberg, Radiocopper in diagnosing liver diseases, *Seminars in Nuclear Medicine*, *2*: 176 (1972).

47. C. M. Weaver, Intrinsic mineral labeling of edible plants: Methods and Uses, *CRC Critical Reviews in Food Science and Nutrition*, *23*: 75 (1985).

48. M. J. Bukovac and S. H. Wittwer, Absorption and mobility of foliar applied nutrients, *Plant Physiol.*, *32*: 428 (1957).

49. T. L. Starks and P. E. Johnson, Techniques for intrinsically labeling wheat with ^{65}Zn, *J. Agric. Food Chem.*, *33*: 691 (1985).

50. L. Hallberg, Bioavailability of dietary iron in man, *Ann. Rev. Nutr.*, *1*: 123 (1981).

51. J. D. Cook, M. Layrisse, C. Martinez-Torres, R. Walker, E. Monsen, and C. A. Finch, Food iron absorption measured by an extrinsic tag, *J. Clin. Invest.*, *51*: 805 (1972).

52. L. Hallberg and E. Bjorn-Rasmussen, Determination of iron absorption from whole diet: A new two-pool model using two radioiron isotopes given as haem and non-haem iron, *Scand. J. Haematol.*, *9*: 193 (1972).

53. E. Bjorn-Rasmussen, L. Hallberg, and R. Walker, Food iron absorption in Man. II. Isotopic exchange of iron between labeled foods and between a food and a iron salt, *Am. J. Clin. Nutr.*, *26*: 1311 (1973).

54. C. Martinez-Torres and M. Layrisse, Iron absorption from veal muscle, *Am. J. Clin Nutr.*, *24*: 531 (1971).
55. J. D. Cook, V. Minnich, C. V. Moore, A. Rasmussen, W. B. Bradley, and C. A. Finch, Absorption of fortification iron in bread, *Am. J. Clin. Nutr.*, *26*: 861 (1973).
56. D. D. Derman, M. Sayers, S. R. Lynch, R. W. Charlton, T. H. Bothwell, and F. Mayet, Iron absorption from a cereal diet containing cane sugar fortified with ascorbic acid, *Brit. J. Nutr.*, *38*: 261 (1977).
57. K. T. Smith, Effects of chemical environment on iron bioavailability measurements, *Food Technol.*, *37*: 115 (1983).
58. N. R. Meyer, M. A. Stuart, and C. M. Weaver, Bioavailability of zinc from defatted soy flour, soy hulls, and whole eggs as determined by intrinsic and extrinsic labeling techniques, *J. Nutr.*, *113*: 1255 (1983).
59. S. M. Ketelsen, M. A. Stuart, C. M. Weaver, R. M. Forbes, and J. W. Erdman, Bioavailability of zinc to rats from defatted soy flour, acid-precipatated soy concentrate and neutralized soy concentrate as determined by intrinsic and soy concentrate as determined by intrinsic and extrinsic labeling techniques, *J. Nutr.*, *114*: 536 (1984).
60. G. W. Evans and P. E. Johnson, Determination of zinc availability in foods by the extrinsic label technique, *Am. J. Clin. Nutr.*, *30*: 873 (1977).
61. R. Schwartz, H. Spencer, and J. J. Welsh, Magnesium absorption in human subjects from leafy vegetables, intrinsically labeled with stable ^{26}Mg, *Am. J. Clin. Nutr.*, *39*: 571 (1984).
62. M. Janghorbani, N. W. Istfan, J. O. Pagounes, F. H. Steinke, and V. R. Young, Absorption of dietary zinc in man: Comparison of intrinsic and extrinsic labels using a triple stable isotope method, *Am. J. Clin. Nutr.*, *36*: 537 (1982).
63. R. P. Sirichakwal, V. R. Young, and M. Janghorbani, Absorption and retention of selenium from intrinsically labeled egg and selenite as determined by stable isotope studies in humans, *Am. J. Clin. Nutr.*, *41*: 264 (1985).
64. M. J. Christensen, M. Janghorbani, F. H. Steinke, N. W. Istfan, and V. R. Young, Simultaneous determination of absorption of selenium from poultry meat and selenite in young men: application of a triple stable isotope method, *Br. J. Nutr.*, *50*: 43 (1983).
65. L. May, E. R. Morris, and R. Ellis, Chemical identity of iron in wheat by Mossbauer spectroscopy, *J. Agric. Food Chem.*, *28*: 1004 (1980).
66. O. E. Olson, E. J. Novacek, E. I. Whitehead, and I. S. Palmer, Investigation on selenium in wheat, *Phytochemistry*, *9*: 1181 (1970).

67. K. C. Weiss and M. C. Linder, Copper transport in rats involving a new plasma protein, *Am. Physiol. Soc.*, *249*: E77 (1985).
68. G. H. Starich and C. J. Blincoe, Properties of a chromium complex from higher plants, *J. Agric. Food Chem.*, *30*: 458 (1982).

Index